Newton's *Principia*: The Central Argument

Newton's *Principia*
The Central Argument

Dana Densmore

Translations and diagrams by
William H. Donahue

Third Edition
Completely revised and redesigned
New sections added

Green Lion Press
Santa Fe, New Mexico

Manufactured in the United States of America.

Published by Green Lion Press,
1611 Camino Cruz Blanca, Santa Fe, New Mexico 87505-0353 USA.
Telephone (505) 983-3675; FAX (505) 989-9314

mail@greenlion.com
www.greenlion.com.

Green Lion Press books are printed on acid-free paper. Both softbound and clothbound editions have sewn bindings designed to lie flat and allow heavy use by students and researchers. Clothbound editions meet the guidelines for permanence and durability of the Committee on Production Guidelines for Book Longevity of the Council on Library Resources.

Printed and bound by Sheridan Books, Inc., Chelsea, Michigan.

Third Edition, 2003, completely redesigned, revised, and expanded.

Cover illustration by Josh Foster.

Cataloging-in-Publication Data:

Dana Densmore
Newton's Principia: the central argument

Excerpts of text of Isaac Newton
Notes and commentary by Dana Densmore
Translation and diagrams by William H. Donahue

Includes translation, introductions, notes, commentary, glossary, bibliography, and index.

ISBN 1-888009-24-1 (cloth binding with dust jacket)
ISBN 1-888009-23-3 (sewn softcover binding)

1. Newton, Isaac, 1642-1727, Principia. 2. History of Mathematics. 3. History of Science. 4. Gravitational theory. 5. Planetary theory. 6. Physics.
I. Densmore, Dana 1945–. II. Donahue, William H., 1943–. III. Title.

QA803.A45 2003

Library of Congress Card Number 2003094208

This guidebook is dedicated to Eva Brann, who in 1963 first introduced me to Isaac Newton and the thrilling adventure of *Principia*; to my mother, Donna Allen, who taught me to love hard work and to dance in the joy of discovery; and to my father, Russell Allen, whose penetrating eye was always looking through the surfaces of things and the easy answers.

Contents

Newton's *Principia*

Foreword by Curtis Wilson

Newton's *Principia* must surely be the most unread famous book in the Western World. Even in the seventeenth century few essayed to read it, and one such, the philosopher John Locke, had to ask Newton for help in finding his way through the main argument. In the eighteenth century, the foremost mathematicians of Europe adopted the Leibnizian symbolism for the calculus, and proceeded to translate basic propositions from Newton's book into an algebraic form. Mathematical physics became algorithmic and, to the uninitiate, an unapproachable mystery, for high priests only.

To be sure, mathematical physics is in fact a wonder. Real education has more to do with learning to recognize the wonderful than with dispelling wonder. For physicists as for nonphysicists, there is a special gain in coming to be educatedly present at the moment in which, as Ernst Mach put it, "the uncommon incomprehensibility [of universal gravitation] became a common incomprehensibility." For this, one must follow Newton's central argument questioningly; appreciate its force, its strangeness for his contemporaries, its potentiality for the future. A new method here emerges. Willynilly, despite Newton's expressed wish, it was to prove pursuable independent of metaphysical or theological persuasion. What we call *science*, independent of philosophy, here came into existence.

Dana Densmore has provided the guide that is needed if one is to follow the central argument of Newton's *Principia* with comprehension and assurance. Putting students on their mettle, she engages them in their own reading of the text, but provides detailed help wherever need for it may be felt. The study is demanding, but open to all with an elementary background in Euclidean geometry. In restatements of and comments on the propositions, Densmore probes the implications of Newton's often cryptic formulations. Some lapses in Newton's argumentation there are, and she does not blink at pointing them out. Among them, Newton's tacit assumption of attraction in III.7, which ostensibly establishes universal gravitation, is worth dwelling on. It shows why the argument for universal gravitation must be open-ended: truths about the real world cannot be proved but only tested or verified.

To dwell long enough and understandingly enough with Newton's text to gain a genuine sense of where the author is coming from and whither he is going: that is what the liberal artist will want to do. Dana Densmore's guide makes it possible.

Curtis Wilson
Annapolis, Maryland
July 1995

Translator's Preface

Over the years of gathering notes for this guidebook, Dana Densmore frequently called upon me to consult the Latin text to elucidate Newton's meaning or to track down possible errors. Our experience suggested that a new translation of the guidebook selections was called for.

Despite the general excellence of Andrew Motte's version, it is still somewhat too free, and Motte's eighteenth-century English is not always clear to a modern reader. Motte also sometimes added or altered words. An especially important example of this is at the end of the General Scholium, where he inserted the words "electric and elastic," which, although it did come from Newton, never appeared in any edition of *Principia.*

Motte's original version is now available, and despite its flaws, it is much to be preferred to Florian Cajori's revision, which for many years was the only translation of *Principia* in print. Of this unfortunate attempt, I. Bernard Cohen notes: "Cajori's modernization of Motte's translation introduces so many infelicities that it may serve as a cautionary object-lesson in not tampering with older translations. Again and again scholars have been misled by Cajori's version, which must always be used with the greatest caution and always checked against the original."*

Cohen himself, together with Anne Whitman, recently brought out a complete new translation, done with the help of his lifetime of experience studying Newton, and with the input of a number of prominent Newton scholars. Cohen's translation is a significant improvement over both Motte's and Cajori's revision of Motte's, and delicately balances accuracy against clarity and fluency. Occasionally, this has resulted in rephrasing Newton's language, changing plurals into singulars (or the other way around), substituting verbs for nouns, or replacing an odd or archaic turn of phrase with something more idiomatic. This rephrasing has the advantage of making the argument much less obscure, but it also glosses over some of the quirks of Newton's text, attention to which is essential to a careful reading. Newton's mathematical terminology has in some instances been replaced by more modern words, distorting the argument in subtle ways.

The present translation has a different aim from the Cohen/Whitman version. Although I have tried to make it consistent with modern English usage, Dana Densmore and I wanted the translation to serve as a window to

---

*I. Bernard Cohen, *Introduction to Newton's 'Principia'* (Cambridge, MA: Harvard University Press, 1971), p. xvi note 17.

the Latin text, as much as possible. Even if it seems to make no difference whether Newton wrote of "force" (singular) or "forces" (plural) in stating a proposition, I have tried to represent accurately what he wrote. And since the reader has Densmore's excellent notes and expanded proofs to help him or her past the awkward places, there is less need for my translation to interpret, and a corresponding opportunity to represent the text more exactly.

A clear example showing how the present translation differs from Motte's and Cohen's is provided by the translation of the title of the "Rules" at the beginning of Book III. The Latin text reads, "Regulae Philosophandi." "Regulae" are rules: there is no problem here. But "Philosophandi" presents a puzzle. It is the genitive case of the gerund of the late Latin verb "philosophari," which means to do philosophy, or to philosophize. Literally, then, the title would seem to mean, "Rules of philosophizing," or "Rules of doing philosophy." But from the beginning, translators had trouble Englishing the title in this simple and straightforward way. Motte had "Rules of Reasoning in Philosophy," presuming that these rules could not apply to philosophizing *per se.* Cajori, for once, declined to meddle with Motte. Cohen, however, could not stand to leave "philosophy" unglossed, supposing that it must refer to *natural* philosophy, and not to ethics, or metaphysics, or any of the other branches of philosophy. But he went farther, evidently feeling that Motte had been right to insert some auxiliary activity, but judging that "reasoning" was inadequate. And thus, in the Cohen version, we have "Rules for the Study of Natural Philosophy." This seems to me first of all out of keeping with what Newton is doing here, which is surely not *studying* natural philosophy, but making it all anew. But worse, it misleads the reader by substituting a gloss for the plain meaning of Newton's words. They may be puzzling, but they are his, and I intend to leave them as he wrote them.

In translating the proofs, I have given particular attention to Newton's technical terminology. Newton did not use twentieth-century mathematical language, and any attempt to make him do so must end poorly. Newton's "sagittae" are not versed sines (as in Motte-Cajori). And although Newton moved freely back and forth between the Euclidean language of proportion and the algebraic language of equation, the translator must beware of rewriting Newton to make him more algebraic. For example, although Newton treats duplicate ratio as equivalent to ratio of squares, a faithful translation should not presume to "correct" Newton by substituting the latter for the former wherever it appears. Nor should the geometric "square" or "cube" be expressed as an algebraic exponent (second or third power), even though it is usually safe to transform the one into the other. Compound ratios, *ex aequali* proportions, and other Euclidean relations should also be maintained, in order to preserve the essentially geometrical nature of Newton's approach to physics On these points, the present translation differs from Cohen, and is closer to Motte (though the latter does use exponents in place of squares and cubes).

Newton's practice of giving only sketches instead of proofs also creates difficulties for the translator. The correct English equivalent for a term often depends upon the context in which it is used, and in the *Principia* the context is usually implicit. Accordingly, to find the right word it was often necessary to know how the sketch proof in which the word appeared would be expanded into a complete proof. For example, in the Corollary to Lemma 4, the text reads, in part, "partes illae... datam obtineant rationem...," which would seem to mean, "those parts... maintain a given ratio...." However, scrutiny of the steps in the proof show that "obtineant" must have a meaning closer to its English cognate "obtain" than to the classically sanctioned "maintain". Both Motte and Cohen/Whitman miss this entirely, reading "obtineant" as "have" or "maintain," respectively. A similar error occurs in II.24, where the quotient of the times and velocities varies as the squares of the times: Motte-Cajori suggests that the times vary as the squares of the times, which is, of course, absurd. Motte himself, as well as Cohen/Whitman, got this right.

In the translation of mathematical terms and procedures, more is at stake than simply style, particularly for Newton. Although he had developed the calculus in an algebraic form, he deliberately chose to use an idiosyncratic revision of ancient geometrical forms and arguments in the *Principia*. Much has been written about this choice, and this is not the place to review the various opinions that have been expressed. However, it would not be out of place to quote Newton's own remarks on the subject, from the preface to the first edition:

> But since artisans are used to working with little accuracy, *mechanics* as a whole has come to be distinguished from *geometry* in that whatever is accurate is ascribed to *geometry*, and whatever is less accurate, to *mechanics*. Nevertheless, the errors do not belong to the art, but to the artisans. He who works less accurately is a more imperfect mechanic, and if anyone could work with perfect accuracy, he would be the most perfect mechanic of all. For the description of both straight lines and circles, upon which *geometry* is founded, belongs to *mechanics*.

There is more than a hint here that this book will be treating the work of "the most perfect mechanic of all." We live in a world governed by mechanical principles, which is truly geometrical because it has been created by that Perfect Mechanic. It is therefore no mere stylistic whim that the *Principia* was written in a language that as closely as possible reflects that mechanic's creation.

A translator should therefore beware of departing from Newton's terminology and procedures on the grounds that these are only matters of style. This is not to say that Newton is strictly Euclidean. He takes liberties in his treatment of ratios and proportions that would have been unacceptable to the Greeks. However, it is not for the translator to decide where to follow

Newton's usage and where to ignore it. Newton must be allowed to speak for himself as much as possible.

On the subject of accuracy, there was an error in my translation (corrected in this edition) of the enunciation of III.7. The origin of this error can serve as an object lesson on the danger of habitually using a translation known to be inaccurate. Newton's Latin begins, "Gravitatem in corpora universa fieri,...". Since "corpora universa" is in the accusative case, the preposition "in" must mean motion or direction towards the bodies. So this phrase should be translated, "That gravity [or, heaviness] takes place [or, happens] towards bodies universally,...". Motte wrote, "That there is a power of gravity tending to all bodies,..." which transforms gravity from the heaviness of things into a presumably separate power, but at least gets the other part right. Cajori evidently didn't think "tending" was quite right for "fieri" (with some justification), but in the process forgot about the accusative, and so wrote, "That there is a power of gravity pertaining to all bodies,..." thus turning gravity into an essential power that resides in the attracting bodies—exactly what Newton was trying not to say.

When I came to that proposition, I was so used to Cajori's reading that I missed the accusative and wrote, "That gravity is given in bodies universally,...". Evidently the same thing happened to Cohen and Whitman, whose version reads, "Gravity exists in all bodies universally...." I was unaware of the error until François De Gandt, no doubt less tainted by the Cajori translation, pointed it out.

One or two other errors in my original translation, less drastic than this, have been caught by alert readers and corrected in the present edition.

Considerable thought has been devoted to the selection of the text upon which this translation is based. The *Principia* appeared in three editions during Newton's lifetime, and was greatly altered and expanded for each. Many of these changes are valuable clarifications and emendations of arguments that were at first obscurely stated. Others, however, represent new thoughts and alternative approaches that do not bear directly upon the original argument. Dana Densmore's wish was to keep the clarifications (such as the reworking of Book I Proposition 6) but to avoid complicating the task of reading the *Principia* by including extraneous matter.

The present translation is based upon the third edition, with occasional reversions to the first edition text. All the alternative proofs (headed "the same, otherwise" in Motte's translation) are omitted. (For more on this decision, see Preliminaries, Section 1, below). Other changes from the text of the third edition are noted as they occur.

William H. Donahue
June, 2003

Acknowledgements

This guidebook is indebted to students, colleagues, and many other thoughtful readers who raised questions and gave me helpful comments and criticisms at every stage of its development over the last fifteen years.

Originally it consisted of my own notes for my classes, then xeroxed handouts for my students, then as a book it went through two preliminary printed editions used in classes at St. John's College, and at this point it was offered to Rutgers University Press for its Masterworks in Discovery series. It was later withdrawn from Rutgers when it became clear that this wasn't the best way to present the book and in 1995 the first Green Lion edition was published. This first Green Lion edition sold out in the first nine months; a second edition was issued in 1996 with only minor additions and corrections. The present third Green Lion edition constitutes a major redesign of the book, with much more extensive revisions and additions. It has been newly typeset using a color-capable page layout system.

In the earliest phases of filling in the steps of Newton's sketched proofs, I am most indebted to my students, with whom, of course, the process was a shared adventure, and to my husband, William H. Donahue, who not only did the translations for me but with whom I discussed every thorny question.

The students in my junior mathematics tutorials were the original Muses of this guidebook. They showed me what was needed and challenged my inadequate formulations. I make particular acknowledgement of, and warmly thank, the students in my junior mathematics tutorial of 1991-92, who worked through the first printed draft of the guidebook and offered me so many insightful suggestions. Later, when I was preparing the book for the Green Lion, my student Josh Foster responded to my expressed cover idea of a hand holding up an apple to the moon by presenting me with the beautiful illustration we are still using.

William H. Donahue was unsparing in providing consultations during the years of developing the notes for the guide and during the years of its formalization and revisions. He helped me enormously in understanding what Newton was doing and with the background and context. When I had the steps to the proofs on my loom, I carried many a knot and broken thread to his desk and always came away ready to weave anew. I am grateful also for his generosity in taking time from his own scholarly work to make the fine new translation and to draw all the diagrams.

Curtis Wilson, another generous Newtonian scholar, also gave of his time, energy, and depth of understanding and insight. He painstakingly reviewed several early drafts, writing out extensive and detailed comments. His support and encouragement have meant a great deal to me during the genesis of this book. He contributed to its soundness both by pointing out errors and by alerting me to context and ramifications of the propositions. In addition, he showed me many intriguing by-ways that enriched my thinking about Newton's project.

I am indebted in more ways that can be expressed to St. John's College in Santa Fe and Annapolis. Eva Brann, to whom this book is dedicated (in the honorable company of the very greatest of my teachers), introduced me to Newton's project and first invited me to engage in it as if I were equal to the task. Later, when I approached this text again not as a student but as a faculty member myself, colleagues Robert S. Bart and Ralph Swentzell were particular inspirations to me.

Special thanks go to the many faculty and students who used the guidebook in junior year mathematics tutorials and took the time to let me know about errors, inconsistencies, and obscurities. I am also most grateful for the enthusiasm and encouragement of faculty and students on both campuses who went out of their way to tell me how this guide enriched their reading of Newton.

While the book was under contract with Rutgers suggestions from that process contributed to the usefulness of the book. Special appreciation goes to Harvey Flaumenhaft, series editor for Masterworks of Discovery, and to Karen Reeds, Science Editor.

In production of the text for the first Green Lion edition, I had the support of William H. Donahue who set up the typesetting programs on my computer and provided advice and patient help with their mysteries and idiosyncracies. We used Paul Anagnostopoulos's ZzTeX book layout system and Y & Y Software's integrated TeX typesetting system for the first two editions; diagrams were drawn with CorelDraw.

For the corrections in the second edition I am grateful to those who read this guide or used it during the 1995–96 school year and made suggestions, and especially to Adam Schulman and his hawkeyed students.

For the third edition I moved the text out of TeX and into QuarkXPress. The book has been redesigned to separate my text from Newton's by using color for Newton's text as well as a distinctive type font. (Previously I had used boldface with a different type font to distinguish Newton's text, but it didn't distinguish it sufficiently to satisfy me; further, I found the overall look busy.) The old font for Newton text had been originally dictated by Rutgers; the one I chose for the third edition, Caslon Antique, is beautiful and appropriate to an old text, thus removing an aesthetic irritant. For my notes and expanded proofs I use the font Stone Sans.

Text for the third edition was set using QuarkXpress; mathematical sections were set in TeX using Blue Sky Textures; diagrams were prepared using both Adobe Illustrator and CorelDraw.

Over the course of seven years of using the second edition, I had noticed places I could have said things better, and many readers reported their thoughts: what was helpful and where they had difficulties as well as their suggestions for additions and corrections. The result is a major revision with many improvements. Among these helpful readers I would like to mention particularly David Bolotin, William Decker, William Donahue, Howard Fisher, Mera Flaumenhaft, François De Gandt, Niccolò Guicciardini, Nicholas Huggett, and Joe Sachs. My grateful thanks to all.

In production of this third edition, I had the considerable support of the Green Lion team, especially in redoing all the equations and diagrams, but also in the process of improving and correcting the notes and expanded proofs. William H. Donahue continued to consult on difficult bits of Newton's text, as he has since the early stages of this book. (He contributed the careful analysis of I.66 and its corollaries in Appendix B.) He also converted and redrew diagrams and turned the more complicated TeX equations into EPS pictures for inclusion in the Quark. Associate editor Howard J. Fisher did the initial conversion of the text from TeX to Quark, worked on diagram conversion, rekeyed many of the equations, and prepared the new index. His alertly engaged proofreading, and his advice about every aspect of the revision, was most valuable. I'm grateful to be part of such a terrific team.

William Davis, Janet Dougherty, and Rachel Pagano helped us out in the thick of production by proofreading large chunks of text to ensure that the newly converted equations still said what they used to say and appeared where they were supposed to appear. Thanks for that careful checking!

And finally, I thank and acknowledge all of *you,* dear readers, who have taken on this thrilling but very difficult piece of work of actively following through Newton's development as he presented it. Thinking about you has inspired me and kept me going over the demanding process of preparing this revision.

Dana Densmore
July 2003

Preliminaries

Section 1: The Grand Sweep of *Principia* and its Central Argument

In January 1684 a conversation took place between Robert Hooke, Edmund Halley, and Christopher Wren at a meeting of the Royal Society in London regarding the possibility of deriving Kepler's laws of planetary motion from physical forces.*

Hooke claimed to be able to demonstrate all the laws of celestial motion by assuming a power inversely as the square of the distance between the celestial bodies.

This "inverse square" relation had been observed in connection with the spreading of light and the action of magnets, and many people, including Wren and Halley, were speculating that it might apply to celestial actions. Wren had discussed the possibility of an inverse square force law with Isaac Newton as early as 1677.

Wren responded to Hooke's claim in that conversation by offering a prize for anyone who could produce a demonstration that an inverse square force law would lead to the motions of the planets described by Kepler. No such demonstration was forthcoming from Hooke, at least nothing that satisfied Wren, and the question stood.

In August of that year, Halley visited Newton at his home in Cambridge and mentioned the challenge, asking Newton whether he knew what sort of orbit an inverse square force law would produce. Newton answered that it was an ellipse, and that he had demonstrated it. Halley, excited, urged him to send the demonstration to him as soon as he could.

The first version Newton sent was a nine-page document entitled *De motu corporum in gyrum* (On the Motion of Bodies in Orbit), submitted to the Royal Society in November 1684. It not only demonstrated the planetary ellipses but also showed how all of Kepler's laws may be seen as consequences of physical forces.

It was obvious to Halley that this was a momentous contribution to placing the mathematics of planetary motion on a sound physical

---

* The Royal Society was the most influential forum for scientific ideas in England at that time. It registered discoveries, discussed and debated them, and often published important works. For a fuller account of the writing of *Principia,* see Richard S. Westfall, *Never at Rest: A Biography of Isaac Newton.* Cambridge: Cambridge University Press, 1984), to which work this summary is indebted. Of the several recent biographies of Newton, this is the one that gives the most emphasis to the scholarly side of Newton's life. Background on the writing of *Principia* can also be found in the introductory volume to the Latin edition used in this translation: I.B. Cohen, *Introduction to Newton's 'Principia'* Cambridge MA: Harvard University Press, 1971.

foundation. It went beyond the work of Kepler in two primary ways. First, it was universal, not depending on different plans for actions between planets and the sun and actions between planetary matter and the planet itself. Newton was able to show that terrestrial heaviness and the forces that move the planets were a single phenomenon. Second, Newton's system required fewer contrivances and *ad hoc* assumptions than Kepler's. It didn't require reference to imaginary or hypothetical entities. For example, Kepler supposed two powers in the sun, neither of which could be observed other than by their effect on planetary motion. A kind of magnetism had to be supposed for the sun that followed laws different from those obeyed by the magnetism found on earth. The actions Newton relied on could be easily found and tested on earth.

Halley was eager to have the document published. But Newton wanted to develop it further, and urged Halley to hold off on publication until he could re-work and expand it. All his prodigious intellectual energies were consumed with this expansion. The manuscript of Philosophiae naturalis principia mathematica (Mathematical Principles of Natural Philosophy),* was initially delivered to the Royal Society for publication in April 1686.

However, Newton still continued to work on the book, expanding it into three books. The final version of Book III was delivered to Halley for publication a year later in April 1687. Book III applied mathematical demonstrations of the two earlier books of *Principia* to our world and derived from these foundations the principles of universal gravitation and the elliptical planets of orbits (along with many other things, including the motions of the tides and the paths of comets).

Principia was written in a stupendous burst of creative energy. Newton lived during this period like one possessed. He often forgot to eat. "When he has sometimes taken a Turn or two [in the garden], has made a sudden stand, turn'd himself about, run up ye Stairs, ... fall to write on his Desk standing, without giving himself the Leasure to draw a Chair to sit down in."**

In such circumstances, the compass of the *Principia* could not be restrained. Among the many dazzling insights for which he sketched out proofs were: the conceptual equivalent of conservation of kinetic and potential energy; general expressions for orbits under any arbitrary force law; expressions for attractions between finite spheres under any force law; attempts to treat the motion of multiple mutually attracting bodies using approximations and qualitative arguments; the motion of pendulums; the motion of waves in water; the motion of fluids (including a refutation of the

---

* This is the full title of the work most often referred to simply as *Principia* (Latin for "Principles")—the work to which this book is a guide.

** Humphrey Newton, Newton's amanuensis for five years during the writing of *Principia*, quoted by Richard S. Westfall in *Never At Rest*, p. 406.

Cartesian theory of vortices; explanations of tides and of the nutation of the earth's axis and the motion of the moon's nodes; and laws governing the orbits of comets.

Buried within this heap of brilliant propositions is a central jewel, the establishment of universal gravitation and its use to demonstrate the elliptical orbits of the planets, which constitutes the main argument of *Principia*. It is this central jewel of an argument that this guidebook takes us through.

This is not only a supremely important step from the point of view of the history of science, but in addition it is Newton's practical demonstration of his theory of how science could be done in a way that yields certainty, being (as he saw it) purely deductive.

This attempt to give science a logically sound deductive basis constituted a radical departure from Francis Bacon's inductive method, which was very influential at the time. Bacon advocated collecting many and varied instances of the phenomena under study and trying to see patterns among them.

By contrast Newton used minimal experimental data. His main experimental foundations, the "Phenomena" of Book III, were (as we shall see) very far from being pure observations, but they were based on observations and theory generally accepted. Everything was deduced, using mathematical demonstrations, from these few observation-based conclusions about how our world works.

In his "Preface to the Reader" (pages 3–4 below), Newton describes this revolutionary method thus:

> And on that account we present these [writings] of ours as the mathematical principles of philosophy. For the whole difficulty of philosophy appears to turn upon this: that from the phenomena of motion we investigate the forces of nature, and then from these forces we demonstrate the rest of the phenomena. ... In the third book, ..., we present an example of this procedure, in the unfolding of the system of the world. For there, from the celestial phenomena, using the propositions demonstrated mathematically in the preceding books, we derive the forces of gravity by which bodies tend to the sun and the individual planets. Then from the forces, using propositions that are also mathematical, we deduce the motions of the planets, of comets, of the moon, and of the sea. In just the same way it would be possible to derive the rest of the phenomena of nature from mechanical principles by the same manner of argument.

This is the Newtonian version of confirmation theory, which is ingenious and is significantly different from both contemporary and modern theories of scientific explanation.

Kepler had previously shown that the planetary orbits are elliptical. Without using Kepler's laws, building only on his own foundations, Newton successfully used his method to derive the elliptical orbits of planets. This is a test of the method, showing that it arrives independently at true results. Newton's central argument provides this test of the method.

Because this guidebook concentrates upon the central argument, we will be looking here not at the whole of *Principia* but at what Newton explicitly delineates as the core sequence.

This core sequence consists of the following parts of *Principia*. The mathematical foundations for the development of universal gravitation and celestial mechanics are found in the Definitions, Laws of Motion, and some basic hypothetical mathematical propositions, the first seventeen propositions of Book I. That is where we start.

Then we go to Book III, where he introduces what he calls Phenomena, a small number of conclusions about our world. We look at these carefully and satisfy ourselves about them.

Then we follow the primary propositions of the application to our universe in the first thirteen propositions of Book III. (In the process we dip back into Books I and II for a few extra propositions and augment our observation-based data with results from experiments on pendulums.) Here lies the thrilling derivation of universal gravitation: the discovery that the moon is falling just like a rock (or a pendulum bob), that inertial and gravitational mass are quantitatively the same, that every particle attracts every other particle inversely as the square of their distance. Having established these things, we can use them to prove that the planets will move in ellipses. This will be our exhilarating finale.

This is the procedure recommended by Newton in his preface to Book III.

At the end of the study undertaken by this guidebook, we will have established the following from the phenomena of our world using the mathematical tools of the first two books:

- we will have derived the force law governing the movement of the planets;
- we will have found that the same force accounts both for terrestrial gravity and for the turning of the planets into orbit;
- we will have discovered the quantitative identity of inertial and gravitational mass;
- we will have established the principles of universal gravitation;
- and we will have deduced the elliptical orbits of the planets.

This guidebook follows the essentials of Newton's presentation in *Principia*. It is designed to deliver a coherent and complete development of his line of argument appropriate for an academic semester of study.

Newton incorporated many interesting diversions into *Principia* in his great burst of creative energy, diversions that lie outside the central argument, and that are not included in this guidebook. In parts of Book I Newton explores some very interesting attributes of conic sections.* In other parts of Book I, he investigates the instability that results from the fact that our universe contains more than two bodies (he does *not*, as is sometimes carelessly said, "solve the three-body problem"). Book II lays down influential propositions of fluid mechanics. Even in the early propositions in Book I, a series we look at in detail, some propositions, which in themselves are stimulating and entertaining, are omitted as not essential to the line of argument.

For example, in the second edition of *Principia* Newton added some alternate proofs to some of the early propositions of Book I, introduced by the words "the same thing differently" (*idem aliter*). These alternative proofs used a circle of curvature for their proof. Since this seems an interesting approach, why drop them? First, this was not how Newton worked out the proofs of these propositions: it is a quite different alternative approach to the proofs. If Newton had preferred this alternative approach, he could have replaced his original proofs. He had two editions (and at least thirteen years) in which to do so, but he chose to leave the original approach as the primary presentation, with the *idem aliter* proof tacked on afterwards as just another way to look at the problem. So the original is what he wants us to concentrate on. Second, understanding a very different approach in addition to the primary one is an unnecessary obstacle to someone with limited time trying to understand the central argument. There might be many ways to prove a Euclidean proposition, but we don't stop to consider all the other possible approaches as we work through Euclid for the first time.

It is important to remember that *Principia* is not just a cookie jar of entertainment and exercise for the mathematical faculties. It *has* a coherent argument that leads some place in particular, an argument line to which Newton explicitly draws our attention. Knowing that, and if we respect Newton and care to understand his enterprise, we must want to follow that argument to Newton's conclusion. Conveniently, this central argument is just right for one semester of study.

It is greatly tempting, and would be only too easy, to use up all our time for Newton in the early, hypothetical, mathematical demonstrations, and give only a rushed perusal to Book III. Yet Book III is the place where Newton gives his main argument, the place where he actually constructs the edifice for which the early propositions laid out the building blocks. The propositions of the first books are pro-theorems, preparation, for his conclusions about the System of the World.

In the first two books of *Principia*, Newton lays a mathematical foundation for the "philosophical" arguments. (When Newton says "philosophy" he

---

* Conic sections are curves that are found by cutting a cone with a plane (see Glossary).

means natural philosophy, which we would understand as physics and astronomy.) The propositions he lays out here are hypothetical. They do not show how things must be; rather, they show how things will be if certain conditions hold. For example, if it be supposed, as the conditions of a proposition, that a center of forces exists in a certain place, and a body moves around it in some specified path, the proof of the proposition shows what forces would be required to produce that path. Such a path might never actually appear in nature, but that is a matter of complete indifference from a mathematical perspective. These hypothetical propositions are then available to be used, should their conditions be shown to hold, to demonstrate things about our actual universe. This is done in Book III, which he calls "System of the World," for certain of the hypothetical propositions.

Book III, the System of the World, is not an afterthought, not mere application of what has already been proved. Nothing about our world has been proved in the early books. It is in Book III that the breakthrough discoveries are made about gravitation, and it is in Book III that we find that the planets do move in ellipses.

If the impressive insights of Book III look obvious, it is only because we have lived with those conclusions since Newton's time, and it is evidence that we need to give them a more careful reading so that the significance and originality may make themselves felt.

There are two reasons, then, for keeping the careful focus of the guidebook selection. The first is to bring out the coherent line of the central argument. The second is to make sure that there is time for adequately thoughtful consideration of that central line of argument.

Those who have more than a semester to spend on Newton can profitably work through this central line of development and then follow out some of the intriguing side paths (either in the earlier books or later in Book III) on a firm foundation of understanding of the primary endeavor.

All of Book III constitutes the demonstration of Newton's method. The present selection follows his central argument through the derivation of elliptical orbits of planets. Working through this central argument with the help of this guidebook gives a very solid foundation on which to build on one's own, continuing through Book III looking at the moon, the tides, comets, and other celestial motions.

If I had a full year with students, I would continue with Book III, following his advice to dip back into the first two books as needed to support those demonstrations. In particular, I would spend some quality time with the propositions on the moon's motion. This was Newton's most challenging application of his gravitational theory, and ultimately he failed to deduce all the motions of the moon on that basis, having recourse to epicycles as his predecessors Ptolemy and Copernicus had done before him. Despite those extraordinary difficulties, he accomplished remarkable things with his analysis of the forces on the moon and their relation to the existing lunar anomalies.

Section 2: Structure and Use of This Guidebook

This guidebook takes for its premise the idea that we gain the best insight into Newton by reading Newton. If we want to understand his vision and his discoveries, to understand what is really new and thrilling in what he is doing and developing, we do that best by discovering it for ourselves by working through Newton's own demonstrations, step by step as he conceived them.

For most people, however, following him carefully enough to share deeply in that discovery has been problematic. This has been true ever since the publication of *Principia*. First, Newton wrote in Latin; the need for (and our dependence on) translation has increased as ever-fewer students are comfortable working with the Latin text. Translators have not always managed to provide authenticity and accuracy. See the translator's preface to this work for an account of some of the difficulties.

More challenging yet, Newton gave us only sketches of his proofs, and those sketches were insufficient even for many mathematicians of his own time. There have been two responses to this difficulty of access to the inner workings of *Principia*.

One response is that few students actually attempt to construct detailed proofs, looking only at Newton's overall claims and guessing, perhaps, how he may have arrived at them. It should be clear how much one may miss of Newton's own understanding of what he is doing by not following the specific strategies of how he does it. We lose what he uses for foundations, what he assumes and what he proves, what the full generality or particular focus of a proposition turns out to be: things often not evident in the enunciation or noted explicitly in the sketch.*

The other response to the difficulties of access to the text has been for Newton's commentators to offer alternative proofs that show ways to get to Newton's results—ways that for the most part make no pretense of reproducing Newton's strategy, but rather intend to satisfy the reader that Newton's claim is indeed true. This is fine if that is all that is wanted. But I believe that following alternative proofs is not the way to understand Newton; it is certainly not the way to experience the thrill of discovering the power and originality that makes some call this the most exciting and important work in the history of science.

This guidebook is for the student who wants to understand Newton by doing Newton. I let Newton and his propositions stand on their own, without

---

* The *enunciation* is the statement at the beginning of a mathematical proof that sets out what is to be proven or found by the proof. It is followed by the steps of the proof itself, or in the case of the proofs Newton gives in *Principia*, a sketch of the proof that could be made, indicating the strategy without including all the steps.

rearrangement or improvement, even where it may appear that it would be "better" or "clearer" in a different order or with a different proof. Occasionally I will offer as a note such an alternative for the purpose of inviting the student to ponder with me why Newton chose to do it as he did. But the premise of the guidebook is that we understand Newton only in understanding why he proved things as he did, not in preening ourselves on our cleverness in finding ways we would have done it better.

2.1 Who can use this book

2.1.1 The scientifically-minded general reader

This guide gives the intellectually-curious reader, with or without a science, math, or engineering background, everything he or she needs to follow Newton's arguments in his original words, conception, and presentation. We have all acquired the habits of Newtonian thinking, but how many of us can really say what its foundations are? Many people who desired to replace the conventionalities of thought about Newtonian physics with the excitement that comes from genuine understanding have found that understanding in study of the *Principia* with the help of this guidebook.

2.1.2 Scholars and graduate students

This guide is also suitable for scholars and graduate students in history of science and mathematics, and in physics, who wish to study the details of Newton's demonstrations in their original presentation, as opposed to their reconstructions in modern terms.

Great care has been given to the authentic reconstruction of the propositions, adhering exactly to the words Newton wrote and the style in which he composed his arguments. This is something that has never been done before.

A scholar might use the guide as a reference text to check on the overall structure, or on specific details, of Newton's mathematical arguments. In particular, scholars who wish to free themselves from reliance on anachronistic algebraic reformulations will find that this book opens fresh pathways into Newton's actual mathematics.

It can be (and has been) used in a semester-long graduate seminar, as one source text among others in a broader course, or as an individual research project to work through the sequence.

2.1.3 College courses

This guide is designed to be usable as the principal text in a semester course for liberal arts undergraduates, for undergraduate history of science or mathematics students, and for undergraduate physics or mathematics students. It does not require a specialist background either in history of science or in mathematics. The expanded proofs cite the appropriate propositions of plane geometry or conic sections or Galilean kinematics where Newton uses them, quoting or paraphrasing the less familiar ones, including anything not normally covered in high school geometry.

Detailed recommendations for course use, including suggestions about modifying the selection for courses of different lengths, or for students of different abilities, are included in Section 7, Suggestions for Teachers.

2.1.4 High-school level enrichment programs

This guide could be used by gifted high school students, by home-schooling teens with active curiosity about math or science, and by any student desiring or needing further challenge and enrichment.

It goes beyond the usual fare of just providing college-level mathematics or physics texts and awakens wonder and sense of adventure in reconstructing the development of universal gravitation and exploring the foundations of our common sense understanding of Newtonian physics.

Although it can be used by a student with interest and initiative working on his or her own, the process could be guided by a math or science teacher, or by a parent prepared to plunge in with energy and enthusiasm. Sharing the adventure this way can make it much more exciting and enjoyable, and the perspective of another could add richness to the exploration.

In this context one would not need to be limited by the rigidity of the school semester or by the pace of other students. The student could proceed at his or her own speed, dwelling longer where interest dictated, perhaps spending more time on some of the side lines of thought suggested in the text.

2.2 Levels of use

This guidebook can be useful on at least three levels.

2.2.1 A scrupulous reliable translation

This book presents a new translation, done by scholar and translator William H. Donahue for this book. Donahue's previous translations of Kepler's works have been acclaimed not only for their grace in rendering Kepler's notoriously difficult Latin, but for showing that he actually understands Kepler's equally difficult technical arguments (in the case of some of Kepler's mystifying calculations, Donahue was evidently the first to actually succeed in understanding, reconstructing, and verifying them.).

His translations for this guidebook are informed by the painstaking reading of Newton's actual words and thought that this guidebook embodies. A translation by a scholar immersed in Newton's sources and the intricacies of Newton's argument can bring insight to translation choices that a classicist decoding the Latin cannot hope to manage. Every other translation has occasionally gone astray by taking standard translations of Latin words according to the lexicon, but that a close reading of the argument shows cannot be Newton's meaning.

The translation is thus superior to other translations of *Principia,* past and present, in additional ways for the student wishing to do a close reading of Newton's exact presentation, with the clearest insight into the choices Newton made. This translation can be relied on for historical correctness: it does not convert Newton's formulations to more familiar modern notation, thus distorting *Principia*'s characteristic—and intentional—mode of presentation.

This translation is always consciously careful to provide the greatest accuracy to Newton's intention by rendering his formulations and phrases just as he expressed them. If he used a verb, this translation uses a verb; Donahue does not rearrange the sentence and turn the word into a noun. Does it matter? Perhaps not, in many cases. If the translator has interpreted the phrase or sentence correctly, the different English construction may be equivalent.

But *Principia* is full of puzzles, many of which have scarcely been explored. This guidebook points out quite a few that need the most careful investigation, and may involve interpretations different from those made by previous scholars.

The reader or scholar who wishes to think independently about the subtlety of Newton's argument needs the original formulation without interpretation or rearrangement, even if it is the interpretation or rearrangement of another respected scholar. He or she needs to know exactly when Newton used proportions and when he used equations; when he wrote an operation out in words and when he used algebraic-style notation such as a superscript for squaring; when he used a plural for forces and when he used a singular; even, perhaps, when he used a verb instead of a noun in some assertion that was not heretofore recognized as a clue to the solution of a puzzle.

Donahue's translation is exacting and hews to the Latin as scrupulously as a skilled translator can do, maintaining both the original intention and clear and graceful English—or maintaining ambiguity, if ambiguity is present, without presuming to resolve the ambiguity with interpretation. Its degree of authenticity offers a reliable window to Newton's original Latin formulations of his enunciations and arguments.

This guidebook would be a valuable resource even if used only to consult the deeply-informed and scrupulous translation. For those using it this way, all Newton's words appear in a distinctive color and typeface: notes and expanded proofs can be easily passed over wherever they are not needed.

2.2.2 Using only the notes

The next level calls upon only minimal help. The student can read Newton's propositions in the new translation, consult the notes that follow, and then try to work out the full proof independently. Some of the notes outline the significance of the proposition in the overall development, and some alert the reader to pitfalls in the interpretation and proof of the propositions.

I urge every student to try this level, attempting to work out the proofs independently, before checking the expanded proof for help with baffling steps. (See Section 4, Suggestions for Students, for specific suggestions on working out the proofs independently.) Trying to construct the argument oneself is a large part of the fun, and it will make the expanded proof's demonstration, if and when resorted to, more alive and meaningful.

2.2.3 Consulting the expanded proofs

The third level, available when it is needed or desired, is an expansion of Newton's sketch of his proposition. This step-by-step demonstration undertakes to reproduce what Newton would have given as a complete proof had he filled in all the gaps and spelled out the strategy for clearest presentation.

The expanded proof takes each step of Newton's actual sketch (keyed with the words of his step repeated in the color and special type font reserved for Newton's words) and shows how he arrives at it.

The scholar or physicist, who may be familiar with the proof in terms of what it says or what the modern physics equivalent is, may need help, or at least want a second opinion, on how Newton actually managed to get from one step to the next, or how he applied the lemma or the earlier proposition of Principia, or the Euclidean proposition, that he cites in his sketch.

The first-time reader can try to fill in the steps independently, consulting the expanded proof only for the step or steps that remain puzzling at the point at which fun has become frustration, skimming through the expanded proof looking for the words in color for the recalcitrant step.

Although the most active involvement will yield the most satisfying sense of discovery, even the students with the *least* enterprising attitude can still get a lot out the adventure of Newton with this guidebook. When they have done what they feel they can manage with only Newton and the guidebook's notes, they need only go to the expanded proofs, and they will find all the baby steps they could have wished Newton to have put in. This is much more desirable than what the less enterprising students do now, which is to stare at Newton's sketch for five or ten minutes, conclude that they can make nothing of it, and close the book in exasperation, feeling stupid and at best hoping that someone else will explain it in class.

In addition to more active preparation and sense of accomplishment in preparation for all students, there is another payoff to this for students, teacher, and the success of the class that comes when every student can come to class understanding the way the propositions are proved. It becomes possible to spend class time discussing the larger questions, the real questions. The following are examples of questions that can be more deeply addressed if class time is not used up in frustrating thrashing about in the slough of false leads, ungrounded guesses, and dead ends in the proofs. Some of the questions can only be recognized when clarity about Newton's procedures is gained.

- What is the difference between the earlier lemmas that do not compare vanishing quantities and those that do? What exactly can it mean to have a ratio *as the quantities vanish*?

- Exploration of how amazingly open the first proposition of the first book is—how Newton assumes nothing of what we now take for granted about this force except its direction.

- What is a force? Are forces real?

- What is the significance of Newton's choice to work with proportions? Of casting all assertions in terms of relationships?

- Why does he make his proofs geometrical? What is gained by that? Why doesn't he use his calculus notation?

- What does the sequence of mathematical proofs tell us about the nature of gravity? What can it be? What can't it be? Are all the possibilities still open at the end of the sequence? Is it true that he does not "contrive any hypotheses" as he claims in the General Scholium?

- Think about the structure of the book: the idea of a hypothetical tool chest followed by real-world applications.

- Consider Newton's claim that *Principia* presents a demonstrative scientific method that avoids the logical difficulties of inductive arguments. Does it hold up under scrutiny?

Many more such questions turn up naturally; some are included in notes after the translation or the text of the expanded proofs.

2.3 Criteria in constructing expanded proofs

Great care has been exercised in two particular respects in the fashioning of the expanded proofs.

2.3.1 Avoiding anachronism—the need for authenticity

First, care has been taken that the expanded proof be in Newton's terms and style.

What is sought is not just *some* proof of each proposition, but the one that Newton actually had in mind, presented in the way he would have expressed it. This is a reconstruction, of course, not mind-reading or spirit-channeling: one must study the work as a whole and make judgments. But there are criteria. One looks for what is in accord with the approach and context of the whole work, the strategy Newton is using to build up to his ultimate conclusions, and the specific hints and citations he gives in the sketches for the proofs.

This guidebook has as its special contribution to Newtonian scholarship the presentation of Newton's proofs in a historically accurate context and setting. These expansions of Newton's proof-sketches follow Newton's hints scrupulously and are consistent with what Newton has built up previously and with the mathematics of his time.

Careful effort has been made to place the work in its intellectual historical context, particularly with respect to the mathematical tools and the principles of mechanics and planetary theory current in Newton's time. Where he assumes the reader's familiarity with these things, they are offered either in the notes after the translation or in the expanded proofs, as appropriate.

In reconstructing Newton's proofs in *Principia*, we must avoid bringing in any post-Newtonian language or methods of proof. Newton had his own idiosyncratic method of mathematical demonstrations, which is largely geometrical. His proofs, and the steps added in the expanded proofs, depend only on Euclid, the classical writers on conics, Galileo, Huygens, and Newton's own previous laws and demonstrations; and they follow Newton's unique style of demonstration, which is so intriguingly characteristic. We gain much insight into Newton by pondering his style of demonstration, and lose that insight if we toss his method aside for an "improved" modern formulation.

Thus you will not find here material from post-Newtonian commentaries or alternative proofs. Nor is references made to the ways Newton is applied in modern physics. These topics are covered in many other books that have been published from Newton's time to the present.

My own commentary is limited to clarifications of context and occasional queries to stimulate discussion. I am aware of much that has been said by other scholars about these propositions, and there is much that I have noticed in my own study, some of which has been presented at scholarly meetings. But this guidebook is dedicated to presenting Newton in his own terms, not to commenting on him or evaluating him. That is another endeavor and is to be found in other books and in journal articles.

2.3.2 The importance of rigorous proofs

Second, particular care has been taken in fashioning the expanded proofs so that the demonstrations are carefully and deliberately thorough.

Newton himself only sketched his proofs in *Principia*. He was not writing a textbook. There is great danger for us modern students in this, as well as great difficulty. The difficulty is obvious: we don't have Newton's impressive mathematical insight and must struggle mightily simply to follow him. The danger is more subtle. It appears on several levels.

First, because Newton doesn't give us step-by-step proofs, but rather hints and sketches, we might imagine that the proof itself consists in hand-waving and not appreciate the rigor behind the sketch. We may think that the conclusions were suggested to Newton by vague physical intuitions, or even that he was careless about the foundations or strategies for the proofs.

Let me dwell on this for a moment, because I believe that some do Newton an injustice by supposing, where he didn't present an explicit justification for a step in his argument, that he "just assumed" the conclusion or that he "couldn't prove it." It is a mistake to think Newton was only stabbing at his results. His hints often refer very specifically to propositions of Euclid and the writers on conic sections; the trick is to recreate the whole proof for which those propositions are the key. Just because he didn't laboriously write out all the steps doesn't mean he didn't see them. His intuition was powerful, but it was also penetrating.

So why didn't he write out all the steps? Surely he must have known how much easier it would have made things for us who are trying to gain insight into his vision and his work with only his text to guide us.

We must acknowledge that Newton was taken up by that very great vision and great work, whose execution, as it was, took years of all-consuming effort. He had neither the time nor the interest for the writing of a primer or textbook. That job was left for those of us who came after. Instead, Newton was writing for mathematicians, and not just any mathematicians, but the very best.*

The second danger to us in trying to work through *Principia* with only Newton's sketches of the proofs is that if we guess carelessly without

---

* The same Robert Hooke who had claimed to Wren and Halley that he could prove that planets moved by an inverse-square attraction to the sun had suggested this inverse square force law in a letter to Newton. Although Hooke presumably tried, he was evidently unable to come up with any proof for this. Newton not only could but did, and came up with not just a proof for the force law but with the masterpiece of *Principia* universalizing gravitation. Hooke insisted on being credited for first having thought of the force law. Newton, enraged, said that anyone could have a guess, or even a vague intuition. The credit should go to the one who actually proved that an inverse-square force law would result in planets moving in ellipses. Newton's indignation reportedly extended to having said, perhaps more spitefully than accurately, that he wrote his demonstrations in *Principia* so that Robert Hooke's inferior mathematical mind would not be able to understand them.

checking our guesses by constructing a rigorous proof, we may mistake the bases on which the proofs rest. This could lead to a significant misunderstanding about what Newton was doing and on what his conclusions rested.

Finally, we may inadvertently assume what Newton is trying to prove. Our physical intuitions, our very common sense, incorporate the conclusions of Newtonian physics, the principles of universal gravitation for which *Principia* lays the foundations and develops the demonstrations. We may falsely conclude that the proof Newton hasn't specified completely is based on some pieces—we pick the ones that seem needed for a particular proof—of what seems simply obvious to us but that Newton gets to only much further along the line.

Great care is *needed* to start where Newton starts and go with him, proving each step along the way of each proof. Any jump, any intuitive leap, risks pulling in illegitimate assumptions.

For that reason, I sought to make the expanded proofs thorough and rigorous enough that we could be confident that we have assumed nothing. Consequently, they will seem to most readers, at least in places, unnecessarily picky, and I have quite deliberately chosen always to err on that side, having seen over and over how easily students "get the general idea" and get it not quite right or even completely wrong.

The student who wishes to follow Newton's discovery authentically, rather than just to find some apparently plausible way of arriving at a conclusion known since grade school, will do well to self-impose this discipline. A step the mind will blithely intuit without any misgivings can strike the same mind in all its unwarrantedness when written out in black and white.

Section 3: The Context: What Do We Already Know?

Ideally we should assume only what Newton gives us to work with: definitions, axioms, propositions, lemmas, "rules of philosophizing," "phenomena." We may also assume, with Newton, the propositions of Euclid, Apollonius, Galileo, and (in one case) Huygens.

It will be obvious that we may not assume any of the conclusions of celestial mechanics and universal gravitation that Newton gave us.

But what, in the store of what we "know" about these things, did Newton give us, and what was his inheritance? Much of what Newton contributed now seems so obvious that we find it difficult to imagine it not being part of the common sense of all those who came before. There is a tendency, for example, to mistakenly attribute certain intuitions of Newtonian physics to Copernicus, Descartes, Galileo, Kepler, Huygens, and others.

So it will help to review the state of theory and understanding in Newton's time of what we now think of as "gravity." Newton showed that the same force that made rocks fall on earth made the planets orbit the sun and the moon orbit the earth. Previous thinkers, if they had explanations at all, had separate explanations for why heavy bodies fell to earth and for what made the heavenly bodies move as they did in the sky.

3.1 Terrestrial heaviness

Of what was then called gravity we may assume only the phenomenon of terrestrial heaviness. As we observe, heavy bodies lifted above the earth and not supported there will fall back to earth.

Aristotle had said that the heavy (earthy) bodies were moving towards the center of the universe (which the center of the earth was supposed to be) because that was their natural place and when displaced from it they strove to return.

Galileo gave us a law for the acceleration of heavy bodies as they moved towards the center of the earth, an acceleration that he believed to be uniform as well as constant for different distances.

Few managed any speculation of cause; among those who did, the speculations ranged from reworkings of Aristotelian natural place (Copernicus, Galileo), to a theory of an innate and natural tendency of matter to clump together into a sphere (Copernicus, Kepler, Galileo), to a theory postulating some sort of magnetic attraction between the earth and the clumps of earth (Gilbert), to elaborate vortices around the earth that pushed heavier matter down (Descartes, Leibniz, Huygens).

Kepler speculated that each planet was a kind of center for the natural place of its own parts, that, for example, jovial matter had an innate tendency to reunite with Jupiter and saturnine matter with Saturn just as terrestrial matter did with the earth. He believed that the moon was actually terrestrial matter and the attempt of earth and moon to reunite caused the tides. Kepler did not, however, see any such tendency for reunion between the planets and the sun. He had another theory for the motion of planets around the sun.

3.2 Celestial mechanics

Copernicus needed no physical explanation for celestial mechanics: things moved in circles because that motion was natural and most perfect. He reworked Ptolemy because he thought it was more in accord with the nature of the sun than the nature of the earth to be the center of that circular motion.

Descartes and his followers thought the planets moved around the sun because all space was completely full of matter that was in motion in vortices, and the planets were carried around by the vortices. Huygens and Leibniz, while departing from Descartes in the details of the vortex mechanism, shared the same explanation.

Kepler thought the planets were dragged around the sun by magnetic strings reaching out from the sun. The sun rotated (evidenced by the movement of sunspots), and thus the magnetic strings revolved and pulled the planets along.

Galileo saw perpetual inertial motion as circular, so he needed no special explanation for the movement of planets in what he believed, following Copernicus, to be their circular orbits. (He continued in this belief despite Kepler's publication of his discoveries about elliptical orbits.)

Note that none of these thinkers was proposing that the planets moved by the force of gravity. Newton not only proposed that, he proved it, and did so in the propositions we will be studying here.

Section 4: Suggestions for Students

Even if you are approaching *Principia* for the first time, you may want to try to work out the proofs on your own as much as possible, falling back on the expanded proofs only when you get stuck or if you want to check yourself.

This is the recommended use of this guidebook, since it will involve you most actively in Newton's project and make the whole adventure most enjoyable.

The expanded proofs are not intended to substitute for that active involvement. Their intent is to save you from frustration if you can't see where to start or if you are stumped about a particular step. You can also check them to see whether you might have made some illegitimate assumptions.

Students vary greatly in their levels of development of mathematical insight. The existence of the expanded proofs as a resource ensures that all students can satisfy themselves about the integrity of every proof. Every student can come to class prepared to demonstrate and explain every proposition with confident understanding. And none of the excitement or significance of *Principia* will get lost because of frustration or discouragement.

Principia is still hard work, even with all the steps spelled out. It is a challenge, and an exhilarating challenge. But it is one that every student is equal to, if the adventure is approached with an adventurous spirit. Removing the frustration of the mystery-steps at the point when they begin to spoil the fun can help keep up the adventurous spirit.

It is highly recommended, before attempting your own proof, to check the notes that follow Newton's words. These can explain key terms in the proposition, keep you from going down plausible but wrong tracks, and help motivate the effort by explaining the significance of the proposition.

The following are some reminders and warnings to help those who are constructing their own proofs from Newton's sketch without using the guidebook's expanded proofs. Some of these were implied in comments that appeared in earlier sections of this introduction. Some may reappear at dangerous junctures in the development of the propositions.

4.1 He gives the hints, we make the proofs

The first warning is against letting yourself be lulled into thinking that the hints and sketches Newton offers following his enunciations actually provide all that is needed to persuade yourself or others of the conclusion. These hints are not intended to constitute a demonstration in the way that what follows the enunciation in Euclid's *Elements* does. Just because Newton says "x, therefore y" doesn't mean y follows from x with no intervening steps. What it means is that x is a key step that helps you towards y. Y must be proved, as, indeed, x usually must.

You have a choice here of attitudes to take towards all this. One option is to feel resentful or indignant that Newton was so inconsiderate of his readers as not to put in all the steps they might need. This might make you feel justified in turning your back on the whole enterprise, but that is a rather dubious benefit, and no real pleasure.

Another option is to feel honored to be in the select group of real mathematicians to whom Newton is directing his arguments. One can feel the respect he had for his readers (that's *us*!) to expect them to be able to fill in these missing pieces. He isn't going to condescend to us by filling in all the baby steps, as if we were beginners who didn't know how to construct a mathematical proof. We have learned how to construct a proof: we are given certain things by the proposition; certain additional things went before in the form of prior propositions, axioms, and definitions; we look where we're going, at what is to be proved; and we think how to manipulate the pieces of what we have to get there. In short, we *are* the people he is writing for; we are worthy of the respect he is expressing for us.

With this basic confidence, one can feel the sense of adventure and intrigue, the challenge of solving the puzzle. It can be viewed as one views many games and puzzles people buy and play *voluntarily* for *fun*. It can be viewed as one views a good detective story: all the clues are there, now scratch your head and see if you can figure out "whodunit."

All right, suppose the challenge has been accepted. With an adventurous spirit we are undertaking to turn his hints into our proof. How will we recognize what steps we need to fill in his sketch?

Look closely, critically. Write out your own proof on your own paper. Make sure you are satisfied that each line on your paper follows from the previous steps. A step may look plausible—it should, it's true—but *why*? How do we get to that true and plausible conclusion?

Here's where you call on what you know about how to construct a proof.

What has Newton given us so far? That may mean reviewing the lemmas and their corollaries, as well as previous propositions. This review might need to include things you brought out about the relationships established as you proved these lemmas, corollaries, and propositions, things that might not jump out from the enunciation. A running catalog of your own making is very helpful.

Then think back to what you know about geometry and conic sections. What geometric relationships are provided by Euclid and Apollonius?

Now you are ready to generate your own solution to the puzzle. How can all these things be cleverly put together to get to where we need to go? Does this tentative plan fit with what Newton says in the sketch? Have we been careful not to grab a useful conclusion without checking to make sure that we have met all the hypothetical conditions that lead to that conclusion? (The following two subsections will give additional considerations with which to check your proof.)

If you are undertaking to demonstrate this proposition in class, it won't do to abdicate responsibility with "He says..." Everyone will want to know why he says that, why it's true. If you're at the board, it's *your* proof, and you need to be able to say why. Just make sure you're preparing carefully enough to notice where you're stumped before you're on the spot with your classmates. This is not a grim business: when figuring out "why" stops being fun, go check the expanded proof for a leg up to the next step.

4.2 Don't assume what has not yet been proved

The second warning is to avoid bringing in assumptions that are yet to be proved. I recommend committing yourself at the start to clearing your mind of practically everything you know about gravity. This includes the idea of bodies attracting other bodies, mutual attraction, attraction between particles, gravity as attraction, centripetal force as something taking place towards a *body*, gravity as something making planets orbit the sun or moons orbit a planet, and the force of gravity as varying by distance from a center. All these things, and others not here enumerated but that we now have incorporated into our understanding of the physical universe, will be arrived at by our efforts as we work through this book.

We are not entitled to start with them or to pull them in to help us with our demonstrations until we ourselves, with Newton's help, have proved them. (If you catch yourself dragging things in, it might help to review Section 3 of this introduction.)

The significance and the excitement of this project is the very fact that we are finally finding out where all those ideas that we take so for granted come from, and why we have reason to have confidence in them. It is an important investigation and an adventure in self-knowledge.

We don't want to compromise this important investigation by muddying up our assumptions with what we want to prove. Nor do we want to set ourselves up for misunderstanding and frustration.

If we succeed in not bringing in Newton's conclusions as assumptions as we trace our path to those conclusions, we will experience the full excitement of their development and appreciate his impressive contribution.

If we fail to keep those assumptions out, we will feel impatient with Newton, think he is stupid for not realizing what is so obvious to us, and be frustrated that he gives long, difficult proofs for what we already know. This is no state of mind in which to be having fun.

4.2.1 The cause of gravity

I mentioned clearing your mind of the idea of gravity as attraction. Newton makes this a little hard to do by sometimes referring to the operation of gravity as an attraction. However, Newton says repeatedly in *Principia* that he is making no hypotheses about the cause of gravity, just describing its effects. Sometimes, he says, he will speak of a body being impelled towards a center, sometimes of the body being attracted towards the center, but he wishes to leave the mechanism open. He is *not* suggesting that there is something at the center that has a power of physically operating on other bodies with no intervening physical medium. You may be tempted from time to time in the course of a proof to assume something that you deduce from the idea that gravity operates by something actively attracting something else. Resist this temptation. Let's take him at his word and leave the options open. For all we know, the planets may be being pushed from behind towards the center of forces by cosmic gremlins, or at least pushed in by differential pressure of an aether that is thicker farther from the sun, as Newton himself once had speculated.

Consider some of Newton's disclaimers taken from those liberally sprinkled through *Principia*:

In the commentary to the Definitions, he says

"Further, I call attractions and impulses accelerative and motive in the same sense. Moreover, I use the words "attraction", "impulse", or [words denoting] a propensity of any kind toward a center, indifferently and promiscuously for each other: I am considering these forces, not physically, but only mathematically. Therefore, the reader should beware of thinking that by words of this kind I am anywhere defining a species or manner of action, or a cause or physical account, or that I am truly and physically attributing forces to centers (which are mathematical points) if I should happen to say either that centers attract, or that forces belong to centers."

In the scholium after Proposition I.69, he says

"The word 'attraction' I here use generally for any attempt whatever of bodies to approach one another, whether that attempt arise from the action of the bodies (whether of mutually seeking one another or of setting each other in motion by emitted spirits), or whether it arises from the action of the aether, or of air, or of any medium whatsoever (corporeal or incorporeal) in any way pushing bodies floating in it towards each other. I use the word "impulse" [*impulsus*] in the same general sense, as in this treatise I am considering, not species of forces and physical qualities, but mathematical quantities and proportions, as I have explained in the Definitions."

In the General Scholium at the end of Book III, he writes:

"Hitherto I have set forth the phenomena of the heavens and of our sea through the force of gravity, but I have not yet assigned the cause of gravity. . . . The reason for these properties of gravity, however, I have not yet been able to deduce from the phenomena, and I do not contrive hypotheses. For whatever is not deduced from the phenomena is to be called a *hypothesis*, and hypotheses, whether metaphysical or physical, whether of occult qualities or mechanical ones, have no place in *experimental* philosophy."

4.3 Prove it Newton's way

The third reminder is to do the proofs the way Newton would have done them. If it's just a matter of getting to the conclusions, well, we had most of them by third grade and all of them by seventh grade. Any encyclopedia will be happy to summarize them for us.

The enterprise we're engaged in here is to put ourselves in Newton's time and place—in his classroom, as it were—for his presentation of these discoveries. How did he argue for his conclusions? What were his thought processes in proving the propositions? That's what we want to recreate.

Always remember that we're not looking for *any possible* proof. There may be many ways to prove things. We can prove many Newtonian propositions using analytic geometry, for example. These are valid proofs. But that's not the way Newton thought about them; that's not the way the propositions formed themselves in his mind. If it's *Principia* we want to be studying, then we need to study it in a Newtonian style.

There are modern notations that involve conventions and developments that came after Newton; some of them save time and pencil lead; we use them in studies of mechanics. That's fine for modern work or even for applying Newton, as long as we have gone through an explicit conversion to see the transition from the Newtonian language of proportions to the algebraic equations.

---

* See Section 6 for explanation of Newton's use of ratio and proportion.

4.4 Example of the perils of anachronism:

In the study of mechanics we also use things derived from Newton expressed in the modern physicist's way, such as "$f = ma$". That equation has served us well in physics, but don't bring things like that to *Principia*! It will be a source of considerable and recurrent confusion. Let's take the very familiar "$f = ma$" as a case study. What sorts of trouble will we get into by trusting its familiarity and trying to use it to get our feet on the ground in *Principia*?

First, $f = ma$ is an equation. Newton didn't deal in equations; *Principia* throughout is cast in the language of proportions. Nowhere does he say any force is *equal* to any absolute quantity; we always have a comparison of forces in "same ratio"* to some other things, for example, $F_1 : F_2 :: m_1 : m_2$ compounded with $a_1 : a_2$. The equation $f = ma$ was not only not used by Newton, but did not appear in any work until Euler stated it in 1747. To use this equation implies an entire structure of algebraic mathematical physics, in which quantities such as mass and force are expressed as certain multiples of fundamental units, which must be defined. To substitute equations for Newton's proportions is therefore seriously anachronistic, and can even turn Newton's demonstrations into nonsense.

Actually, the proportion we have stated is still not quite right; what Newton actually gives us for that definition of force is "...proportional to change in quantity of motion generated in a given time," that is, Δmv per given time. So what he actually has is:

$$F_1 : F_2 :: \frac{\Delta(m_1 \times v_1)}{\text{given } \Delta t} : \frac{\Delta(m_2 \times v_2)}{\text{given } \Delta t}$$

Now you *can* get from there to "ma". Because m_1 and m_2 are each individually constant as velocity changes, they can be factored out so we can operate on this definition to get

$$F_1 : F_2 :: m_1 \frac{\Delta v_1}{\text{given } \Delta t} : m_2 \frac{\Delta v_2}{\text{given } \Delta t}$$

which translates to mass times average acceleration. Taking this to a limit as $\Delta t \to 0$ to get instantaneous acceleration, we shall have force proportional to ma.

But wait, there's still a problem. What force is this? Newton in his Definitions gives us three different measures of centripetal force: absolute quantity of force, accelerative quantity of force, and motive quantity of force. Which is the one that corresponds to "ma"? It was Definition 8, motive quantity of centripetal force.

But when Newton talks about "force" in the first two books, he doesn't mean motive quantity of force, he means accelerative quantity of force. So if we mentally substitute "*ma*" when we read the word "force" in his lemmas and propositions in Book I, we will find it impossible to follow his argument.

I have dwelt at some length on the confusions that proceed from careless assumptions about what "force" means for two reasons. First, these confusions provide a good example of the dangers of bringing later definitions and formulations into the reading of Newton.

Second, the mistake of assuming that when Newton uses the word "force" one can mentally substitute the modern "mass times acceleration" is made almost universally by beginning readers of *Principia*, even when warnings have been given. It is a persistent confusion that wastes a great deal of time and leads to frustration and misunderstanding. If you can make up your mind now to let Newton define his own terms for *Principia*, I promise you it will work much better for you.

Force is not a word handed down by a deity at the beginning of time complete with a fixed and permanent meaning. The meanings we associate with the word in the physics of the twenty-first century were laboriously worked out over time, with many different conceptions proposed and explored, and many different words used. The century before Newton was a particularly rich time of thinking and experimentation with concepts of force. Our modern definition has emerged as a very useful one, but it is not the only possible one, or even the only useful one.

We are seeing our modern understanding of force beginning to coalesce here in *Principia.* It is an exciting moment—don't take it for granted.

4.5 Anticipations

A final hint is to remember that from time to time Newton refers ahead to what he's going to prove later. He usually puts these anticipations in scholia between the actual propositions, but not always. Some of them show up in his commentaries on his Definitions and Laws of Motion, some even pop up in corollaries. (I have tried to note any cases of this that occur outside of scholia.)

You know, if you have been careful, what has been demonstrated already. If you see something not yet proved, look carefully at it. You may be able to see the evidence that he's presaging what he will prove later. In any case, don't swallow it if it hasn't been proved! Wait and see whether he does prove it before accepting it as a legitimate step in the development.

4.6 Summary

In summary: try the proofs yourself; write them out line by line and insist on getting all the steps; do it the way Newton would have done it; don't assume anything you haven't proved; and, above all, keep up a spirit of adventure and discovery.

Section 5: Newton's Use of Mathematics

In undertaking the adventure of following Newton's process in this product of his thought, we want to avoid the distortion of his thought by projecting modern notions. But the modern notions are what we know. How do we avoid interpreting Newton through modern eyes if those are the eyes we have? The following remarks may help raise our awareness and help us understand why Newton's approach is not just accidentally different from our modern one.

5.1 Geometry

One question that has puzzled readers ever since *Principia* was published is why Newton chose to express his proofs in an idiosyncratic modification of the classic geometry of the ancient Greeks. Newton had already developed his version of the calculus in algebraic form, but chose not to use his "method of fluxions," as he called it, for the *Principia*. Why not?

I'm not going to try to answer that question here; that's for you to explore as part of the adventure of reading *Principia*. But I do want to say a few things about why it is important to read Newton in the mathematical language in which he wrote, rather than translating him into modern notation.

First, Newton, like Kepler before him, had very clearly formulated ideas about the relationship between God, geometry, and the created universe. He expresses these ideas in the "Preface to the Reader" (page 3 below). It is at least arguable that he believed that geometrical proofs exactly match the inner workings of nature as God actually set it up. To throw this out and substitute some alternative means of expression is to reject this astonishing implicit claim out of hand, depriving oneself of the opportunity to consider Newton's take on the ever-perplexing question of the relationship between mathematics and physical reality.

But there were also more down-to-earth reasons why Newton may have preferred geometry. One of these was the discontinuity in the number system: there are infinitely many spatial magnitudes that cannot be expressed in whole numbers or fractions. This fact, which was known to the ancients (it is the subject of Book Ten of Euclid's *Elements*) made it impossible to put the algebraic form of calculus on a rigorous mathematical and logical foundation. It took over a century for number theory to develop to the point where it could encompass the irrationals, and meanwhile strong arguments were raised against the use of "infinitesimals," as Newton's fluxions and Leibniz's differentials were called. The development of a geometrical version of the calculus provided the continuity that the algebraic version lacked. In its geometric form, Newton's notion of limit as expressed in Lemma 1 and the Scholium following Lemma 11 possesses impressive rigor and sophistication.

Further, there is evidence that Newton believed that algebraic expression of physical relationships obscures nature by lacking referential clarity, in contrast with geometry, which always clearly expresses the physical objects. Equations are mechanical and opaque, while geometry gives us a clear visual and intuitive sense of what we are proving. So in presenting *Principia* in geometrical form, Newton is inviting us, not merely to follow an abstract argument, but to experience directly the hitherto invisible forces that make the universe go. Do we want to decline that invitation?

5.2 A warning about ratios and proportions

The practical aspects of Newton's use of Euclidean ratios and proportions are described below in Section 6. Here I think I need to warn students about the dangers of substituting equations for proportions, and especially of reading our shorthand notation "$A \propto B$" as meaning "$A = k \times B$", where k is a constant of proportionality.

A general problem with using equations is that they assume the definition of standard units for all variables, and this assumption presupposes the whole structure of mathematical physics that Newton was beginning to develop. In other words, it's anachronistic. We must always bear in mind, as we study this book, that there were no such things as "grams" or "newtons" that could be plugged into equations. The ideas of force, mass, and so on, were defined in terms of ratios, and their relationships were explored and developed in the course of the book. To treat them as magnitudes given in themselves is to obscure the way these ideas grew out of Newton's work. For further discussion of the way Newton thought of these things, see the notes on Definition 1, below.

However, there are more immediate problems with substituting an equation for a proportion: it can lead to absurdities that are not present in Newton's proof, but are created by the substitution. For example, in I.14, in one step of the proof, Newton states that "QR...is as...SP^2." What he means here is that, when comparing two different places P and p on two orbits,

$QR : qr :: SP^2 : Sp^2$.

The two magnitudes on the left are vanishingly small, while the two on the right remain finite. Nevertheless, each ratio, taken by itself, is well-defined and finite, and the two ratios are the same. No problem. But if we try to turn this into an equation,

$QR = k \times SP^2$,

it obviously becomes nonsense, since the constant k must somehow make something infinitely small (QR) equal to something finite (SP^2). For this reason, you must always keep in mind that a proportion cannot be safely transformed into an equation with a constant multiplier.

That is, $A \propto B$ does not mean the same as $A = k \times B$.

Section 6: Newton's Use of Ratios and Proportions

6.1 Ratio

A ratio is a relationship between two numbers or between two magnitudes. Euclid defines ratio of magnitudes in Book V Definition 3 of the *Elements*:

> A *ratio* is a sort of relation in respect of size between two magnitudes of the same kind.

Note that the magnitudes must be of *the same kind*: areas may be in ratio to areas, lines to lines, velocities to velocities, forces to forces. But Euclid doesn't compare unlike magnitudes, and Newton follows him pretty consistently in this, and doesn't try (for example) to put force in ratio to area. Since ratio is a relation in respect of size, how would we know whether some area were larger, smaller, or equal to some force? How large must an area be before it is twice as large as a certain force? They cannot be directly compared; they have no relation in respect of size.

When we use algebraic equations, on the other hand, we treat what were ratios as fractions, that is, as quotients of numbers. Then we are no longer thinking of the numerator and denominator as magnitudes with a *kind*. Newton sometimes makes use of equations in this way; however, he avoids doing so, because when we do that we no longer have a geometric, and thus visual and intuitive, picture of what we are doing. The manipulation of numbers using algebra is convenient, but it's a bag of tricks that loses for us the reality of the things behind it, at least for the duration of the transformations.

With ratios we speak of the *antecedent* and *consequent* instead of numerator and denominator. In the ratio $A:B$, A is the antecedent and B is the consequent.

6.2 Same ratio and proportion

Let's say we have our given ratio. Now suppose another relationship of like magnitudes with respect to size. This may be the *same ratio* with our given ratio. For example, each consequent may be twice as large as its respective antecedent.

We may call this same relationship "same ratio" even if the like magnitudes of the second ratio are of a different kind from the like magnitudes of the first. For example, in the first ratio, one area may be twice as large as another area; in the second ratio, one force may be twice as large as the another force.

Two ratios that are the same constitute a *proportion* and it is abbreviated as "::". For example, if $a:b$ is the same ratio as $c:d$, then $a:b :: c:d$.

Euclid worked out a number of legitimate manipulations that can be done on the four magnitudes in a proportion, yielding new same ratios.

Antecedent and consequent of each of the same ratios can be *inverted*. If the ratios of antecedents to consequents are the same, the ratios of consequents to antecedents should be the same. Thus if one proportion is discovered or assumed to be true, the "inverted" proportion will also be true. Early commentators called this operation *invertendo*.

They can be *alternated*, such that antecedent is taken in relation to antecedent and consequent to consequent (*alternando*). One must be careful with this: if the second antecedent and consequent are a different kind of magnitude from the first, alternation will produce illegitimate "ratios" between unlike magnitudes. Newton avoids doing this; when he must alternate unlike ratios he switches to algebraic equations. If the ratios are not ratios of magnitudes but of numbers, then we don't have this problem. Since all numbers are homogeneous with one another, a proportion of numbers remains a proportion of numbers even after alternation.

We can also *compose* a ratio (*componendo*). There we add antecedent to consequent in both ratios and set them in ratio to the original consequents. This yields new same ratios.

We can also *separate* ratios (*separando*). In this we take the difference between antecedents and consequents and set them in ratio to the original consequents. This also yields new same ratios.

In *conversion* of ratios of a proportion (*convertendo*), we put the original antecedents in ratio to the difference between the antecedents and consequents.

An *ex aequali* proportion arises when we have a string of two or more proportions such that the same magnitude appears as consequent of one and antecedent of another. Our string may have several of these repeated magnitudes. In algebra we would say that we are multiplying simultaneous equations and "canceling" the intermediate terms. We may cancel enough of them to be left with a resultant proportion of just four simple magnitudes; or we may be left with ratios of multiplied antecedents or consequents.

6.3 Compounding ratios; duplicate and triplicate ratios

A proportion can also express sameness between one ratio and a ratio that is obtained by multiplying two or more other ratios. In Euclidean terms multiplying ratios together is called *compounding ratios*. Euclid doesn't put compound ratio in terms of multiplication, but in terms of relationships between geometrical figures. In Book VI Proposition 23, Euclid gives his geometrical foundation for the operation, letting the magnitudes that are in the ratios be represented by the lengths of sides of equiangular parallelograms. The ratio of the areas of the parallelograms represents the ratio of the compounded ratios.

Newton doesn't try to represent compounding with geometrical figures, despite his overall approach in *Principia* of using geometrical representations rather than ones of algebra or analytic calculus. (One occasionally sees a bit of that geometrical thinking in other places too, as in his referring to the all-important expression representing the ratio of forces in I.6 Corollary 1 as a "solid"). As a rule, however, he treats the operation as simple multiplication, although of course a multiplication of ratios. Taking the shorthand of algebraic equations would have meant abandoning the connection to reality and grounded intuition.

When ratio $C:D$ is *compounded with* ratio $E:F$ the result is the ratio $(C\times E) : (D\times F)$. Thus to express that ratio $A:B$ is same ratio as ratio $C:D$ *compounded with* ratio $E:F$, we may write

$A:B :: (C:D)$ *comp.* $(E:F)$, or

$A:B :: (C\times E) : (D\times F)$.

Sometimes a ratio is proportional to a second ratio compounded with itself. This is called *duplicate ratio.* So $A:B :: C:D$ *compounded with* $C:D$ again; or we say that $A:B$ is the *duplicate ratio* of $C:D$. This is written in various ways for short; in this guidebook I will use the abbreviation *dup*( ). For this example, I will say $A:B :: dup(C:D)$.

Since compounding is algebraically like multiplying, duplicate ratio will lead to what we call squares in algebra. In this example, $A:B :: dup(C:D)$ is, in this algebraic view, the same as $A:B :: C^2 : D^2$. (Euclid defines duplicate ratio in terms of a third proportional, but Newton treats it in a different way, as explained below.)

Triplicate ratio works the same way as duplicate ratio, except that, algebraically, we multiply the ratio times itself twice. Both antecedent and consequent become "cubed" in algebraic terms.

Although Newton was able to use algebra with facility, he chose not to embrace it in *Principia,* and it's instructive to note where he chooses to be Euclidean, where he chooses to be algebraic, and what idiosyncratic middle ground he chooses as his characteristic mode of presentation. (A translation that obscures this—as other present-day translations do in converting duplicate ratios to algebraic squares—does a disservice to the thoughtful reader.) Newton's characteristic way of handling duplicate and triplicate ratio is described in the following paragraphs.

Euclid proved that squares (like any similar figures) are to each other in the duplicate ratio of their sides. Newton, following Apollonius, often makes use of this by substituting the ratio of squares on given lines for the duplicate ratio of the lines. However, it is worth noting that Newton almost always writes "square C" or "sq. C," which can mean either the square constructed on line C or C multiplied by itself. He avoids the superscript notation C^2, although he was familiar with it.

Further, cubes, and other similar solid figures, are in the triplicate ratio of their sides. Therefore, a triplicate ratio can be represented by the ratio of cubes on the two lines. And since a triplicate ratio is a ratio compounded with itself twice, it is equivalent to "cubing" in algebra. Newton writes "cube C" or "cub. C", which can mean either the cube constructed on line C or C multiplied by itself twice. Again, he generally avoids using the superscript "3" to indicate the cubing of a quantity or magnitude.

6.4 Subduplicate ratio

Euclid defines subduplicate ratio in terms of a mean proportional. Geometrically, it could mean that the magnitudes are taken as squares, and we take the ratio of their sides. This is equivalent to taking the square roots of the magnitudes. Newton uses the subduplicate ratio as the ratio of the square roots of the magnitudes.

6.5 Shorthand notation for proportions: the ∝ symbol

Newton frequently works with ratios between corresponding parts in different figures, or different instances of a construction in the same figure as a point moves to its limiting position.

For example, he might have a proportion consisting of two ratios between corresponding parts like this:

$AB : ab :: AD : ad$

This can be read as "*AB* varies as *AD*" and written this way:

$AB \propto AD$

This "∝" notation is always shorthand for a full proportion. Whenever we see it we know that we are actually talking about at least four terms, the first and second magnitudes being in the same ratio as the corresponding third and fourth magnitudes. The ratio of the third and fourth may actually be a complex formula of compounded ratios or multiplied or divided magnitudes (as, indeed, might the first and second). So if we had

$$\frac{AB^2}{Ab^2} = \frac{AG}{Ag} \times \frac{DB}{db},$$

we could say

$AB^2 \propto AG \times DB.$

Section 7: Suggestions for Teachers

7.1 Course duration and scheduling

Section 2 of these Preliminaries offers some guidance about course length and intensity for various categories of students, and about the book's use for individual study.

In a 16-week semester, a good plan is to allow eight weeks for Book I and eight weeks for Book III. As an undergraduate college course, this selection works best as a semester course with classes meeting three times a week for an hour and a quarter or an hour and a half. This allows students to gain and demonstrate mastery of all the propositions and to experience the satisfaction of that mastery as the adventure unfolds. It also allows plenty of time for thoughtful discussion of the larger questions raised. The selection can also be done in a semester with two classes per week with a little less leisure or with some sections read and discussed but not demonstrated. For a course lasting less than a semester, or for one that requires less mathematical application, one could profitably work through just the first four propositions of Book I before turning to Book III.

If time seems to be running out in Book I, try to prove the lemmas and the propositions through Book I Proposition 4, and then move more quickly over the remaining propositions through Proposition 17, skipping the formal demonstrations in class but discussing the propositions, including questions raised in notes.

If time seems to be running out in Book III, the propositions that dip back into the previous books—II.24, I.69, and I.71 through I.75—can be read over, perhaps in the form of the expanded proofs, with careful attention to the notes and Pauses, without being formally proved in class. Try to get at least through III.8: that will provide the culmination of the development of universal gravitation. Then III.13, deriving the planetary ellipses, could be read and discussed to see the application of his method, deducing the motions of the planets from the forces of gravity.

I don't recommend stopping after Book I for two reasons. First, that is not the way Newton recommended his book be read; in his preface to Book III he recommends just our selection. And as he explains in his preface to the whole work, it is intended as a demonstration of his method by applying it to derive the laws governing the other phenomena of our world.

Furthermore, he says at the end of the scholium following the definitions, "[H]ow to determine the true motions from their causes, effects, and apparent differences, and, conversely, how to determine their causes and effects from the motions, whether true or apparent, will be taught more fully in what follows. For it is to this end that I wrote the following treatise."

Thus to stop with his protheorems is to mistake his purpose.

Second, the application to our world is the part that brings the whole project alive. We see how one connects the mathematical tools to the world we can observe and in which we live and have our weight. Devoting all one's time to just the mathematical propositions can leave the student feeling burned out and without any particularly exhilarating result.

In addition, it is only by working through the propositions of Book III included in this selection that we can get to universal gravitation. Book I does not yield this result; even Book I with III.4 does not do this, although the major clue has, of course, been provided.

I also encourage all students of *Principia* to work through the calculations of the Phenomena at the beginning of Book III, to get a hands-on, nitty-gritty sense of how one moves from the hypothetical propositions of the first two books to the demonstration of truths about our world.

Once you have done that, you will never look at Jupiter or its moons the same way again: they will have become intimate and vitally interesting friends.

7.2 Class format

The way to have the most fun with this work, and to get the deepest and most intimate understanding of it, is to have each student prepare proofs for every proposition, and to have students take turns demonstrating the propositions at the board in class.

As the propositions are demonstrated, students, and the teacher, can ask questions, request clarifications, and discuss the method and the assumptions involved in the proof. Sometimes it will be helpful to look back to previous propositions or lemmas, or to definitions or laws.

Class discussions can continue these threads, and also inquire into the larger questions raised. Some of these large questions are listed in Section 2 above. Others are included as "Questions for Discussion" in the Notes and Expanded Proofs.

These discussions will provide fruitful material and inspiration for papers written by the students individually.

I recommend an evening viewing to see Jupiter and its moons while working with Phenomenon 1. I have used a small telescope. A three or four inch reflector, preferably with a good eyepiece, works very nicely. This is the approximate power available to the astronomers on whose work Newton relied. The moons can be seen even with binoculars.

Alternatively, one can connect with a local astronomy club to arrange viewings using their telescopes; one of the students could undertake to give a talk to club members about how the observations were done in Phenomenon 1 and their role in the development of the theory of universal gravitation.

7.3 Encouraging student initiative

The text has been arranged so that it needs to be checked only for the step of the proof that has proved difficult, thus allowing the maximum individual exercise, at least for the first stab at a proposition.

Students can be encouraged to start each proposition with Newton's text and see whether they can figure out how to construct the proof as he sketches it out. The notes after the proposition can be very helpful, in many cases providing information the student would have no way of knowing or figuring out. But students particularly well-prepared with background in history of astronomy and classical mathematics who want the maximum challenge could try their own proofs before checking even the notes.

Where difficulties arise, after having exhausted their own efforts and imagination, the notes and expanded steps of the proof are available to keep the undertaking moving. But if the teacher helps instill sufficient sense of adventure and initiative in students that they make their own attempt at the proof first without the expanded proof, more motivation and engagement can result. The effort may make students invested in finding a proof, and curious how it might be done. This can make the subsequent efforts, when students turn to the expanded proof, more enjoyable and more educational; it can feel less like slogging through someone else's proof.

7.4 Using the helps; and recognizing some dangers

On the other hand, it must be recognized that there are potential problems with only checking the helps when one feels stuck. It will sometimes be the case that the student has been going down an erroneous or fallacious path in making up his or her original proof. In such a case the step in the expanded proof consulted upon hitting a snag may be inconsistent with the presumptions of the student's proof. It may thus not, as I promised, be helpful in providing just the next step needed at this impasse.

There may be things that have been pointed out in the notes or in the expanded proof of a previous proposition that it would be helpful or even essential to know in order to understand why the proof of this proposition must expand as it does, perhaps clarifying how that previous proposition must be invoked in this application.

If everyone is using this guidebook as a resource, then usually some other student will have noted those observations and can bring them up in class. However, I recommend that the teacher look over all the helps as the class moves along, even if he or she is (as I hope) engaging in his or her own fun with the process of proving things independently before checking the helps. These notes and other comments are there to keep things clear and avoid frustrating false paths while suggesting substantive things to wonder about and discuss in class.

Newton's sketches are so elliptical that there are many wrong turnings that can be taken. Many plausible ways of constructing a proof may be illegitimate. The excitement and triumph of seeing—as one imagines—a way to leap one of the divides can blind one to a subtle misapplication of a lemma or previous proposition. There is always lurking, as well, the tendency to use something one "knows" about gravity as a post-Newtonian as a step in one of the proofs, without appreciating that this is not "known" until Newton has completed his demonstration in III.8.

If the teacher checks the notes and expanded proofs, he or she should be able to save the class a wrong turning down a plausible but fallacious path. Class time can be spent considering why some of these other possible proofs look likely but go astray, rather than fruitlessly arguing about them without understanding exactly where the problem lies.

7.5 Avoiding glibness and superficiality

Whether students prepare the demonstrations without consulting the helps, or (at the other extreme) whether they study the full expanded proof, they should be expected to have made the demonstration they bring to class their own. Teacher and fellow students should not permit the one giving the demonstration at the board to fall back on either "Newton says..." or "Densmore says...."

Once the proof is prepared, it is the student's proof, and the student should be expected to be able to satisfy his or her classmates, and the teacher, that the proposed result is legitimately and completely established. Students who are at the board must therefore be satisfied themselves, in their preparation, and able to justify every step. If a previous proposition is cited, they must be able to say just what that proposition proved, and what conditions were on its result.

It is this depth of familiarity with the proposition, and care in its preparation, that results in depth of understanding and satisfaction with the work. It protects from making plausible but possibly fallacious assumptions. And it provides a foundation for discussion of the larger questions—indeed, it is often the careful delving into the actual premises and tactics of the proof that opens many of the larger questions.

Resources

Newton assumes that his readers have familiarity with certain basic mathematical texts. These include, most fundamentally, the *Elements* of Euclid and the classical writers on conic sections, particularly *On Conic Sections* by Apollonius. Newton also assumes an understanding of Galileo's basic propositions about motion.

As explained earlier, the expanded proofs cite the appropriate propositions where Newton uses them, quoting or paraphrasing the less familiar ones, including anything not normally covered in high school geometry. However, if you are attempting to fill in the steps of Newton's propositions independently (without making use of the expanded proofs), it will help to study or review some parts of Euclid, Apollonius, and Galileo in order to know what tools you have to work with.

To make this task easier for those who have not previously studied these authors or whose recollection is shaky, here is a list of the propositions of Euclid, Apollonius, and Galileo that Newton uses in the central argument of *Principia* as presented in this guidebook.

Convenient unabridged editions of Euclid's *Elements* and Apollonius's *On Conic Sections* Books I-III are available from Green Lion Press. Galileo's *Two New Sciences* is available in Stillman Drake's translation (recommended) from Wall & Thompson and from Modern Library Science Series. *Two New Sciences* is also available in the older translation of DeSalvio and Crew from Prometheus Books.

From Euclid's *Elements*:

Book I: Postulates 3 and 4; Propositions 4, 6, 13–15, 23, 26–29, 31–34, 37–41, and 47.

Book II: 11–13.

Book III: 3, 14–18, 21, 29, 31–32, and 36.

Book IV: 5.

Book V: Definition 5; Propositions 7 porism, 8–9, 11–12, 14, and 16–19.

Book VI: 1, 2, 4, 5, 8, 14, 16, 17, 19 por., 20 por., and 23.
Book XI: 1 and 2.

From Apollonius's *On Conic Sections*:

Book I: Definitions 4–8; Propositions 11–13, 15;

Definitions 9 and 11; Propositions 17, 21, 32, 35, 37, 46, 47, 49, and 60.

Book III: Propositions 45 and note, 48 and note, 49, 51, and 52.*

From Galileo's *Two New Sciences*:

Third Day:

"On Equable Motion": Propositions 1, 2, and 4.

"On Naturally Accelerated Motion": Definition of uniformly accelerated motion; Propositions 1 and 2.

Fourth Day:

"On the Motion of Projectiles": Proposition 1.

---

* The cited notes are by R. Catesby Taliaferro; they are included in his translation published by Green Lion Press.

Glossary

A fortiori means "the case is even stronger for saying that..."

Alternando. See Section 6 of the Preliminaries.

Anomaly. In Ptolemaic astronomy, the discrepancy between a planet's observed position (with respect to longitude) and the position it would occupy if it were moving with uniform circular motion about a known center. But in the seventeenth century, and for Newton, anomaly was the the angular distance a celestial body has progressed along its orbit measured from the point of greatest distance from the center of forces. For a planet, the angle is measured from aphelion, with sun as center; for the moon, the angle is measured from apogee, with earth as center.

Aphelion. In the path of a body orbiting the sun, the aphelion is the point furthest from the sun. The aphelia are "at rest" if the body returns to the same point at each aphelion.

Apogee. In the path of a body orbiting the earth, apogee is the point furthest from the earth.

Apsides. The line of apsides is the line between the nearest approach of the orbiting body to the center of forces and the most distant point. (For a body orbiting the sun, the line of apsides is the line between perihelion, the point on the path closest to the sun, and aphelion, the point on the path farthest from the sun. The line of apsides of the moon is the line between perigee, nearest approach to the earth, and apogee, greatest distance from the earth in the moon's orbit.) In an elliptical path, with the center of forces at the focus, this line will be the major axis.

Invertendo. See Section 6.2 of the Preliminaries.

Componendo. See Section 6.2 of the Preliminaries.

Compounded ratio. See Section 6.3 of the Preliminaries.

Conic section. Conic sections are curves that are found by cutting a cone with a plane. A plane parallel to the axis results in a hyperbola; a plane parallel to a side of the cone results in a parabola; a plane cutting through both sides results in an ellipse (if this plane is parallel to the base of the cone it will be a circle); degenerate conic sections such as point and straight line are also possible with other placements of the cutting plane.

Conjunction. In conjunction, the moon is on the same side of the earth as the sun, and is seen as "new." An inner planet in conjunction will be seen as new if it is between the earth and the sun; it will be seen as full if it is beyond the sun. An outer planet in conjunction will be on the same side of the earth as the sun but will be beyond the sun and will be seen full phase.

Conjugate diameters in an ellipse or hyperbola. Take any diameter. Take the tangent where this diameter meets the curve. Then draw a line through the center of the conic section parallel to this tangent. This line will be the *conjugate* diameter (See Apollonius *Conics* Book I Definitions 5 and 6, and I.47.)

One set of conjugate diameters of the ellipse is the major and minor axes. These are the only ones that are perpendicular unless it's a circle, where all sets of conjugate diameters are mutually perpendicular. (See Apollonius *Conics* Book I Definitions 7 and 8.)

Diameter of Ellipse or Hyperbola. A line through the center of the ellipse or hyperbola meeting the curve in both directions. (See Apollonius *Conics* Book I Definition 4.)

Diameter of Parabola. Any line that intersects the curve exactly once. All such lines turn out to be parallel. (See Apollonius *Conics* Book I Definition 4.)

Duplicate ratio. See Section 6.3 of the Preliminaries.

Elongation. The angular distance between a planet and the sun; or between any celestial body and another about which it revolves. Although elongations are observed and measured from earth, we may (by a suitable calculation) translate any such measurement into the angle that would be observed from the sun, in which case it is called "heliocentric elongation." In Phenomenon 1, for example, Newton reports the heliocentric elongations of Jupiter's moons with respect to Jupiter.

Enunciation. The statement at the beginning of a mathematical proof that articulates what is to be proven or found by the proof. It is followed by the steps of the proof itself, or in the case of the proofs Newton gives in *Principia*, a sketch of the proof that could be made, indicating the strategy without including all the steps.

Ex aequali. See Section 6.2 of the Preliminaries.

Gibbous. Of the phase of the moon or a planet, greater than half phase and less than full.

In infinitum. Of a process, continued without limit.

Lemmita. A small lemma necessary for the expanded proof, not explicitly given by Newton.

Natural philosophy. A term that up until Newton's time had been used for the Aristotelian study of motion and change. After Newton used the term for this book, it was never the same again, and was soon synonymous with what we call physics or astronomy.

Opposition. In opposition, the moon or planet is on the opposite side of the earth from the sun, and is seen as full.

Ordinate. Drawn from a point on the curve to a particular diameter parallel to the tangent where that diameter meets the curve. (See Apollonius Definition 5.)

Parallax. Parallax is the apparent angular displacement of the observed body that results from change in the position of the observer. See Note 5 in III.4 (page 364).

Perigee. In the path of a body around the earth, perigee is the point closest to the earth.

Perihelion. In the path of a body around the sun, the perihelion is the point closest to the sun.

Planets. Traditionally the "seven planets" were Mercury, Venus, Mars, Jupiter, Saturn, the sun, and the earth's moon. When Newton says "planets," he may include the sun or earth's moon; judge by the context. Newton also at times refers to the moons of Jupiter and Saturn as "planets." Sometimes he makes the distinction by calling the moons "secondary planets" as opposed to "primary planets." Note that traditionally, and in Newton, the earth is not called a planet, although there are places where Newton states general conclusions about "planets" which we have proved apply to the earth.

Principia. Latin for *Principles*. In this guidebook and generally, short for *Philosophiae naturalis principia mathematica* (Mathematical Principles of Natural Philosophy), the full name of the work to which this book is a guide.

Reductio ad absurdum. A type of proof in which all possibilities other than the one that we want to establish are shown to involve a contradiction. That is, all other cases are reduced to some absurdity.

Refraction. Refraction is the apparent displacement of a celestial body owing to the bending of light by the earth's atmosphere. For example, as light from the moon enters the earth's atmosphere, a denser medium, it is bent downward towards the vertical. See Note 5 in III.4 (page 365).

Sagittae (singular, *sagitta*) of arcs lie on lines that originate from a point on the concave side of the curve, and, passing through the points of bisection of the chords to those arcs, extend to reach the curve. The actual sagitta is the segment of this line lying between the chord and the arc. (The word *sagitta* is Latin for "arrow.")

Sesquiplicate ratio. See Section 6 of the Preliminaries for discussion of the powers of ratios. The triplicate ratio is a ratio of cubes; the subduplicate ratio is a ratio of square roots. The sesquiplicate ratio is the subduplicate of the triplicate; that is, a ratio of two quantities each taken to the 3/2 power.

Separando. See Section 6.2 of the Preliminaries.

Syzygies. "Syzygies" means "places of being yoked together," namely, conjunction and opposition, when the sun, earth, and moon are in line.

Q.E.D. is an abbreviation for *quod erat demonstrandum*, "that which was to have been demonstrated."

Q.E.F. is an abbreviation for *quod erat faciendum*, "that which was to have been done."

Q.E.I. is an abbreviation for *quod erat inveniendum*, "that which was to have been found."

Q.E.O. is an abbreviation for *quod erat ostendendum*, "that which was to have been shown."

Quadrature. 1. The angle between two bodies measured from a chosen third point is a right angle. For example, a planet or the moon is positioned such that the angle between the planet and the sun, observed from the earth, is a right angle.

2. Newton specifies that the solution to some problems requires knowledge of "the quadratures of curvilinear figures." In practice, this amounts to being able to compute the area contained by a curve (representing the force law), an axis, and two ordinates drawn parallel from the axis to the curve. This kind of problem (which we would nowadays call "integration") does not usually have a geometrical solution, but was well within the capabilities of Newton's calculus, which he called "the method of fluxions." Thus in those few places in *Principia* where Newton states that the quadratures of curvilinear figures need to be provided, he is warning us that we are to some extent stepping outside the bounds of the strict geometrical method.

Vortex, Vortices. Literally, "whirlpool." This word was used by Descartes and his followers for the swirls of invisible matter that they believed filled all space and carried the planets around the sun, the moon around the earth, and perhaps other planets around other stars.

PHILOSOPHIÆ NATURALIS PRINCIPIA MATHEMATICA.

Autore *JS.* NEWTON, *Trin. Coll. Cantab. Soc.* Matheseos Professore *Lucasiano*, & Societatis Regalis Sodali.

IMPRIMATUR.
S. PEPYS, *Reg. Soc.* PRÆSES.
Julii 5. 1686.

LONDINI,

Jussu *Societatis Regiæ* ac Typis *Josephi Streater.* Prostat apud plures Bibliopolas. *Anno* MDCLXXXVII.

Newton's Preface to the Reader

The ancients, as Pappus wrote, made mechanics of the highest value in the investigation of natural matters, and more recent writers, having dismissed substantial forms and occult qualities, have made an approach to referring the phenomena of nature back to mathematical laws. It has accordingly seemed fitting in this treatise to develop mathematics insofar as it looks to philosophy. Now the ancients established two branches of mechanics: rational, which proceeds accurately by demonstrations; and practical. To practical mechanics all the manual arts look, and from here its name "*mechanica*" is borrowed. But since artisans are accustomed to work with little accuracy, it happens that mechanics as a whole is so distinguished from geometry, that whatever is accurate is referred to geometry, and whatever is less accurate, to mechanics. The errors, however, belong to the artisan, not the art. One who works less accurately is a more imperfect mechanic, and if any could work with perfect accuracy, this would be the most perfect mechanic of all. For the drawing of both straight lines and circles, upon which geometry is founded, belongs to mechanics. Geometry does not teach how to draw these lines, but requires [*postulat*] that they be drawn. For it requires that the beginner learn to draw them accurately before crossing the threshold of geometry, and then teaches how problems are solved by these operations. To draw straight lines and circles are problems, but not geometrical problems. The solution of these is required of mechanics, and once the solutions are found, their use is taught in geometry. And it is the glory of geometry that so much is accomplished with so few principles that are obtained elsewhere. Thus geometry is founded upon mechanical procedure, and is nothing else but that part of universal mechanics that accurately sets forth and demonstrates the art of measuring. Further, since the manual arts are chiefly concerned with making bodies move, it happens that geometry is commonly related to magnitude, and mechanics to motion. In this sense, rational mechanics will be the science of the motions that result from any forces whatever, and of the forces that are required for any motions whatever, accurately set forth and demonstrated. This part of mechanics was developed into five powers by the ancients, looking to the manual arts, since they considered gravity (which is not a manual power) not otherwise than in the weights that were to be moved by those powers. We, however, are interested, not in the arts, but in philosophy, and write of powers that are not manual but natural, treating mainly those matters pertaining to gravity, levity, elastic force, the resistance of fluids, and forces of this kind, whether attractive or impulsive. And on that account we present these [writings] of ours as the mathematical principles of philosophy. For the whole difficulty of philosophy appears to turn upon this: that from the phenomena of motion we may investigate the forces of nature, and then from these forces we may demonstrate the rest of the phenomena. And to this end are aimed the general propositions to which we have given careful study in the first and second books. In the third book, on the other hand, we present an example of this procedure, in the unfolding of the system of the world. For there, from the celestial phenomena, using the propositions demonstrated mathematically in the preceding books, we derive the forces of gravity by which bodies tend to the sun and the individual planets. Then from the forces, using propositions that are also mathematical, we deduce the motions of the planets, of comets, of the moon, and of the sea. In just the same way it would be possible to derive the rest of the phenomena of nature from mechanical principles by the same manner of argument. For I am led by many reasons to strongly suspect that all of them can depend upon certain forces by which the particles of bodies, by causes not yet known, either are impelled towards each other mutually and cohere in regular shapes, or flee from one another and recede. These forces being unknown, philosophers have hitherto probed nature in vain. It is my hope, however, that the principles set forth here will shed some light either upon this manner of philosophizing, or upon some truer one.

[The rest is omitted, since it deals with details of publication that do not bear directly upon the argument.]

Definitions

Quantity of Matter Defined

Definition 1

The quantity of matter is the measure of the same arising from its density and magnitude conjointly.

Air of double density, in a space that is also doubled, is quadrupled; in a tripled [space], sextupled. The same is to be understood of snow and powdered substances condensed by compression or liquefaction. And the same account is given of all bodies which are condensed in various ways through various causes. In this I do not take account of the medium (if any) freely pervading the interstices of the parts. Further, in what follows, by the names "body" or "mass" I everywhere mean this quantity. It is apprehended through an individual body's weight. For it is found by experiments with pendulums carried out with the greatest accuracy to be proportional to weight, as will be shown hereafter.

Notes on Definition 1

- **Measure and Proportionality: a Question of Relationships**

"The quantity of matter is the measure of the same arising from its density and magnitude conjointly."

For a present-day student it may not be clear here from Newton's wording that this definition is giving us a proportion, not an equation. When he speaks of the measure "arising," this is one indicator of it being a proportion. The stronger and more explicit indicator is the word "conjointly." Another way to express this would have been that the quantity of matter varies as the density and magnitude conjointly; that is, that two quantities of matter are in the same ratio as the ratio of the densities compounded with the ratio of the volumes.

That Newton is thinking in terms of a proportion is confirmed by the first words in the commentary: "Air of double density, in a space that is also doubled, is quadrupled; in a tripled [space], sextupled."

Newton is not telling us here that we will get a measure of mass (in grams or some such unit) if we multiply a body's density times its bulk. Rather, he is telling us how to compare two or more bodies. The ratio of the masses is the same as the ratio of respective densities compounded with the ratio of their respective bulks or volumes.

(Compounding is a way of multiplying ratios. Section 6 of the Preliminaries discusses in detail this and other terms in Newton's use of ratio and proportion. Section 5 of Preliminaries discusses the significance of Newton's use of the language of proportion rather than the language of algebra. If the subtleties of the classical Euclidean or Eudoxian use of proportions and the differences between that and our present-day use of equations are not familiar to you, those two sections should be helpful.)

What does Newton mean by "measure"? How does he measure things? In *Principia*, he measures things by looking at relationships, that is, he compares two things of the same sort and finds that they are in the same relationship as the ratio (or compounded ratios) of some other pairs of things, always comparing things of the same sort.

Unlike what we may be accustomed to in physics textbooks, in *Principia* we very seldom get statements that any one thing is *equal to* some particular other thing or combination of other things, for example that a particular mass can be calculated by multiplying a particular density times a particular volume. This language of equations, so familiar to us and so convenient, is almost completely absent from *Principia*.

This is in part because in Newton's time the use of algebra was just starting to establish itself against the traditional use of proportions. But it was also a choice for Newton. He could have used algebra; indeed, elsewhere (and even in a few places in *Principia*) he did use algebra and the analytic calculus he had himself developed. But he evidently felt that what he wanted to represent was better conveyed by the language of proportions.

Though cumbersome, the insistence on sticking with proportions did two things that are lost in converting to algebra and equations (where that conversion would have been possible).

First, it stayed clear about the meanings of the quantities being worked with. For example, we understand that one distance might be twice another distance, but whatever could be the meaning of dividing a distance by a force?

Second, it kept things in terms of relationships. One might want to speculate about why that might have been a value to Newton, and to watch as one works through *Principia* for other ways in that Newton seems to be seeing relationships as the way to understand foundational things. Keep this in mind for pondering the meaning of the Third Law of Motion and the understanding of the working of gravity itself.

• Nature of a Definition

A student once asked, "Why does Newton define quantity of matter in terms of density when it seems that we know how to measure quantity of matter directly, but we don't know how to measure density directly?"

This question raises several interesting issues, two of which may be helpful to explore here.

First, the question confuses a definition with a procedure for measuring the defined item in a laboratory. Definitions in a mathematical exposition describe how terms will be used and do not establish that the thing defined exists. And if the thing defined does exist, the definition doesn't provide the construction or method of finding it. That may need to come in a later proposition, or perhaps in a postulate. How one would go about assigning a number to the quantity of matter in a given thing in the world is not the concern in the definition (or generally in *Principia*). Rather, he is laying out here the relationship between quantity of matter, density, and size.

Another sort of difficulty with this question is that it claims that we know how to measure quantity of matter directly. Do we? We have balance scales that compare weights and spring scales that measure the force we call weight. But weight and quantity of matter, force and stuff, are different things, and we don't yet know the relationship between them. That's one of the important things worked out in *Principia* (see the next note).

This first definition that opens *Principia* invites us to start thinking about just what mass (or quantity of matter) could be on its own. **Question for Discussion.** Imagine that we have no Newtonian physics, and we're trying to sort out these things we're going to be working with. What might quantity of matter be? How would we want to define it, how would we give people a picture of what we will mean when we use the words? Then consider Newton's answers to those questions in this definition.

We have a tendency, living in a post-Newtonian world, to think of mass as something absolute, the concept created along with the matter itself, and that anyone with any other concept or definition must be ignorant or confused. But many ways were open to Newton in conceptualizing what we now call mass. The way he did decide to think about it, the way he defined it, turned out to be very useful, so useful that it's now hard to imagine any other way of thinking about it.

The way to think about these definitions is not to ask whether we would have picked different words (or equations) to describe what we as his beneficiaries understand these things to be. That might be appropriate if he had been given the understanding modern physics uses and was just writing a textbook to codify it. Then we (or some other modern textbook we have read) might have different or better ways of expressing this understanding.

But our opportunity here is to do something much more thrilling and engaging than fine-tuning a textbook. Rather, we can step back and stand with Newton at the moment of formulating what a useful way of thinking about these things would be.

- **Quantity of Matter and Weight**

"[Mass] is apprehended through an individual body's weight. For it is found by experiments with pendulums carried out with the greatest accuracy to be proportional to weight, as will be shown hereafter."

As Newton will be doing frequently in his commentaries to the definitions and the laws of motion, he is here looking ahead to what will emerge from this work as a whole, and in particular to what will prove to be true when we pull in information about how things work in our world. In this case, as we embark on this investigation, we don't know the relationship between mass and weight, and these experiments he is referring to are part of the development that *Principia* is about to lead us through. That mass, or quantity of matter, can be "apprehended through weight" must be shown, and it is indeed finally shown, but not until Book III, which makes the various applications to our actual world. The experiments Newton mentions here are cited, and the relationship of mass and weight established, in Book III Proposition 6.

• **Mass and Quantity of Matter:** The word this guidebook translates as "mass" is Latin "*massa.*" *Massa* is a large irregular lump of something, a bundle or heap. This is in contrast to our modern use of the word mass, which is more technical and abstract. The technical and abstract term for Newton, corresponding to our modern use of the word mass, was "quantity of matter."

The translation in this guidebook always distinguishes *massa,* mass, from *quantitas materiae,* quantity of matter.

You won't get into any trouble in reading *Principia* if you mentally substitute "mass" for "quantity of matter." But you might want to notice when Newton uses each term, keeping in mind that, when he used the word mass as opposed to the term quantity of matter, he was likely picturing something in its individuality as a lump of matter.

Quantity of Motion Defined

Definition 2

The quantity of motion is the measure of the same arising from the velocity and the quantity of matter conjointly.

The motion of the whole is the sum of the motions in the individual parts, and therefore in a body twice as big, with an equal velocity, it is doubled, and with a doubled velocity it is quadrupled.

Note on Definition 2

This definition is expressed as a proportion, not as an equation. See the first note to Definition 1.

Again, we are invited to think about what quantity of motion might be. **Question for Discussion.** What is motion itself? How might one measure it (not necessarily in the laboratory, but in thought)? One might have chosen to think of motion as the movement itself, maybe measured by the distance traveled in a particular time, or even just the total distance traveled. (In casual speech, we do sometimes use the word in both these ways.)

Newton chooses in *Principia* to use a definition of motion as the speed and the quantity of matter conjointly—that is, the ratio of their corresponding speeds compounded with the ratio of their corresponding quantities of matter. **Question for Discussion.** What does this tell us about the way he is thinking of motion? How is his picture different from other definitions someone might have made?

Kinds of Force (Definitions 3–8)

Newton defines two types of force for us in this section. The first is *vis insita,* or inherent force, given in Definition 3. The second is *vis impressa,* impressed force, given in Definition 4. In Definition 5 he goes on to describe a particular type of impressed force, "centripetal force," the investigation of which is central to this book. He then further distinguishes three different measures of this sort of impressed force, the absolute, the accelerative, and the motive quantities of centripetal force, in Definitions 6–8 and the following commentary.

Inherent Force (Inertia, *vis insita*) Defined

Definition 3

The inherent force of matter is the power of resisting, by which each and every body, to the extent that it can, perseveres in its state either of resting or of moving uniformly in a straight line.

This is always proportional to its body, and does not differ in any way from the inertia of mass, except in the mode of conception. Through the inertia of matter it comes to be that every body is with difficulty disturbed from its state either of resting or of moving. Whence the inherent force can also be called by the extremely significant name, "force of Inertia." A body exercises this force only in the alteration of its status by another force being impressed upon it, and this exercise falls under the diverse considerations of resistance and impetus: resistance, to the extent that a body resists an impressed force in order to preserve its state, and impetus, to the extent that the same body, in giving way with difficulty to the force of a resisting obstacle, endeavors to change the state of that obstacle. Common opinion attributes resistance to things at rest and impetus to things in motion, but motion and rest, as they are commonly conceived, are distinguished from each other only with respect [to each other], nor are those things really at rest which are commonly seen as if at rest.

Notes on Definition 3

• This is the definition of inherent force, saying, in the way of definitions, how the term will be used. The existence of inherent force is postulated in Law 1 where it is asserted that every body continues in its state of motion or rest unless driven to alter that state by an impressed force. Definition 3 and Law 1 are best read in conjunction by anyone attempting to understand this phenomenon.

• The word inertia was not a familiar physical term when Newton used it in this book. It had first been used in a physical sense by Kepler, for whom it meant a body's tendency to come to rest. Newton borrowed the word, but changed its meaning to fit his physics. The basic meaning is "laziness" or "sluggishness."

• We must be especially careful not to project modern textbook definitions or concepts of inertia and force onto Newton's definitions. We're in another world as we read *Principia,* a world we have entered by time travel. Only careful alert exploration will reveal how much of what we find here is what we're accustomed to at home and how much will be different in interesting and even instructive (but possibly subtle) ways. Concepts evolve over time. Words change their meaning. Formulas of modern physics that claim to be "Newtonian" may in fact use different definitions of the same terms. There is no one eternal meaning of such terms as force. Rather, these are concepts that evolve over time. We're at one snapshot of time—a very important moment, but one more than three centuries in the past.

Question for Discussion. What is our own understanding of force? I don't mean to ask what formula we learned in high school physics. What is this thing in the world, in our experience, that we call force? We can ask how we use the word when we're talking about our own personal experience of the world, when describing our everyday experience with material objects and when using the world analogously for psychological, social, or political phenomena. Do we have a direct experience of this thing (or maybe a set of things having something in common) that leads us to a concept prior to any definitions or formulas we might learn in school?

• Returning now to Newton, and looking at the wording of this definition, we might wonder what sort of force this could be. He says that it is that something that is proportional only to body (note that he says in Definition 1 that when he uses the terms body and mass he means quantity of matter). If inherent force is proportional "only" to quantity of matter it is not, presumably, proportional to anything else (and he doesn't mention anything else). But how is it that something proportional only to body comes to act as a force, and not only act in some force-like way, but be *defined* as force?

One notes that not only does he define this using the word force (*vis*), but he goes out of his way to emphasize that it is important to him to so classify it, calling it an "extremely significant" name:

"Through the inertia of matter it comes to be that every body is with difficulty disturbed from its state either of resting or of moving. Whence the inherent force can also be called by the extremely significant name, 'force of Inertia.'"

Question for Discussion. It is worth speculating before we move on about the way Newton is understanding this force of inertia or inherent force, considering in what sense he is seeing it as a force (*vis*), with the possible insights such thinking might yield about what force means for him. A look ahead at the impressed force of Definition 4 might give something to go on. Is there a way in which inherent force can be seen acting in a similar way to the way he says impressed force acts? How is it different? He calls *vis insita* a "power of resisting." How would that be measured?

• If it seems that the more carefully you read and question the text the more puzzles you find, don't feel discouraged or yield to a temptation to fall back on modern formulas or understandings. You don't expect to have figured out "whodunit" in the first pages of a mystery novel. Rather, let that experience of puzzlement awaken a sense of wonder and give you something to ponder as you go along through *Principia*. At this early point we really don't have a great deal to go on. We haven't seen how he will be speaking about force in propositions or even in the Laws of Motion. We will need to stay alert for how he applies these terms in what follows and continue to feel out his vision and intent.

• "...motion and rest, as they are commonly conceived, are distinguished from each other only with respect [to each other], nor are those things really at rest which are commonly seen as if at rest."

Notice Newton's assertion here of the relativity of motion and rest, "as they are commonly conceived." That is, from our vantage point, we can't tell whether things are absolutely moving or at rest: we can only say whether they are moving or at rest with respect ourselves or with respect to one other. He will take up this question of relativity of motion and whether there is such a thing as knowable absolute motion in the scholium following these definitions.

Impressed Force (*vis impressa*) Defined

Definition 4

Impressed force is an action exerted upon a body for changing its state either of resting or of moving uniformly in a straight line.

This force consists in the action alone, and does not remain in the body after the action. For the body continues in each new state through the force of inertia alone. Moreover, impressed force has various origins, such as from impact, from pressure, from centripetal force.

Notes on Definition 4

- This is the definition of impressed force, saying, in the way of definitions, how the term will be used. The Second Law of Motion describes the operation of impressed force; you may wish to read Definition 4 and Law 2 in conjunction, or at least refer back to this definition when you study the Second Law.

- Newton asserts in this definition that impressed force does not remain in the body after the action of impressing.

He is here contradicting the impetus theorists who believed that the force applied to a body, an impetus, remained in the body. This impetus accounted for the body's continuing in motion after the force was no longer being applied.

His new view replaces traditional impetus theories that were current at least from the sixth century. Earlier versions of the theories tended to see the impetus that was added in launching a projectile as gradually expended in the motion. As the impetus was expended, the projectile slowed down; once it was used up the projectile came to rest again.

In the fourteenth century Jean Buridan asserted that impetus from impressed force remained in the body, only diminishing from external resistance or internal tendency to motion in a contrary direction. Aside from being impressed from an external force, it could be augmented in falling, each increment of acquired speed adding impetus. This view of the impetus as physically present in the body was influential through Newton's time.

Closer to Newton's time, the view of force as something that resides in a body and can be transferred to other bodies by impact was propounded by Descartes and his followers. The Cartesians were trying to identify things that were conserved in physical interactions, and the efforts of Cartesians such as Huygens and Leibniz led to the principles of conservation of momentum and conservation of energy (which they called living force, *vis viva*). Newton, however, proposes here a definition of impressed force that is more closely related to observation and more easily expressed geometrically.

Newton seems to retain something like the older view of impetus in his concept of inherent force (*vis insita*), as presented in Definition 3: both traditional impetus and inherent force remain in the body. Note that he says in his comments on Definition 3 that people are accustomed to think of inertia being in bodies at rest and impetus in moving bodies. Newton replaces both older ideas with his inherent force, saying that motion and rest, "as they are commonly conceived," are distinguished only relative to each other. Impressed force, then, is something new, acting on the body but not remaining with it.

Question for Discussion. If impetus is the same as inertia, it is in the body; it is the body's inherent force. How would you articulate the difference between that force which seems to be in the body (inherent force) and the force which ceases when the impressing action ceases? Is there a transformation of that impressing force from an external action to an internal force? How exactly would you say Newton has altered impetus theory, or has he?

• **Question for Discussion.** Impressed force is called an "action." What does Newton mean by an action? He says in his commentary that impressed force can have its origin in pressure. So apparently this "action" can be operating without any resulting motion if there are contrary forces balancing each other out. If you had to write a definition of "action" that would work for Newton's impressed force, what would you say about it?

• Inherent force and impressed force seem to be the two basic kinds of force. Centripetal force, which will occupy his attention in the next three definitions (and for most of this book) is said here to be an example of impressed force. How that force is impressed in the case of centripetal force is not and cannot be specified in this book. But as you read and ponder this definition and the Second Law of Motion you might want to keep somewhere in mind that whatever the mechanism of gravity could be, the force of gravity will turn out to be an impressed force under this definition.

• **Question for Discussion.** Have we gotten any more insight into what a force in itself is? How are inherent force and impressed force both sorts of force? How are they different? Is inherent force an "action" in any sense? Presumably in the case of impressed force the action is the action of the impressing force on a body. But is there a sort of negative action in a body's resistance to changing its motion?

Notice also the interesting complementarity in the way Newton defines these two kind of force. The inherent force keeps a body at rest or moving in a straight line. Impressed force is an action that changes a body's state of being at rest or moving in a straight line.

Centripetal Force Defined

Definition 5

Centripetal force is that by which bodies are pulled, pushed, or in any way tend, towards some point from all sides, as to a center.

Of this kind is gravity, by which bodies tend to the center of the earth; magnetic force, by which iron seeks a magnet; and that force, whatever it might be, by which the planets are perpetually drawn back from rectilinear motions and are driven to revolve in curved lines. A stone, whirled around in a sling, attempts to depart from the hand that drives it around, and by its attempt stretches out the sling, doing so more strongly as it revolves more swiftly, and as soon as it is released, it flies away. The force contrary to that attempt, by which the sling perpetually draws the stone back to the hand and retains it in its orbit, I call "centripetal," because it is directed towards the hand as to the center of the orbit. And the account of all bodies that are driven in a gyre is the same. They all attempt to recede from the centers of the orbits, and in the absence of some force contrary to that attempt, by which they are pulled together and kept in their orbits, and which I therefore call "centripetal," they will go off in straight lines with uniform motion.

If a projectile were deprived of the force of gravity, it would not be deflected towards the earth, but would go off in a straight line toward the heavens, doing so with a uniform motion, provided that the resistance of the air be removed. It is drawn back by its gravity from the rectilinear path and is perpetually bent towards the earth, more or less according to its gravity and the velocity of motion. Where its gravity is less in proportion to the quantity of matter, or where the velocity with which it is propelled greater, it will deviate correspondingly less from the rectilinear path, and will travel farther.

If a lead ball, propelled by gunpowder from the summit of some mountain in a horizontal line with a given velocity, were to travel in a curved line for the distance of two miles before it fell to earth, it would travel about twice as far with double the velocity, and about ten times as far with ten times the velocity, provided that the resistance of the air be removed. And by increasing the velocity, the distance to which it is propelled may be increased at will, and the curvature of the line which it describes may be diminished, so that it would finally fall at a distance of ten or thirty or ninety degrees, or it might even go around the whole earth, or, at last, go off towards the heavens, continuing on *in infinitum* with the motion with which it departed. And by the same account, by which a projectile may be deflected into an orbit by the force of gravity and may go around the whole earth, the moon too, whether by the force of gravity (provided it be heavy) or by another force of whatever kind, by which it is urged towards the earth, can be always pulled back towards the earth from its rectilinear path, and deflected into its orbit; and without such a force the moon cannot be held back in its orbit.

This force, if it were less than required, would not sufficiently deflect the moon from the rectilinear path; and if greater than required, would deflect it more than sufficiently, and would lead it down from its orbit towards the earth. It is indeed requisite that it be of exactly the right magnitude, and it is for the Mathematicians to find the force by which a body can be accurately kept back in any given orbit you please with a given velocity, and in turn to find the curvilinear line into which a body departing from any given place you please with a given velocity would be deflected by a given force.

Further, the quantity of this centripetal force is of three kinds: absolute, accelerative, and motive.

Notes on Definition 5

- **Gravity as centripetal force.**

"Of this kind is gravity, by which bodies tend to the center of the earth ..."

One perhaps doesn't need a formal proof to recognize that the phenomenon of terrestrial heaviness seems to illustrate a tendency towards a center. If looking around at falling bodies doesn't suggest this, reading Aristotle makes the idea a familiar one.

We must be careful, though. In our day we understand gravity to be more than terrestrial heaviness; we understand it to be the force that moves the heavenly bodies in their curved paths, drawing them out of their tangential motion. If we interpret the statement just quoted from the commentary to this definition as including that too, then we have gotten ahead of ourselves. That assertion must be proved and Newton doesn't assume it until he has proved it. The foundation for the proof is being laid through the first two books and is formally arrived at in Book III Proposition 5.

"And by the same account, by which a projectile may be deflected into an orbit by the force of gravity and may go around the whole earth, the moon too, whether by the force of gravity (provided it be heavy) or by another force of whatever kind, by which it is urged towards the earth, can be always pulled back towards the earth from its rectilinear path, and deflected into its orbit; and without such a force the moon cannot be held back in its orbit."

Notice that Newton is not assuming that the moon is held in orbit by gravity. He says "...whether by the force of gravity (provided it be heavy) or by another force of whatever kind...." Whether the moon is "heavy," in the terrestrial sense of possessing weight, must be shown. Newton intends to prove this; the proof is completed in III.4.

• **Centrifugal force and centripetal force.**

"A stone, whirled around in a sling, attempts to depart from the hand that drives it around, and by its attempt stretches out the sling, doing so more strongly as it revolves more swiftly, and as soon as it is released, it flies away. The force contrary to that attempt, by which the sling perpetually draws the stone back to the hand and retains it in its orbit, I call "centripetal," because it is directed towards the hand as to the center of the orbit."

In the first sentence of this quotation, Newton is alluding to the force that the Cartesians (most notably Huygens) called "*vis centrifuga*," centrifugal force. This force is a tendency away from the center and is central to their dynamic analysis of all circular motion, including that of the orbits of planets. Newton is shifting our viewpoint here from a supposed tendency away from the center to a tendency towards the center, and has made up the word "centripetal" to refer to that latter tendency.

Absolute Quantity of Centripetal Force Defined

Definition 6

The absolute quantity of centripetal force is the measure of the same, greater or less in proportion to the efficacy of the cause propagating it from the center through the encircling regions.

As the magnetic force is greater in one magnet, less in another, in proportion to the size [*moles*] of the magnet or the intensity of the power [*virtus*].

Notes on Definition 6

• We must again remember that this is a definition of how the term will be used. It is not proof or even assertion that this thing exists. That must be shown. Centripetal force may not be present in the encircling regions around the center; it may be there in the spaces but not propagated from the center. In the last paragraph of the commentary after these last three definitions, Newton will warn that he is considering the forces mathematically, not physically, and that

> ...the reader should beware of thinking that...I am anywhere defining a species or manner of action, or a cause or physical account, or that I am truly and physically attributing forces to centers (which are mathematical points) if I should happen to say either that centers attract, or that forces belong to centers.

• Absolute quantity of centripetal force is not mentioned again until I.17, and then it is, as always prior to Book III, hypothetical. That is, I.17 says that if we suppose there to be given a certain absolute quantity of force, along with other conditions, he will show us how to determine a resulting curve.

• Notice again that we are given not an equation but a proportion. See the note to Definition 1.

• **Question for Discussion.** Does the idea of force propagated out into surrounding spaces sound occult to you? If it doesn't, should it? How could this absolute quantity of force be measured? How would you propose it be measured?

Keep these speculations in mind as you read the following definitions and when absolute force reappears in I.17.

Accelerative Quantity of Centripetal Force Defined

Definition 7

The accelerative quantity of centripetal force is the measure of the same, proportional to the velocity which it generates in a given time.

Thus the power of the same magnet is greater at a less distance, less at a greater one; or the gravitating force is greater in valleys, less on the peaks of high mountains, and less still (as will become clear hereafter) at greater distances from the earth's globe. At equal distances, however, it is everywhere the same, because it equally accelerates all falling bodies (heavy or light, great or small) once the resistance of the air is removed.

Notes on Definition 7

• We again have a proportion. We are instructed that we are to compare two centripetal forces by comparing the velocity each generates in a given time. We are not being told how to calculate the magnitude or number of a particular force or of its accelerative quantity.

• Note that neither Definition 7 nor Definition 8 assumes or defines an instantaneous force. He says "in a given time." This time could be taken to the limit (gradually reduced towards zero), thus yielding a rato of forces at particular points, but it could also be finite.

Look out for assuming that accelerative quantity of force must be the second derivative of distance. Aside from the fact that this definition gives us, as noted in the previous bullet, a ratio of forces and not a value for a particular force, there is another problem with making that conscious or unconscious translation (a translation that can and has led to serious misunderstandings of some of his proofs even by scholars who should know better). Modern physics has become very comfortable, even blasé, about operating on infinitesimals, but Newton is very careful in his work with vanishing quantities. He doesn't throw around ratios of infinitesimals, but rather takes ratios of finite magnitudes and looks at what happens to those ratios as time, or as one of the magnitudes, shrinks towards zero.

As you work through his Lemmas in the section coming up, you will see how careful he is, and how he manages to establish a firm grip on the notoriously slippery creatures that vanishing quantities are. (But be warned: you will only see how careful he is being if you can manage not to look through modern eyes. Since this is easier said than done, I recommend particular attention to my Pause Before Lemma 6, where I explore the subtleties of his handling of ratios of vanishing quantities.)

- Be careful about Newton's commentary. The italicized part is his actual definition. The commentary following it is looking ahead at what he will develop formally. His readers know that experiments indicated the empirical truth of these claims about the magnet and the power of the gravitational force on the earth's surface, but note that the definition says nothing about the force varying as the distance. As you begin Newton's exposition in the next section, Section 2, "On the Finding of Centripetal Forces" (it begins with Proposition 1), you will be astonished by how little he assumes. All this discussion in these commentaries to the Definitions and Laws, and the scholia after the Definitions and Laws, will be ignored. That gravitational force varies with the distance will be brought out with the greatest care, and not established until Book III. Nor will he assume the equal acceleration of light and heavy bodies at equal distances without a formal demonstration; this too is established finally in Book III.

- If you are thinking that accelerative quantity of force, measured only by the change in velocity it effects, is not the "proper" definition of force, you are no doubt being overly influenced by what you learned in your physics class. Set that aside. (For further explanation of why this sort of thing must be set aside, see the discussion entitled "Example of the perils of anachronism" in Section 4 of the Preliminaries, which talks about why $f = ma$ is not Newton).

Let Newton unfold his system to you. He has good reasons for what he does. If you think you want to be dealing with mass right from the start, you don't know what you would be getting yourself in for. Watch and see how Newton does it. When he finally brings mass in, in Book III, first in the attracted body and then in the attracting body, he will have carefully built up a very complex system adding one layer of complexity at a time. You will appreciate the brilliance (and kindness) of his exposition.

• **Question for Discussion.** We asked in thinking about the previous definition how we might measure the absolute quantity of force, "the efficacy of the cause propagating it from the center through the encircling regions." Would this definition offer us a way to measure it? Newton is suggesting that we look at how much velocity is generated in a given time. If we look at those velocities in different places, perhaps we could say that we were measuring the efficacy of the force at those places. What do you think?

Motive Quantity of Centripetal Force Defined

Definition 8

The motive quantity of centripetal force is the measure of the same proportional to the motion which it generates in a given time.

Thus the weight is greater in a greater body, less in a lesser one, and in the same body it is greater near the earth, less in the heavens. This quantity is the whole body's centripetency, or propensity towards the center, and (if you will) weight, and is always known through the force contrary and equal to it, by which the descent of the body can be prevented.

Notes on Definition 8

• We again have a proportion. The definition says that motive forces are proportional to their respective motions generated in a given time.

A shorthand way of saying this is "motive force varies as motion generated in a given time." But if you do use this shorthand, it is important to understand that, if you're talking about Newton, your shorthand formulation means comparing two (or more) motive forces by comparing the amount of motion (defined in Definition 2) each has generated in a particular time segment.

We are instructed in the way to compare two centripetal forces. We are not being told how to calculate the magnitude or number of a particular force or of its motive quantity.

• **"a given time."** This is not a definition of an instantaneous force; see the second bulleted note under Definition 7.

• **Weight.** The argument that weight is an example of motive force comes in the paragraphs that follow in Newton's text, in his commentary on Definitions 6–8.

• **Motive Force in Contrast to Accelerative Force.** Recall the definition of accelerative quantity of force (Definition 7). Those forces are proportional to the respective velocities generated in a given time. In contrast, the motive quantities of force are proportional to the respective quantities of motion (Definition 2) generated in a given time. The difference here is the introduction of the mass of the attracted body.

The mass of the attracted body (the body on which the centripetal force is exerted) will not become relevant to the development of universal gravitation until Book III Proposition 6; in fact it won't even be mentioned in the central argument until then. There is much groundwork to be laid just looking at accelerative force before that next level of complication can be brought in. (And it will turn out, of course, that bringing in the mass of the attracted body is itself only a step on the way: to complete the line of argument we will have to consider also the mass of the attracting body.)

• **Something about $f = ma$.** The reader may have recognized motive force as the sort of force modern physics has in mind when it uses the equation $f = ma$. We've reminded ourselves in the first note above that Newton doesn't give us an equation here, but rather a proportion. For more warning about the pitfalls of moving into *Principia* with $f = ma$ dancing in one's head, please see the passage in Section 4 of Preliminaries that takes this up as a case study in anachronism.

Newton's Commentary to Definitions 6–8

For the sake of brevity, these quantities of force may respectively be called motive forces, accelerative forces, and absolute forces, and for the sake of distinction, may be said to trace their origins respectively to the bodies seeking the center, to the places of the bodies, and to the center of forces. That is, motive force traces its origin to the body, as if the force were the endeavor of the whole toward the center, composed of the endeavors of all the parts; accelerative force traces its origin to the place of the body, as if it were a sort of efficacy, spread out from the center through the individual places on the circumference, for moving bodies that are in those places; and absolute force traces its origin to the center, as if it were endowed with some cause without which the motive forces would not be propagated through the regions on the circumference, whether that cause be some central body (such as is a magnet at the center of the magnetic force, or the earth in the center of the gravitating force) or something else that does not appear. This concept is strictly mathematical, for I am not now considering the causes and physical seats of the forces.

The accelerative force is accordingly to the motive force as speed is to motion. For the quantity of motion arises from the speed and the quantity of matter, and the motive force arises from the accelerative force and the quantity of the same matter conjoined. For the total action of the accelerative force on the individual particles of a body is the motive force of the whole. Whence, near the surface of the earth, where the accelerative gravity or the gravitating force is the same in all bodies, the motive gravity, or weight, is as the body; but if an ascent be made to regions where the accelerative gravity is less, the weight gradually decreases, and will always be as the body and the accelerative gravity conjoined. Thus in regions where the accelerative gravity is less by a factor of two, the weight of a body that is smaller by a factor of two or three will be less by a factor of four or six.

Further, I call attractions and impulses accelerative and motive in the same sense. Moreover, I use the words "attraction," "impulse," or [words denoting] a propensity of any kind toward a center, indifferently and promiscuously for each other: I am considering these forces, not physically, but only mathematically. Therefore, the reader should beware of thinking that by words of this kind I am anywhere defining a species or manner of action, or a cause or physical account, or that I am truly and physically attributing forces to centers (which are mathematical points) if I should happen to say either that centers attract, or that forces belong to centers.

Notes on Commentary to Definitions 6–8

- **Disclaimers on Cause of Gravity.** Pay special attention to the whole last paragraph above, as well as to Newton's earlier statement, "this concept is strictly mathematical, for I am not now considering the causes and physical seats of the forces," which also disavows any claim that the power comes from the center or a central body.

You will want to keep these Newtonian warnings in mind, because Newton will indeed talk about centers and attractions and impulses, and one could indeed get a distorted idea, not only of his physical picture but the structure of his argument, if one were to think he was attributing physical reality to things that are not seen so by him. It is a vulgar modern idea that bodies act on other bodies at a distance; Newton found the idea absurd.

None of this, however, prevents you from deeply pondering, as you work through this book, what the cause of gravity might be, and to ponder as well what Newton's speculations might have been.

Scholium

Hitherto it has seemed appropriate to explain the less familiar terms, [and] the sense in which they are to be taken in what follows. Time, space, place, and motion, are very familiar to everyone. It should nevertheless be noted that these are not commonly conceived of otherwise than from their relation to sensible objects. And from this there arise certain prejudices, for the removal of which it is useful for these same [terms] to be distinguished into absolute and relative, true and apparent, mathematical and common.

I. Absolute, true, and mathematical time, in itself and by its nature without relation to anything external, flows uniformly, and by another name is called "duration." Relative, apparent, and common [time] is the perceptible and external measure (whether accurate or varying in rate) of any duration you please by means of motion, which is commonly used in place of true time, such as an hour, a day, a month, a year.

II. Absolute space, by its nature, without relation to anything external, always remains similar and motionless. Relative [space] is any movable measure or dimension you please of this space, which [measure] is defined by our senses through its position with respect to bodies, and is commonly taken in place of motionless space, such as the dimension of subterraneous, aerial, or celestial space defined through its position with respect to earth. Absolute and relative space are the same in form and size, but do not always remain the same in number. For if the earth, for example, were to move, the space of our air, which relatively and with respect to our earth always stays the same, will be now one part of absolute space in which the air moves across, now another part of it, and thus, absolutely, it will perpetually change.

III. Place is the part of space which a body occupies, and is absolute or relative according to the space. It is a part, I say, of space, not the location of a body, or the enclosing surface. For the places of equal solids are always equal; the surfaces, however, are nearly always unequal because of the dissimilarity of figures. Locations, on the other hand, do not, properly speaking, have quantity, nor are they so much places as properties of places. The motion of the whole is the same as the sum of the motions of the parts: that is, the translation of the whole from its place is the same as the sum of the translations of the parts from their places. Consequently, the place of the whole is the same as the sum of the places of the parts, and for that reason it is internal and in the whole body.

IV. Absolute motion is the translation of a body from absolute place to absolute place; relative [motion is the translation of a body] from relative [place] to relative [place]. Thus in a boat which is carried with sails set, the relative place of a body is that region of the boat in which the body is, or that part of the whole concavity which the body fills, and which to that extent moves along with the boat; and relative rest is the body's continuing to remain in the same region of the boat or part of the concavity. But true rest is the body's continuing to remain in the same motionless part of that space in which the boat itself along with its concavity and all its contents moves. Whence if the earth is really at rest, a body which is relatively at rest in the boat will move truly and absolutely with that velocity with which the boat moves upon the earth. If on the contrary the earth also moves, the true and absolute motion of the body arises partly from the true motion of the earth in motionless space, partly from the relative motion of the boat upon the earth; and if the body also moves relatively in the boat, its true motion arises partly from the true motion of the earth in motionless space, partly from the relative motions of the boat upon earth and of the body in the boat; and from these relative motions there arises the body's relative motion upon earth. Thus, if that part of earth where the boat is placed be really moved eastward with a velocity of 10,010 units, and the boat be carried by the sails and the wind westward with a velocity of ten units, while the boatman walk on the boat towards the east with a velocity of one unit, the boatman will move truly and absolutely in motionless space with 10,001 units of velocity eastward, and relatively upon earth towards the west with nine units of velocity.

Absolute time is distinguished from relative in astronomy by the equation of common time. For the natural days are unequal, but are commonly taken as if equal for a measure of time. Astronomers make a correction for this inequality, so that they may measure the celestial motions from a truer time. It is possible that there is no uniform motion by which time may be measured accurately. All motions can be accelerated and retarded, but the flow of absolute time cannot be changed. The duration or perseverance of the existence of things is the same, whether the motions are fast or slow or none. Furthermore, these are rightly distinguished from their perceptible measures, and are reckoned from them through the astronomical equation. Moreover, the need for this equation in determining phenomena is established both by the experimental evidence of the pendulum clock and by the eclipses of Jupiter's moons.

As the order of the parts of time is unchangeable, so also is the order of the parts of space. If these move from their places, they will also (so to speak) move from themselves. For times and spaces are in a way the places of themselves and of all things. Everything without exception is located in time according to order of succession, in space according to order of place. It is of their essence that they be places, and it is absurd for primary places to move. These are therefore absolute places, and translations from these places are alone absolute motions.

However, because these parts of space cannot be seen and cannot be distinguished from each other by our senses, we introduce perceptible measures in their stead. For from the positions and distances of things from some body, which we see as motionless, we define all places universally, and thereafter we also estimate all motions as well with respect to the places mentioned previously, insofar as we conceive the bodies to be carried away from them. We thus use relative places and motions in place of absolute ones, and this is not an inconvenience in human affairs. In philosophical matters, however, an abstraction from the senses must be made. For it can happen that there is no body really at rest, to which places and motions may be referred.

Further, absolute and relative motions are distinguished from each other through their properties and causes and effects. It is a property of rest that bodies really at rest are at rest among themselves. Therefore, since it is possible that some body in the regions of the fixed stars, or far beyond, be absolutely at rest, while it cannot be known from the position of bodies with respect to each other in our regions whether one of them may preserve its given position with respect to that distant [body], true rest cannot be defined from the position of the latter [bodies] among themselves.

It is a property of motion that the parts, which preserve given positions with respect to the wholes, participate in the motions of the same wholes. For all parts of bodies moving in curves strive to recede from the axis of motion, and the impetus of bodies moving forward arises from the conjoined impetus of the individual parts. Therefore, when the surrounding bodies are moving, those are moving which are at rest among the surrounding ones. And for that reason, true and absolute motion cannot be defined through translation away from nearby bodies, which are viewed as if they were resting bodies. For the external bodies ought not only to be viewed as if they were resting, but also to be truly at rest. Otherwise, all the surrounded [motions], other than translation away from the nearby surrounding [bodies], will also participate in the true motions of the surrounding [bodies], and when that translation is removed, they are not truly at rest but will only be viewed as if they were resting. For the surrounding are to the surrounded as the exterior part of the whole is to the interior part, or as the shell to the kernel. And when the shell is moved, the kernel too is moved without translation away from the surrounding shell, as a part of a whole.

Related to the preceding property is that when the place is moved the thing in the place is moved along with it: thus a body that is moved away from a moved place, also participates in the motion of its place. Therefore, all motions which take place away from [i.e., with respect to] moving places are only parts of the whole and absolute motions, and every whole motion is compounded of the motion of a body from its prime place, and the motion of that place from its place, and so on, until it comes to a motionless place, as in the example of the boatman mentioned above. Hence, whole and absolute motions can be defined only through motionless places, and I have consequently related them above to motionless places, and relative [motions] to movable [places]. However, only those places are motionless which from infinity to infinity all preserve given positions with respect to each other, and moreover remain forever motionless, and constitute that space which I call immovable.

The causes by which true and relative motions are distinguished from each other are the forces impressed upon bodies for generating motions. True motion is neither generated nor changed except by forces impressed upon the moved body itself. But relative motion can be generated and changed without forces being impressed upon this body. For it suffices that they be impressed only upon other bodies to which it is related, so that, when they give way, that relation, of which the relative rest or motion of this [body] consists, is changed. Again, true motion is always changed by forces impressed upon the moved body, but relative motion is not necessarily changed by these forces. For if the same forces were impressed upon other bodies as well, to which there is a relation, in such a way that the relative position be preserved, the relation in which the relative motion consists will be preserved. Therefore, every relative motion can be changed while the true motion is preserved, and can be preserved while the true motion is changed, and for that reason true motion consists not at all in relations of this sort.

The effects by which absolute and relative motions are distinguished from each other are the forces of receding from the axis of circular motion. For in purely relative circular motion, these forces are none, but in true and absolute [circular motion] they are greater or less according to the quantity of motion.

Suppose that a pail should hang from a very long cord, and be driven continually in a circular path, until the cord becomes somewhat stiff from twisting, and [the pail] next be filled with water, and be at rest along with the water, and then be driven by some sudden force in a circular path with a contrary motion, persevering in that motion for a long time as the cord untwists. At the beginning, the surface of the water will be flat, as it was before the motion of the vessel, but after the vessel, by a force gradually impressed upon the water, makes it too begin to rotate perceptibly, it will itself gradually withdraw from the middle, and will climb up to the sides of the vessel, adopting a concave form (as I have myself experienced); and, with an ever increasing motion, will climb more and more, until it make its revolutions in an equal time with the vessel, and come to rest relative to it. This climbing is an indicator of a striving to withdraw from the axis of motion, and through such a striving the true and absolute circular motion of the water, here completely contrary to relative motion, comes to be known and is measured. At first, where the relative motion of the water in the vessel was greatest, that motion did not arouse any striving to withdraw from the axis: the water was not seeking the circumference by climbing up the sides of the vessel, but stayed flat, and for that reason its true circular motion had not yet begun. Later, however, where the relative motion of the water decreased, its ascent up the sides of the vessel was an indicator of the striving to withdraw from the axis, and this striving showed its true circular motion, which was ever increasing, and was finally made greatest where the water came to rest relatively in the vessel. Therefore, this striving will not depend upon the translational motion of the water with respect to the surrounding bodies, and consequently, true circular motion cannot be defined by such translations.

The true circular motion of any revolving [body] is unique, corresponding to a unique striving which is its own, as it were, and is commensurate with the effect. Relative motions, however, are countless, in accord with the various relations with external [bodies], and, like relations, they are entirely bereft of true effects, except insofar as they participate in that true and unique motion. Hence also, in the system of those people who would have our heavens rotate in an orb beneath the heavens of the fixed stars, and bear the planets along with them, the individual parts of the heavens, and the planets which are indeed relatively at rest in the heavens nearest them, in truth move. For they change their positions with respect to each other (unlike what happens in those truly at rest), and, being carried along with the heavens, they participate in their motion, and, like the parts of all revolving things, strive to recede from their axes.

Relative quantities are therefore not those quantities themselves whose names they display, but are those perceptible measures of them (whether true or erroneous) which are commonly used in place of the measured quantities. And if the meanings of words should be defined from usage, then by those names "time," "space," "place," and "motion," these perceptible measures should properly be understood, and the discussion will be out of the ordinary and purely mathematical if the quantities measured be understood here. Furthermore, they do violence to sacred scripture who interpret these words as concerning measured quantities there. Nor are mathematics and philosophy any the less defiled by those who confuse the true quantities with their relationships and common measures.

To recognize the true motions of individual bodies and to distinguish them in fact from the apparent ones, is indeed extremely difficult, for the reason that the parts of that motionless space in which the bodies truly move do not flow in to the senses. Nevertheless, the cause is not entirely hopeless. For arguments can be taken, partly from the apparent motions which are the differences of true motions, partly from the forces which are the causes and effects of the true motions.

As, if two globes, joined together at a given distance from each other by a cord between them, were to revolve about their common center of gravity, the striving of the globes to recede from the axis of motion would come to be known from the tension of the cord, and from that the quantity of circular motion might be computed. Then if any equal forces you please were to be simultaneously impressed upon alternate faces of the globes, to increase or decrease the circular motion, the increase or decrease of the motion would come to be known from the increased or decreased tension of the cord, and finally, it would be possible from that to find the faces of the globes upon which forces should be impressed in order most greatly to increase the motion, that is, the aftermost faces, or those which are the following ones in the circular motion. And when the faces which are following are known, and the opposite faces which are leading, the direction of motion would be known. In this way, both the quantity and the direction of this circular motion might be found in whatever immense void you please, where nothing external and perceptible were to exist with which the globes might be compared. Now, if there were to be set up in that space some distant bodies maintaining a given position among themselves, such as are the fixed stars in the regions of the heavens, it would indeed be impossible to tell from the relative translation of the globes within the bodies whether the motion should be attributed to the former or the latter. But if attention were paid to the cord, and it were ascertained that its tension were exactly that which the motion of the globes would require, it would be permissible to conclude that the motion belonged to the globes, and that the bodies were at rest, and, finally, from the translation of the globes within the bodies, to determine the direction of the motion. But how to determine the true motions from their causes, effects, and apparent differences, and, conversely, how to determine their causes and effects from the motions, whether true or apparent, will be taught more fully in what follows. For it is to this end that I wrote the following treatise.

Axioms, or Laws of Motion

Law 1

Every body continues in its state of resting or of moving uniformly in a straight line, except insofar as it is driven by impressed forces to alter its state.

Projectiles continue in their motions except insofar as they are slowed by the resistance of the air, and insofar as they are driven downward by the force of gravity. A top, whose parts, by cohering, perpetually draw themselves back from rectilinear motions, does not stop rotating, except insofar as it is slowed by the air. And the greater bodies of the planets and comets preserve their motions, both progressive and circular, carried out in spaces of less resistance, for a longer time.

Note on Law 1

- This law in effect asserts the existence of *vis insita,* inherent force or inertia, defined in Definition 3. Now it can be used, as it is in Proposition 1.

Law 2

The change of motion is proportional to the motive force impressed, and takes place following the straight line in which that force is impressed.

If some force should generate any motion you please, a double [force] will generate a double [motion], and a triple [force] a triple [motion], whether it has been impressed all at once, or gradually and successively. And because this motion is always directed in the same way as the generating force, if the body was previously in motion, then this [impressed] motion is either added to its motion (if they have the same sense) or subtracted (if contrary), or joined on obliquely (if oblique) and compounded with it according to the determination of the two.

Notes on Law 2

- See Definition 4. More is added to the definition of impressed force through this law. We are now told that impressed force is a motive force.

Note that we are told this about impressed forces generally, and by Definition 4 we know that impressed forces can be of various kinds, only one of which is centripetal force. When we read "motive force" here, therefore, the words are not elliptical for "motive quantity of centripetal force" (Definition 8). Both might have the same measure (both being proportional to quantity of motion), but motive force may cover forces which are not centripetal.

- "...whether it has been impressed all at once, or gradually and successively."

Motive force is defined as taking place "in a given time." Thus the impressed force whose action is the subject of this law takes place as Newton describes in his commentary (quoted above). We note that Newton does not envision impressed force as an instantaneous action.

Law 3

To an action there is always a contrary and equal reaction; or, the mutual actions of two bodies upon each other are always equal and directed to contrary parts.

Whatever pushes or pulls something else is pushed or pulled by it to the same degree. If one pushes a stone with a finger, his finger is also pushed by the stone. If a horse pulls a stone tied to a rope, the horse will also be equally pulled (so to speak) to the stone; for the rope, being stretched in both directions, will by the same attempt to slacken itself urge the horse towards the stone, and the stone towards the horse, and will impede the progress of the one to the same degree that it promotes the progress of the other. If some body, striking upon another body, should change the latter's motion in any way by its own force, the same body (because of the equality of the mutual pushing) will also in turn undergo the same change in its own motion, in the contrary direction, by the force of the other. These actions produce equal changes, not of velocities, but of motions — that is, in bodies that are unhindered in any other way. For changes in velocities made thus in opposite directions, are inversely proportional to the bodies, because the motions are equally changed. This law applies to attractions as well, as will be proven in the next Scholium.

Note on Law 3

- Newton, in the last sentence above, extends the workings of this law to apply to "attractions" as well as conventional mechanical pushes and pulls. He says that this will be proved in the scholium which follows the corollaries.

This final sentence in the commentary to Law 3 was added in the second edition of *Principia*, along with the two thought experiments in the scholium that constitute the promised proof. Consider Law 1 and Corollary 4 as you evaluate these thought experiments. After considering what Newton says, you may wish to read my note which discusses this extension of Law 3. That note follows the scholium.

• Of course, a law of motion is a law, not a proposition. It doesn't need to be proved. It is like a postulate, but not a hypothetical postulate of the form "suppose we assume this, then what would follow?" It resembles more a statement of something sufficiently clear from experience as not to need a proof. It may be a new formulation, a new way of thinking about things, or even a new observation; but it is asserted with no fear that contradictory instances will be found.

This is clearly the case with the earlier parts of the statement of Law 3, the ones involving pushes and pulls of a direct mechanical sort. The case with attractions has a somewhat different status. We don't have hands-on experience of the workings of third law interactions in the case of attraction.

We may even have the sense of attractions being independent of third law reactions or reciprocation, as they tend to be for the attractions arising within the souls of living beings. (By soul I mean that element in living beings—sometimes called psyche or consciousness or even "heart"—that makes them more than inert matter.)

And we start to wonder what "attractions" between lumps of *matter* could be. Since they don't have souls, it must be some push or pull mechanism, but what? Depending on the mechanism, the third law might or might not apply.

For this reason Newton feels that he needs to offer a proof specifically for attractions. He says it will be proved (Latin *probabitur*) in the scholium, but what he offers is more like a reassurance that our universe does indeed work as if gravitational attractions were obeying the third law.

Corollaries to the Laws of Motion

Note on Corollaries to the Laws

The note after the scholium following these corollaries will raise a question about the relationship between Law 3 and Corollary 4, so you will want to be alert to what emerges about that relationship as you read the corollary.

Newton presents three laws and then gives us six corollaries. But to *what,* exactly, are they corollaries? One might wonder whether they are all corollaries to Law 3, all corollaries to all the laws, or connected to the laws in various other ways.

It is worth noting in each case what the corollary follows from and what Newton says it depends on. You will probably want to think that through and sort it out for yourself.

If you're referring back from a later time or a later point in the book, here's the summary.

Corollary 1 depends on Laws 1 and 2. Corollary 2 extends Corollary 1 and also depends on the first two laws. Corollary 3 extends the previous corollaries and depends on Law 3 and Law 2 and on Corollary 2.

Corollary 4 extends Corollary 3 and depends on Lemma 23 and its corollary. (In *Principia* these appear after Proposition 26 of Book I; I have included their translation as a note to Corollary 4.) It also apparently depends on Law 3—see the note to Corollary 4 for discussion of this dependence.

Corollary 5 extends Corollary 4 and also depends on Laws 2 and 3. Corollary 6 extends the previous corollaries and depends on Law 2.

Corollary 1

A body [urged] by forces joined together, describes the diagonal of a parallelogram in the same time in which it describes the sides separately.

Suppose that, in a given time, by the force M alone, impressed at place A, a body would be carried with uniform motion from A to B, and by the force N alone, impressed at the same place, it would be carried from A to C. Let the parallelogram $ABCD$ be completed, and that body will be carried by both forces in the same time on the diagonal from A to D. For because the force N acts along the line AC parallel to BD, this force, by Law 2, will not at all change the velocity of approach to that line BD generated by the other force. Therefore, the body will arrive at the line BD in the same time whether force N be impressed or not; and therefore, at the end of that time, it will be found somewhere on that line BD. By the same argument, at the end of the same time, it will be found somewhere on the line CD, and for that reason it must necessarily be found at the intersection of the two lines D. But, by Law 1, it will proceed with a rectilinear motion from A to D.

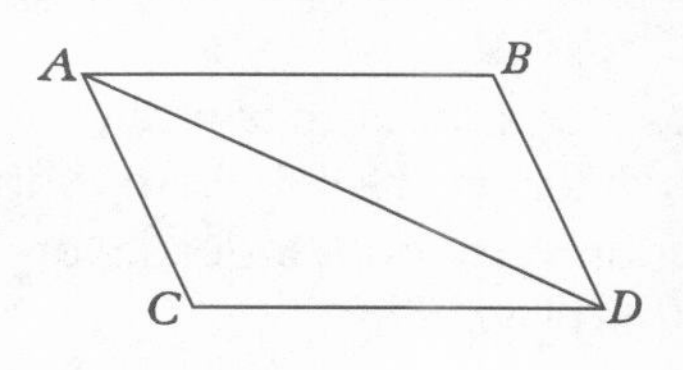

Corollary 2

And hence is evident the composition of a direct force AD from any oblique forces you please AB and BD, and, in turn, the resolution of any force you please AD into any oblique ones whatever AB and BD. This *composition and resolution is, moreover, abundantly confirmed from mechanics.*

[Newton does not offer a proof of Corollary 2. Instead, he gives an example of its application to "wheels, winches, pulleys, levers, taut cords, and weights"—the basic elements out of which, he asserts, all machines are compounded. In this way he shows that the principle of composition and resolution of forces pervades the whole doctrine of mechanics.

As his example he describes a delightfully elaborate contraption that has come to be known as "Newton's wheel," though it involves far more than a wheel. But the analysis is long and complicated and does not pertain to Newton's central argument (if indeed it applies anywhere in Principia) and so is omitted here.]

Corollary 3

The quantity of motion that is obtained by taking the sum of the motions made in the same direction, and the difference of those made in opposite directions, is not changed by action of the bodies among themselves.

For action and the reaction contrary to it are equal by Law 3, and consequently, by Law 2, bring about equal changes in motions in opposite directions. Therefore, if the motions take place in the same direction, whatever is added to the motion of a body that is speeding away will be subtracted from the motion of a following body, so that the sum remains the same as before. On the other hand, if bodies move in opposite directions, the subtraction from the motions will be equal on both sides, and consequently the difference of motions made in contrary directions will remain the same.

As, for example, let the spherical body A be three times as large as the spherical body B, and let it have two units of velocity, and let B follow in the same straight line with ten units of velocity: therefore, the motion of A is to the motion of B as six to ten. Let motions be supposed for them that are of six units and ten units, and the sum will be [a motion] of sixteen units. Therefore, in the bodies' collision, if body A should gain three or four or five units of motion, body B will lose the same number of units, and accordingly, after rebounding, body A will proceed with nine or ten or eleven units, and B with seven or six or five, the sum of the units always being sixteen, as before. If body A should gain nine or ten or eleven or twelve units, so that after the collision it should move forward with fifteen or sixteen or seventeen or eighteen units, body B, in losing as many units as A gains, will either move forward with one unit, having lost nine units, or will be at rest, having lost its forward motion of ten units, or will go back with one unit, having lost its motion and (so to speak) one unit more, or will go back with two units because a forward motion of twelve units has been taken away. And so the sums of the motions in agreement, $15 + 1$ or $16 + 0$, and the differences of the contrary motions, $17 - 1$ and $18 - 2$, will always be sixteen units, as they were before colliding and rebounding. Further, when the motions with which the bodies proceed after rebounding are known, the velocity of either will be found by supposing that it is to the velocity before rebounding, as the motion afterwards is to the motion before. As in the last case, where the motion of body *A* was six units before rebounding and eighteen units afterward, and the velocity was two units before rebounding, it will be found that its velocity after rebounding is six units, by saying that the six units of motion before rebounding are to the eighteen units of motion afterward, as the two units of velocity before rebounding are to six units of velocity afterward.

Suppose the bodies are either not spherical or are moving in different straight lines and encounter each other obliquely, and the motions after their rebound are required. One must know the position of the plane to which the colliding bodies are tangent at the point of collision; then, the motion of each body (by Corollary 2) must be divided into two, one perpendicular to this plane, the other parallel to it. And, because the bodies act upon each other along a line perpendicular to this plane, the parallel motions are to be kept the same after rebounding as before, and to the perpendicular motions, equal changes in opposite directions must be given, so that the sum of the agreeing [motions] and the difference of contrary ones stays the same as before. From rebounds of this kind, there also usually arise circular motions of the bodies about their own centers. But in what follows I do not consider these cases, and it would be exceedingly long to demonstrate everything relating to them.

Corollary 4

The common center of gravity of two or more bodies does not change its state either of motion or of rest by actions of the bodies among themselves, and for that reason the common center of gravity of all bodies acting mutually upon one another (external actions and hindrances being excluded) either is at rest or moves uniformly in a straight line.

For if two points move forward with uniform motion in straight lines, and their distance be divided in a given ratio, the dividing point either is at rest or moves forward uniformly in a straight line. This is demonstrated below in Lemma 23 and its corollary, if the motions of the points take place in the same plane, and it can be demonstrated by the same procedure if those motions do not take place in the same plane. Therefore, if any number of bodies move uniformly in straight lines, the common center of gravity of any two you please either is at rest or moves forward uniformly in a straight line, because the line joining the centers of these bodies moving forward uniformly in straight lines is divided by this common center in a given ratio. Likewise also, the common center of these two and any third you please either is at rest or moves forward uniformly in a straight line, because the distance of the common center of the two bodies and of the center of the third body is divided by it in a given ratio. And in the same manner, the common center of these three and of any fourth you please either is at rest or moves forward uniformly in a straight line, because the distance of the common center of the three and of the center of the fourth is divided by it in a given ratio, and so on *in infinitum*. Therefore, in a system of bodies that are entirely void of reciprocal actions among themselves and of all others impressed upon themselves from outside the system, and that accordingly each move uniformly in individual straight lines, the common center of gravity of all of them either is at rest or moves uniformly in a straight line.

Further, in a system of two bodies acting upon each other, since the distances of the centers of each from the common center of gravity are inversely as the bodies, the relative motions of the same bodies, whether of approaching that center or of receding from the same, will be equal among themselves. In this way, by equal alterations in the motions made in opposite directions, and thus by actions of these bodies among themselves, that center is neither moved forward nor retarded, nor suffers any alteration in its state as regards motion or rest. But in a system of many bodies, the common center of gravity of any two of them acting upon each other mutually does not change its state in any way because of that action, and the common center of gravity of the rest of them, among which that action does not intervene, is unaffected by it. Moreover, the distance of these two centers is divided by the common center of all the bodies into parts inversely proportional to the total sums of the bodies of which they are the centers, and accordingly, as those two centers preserve their state of moving or being at rest, the common center of all the bodies also preserves its state. Therefore, it is evident that that common center of all never changes its state as regards motion or rest because of the actions of two bodies between themselves. But in such a system, the actions of all the bodies among themselves are either between two bodies, or are compounded of actions between two bodies, and for that reason never introduce a change in the state of motion or rest in the common center of all the bodies. Consequently, since, where the bodies do not act upon each other, that center either is at rest or progresses uniformly in some straight line, it will proceed in the same way regardless of the actions of the bodies among themselves, whether to be always at rest, or always to progress uniformly in a straight line, unless it be dislodged from that state by forces impressed upon the system from outside. Therefore, the law for a system of many bodies is the same as for a solitary body, as regards perseverance in the state of motion or rest. For forward motion, whether of a solitary body or of a system of bodies, should always be reckoned by the motion of the center of gravity.

Notes to Corollary 4

- **Center of gravity:**

"...the line joining the centers of [two] bodies moving forward uniformly in straight lines is divided by this common center in a given ratio."

This is Newton's implicit definition of "common center," or center of gravity, of any number of bodies; the "given ratio" he refers to is the ratio of the bodies' respective quantities of matter. Thus, in general, the respective distances of any number of bodies from their common center of gravity are inversely as the quantities of matter in those bodies. In the case of two bodies, the center of gravity lies on the line between them, and the quantity of matter of one body times its distance from the center equals the quantity of matter of the other times its distance.

Classically, the center of gravity would have been defined in terms of the ratio of bodies' *weights*, not their quantities of matter. Newton is here appropriating the classical term to suit his own purposes; but in Proposition III.6 he will demonstrate that (at a given location) bodies' weights are proportional to their quantities of matter, so the calculated location of his "center of gravity" will not differ from that of the classical one.

- "For if two points move forward with uniform motion in straight lines, and their distance be divided in a given ratio, the dividing point either is at rest or moves forward uniformly in a straight line. This is demonstrated below in Lemma 23 and its corollary.... Therefore, in a system of bodies that are entirely void of reciprocal actions among themselves and of all others impressed upon themselves from outside the system, and that accordingly each move uniformly in individual straight lines, the common center of gravity of all of them either is at rest or moves uniformly in a straight line."

This is the first case of Corollary 4, where there is no interaction between or among the bodies. It depends on Lemma 23 and its corollary; these are included in the last note to this corollary.

- "Further, in a system of two bodies acting upon each other, since the distances of the centers of each from the common center of gravity are inversely as the bodies, the relative motions of the same bodies, whether of approaching that center or of receding from the same, will be equal among themselves. In this way, by equal alterations in the motions made in opposite directions, and thus by actions of these bodies among themselves, that center is neither moved forward nor retarded, nor suffers any alteration in its state as regards motion or rest."

This is the case of bodies that do interact. How will their interaction modify the previous results? Newton says that the interaction will produce "equal alterations in the motions." These alterations, moreover, will be "in opposite directions."

It is important to appreciate that by the "motion" of a body Newton here means the *quantity of motion:* not just the speed but the product of the body's speed and its quantity of matter. Thus if by their interaction the bodies experience "equal alterations" in their quantities of motion, each will change its speed by an amount that is inversely as the mass of the body. During any finite time, therefore, the bodies will move additional distances that are inversely as the bodies—and in opposite directions—compared to what they would have moved had there been no interaction.

But by definition, the distances of the bodies from their common center of gravity will be "inversely as the bodies" independent of any interaction. So when these distances are compounded with the additional distances attributable to the bodies' interaction, the resulting distances will still be inversely as the bodies; therefore any interaction between the bodies will not necessitate any change in the condition of motion (or rest) of the common center of gravity.

What is the basis of Newton's statement that any interaction between the bodies will produce equal alterations in opposite directions? Newton does not explicitly invoke the Third Law here, but the behavior he is describing is just what the third law would require. And the terms "equal" and "opposite" further suggest that he is appealing to that law. Can you think of any other reason the interactions might produce equal alterations in opposite directions or justification for that statement?

If the proof of this corollary does indeed depend on Law 3, keep this in mind as you think about the proofs Newton will give that Law 3 applies to attractions. Newton's proofs of this are given in the scholium following these corollaries. A note following the scholium will raise some additional questions.

- The following lemma and its corollary appear after Book I Proposition 26. I am including them here since Newton cites them in the sketch of this Corollary.

Lemma 23

If two straight lines AC, BD, given in position, be bounded at given points A, B, and have a given ratio to each other, and the straight line CD, by which the indeterminate points C, D are connected, be cut in a given ratio at K: I say that the point K will be located on a straight line in a given position.

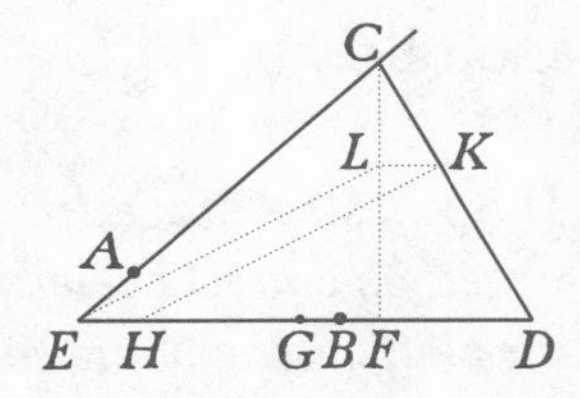

For let the straight lines AC, BD meet at E, and on BE let BG be taken to AE as BD is to AC, and let FD be always equal to the given [line] EG, and, by construction, EC will be to GD, that is, to EF, as AC is to BD, and therefore [will be] in a given ratio; and therefore, the triangle EFC will be given in shape. Let CF be cut at L so that CL may be to CF in the ratio of CK to CD, and, because of that given ratio, triangle EFL will also be given in shape; and further, the point L will be located on the straight line EL in a given position. Join LK, and triangles CLK, CFD will be similar; and, because FD is given, and the ratio of LK to FD is given, LK will be given. Let EH be taken equal to this, and $ELKH$ will always be a parallelogram. Therefore, the point K is located on the side HK of that parallelogram given in position.

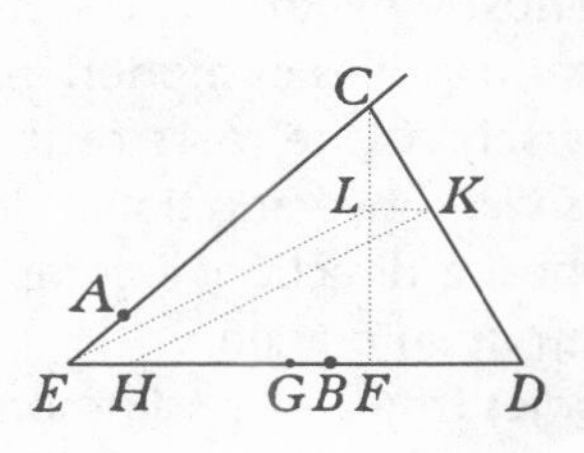

Q.E.D.

Lemma 23 Corollary

Because the figure $EFLC$ is given in shape, the three straight line EF, EL, and EC, that is, GD, HK, and EC, have given ratios to each other.

Corollary 5

The motions of bodies contained in a given space are the same among themselves, whether that space be at rest, or whether it move uniformly in a straight line without circular motion.

For the differences of motions tending in the same direction, and sums of those tending in opposite directions, are the same at the beginning in both cases (by supposition), and from these sums or difference arise the collisions and impulses with which the bodies strike each other. Therefore, by Law 2, the effects of the collisions will be equal in both cases, and therefore the motions among themselves in the one case will remain equal to the motions among themselves in the other. The same is confirmed by a most lucid experiment. All motions on a ship relate to each other in the same way whether the ship be at rest or move uniformly in a straight line.

Corollary 6

If bodies be moved in any manner whatever among themselves, and be urged by equal accelerative forces along parallel lines, everything goes on moving in the same manner among themselves as if they had not been impelled by those forces.

For those forces, in acting equally (in proportion to the quantities of the bodies to be moved) and along parallel lines, will move all the bodies equally (as regards velocity), by Law 2. Therefore, their positions and motions among themselves will not change.

Scholium After the Laws of Motion

Scholium

Up to this point, I have presented principles accepted by mathematicians and confirmed by manifold experience. By the first two laws and the first two corollaries, Galileo found that the descent of heavy bodies is in the duplicate ratio of the time, and that the motion of projectiles takes place in a parabola, in agreement with experience, except to the extent that those motions are retarded by some small amount by the resistance of air. When a body falls, uniform gravity, acting equally in the individual equal particles of time, impresses equal forces upon that body, and generates equal velocities; and in the whole time impresses a whole force and generates a whole velocity proportional to the time. And spaces described in proportional times are as the velocities and the times conjointly, that is, in the duplicate ratio of the times. And when a body is projected upwards, uniform gravity impresses forces and subtracts velocities proportional to the times, and the times of ascending to the highest altitudes are as the velocities to be subtracted, and those altitudes are as the velocities and the times conjointly, or in the duplicate ratio of the velocities. And the motion of a body projected along any straight line you please, arising from the projection, is compounded with the motion arising from gravity.

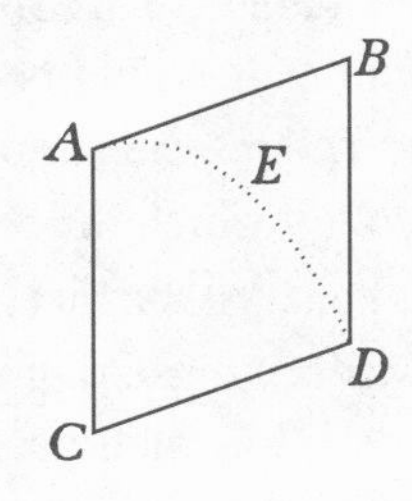

As, for example, a body A might, by the motion of projection alone, describe the straight line AB in a given time, and by the motion of falling alone might in the same time describe the height AC. Let the parallelogram $ABDC$ be completed, and at the end of the time that body in its compound motion will be found at place D, and the curved line AED, which that body will describe, will be a parabola which the straight line AB touches at A, and whose ordinate BD is as sq. AB. From the same laws and corollaries depend matters demonstrated of the times of swinging pendulums, with the support of everyday experience of clocks.

From these same things and the Third Law, Sir Christopher Wren, Dr. John Wallis, and Christian Huygens, easily the chief geometers of recent times, have found the rules of collisions and rebounds of two bodies, and communicated them to the Royal Society at nearly the same time, fully in agreement with each other (as regards these laws); and in fact first Wallis, then Wren and Huygens, produced the discovery. But the truth was also confirmed by Wren before the Royal Society by an experiment of pendulums, which the most illustrious Mariotte also soon saw fit to expound in an entire book. In fact, to make this experiment match the theories exactly, one must take into account the resistance of air, as well as the elastic force of the colliding bodies.

Let spherical bodies A, B hang by parallel and equal strings AC, BD, from centers C, D. With these centers and radii, let semicircles EAF, GBH be described, bisected by radii CA, DB. Let body A be pulled to any point you wish R of arc EAF, and (body B being removed) let it be sent down from there, and after one swing let it return to point V. RV is the retardation from the resistance of the air. Let ST be made one fourth of this RV, located in the middle, in such a way that RS and TV are made equal, and let RS be to ST as 3 to 2. And this ST will show the retardation in the descent from S to A, approximately. Let body B be restored to its place. Let the body A fall from point S, and its velocity at the point of rebound A will be as great as if it had fallen in a vacuum from the place T, without perceptible error. Therefore, let this velocity be set out by the chord of the arc TA. For it is a proposition very well known to geometers that the velocity of a pendulum at the lowest point is as the chord of the arc which it describes in falling. After rebound, let the body A carry through to place s, and body B to place k. Let body B be removed and let the place υ be found, such that if body A be sent down from it and return after one swing to point r, st would be one fourth of $r\upsilon$ located in the middle, in such a way that rs and $t\upsilon$ are made equal, and by the chord of the arc tA let there be expressed the velocity which the body A had at place A immediately after rebounding.

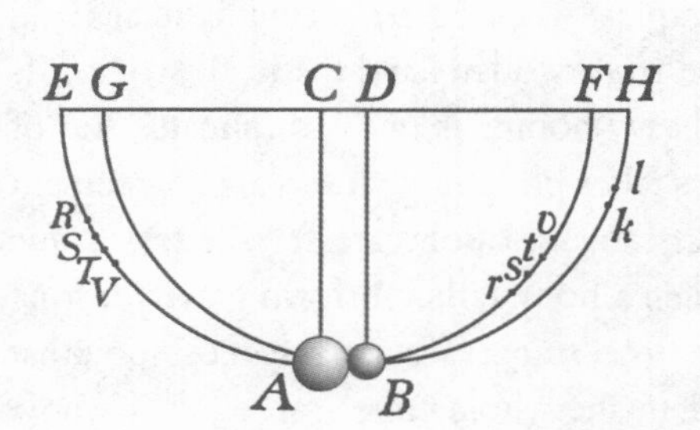

For t will be that true and correct place to which the body A, with the resistance of air removed, ought to have ascended. By a like method the place k is to be corrected, to which the body B ascends, and the place l is to be found, to which that body ought to have ascended in a vacuum. In this way, everything may be experienced exactly as if we were set up in a vacuum. Finally, body A is to be multiplied (so to speak) by the chord of the arc TA, which represents its velocity, so that its motion at place A immediately before rebounding may be had; then by the chord of the arc tA, so that its motion at place A immediately after rebounding may be had. And likewise, the body B is to be multiplied by the chord of the arc Bl, so that its motion immediately after rebounding may be had. And by a similar method, where two bodies be sent down simultaneously from different places, the motions of the two are to be found before rebounding as well as after, and then at last the motions are to be compared among themselves and the effects of the rebound concluded. In testing the matter in this way with pendulums of ten feet, and with bodies both equal and unequal, and making the bodies collide from very large intervals, such as eight or twelve or sixteen feet, I always found without an error of three inches in the measurements, where the bodies encountered each other directly, that the changes of motions introduced into the bodies in opposite directions are equal, and therefore, that the action and the reaction are always equal. For example, if body A struck body B, at rest, with nine units of motion, and, losing seven units, continued with two after rebounding, body B moved back with those seven units.

If the bodies went towards each other, A with twelve units and B with six, and A went back with two, B went back with eight, a subtraction of fourteen units being made on both sides. Let twelve units be subtracted from A, and there will remain none; let two more units be subtracted, and a motion of two units in the opposite direction results. Likewise, by subtracting fourteen units from the motion of body B of six units, there result eight units in the opposite direction. If the bodies went in the same direction, A faster with fourteen units, and B slower with five units, and after rebound A continued with five units, B continued with fourteen, a transfer of nine units being made from A to B. And so for the rest. By the meeting and collision of bodies, the quantity of motion was never altered, which was reckoned from the sum of agreeing motions and the difference of contrary ones. For I would attribute an error of one or two inches in the measurements to the difficulty of carrying out each step accurately enough. It was difficult to let the pendulums go simultaneously so that the bodies should meet one another at the lowest place AB; and to mark the places s, k, to which the bodies ascended after collision. But also, in the pendulous bodies themselves, the unequal density of parts, and the irregular texture arising from other causes, introduced errors.

Further, so that no one should object to the rule, for the testing of which this experiment was devised, that it presupposes bodies that are either absolutely hard, or at least perfectly elastic, of which sort none are found in natural compounds, I add that the experiments just described work in soft bodies as well as in hard; that is, that nothing depends upon the condition of hardness. For if that rule is tested in bodies that are not perfectly hard, it will only be necessary to decrease the rebound in a certain proportion according to the quantity of the elastic force. In the theory of Wren and Huygens, absolutely hard bodies return from each other with the velocity of collision. That will be stated with greater certainty of perfectly elastic bodies. In imperfectly elastic bodies, the velocity of return is to be diminished along with the elastic force, because that force is (as much as I can tell) certain and determinate (except where the parts are damaged in the collision, or undergo some stretching as if under a hammer), and makes the bodies return from each other with a relative velocity that is to the relative velocity of collision in a given ratio. I have tested this in balls of tightly packed and strongly bound wool, thus. First, by releasing pendulums and measuring the rebound, I found the quantity of the elastic force. Next, by this force, I determined the rebounds in other cases of collision, and the experiments answered to them. The balls always recoiled from each other with a relative velocity which would be to the relative velocity of collision as about 5 to 9. Balls of steel recoiled with nearly the same velocity; others, of cork, with a little less; but in glass balls the ratio was about 15 to 16. And in this way the Third Law, as to impacts and rebounds, is confirmed in a theory which is entirely in accord with experience.

In attractions, I show the matter briefly thus. There being any two bodies you please A, B, that mutually pull each other, conceive of any obstacle you please interposed, by which their coming together may be impeded. If one of the bodies A is pulled more towards the other body B than that other B [is pulled] towards the former A, the obstacle will be urged more by the pressure of the body A than by the pressure of the body B, and accordingly will not remain in equilibrium. The stronger pressure will prevail, and will make the system of the two bodies and the obstacle move in a straight line towards the parts in the direction of B, and go off *in infinitum* in free spaces with a motion always accelerated. Which is absurd and contrary to the First Law. For by the First Law the system will be obligated to persist in its state whether of resting or of moving uniformly in a straight line, and accordingly the bodies will urge the obstacle equally, and for that reason will be equally pulled towards each other. I have tried this on a magnet and iron. If these two be placed by themselves in separate vessels that touch each other, in still water they will float together; neither will propel the other, but by an equality of attraction on both sides they will maintain equal strivings towards each other, and at length, being set in equilibrium, will come to rest.

Thus also, gravity between the earth and its parts is mutual. Let the earth FI be cut by any plane you please EG into two parts EGF and EGI, and the weights of these towards each other mutually will be equal. For if by another plane HK, which is parallel to the former EG, the greater part EGI be cut into two parts $EGKH$ and HKI, of which HKI is equal to the former part cut off EFG, it is manifest that the middle part $EGKH$ will by its own weight be inclined towards neither of the extreme parts, but will be suspended between the two in equilibrium, so to speak,

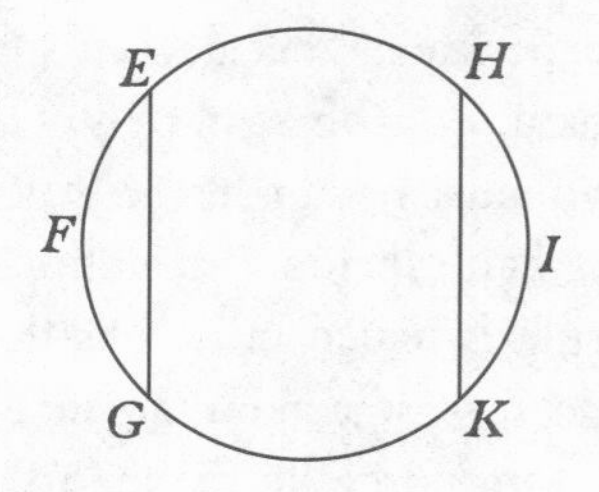

and will be at rest. But the extreme part HKI will press on the middle part with its whole weight, and will urge it towards the other extreme part EFG, and therefore the force by which the sum EGI of the parts HKI and $EGKH$ tends towards the third part EGF is equal to the weight of the part HKI, that is, to the weight of the third part EGF. And on this account the weights of the two parts EGI, EGF to each other mutually are equal, just as I wished to show. And unless those weights were equal, the whole earth, floating in the free ether, would give way to the greater weight, and in fleeing from it would go off *in infinitum*.

As bodies have the same strength in collision and rebound, whose velocities are inversely as the inherent forces, likewise, in making mechanical instruments move, agents have the same power and support each other by contrary efforts, whose velocities, reckoned along the direction of the forces, are inversely as the forces.

Thus, weights will be equivalent in moving the arms of a balance, which, when the balance swings, are inversely as their velocities upwards and downwards; that is, weights, if they ascend and descend directly, will be equivalent, which are inversely as the distance from the balance's axis of the points from which they are suspended. If, however, they ascend or descend obliquely, impeded by inclined planes or other obstacles set in the way, [those weights] will be equivalent, which are inversely as the ascents and descents, insofar as they are made along the perpendicular: this is because gravity is directed downward.

Similarly, in a winch or block and tackle, the force of the hand directly pulling the rope, which is to the weight, ascending either directly or obliquely, as the velocity of perpendicular ascent is to the velocity of the hand pulling the rope, will support the weight.

In clocks and similar instruments, which are constructed of linked wheels, the contrary forces for furthering and hindering the motion of the wheels, if they are inversely as the velocities of the parts of the wheels upon which they are impressed, will support each other.

The force of a screw for pressing a body is to the force of the hand driving the handle around, as the circular velocity of the handle at that part where the hand urges it, is to the forward velocity of the screw towards the pressed body. The forces by which a wedge urges the two parts of a split timber are to the force of the hammer on the wedge, as the forward motion of the wedge along the direction of the force impressed upon it by the hammer, is to the velocity by which the parts of the timber give way to the wedge, along lines perpendicular to the faces of the wedge. And the account of all machines is comparable.

The efficacy and usefulness of these consists in this alone, that in decreasing the velocity we may increase the force, and vice versa. From this comes the solution, in every kind of suitable device, of the problem, "to move a given weight with a given force," or otherwise, to overcome a given resistance with a given force. For if machines be so arranged that the velocities of the agent and the resistance are inversely as the forces, the agent will support the resistance, and with a greater disparity of velocities, will overcome the same. Certainly, if the disparity of velocities be so great that all resistance is overcome, that normally arises from the wearing away of bodies that are contiguous and rubbing each other, as well as from the cohesion of [bodies that are] continuous and that must be separated from each other, and [from] the weights of [bodies that are] to be lifted, once all resistance is overcome, the force left over will produce an acceleration of motion proportional to itself, partly in the parts of the machine, partly in the resisting body. Nevertheless, to treat of mechanics is not the purpose of this. In these remarks, I have intended only to show how broadly evident and how certain the Third Law of Motion is. For if the action of an agent is reckoned from its force and velocity conjointly, and the reaction of the resistance is likewise reckoned conjointly from the velocities of its individual parts and the forces of resistance arising from their wearing away, cohesion, weight, and acceleration, the action and reaction, in every use of instruments, will always be equal to each other. And insofar as an action is propagated through the instrument and is ultimately impressed upon every resisting body, its final direction will always be contrary to the direction of the reaction.

Note on Law 3 and Scholium to the Laws

The following notes raise a few questions about Newton's application of Law 3 to gravitational attractions as supported by the two thought experiments in the preceding scholium. You might want to reread or refer back to Law 3 and its commentary (including the last sentence in which he says that that he proves in the scholium that Law 3 applies to attractions as well as mechanical pushes and pulls) and to the thought experiments on pages 42 and 43 of the scholium. The two thought experiments are given in the two paragraphs beginning with the words "In attractions, I show the matter briefly thus."

• **What are "attractions"?** I put quotation marks around attractions, because it is not clear exactly what attractions are when we are talking about the physical world and not about movements of the soul. The thought experiments involve gravitational attractions, so perhaps we can at least narrow the focus down somewhat. But we remember Newton's emphasis in his commentary after the Definitions that when he speaks of attraction he only means it mathematically, not physically, and says "I use the words 'attraction,' 'impulse,' or words denoting a propensity of any kind toward a center, indifferently and promiscuously for each other… Therefore, the reader should beware of thinking that by words of this kind I am anywhere defining a species or manner of action, or a cause or physical account."

This means that the physical cause or mechanism for what he calls attraction may be impulse(s) from behind the attracted body—differential ether pressure, for example. If so, the physical action would not be between the two bodies: there would indeed be Law 3 interactions but they would be between each body and the ether particles impelling it.

• **Questions for Discussion:** What do you make of Newton's two thought experiments showing that Law 3 applies to "attractions"? Has the assertion made in the last sentence of his commentary to Law 3 that the law applies been, as he said, "proven"? And, considering that this is a law and not a proposition, is proof even the right term? Laws are laws of nature. They are like postulates. And isn't Newton, in his exposition, in fact proceeding as if they are postulates? One might at least get the impression that he is calling on our experience of the world in observing that the earth, or the system of two bodies with an obstacle between them, remains at rest and doesn't go accelerating out in some direction to infinity.

And yet, what about his formal invoking of Law 1? He says that the bodies don't fly off because that would violate Law 1. So what is really at issue here, Law 3 or Law 1? And how exactly does the acceleration of the two bodies, in the case he declares absurd, violate Law 1? Law 1 only applies to bodies in the absence of an impressed force. But gravitational attraction, centripetal force, is an impressed force, as Newton tells us in Definition 4.

• Since, as we observed in a note after Law 3, a law of motion is a law, not a proposition, and doesn't need to be proved, Newton rightly saw no need to offer a proof of the direct mechanical pushes and pulls in the main body of this discussion of the third law, although he does make an argument for it in the last section of the scholium.

But Newton might have tried to "prove" that earlier part of Law 3 the same way he does for attractions, the proof looking very similar to the proofs about attractions in the scholium. The method of proof has an implied invocation of Corollary 4 of the Laws. In the "attraction" thought experiments, the actions of the bodies (or parts of the earth) among themselves should not, by Corollary 4, change the common center of gravity of the system. But if the bodies accelerated out together in some direction, the common center of gravity would be accelerating out in that direction.

We could see the same argument with actions which are not classed as attractions. If I have rowed a boat up to the dock, so that the bow is touching the dock and the stern extending out into the lake, and then walk up to the bow with the idea of climbing onto the dock, I find that as I walk towards the bow of the boat, the boat, by Law 3, will move out into the lake. This is a Law 3 mutual action between me and the boat. Furthermore, I can prove that it must be so. Before I begin walking, the boat and I have a common center of gravity in the space defined by the dock and the lake. If the boat did not move back as I moved forward, the common center of gravity of the boat and me would have moved closer to the dock, contrary to Corollary 4.

But if Law 3 is just an application of Corollary 4, why do we need a separate law?

Moreover, it looks like we have yet another sort of puzzle. On the one hand, in our discussion following Corollary 4, we saw that the proof of Corollary 4 appears to rely on Law 3. On the other hand, in the case of Newton's explicit arguments that Law 3 applies to attractions in the scholium we just read, he seems to be calling on Corollary 4 in order to reach that conclusion. What do you make of this seeming tangle of mutual dependency?

Or perhaps Newton should not have led us into this sort of thinking by claiming to give a proof in the first place. Instead of the "proofs" in the scholium, he might have just said that the applicability of the third law to attractions was an assumption on which he intended to rely and whose legitimacy would be demonstrated by the results of his whole system.

Book I

On the Motion of Bodies ❦ First Book

SECTION 1

On the method of first and ultimate ratios, by means of which what follows is demonstrated.

Lemma 1

Quantities, as well as the ratios of quantities, which in any finite time you please constantly tend towards equality, and before the end of that time approach nearer to each other than by any given difference you please, become ultimately equal.

If you deny it, let them become ultimately unequal, and let their ultimate difference be D. Therefore, they cannot approach nearer to each other than by the given difference D, contrary to the hypothesis.

Notes on Lemma 1

• Lemmas are derivations of relationships that can then be called on succinctly as parts of later proofs, without having to be reproven each time. In the case of this series of eleven, the lemmas present Newton's calculus as he will use it in *Principia*.

• **Technical Meaning of "Ultimate":**

Lemma 1 is a method of *finding what the limit is*. We then say that the thing converging towards that limit, and the limit itself, are "ultimately" equal. No claim is made that a particular convergence *reaches* the limit.

The following lemmas, and all of *Principia*, are based on this lemma and its procedure. It is crucial to understand what Newton means by "ultimately equal," the conclusion we may make if the conditions are met.

When Newton says that a quantity "ultimately" equals another, he doesn't mean that it becomes equal to the other "eventually," as we might understand of that adverb in common speech. It may never get there at all. His thinking, and the applications made of this lemma, include quantities that can never actually become equal to the others that they approach, at least not without ceasing to be what they are.

Sometimes these may have vanished at that point (the Pause Before Lemma 7 looks further into this); quantities that no longer exist are no longer what they were.

In the next two lemmas we are dealing with areas bounded by connected straight line segments that get closer and closer to coinciding with areas bounded by a curved line. Newton will say that "ultimately" the rectilinear areas will coincide with the curvilinear areas. But as long as they are what they are—rectilinear areas—their sides can never coincide with arcs of a curve.

What does happen is that the segments can be made to approach the curve more and more closely, and the areas bounded by the straight line segments and the areas bounded by the curve can become closer and closer to being equal.

As they become closer and closer, they approach a *limit*. The "limit" is a technical term that means some value or relationship towards which something constantly changing is headed, but that it may never actually reach (sometimes this is phrased as "will never reach in a finite time"), and that in any case it will never go beyond.

Here's an example. Suppose you had a quart of ice cream in the freezer. For dessert you eat half of it. Later, remembering how good it tasted, you go back to the freezer; but knowing that you already had dessert, you eat only half of what is left. A little later, temptation strikes again, but not wanting to feel like a glutton, you again eat only half of what is left. Let's say that this yielding to temptation, but with the same moderation, goes on *in infinitum*. There are many problems with imagining actually completing the transfer of the ice cream from the carton to your stomach by this process. But there is no question at all about the *limit* that is being approached. The limit is one quart.

You are getting closer and closer to having consumed a quart of ice cream, and by this process you will never consume *more* than a quart. You are "tending constantly" towards having eaten a quart of ice cream, and you are diminishing the difference between the original quart and the amount eaten below any imagined difference however small. What Newton would say is that you "ultimately" will have consumed a quart. The amount consumed will be "ultimately equal" to the amount originally in the carton.

Ultimate then refers to the relationship among the variables being approached at a limit. The process need not reach that limit, or, if it does, the magnitudes might not exist there. But the limit itself, and the relationship among the quantities being approached at that limit, may still be quite clear or at least determinable by geometric demonstration.

• **Lemma 1 as a definition:** The word "ultimately" takes on two different guises in the lemma. When we first encounter it, it is subordinate to the adjective "equal," which it modifies. The force of the demonstration is to show that if certain conditions are satisfied, the quantities or ratios in question cannot be unequal, and must therefore be (in some sense) equal. This is Lemma 1 as a proof.

At this point, however, before our very eyes, the word "ultimately" ceases to be a mere modifier and becomes a technical term referring to a relationship among quantities or ratios that satisfies the two conditions. Thus, throughout the *Principia,* the words "ultimately equal" mean that the quantities or ratios

1. constantly tend towards equality; and
2. approach nearer to each other than by any given difference.

This is Lemma 1 as a definition.

We shall have occasion, in what follows, to refer back to this lemma in ascertaining what Newon means when he establishes "ultimate" relationships.

• The parenthetical phrase "as well as ratios of quantities" is included to deal with the situation in which the quantities themselves are vanishing—not just the difference between them. This phrase would not be needed if Newton only meant to talk about diminishing differences between finite quantities. The statement about what happens to the ratio also applies to quantities that are not vanishing: when finite quantities are approaching equality with each other (that is, their *difference* is vanishing) their ratio also approaches a ratio of equality.

The first five lemmas are concerned with quantities that are not vanishing, although the difference between them is understood to vanish. Lemma 6 involves a single angle vanishing. We get a trickier situation when the quantities being compared are themselves vanishing, and we must deal with some of those situations beginning with Lemma 7.

In the Pause Before Lemma 7 we will rethink Lemma 1 in relation to vanishing quantities. For now, we'll keep it simple by considering the primary intention of Lemma 1 and its uses in the lemmas that directly follow.

Expansion of Newton's Sketch of Lemma 1

"Quantities, as well as the ratios of quantities, which in any finite time you please constantly tend towards equality, and before the end of that time approach nearer to each other than by any given difference you please, become ultimately equal."

Given:

(1) Two [nonvanishing] quantities tend constantly towards equality;

(2) The quantities, in a finite time, approach nearer to each other than any given difference, however small.

To Prove:

The two quantities become "ultimately equal," that is, they have a limit of equality.

Proof:

"If you deny it, let them become ultimately unequal, and let their ultimate difference be *D*. Therefore, they cannot approach nearer to each other than by the given difference *D*, contrary to the hypothesis."

Suppose that the quantities did not become equal at the limit. Let there be at the limit some difference *D*. They therefore cannot have approached nearer to one another than the difference *D*, which is contrary to the second "given."

Therefore our supposition is absurd; that leaves only the possibility that they do become equal at the limit.

Lemma 2

If in any figure you please AacE, contained by the straight lines Aa, AE, and the curve acE, there be inscribed any number of parallelograms Ab, Bc, Cd, and so on, contained beneath equal bases AB, BC, CD, and so on, and by the sides Bb, Cc, Dd, and so on, parallel to the side Aa of the figure; and if the parallelograms aKbl, bLcm, cMdn, and so on, be completed, and if afterward the breadth of the latter parallelograms be diminished and their number be increased in infinitum, *I say that the ultimate ratios which the inscribed figure AKbLcMdD, the circumscribed figure AalbmcndoE, and the curvilinear figure AabcdE, have to each other are the ratios of equality.*

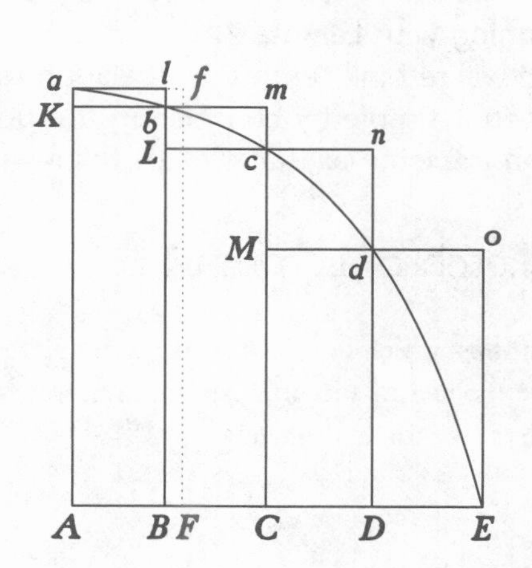

For the difference of the inscribed figure and the circumscribed [one] is the sum of the parallelograms Kl, Lm, Mn, Do, that is (because of the equal bases of them all), the rectangle beneath the base Kb of one of them and the sum of the altitudes Aa, that is, the rectangle $ABla$. But because the breadth AB of this rectangle is diminished in infinitum, the rectangle becomes less than any given [magnitude] you please. Therefore (by Lemma 1) the inscribed figure and the circumscribed [one], and much more so the intermediate curvilinear [figure], become ultimately equal.

Q.E.D.

[Q.E.D. is an abbreviation for *quod erat demonstrandum*, "that which was to have been demonstrated."]

[**Note on Newton's diagram:** The engravings for Newton's diagrams were expensive, and he made do with as few as possible, making different propositions share the same drawing even when their constructions are different. Newton gives this diagram for use with both Lemma 2 and Lemma 3. The extra lines and letters may make it a bit confusing for Lemma 2; the diagram is simply incorrect for Lemma 3. Construct your own diagrams for these two lemmas based on the steps of the particular proofs. Or consult the diagrams I supply with the expanded proofs.]

Notes on Lemma 2

• Our interest is in learning something about the area under a curve. Measuring an area with a curved boundary has always been a difficult problem because of the incommensurability of curved areas and rectilinear areas. We measure area in "square feet" or "square millimeters," but curved areas can't be broken into square anythings.

• The proof Newton gives here depends upon the segment of curve in question being as he has drawn it in the diagram he supplied, that is, with a negative slope through its length (curving down monotonically from a high point). It would work equally well with a segment of curve with positive slope through its length (curving up from a low point). To deal with a more complicated segment of curve, such as one that went first up then down, or vice versa, we would break the larger curve into pieces each of which was monotonically increasing or decreasing and use this demonstration on each part.

Expansion of Newton's Sketch of Lemma 2

Given:

"If in any figure you please $AacE$, contained by the straight lines Aa, AE, and the

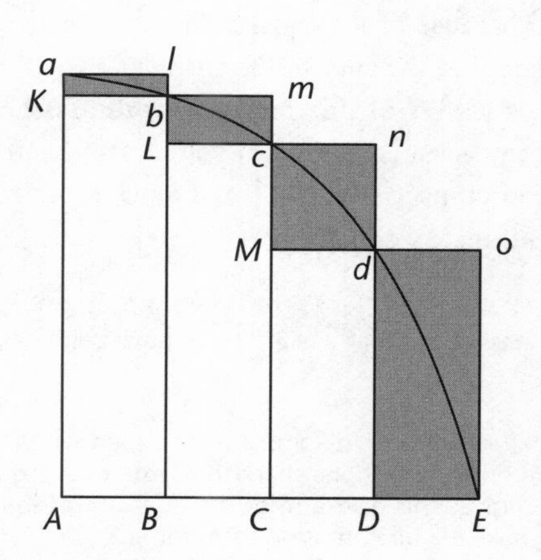

curve acE, there be inscribed any number of parallelograms Ab, Bc, Cd, and so on, contained beneath equal bases AB, BC, CD, and so on, and by the sides Bb, Cc, Dd, and so on, parallel to the side Aa of the figure; and if the parallelograms $aKbl$, $bLcm$, $cMdn$, and so on, be completed, and if afterward the breadth of the latter parallelograms be diminished and their number be increased *in infinitum*..."

1. A figure bounded by straight lines on two sides and a curve on the third side;

2. Parallelograms inscribed in and circumscribing the curve on equal bases;

3. Suppose the bases to be increased and the parallelograms' breadth diminished without limit.

To Prove:

"I say that the ultimate ratios which the inscribed figure $AKbLcMdD$, the circumscribed figure $AalbmcndoE$, and the curvilinear figure $AabcdE$, have to each other are the ratios of equality."

Proof:

"For the difference of the inscribed figure and the circumscribed [one] is the sum of the parallelograms Kl, Lm, Mn, Do, that is (because of the equal bases of them all), the rectangle beneath the base Kb of one of them and the sum of the altitudes Aa, that is, the rectangle $ABla$."

The proof is simple and elegant. We divide the base of the area under the curve into equal segments and construct inscribed and circumscribing parallelograms on those bases. We first note that the area is approximated by the parallelograms, and the more parallelograms there are the closer is the approximation. But the area under the curve is always greater than the sum of the areas of the inscribed parallelograms, and less than the sum of the areas of the circumscribing parallelograms.

The proof requires us to consider the staircase of small parallelograms that are the difference, on each base, between the inscribed and circumscribing parallelograms. We notice that they could all slide over and neatly fill the first parallelogram. Thus the difference between the sum of the inscribed and the sum of the circumscribing parallelograms is equal to the area of the first, or tallest, circumscribing parallelogram *ABla*.

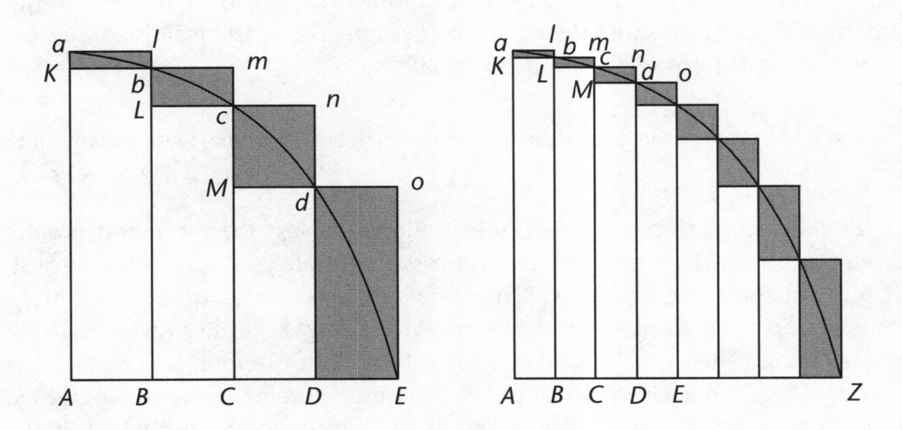

"But because the breadth *AB* of this rectangle is diminished *in infinitum*, the rectangle becomes less than any given [magnitude] you please."

Now we imagine the number of divisions of the base of the curvilinear area increased. For example, we might double the number of divisions. We must then draw new sets of circumscribed and inscribed parallelograms. These parallelograms will be narrower, having, in our example, half the previous breadth. There will again be an offset stack of difference-parallelograms, which can be again slid over to just cover the first or tallest circumscribing parallelogram, which, now being half the breadth it was before, consequently possesses half the area. Thus the difference between the circumscribing and inscribed parallelograms is half what it was. Again double the number of bases, construct new narrower inscribed and circumscribing parallelograms, and note that the area of the difference is again cut by half. (It need not have been half; that was just our example here. Any consistent scheme of reducing the size of the parallelograms would do.)

Repeat this process. Because the difference between the areas of the two sets of parallelograms (the circumscribing and the inscribed) is continually decreasing, the summed areas are "tending constantly towards equality" in the terms of Lemma 1. Furthermore, if we were to propose a very small area, as tiny as we pleased, someone could cut the parallelogram bases down until the area of the first circumscribing parallelogram (and therefore the differences between the sets of parallelograms) was smaller than that tiny area. And it could be done in a finite time, whatever method was being used to increase the number of bases.

Note that in setting this up, Newton is setting up the conditions of Lemma 1.

"Therefore (by Lemma 1) the inscribed figure and the circumscribed [one] ... [become ultimately equal] ..."

We have already established the conditions for Lemma 1 for nonvanishing quantities. Consequently we say by Lemma 1 that the inscribed and circumscribing figures are "ultimately equal."

"... and much more so the intermediate curvilinear [figure], become[s] ultimately equal. Q.E.D."

But we noted that the area under the curve is less than one and greater than the other. The smallest area becomes "ultimately equal" to the largest; it must at the same moment, if not previously, become equal to the intermediate area. One could say the same for the largest on its way to its "ultimate equality" with the smallest: it must then, if not before, be equal to the intermediate. Thus either set of parallelograms, if their number be increased and their breadth decreased *in infinitum*, will approach a limit which is the area under the curve.

Conclusion

Newton's sketch ends with proving ultimate equality, which is all that is needed for nonvanishing quantities. But his enunciation says that he will prove that the ultimate *ratios* are ones of equality. To be complete the proof should end with what the enunciation asserts. If we intend to use this lemma as enunciated in our later propositions we must make sure that we can take it all the way.

"I say that the ultimate ratios which the inscribed figure $AKbLcMdD$, the circumscribed figure $AalbmcndoE$, and the curvilinear figure $AabcdE$, have to each other are the ratios of equality."

As the difference between the two areas diminishes, the ratio between those two areas tends constantly towards equality. As the difference decreases *in infinitum* the ratio will become closer to equality than any given difference. As the difference vanishes, that limit is a ratio of equality.

Since the limit of both is the area under the curve, the ultimate ratios that the inscribed figure, the circumscribed figure, and the curvilinear figure have to each other are the ratios of equality.

Q.E.D.

Note: Neither set of parallelograms will ever actually reach *coincidence* with the curved area as long as they are actually parallelograms, that is, as long as there is any finite breadth to the base. But the area under the curve is the limit they both approach. This means that (1) they will never go past that limit; (2) they will get continually closer to that area; and (3) they will get closer than any given difference.

Lemma 3

The same ultimate ratios are also ratios of equality, where the breadths of the parallelograms AB, BC, CD, and so on, are unequal, and all are diminished in infinitum.

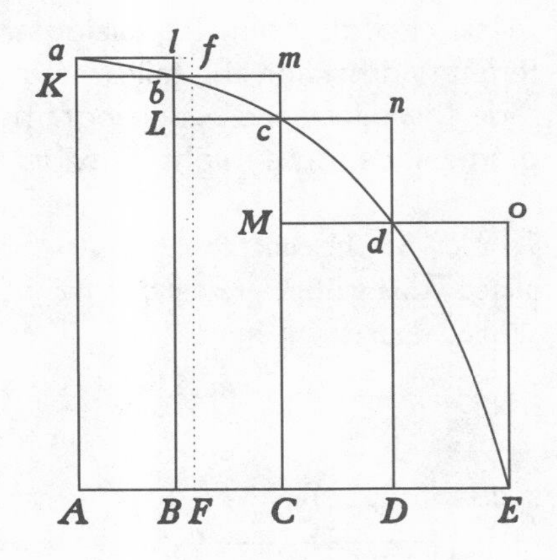

For let *AF* be equal to the greatest breadth, and let the parallelogram *FAaf* be completed. This will be greater than the difference of the inscribed figure and the circumscribed figure; and when its breadth *AF* is diminished in infinitum, it becomes less than any given rectangle.

Q.E.D.

[**Note on Newton's diagram:** This diagram was given with Lemma 2 and is evidently intended to serve for both. As noted where it first appeared in Lemma 2, it is incorrect for Lemma 3, since *AF* is supposed to be the width of the widest parallelogram. Since the diagram has all the bases equal, *AF* is wider than any shown. Use imagination or consult the diagrams in the expanded proof below.]

Notes on Lemma 3

• In order to be able to invoke Lemma 1, we need to have a repeatable procedure for dividing the base of the curvilinear area into these unequal parallelogram bases. Such a procedure might be a simple one, such as dividing the whole base of the curvilinear area in some ratio, perhaps some simple whole number ratio like 2:3. Those two line segments (one being 1/3 and the other being 2/3 of the original base) would be the bases of the first set of parallelograms. This is repeated by dividing each of those unequal bases again in the same ratio. Or the division algorithm might be more complicated. The important thing is that it be a procedure that can be repeated, a process that will operate every time on every base of the previous iteration, so that the breadth of the widest base will diminish constantly.

• See the second note on Lemma 2: this proof assumes a diagram like the one Newton gives, but could be applied to a more general curve as suggested in that note.

Expansion of Newton's Sketch of Lemma 3

In Lemma 2 all the parallelograms had the same breadth, and thus the stack of parallelograms making up the differences could be slid into the tallest circumscribing parallelogram to fill it exactly. Here we may have different widths, and the tallest might not be the widest. Thus we can't simply slide the difference-parallelograms into the tallest circumscribing parallelogram, since some might be wider and we don't know by how much.

"For let AF be equal to the greatest breadth, and let the parallelogram $FAaf$ be completed. This will be greater than the difference of the inscribed figure and the circumscribed figure; ..."

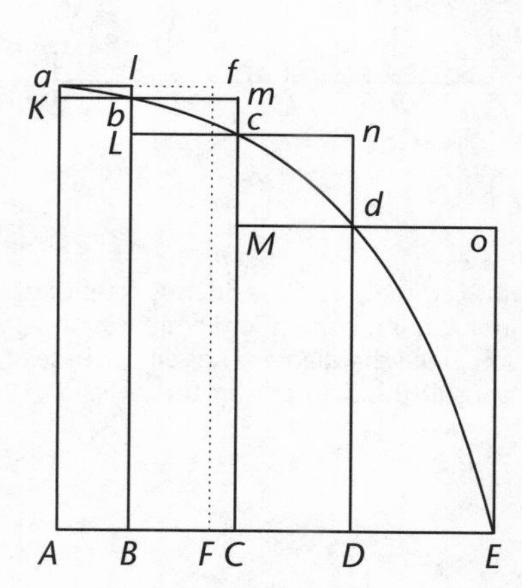

Newton's strategy is to take the tallest (in this diagram parallelogram *aB*) and cut off a base (*AF*) equal to the widest base (the base of parallelogram *dE*), creating a new parallelogram (parallelogram *aF*) into which all the difference-parallelograms may now be slid, with all of them guaranteed to be inside. Of course, some will be narrower, and so the sum of the areas of the difference-parallelograms will now fall short of the area of this constructed parallelogram, but that will only make our demonstration stronger.

We will prove that the area of the constructed parallelogram *aF* can, by increasing the number of divisions of the base of the curvilinear area, be made as small as we please, smaller than any given area however tiny, following the conditions of Lemma 1.

"...and when its breadth AF is diminished *in infinitum*, it becomes less than any given rectangle.

Q.E.D."

As we increase the number of divisions of the base by whatever process we have selected or been given, the widest base will also become continuously smaller. The maximum height under the curve will remain the same. Thus the area of a constructed parallelogram with the maximum height and the greatest breadth will become continuously smaller, eventually (in a finite time) smaller than any tiny area we posit. And since the actual sum of the areas of the difference-parallelograms, or the total difference between the areas of the circumscribing and the inscribed parallelograms, will be even *smaller* than that constructed parallelogram, the case is even stronger for saying that the difference will become smaller than any tiny area we posit.

To complete the proof we must add the final steps, following the pattern of the argument Newton gave us for Lemma 2.

We have fulfilled the conditions for Lemma 1 and so may say that the areas which are the sums of the inscribed and circumscribing parallelograms are ultimately equal.

The area under the curve, greater than the area of the inscribed, and less than the area of the circumscribing, parallelograms, must "much more so" become equal to both at their limits.

Furthermore, they tend constantly towards ratios of equality and, as the difference between them vanishes, the ultimate ratios are ratios of equality.

Q.E.D.

Lemma 3 Corollary 1

Hence, the ultimate sum of the evanescent parallelograms coincides on all sides with the curvilinear figure.

Expansion of Newton's Sketch of Corollary 1

And the areas of these circumscribing and inscribed parallelograms are not only ultimately equal, they are ultimately coincident with the curvilinear figure, that is, the curvilinear area, which is their limit.

Since the limit of the changing rectilinear areas is the curvilinear area, we must say that at the limit the staircases would coincide with the curve. This is not the same as saying that the *length* of the staircases approaches the *length* of the curve. For now, just a warning. This question of the relationship of rectilinear and curvilinear lengths is explored in more detail in the Notes on Corollary 4 and in the Pause After Lemma 3 Corollaries, where the stairstep example is taken up specifically in the last example of that Pause.

Lemma 3 Corollary 2

And much more does the rectilinear figure enclosed by the chords of the evanescent arcs ab, bc, cd, and so on, coincide ultimately with the curvilinear figure.

Expansion of Newton's Sketch of Corollary 2

Instead of the inscribed parallelograms, we now place a figure on each base which meets the curve on the two sides and is topped by the chord. (In this diagram these would be rectilinear figures *abBA*, *bcCB*, and so on.) The sum of these inscribed figures is closer in area to the area under the curve than was the sum of the areas of the inscribed parallelograms by the sum of the triangles *aKb*, *bLc*, etc.

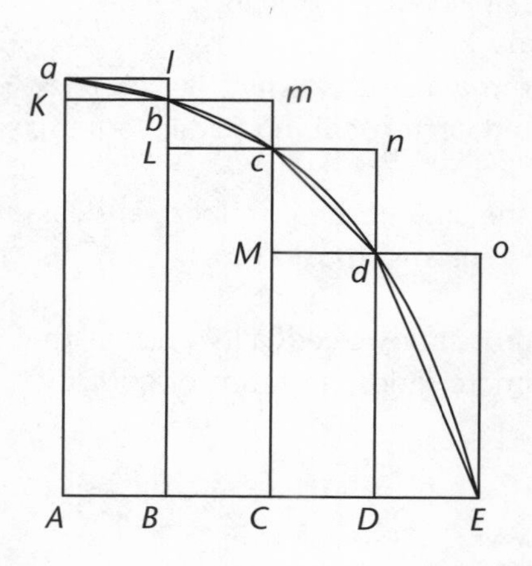

The figure bounded by the chords is always closer than the area of the inscribed parallelograms to the area under the curve. And the area of the inscribed parallelograms approaches a limit of equality with the area under the curve, by Lemmas 2 and 3. Therefore the figure enclosed by chords must even more so approach equality with the curved figure.

Lemma 3 Corollary 3

And also the circumscribed rectilinear figure which is enclosed by the tangents of the same arcs.

Note on Corollary 3

We're being invited to consider a tangent to this curve, but this curve is not necessarily a circle (to which Euclid gave us a definition for a tangent) or even a conic section (for which Apollonius described tangents). Newton will give us a general definition of the tangent to any continuous curve, but not until Lemma 6. Meanwhile we may use our intuition based on the circle to picture tangents as straight lines that touch the curve but do not intersect it.

Expansion of Newton's Sketch of Corollary 3

Tangents have been drawn touching the curve at *a*, *b*, and so on. Let them intersect at *p*,*q*,*r*,*s* to enclose an area *apbqcrdsEA*. The figure *apbqcrdsEA* bounded by the tangents is larger than the curvilinear area *abcdEA* but smaller than the sum of the circumscribing parallelograms *albmcndoEA*; therefore, since the circumscribing parallelograms reach equality with the area under the curve, even more so must the figure bounded by the tangents reach a limit of equality with the curvilinear area as the bases are continuously diminished.

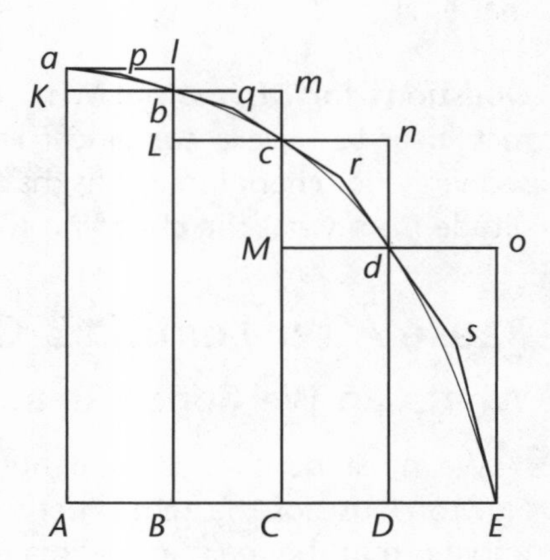

Lemma 3 Corollary 4

And therefore, these ultimate figures (as regards their perimeters acE) are not rectilinear, but curvilinear limits of rectilinear figures.

Note on Corollary 4

Curve as Limit of the Figures:

The term "figure," which he uses in this corollary, refers to the area contained by the perimeter or boundary. (Euclid, Book I Definition 14: "A *figure* is that which is contained by any boundary or boundaries.") Rectilinear figures are areas bounded by straight lines; curvilinear figures are areas that are bounded at least in part by curved lines.

"These figures" presumably are the ultimate figures of the lemma and previous corollaries. This corollary summarizes that the curvilinear figure *acE* is the limit of all four of the rectilinear figures we have been looking at: the circumscribing and inscribed parallelograms, the figure made with the tangents, and the figure made with the chords.

Newton tells us that "these ultimate figures are not rectilinear, but curvilinear limits of rectilinear figures." We have here areas bounded by straight lines which get closer and closer to a curved area. The limit is a curved area. Thus, by the technical meaning of "ultimate," the ultimate figure in this process of multiplying the bases is a curvilinear figure.

This corollary reiterates that when he speaks of the "ultimate figure" of the parallelograms, Newton means the figure bounded by curve *acE*.

The limit is itself not a rectilinear area, even though the figure as it approaches the limit will *always* be rectilinear. The figure which is the limit is something really different from every version of the figure that approaches that limit.

Questions for Discussion: What can it mean for something that is always rectilinear to have a curvilinear limit? Can you think of other examples of sequences or changing things that have a limit that is a different sort of magnitude from what the changing thing always is?

Pause After Lemma 3 Corollaries

What Can We Conclude about Lengths of Perimeters?

We must be careful about jumping to unwarranted conclusions from Newton's use of the term "perimeter" in Corollary 4. He has so far been proving that the *areas* are ultimately equal, not that the perimeters become ultimately equal.

Length of Curve as Limit of Lengths of Chords

We might like to conclude from these corollaries that the ultimate length of the figure made up of the chords is the length of the curve.

For one thing, it would give us a length of the curve in terms of the lengths of straight lines; previously these two kinds of length have been incommensurable and only able to be compared even in terms of greater or less by axiom. (Even that a curve is *longer* than a straight line between the same two end points could not be proved, and could only be supposed as an axiom by Archimedes.)

However, although we might like Newton to have given us that, it does not seem that he has, nor does it seem to me that he intended to assert such a conclusion or that he believes that he has proved anything about the relationship of lengths. One piece of evidence that he doesn't think he has proved it here is the fact that he undertakes just such a proof, explicitly, later (in Lemma 7). Had he believed that Lemma 3 in some sense proved this already, Lemma 7 would not have been necessary.

In addition, a close look at what is established by Lemma 3 does not support the idea that even such an assertion is being made, let alone a proof.

Corollary 1 has pointed out that the ultimate figure which is the sum of the evanescent circumscribing and inscribed parallelograms coincides on all sides with the curvilinear figure. This explicitly draws our attention to the fact that there is coincidence of the perimeter of the curve and the ultimate figure (the limit towards which the sum of the parallelograms tends). But the fact that the *area* has approached the limit of the curvilinear area according to the conditions of Lemma 1 is not enough to prove that the length of the perimeter has approached the length of the curve according to the conditions of Lemma 1.

Corollary 4 concerns itself with the figures, or areas, with the perimeter being specified only to identify the curve that is the limit. It does not and could not make the claim that the length of the *perimeters* converge according to Lemma 1 (something one might mistakenly conclude by interpreting "figure" as the boundary rather than that which is contained by the boundary).

For such a striking and important advance in mathematics as nailing down the relationship between rectilinear and curvilinear lengths, Newton and we ourselves would both want an explicit and convincing proof.

But, one might ask, could it not be easily proved here? Or might it not be so obvious as not to need proof? Does it not follow from Corollaries 1 and 4 that the curved perimeter must also be the limit towards which the perimeters of the four kinds of rectilinear figures tend? If the perimeters coincide *at* the limit, must they not have *approached* that limit?

Newton will prove in Lemma 7 that the arc, chord, and tangent will ultimately coincide, using Lemma 1. So it can be shown using Lemma 7 that the perimeter of the figure that includes the chords of diminishing arcs (Corollary 2), and the perimeter of the figure that includes tangents to diminishing arcs (Corollary 3), *do* both "ultimately" coincide with the curve.

Perimeters That Fail the Conditions of Lemma 1: The Stairstep

But even when this has been proved for the perimeters made of the chords and the tangents, we would be misled if we supposed from this that the lengths of *all* the perimeters approach the length of the curve.

On the contrary, the perimeters of the figures which include the stairstep shape of the inscribed and circumscribing parallelograms of Lemmas 2 and 3 do not approach the length of the figure bounded by the curve. Someone might protest that Corollary 1 tells us that the ultimate sum of parallelograms, the figure that is the limit, will coincide on all sides with the curve. And we know by Lemma 3 that the area of the figure approaches the area under the curve. But consider: in that approach of areas, has the perimeter *approached* the perimeter in the sense of Lemma 1?

The figures made up of the sums of inscribed and circumscribing parallelograms are worth considering because they provide an example that *doesn't* meet the conditions of Lemma 1 when we multiply the number of bases and decrease their breadth *in infinitum*. As we continue through the lemmas and the propositions to come, it will be important to remember that it is *not* the case that "everything is equal at the limit," or that everything that is equal or coincident after the limit has been taken even moves towards that limit.

Consider a single step *aw*, *wE* such that the two line segments terminate at the ends of the curve segment. It has a particular length $aw + wE$, and that length bears some particular ratio to the length of the curved segment.

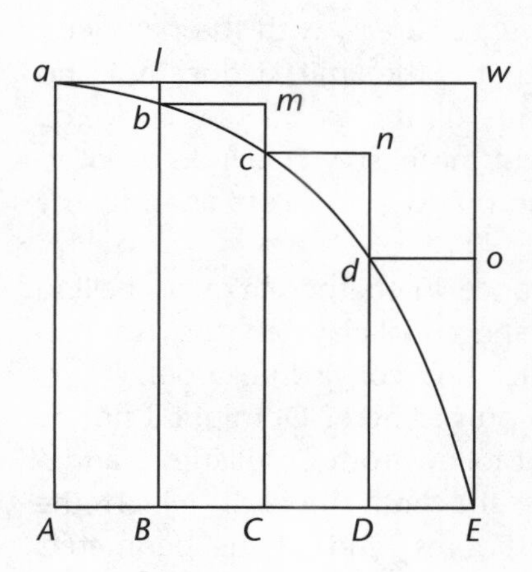

If we double the number of steps, we find that the length is still the same (it is found by adding up all the horizontal and vertical distances) and the ratio is the same. As we continue to increase the number of steps (which happens as we increase the number of parallelograms) the difference in length between the straight line segments and the curve does not change. (The sum of lengths $al + bm + cn + do = aw$; while $lb + mc + nd + oE = wE$.)

Thus we do not meet the condition that the lengths "tend constantly towards equality." Even as we increase the number of parallelograms *in infinitum* the length of the stairstep perimeter will remain just what it was at the beginning. Thus the rectilinear perimeter is *not* "ultimately equal" to the curve.

Fortunately Newton is not making such a claim in the fourth corollary. He specifies the perimeter *acE* to identify the curve that bounds the area towards which the rectilinear areas tend.

Question for Discussion: Corollary 1 asserted, and we acknowledged, that the curvilinear area is the limit which the rectilinear areas approach as the number of parallelograms increases *in infinitum*. But we can see that the length of the staircase does not approach the length of the curve as a limit. What do we make of this seeming paradox?

Lemma 4

If in two figures AacE, PprT, there be inscribed two series of parallelograms (as before), and if the number of the two be the same, and if, where the breadths are diminished in infinitum, *the ultimate ratios of the individual parallelograms in one figure to the [corresponding] individual parallelograms in the other, be the same, I say that the two figures AacE, PprT, are to one another in that same ratio.*

For as individual parallelogram is to individual parallelogram, so, *componendo*, would the sum of them all be to the sum of them all, and thus the figure to the figure; because,

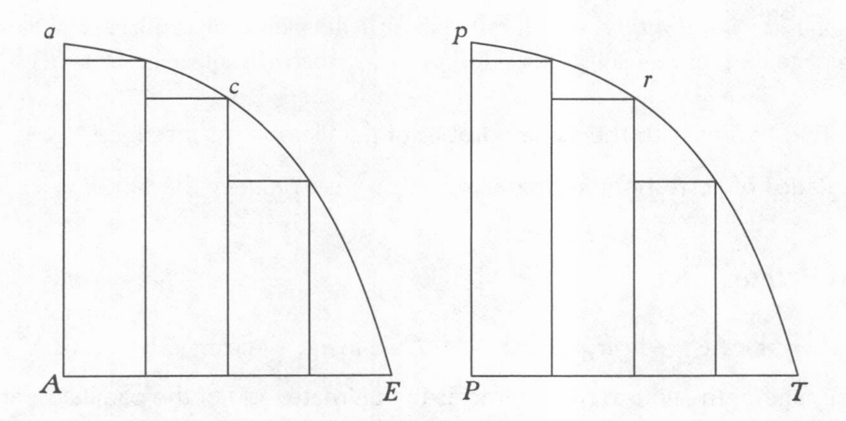

by Lemma 3, the former figure stands to the former sum, and the latter figure to the latter sum, in the ratio of equality.

Q.E.D.

Notes on Lemma 4

• For Newton's use of ratio and proportion generally, and *componendo* in particular, see Section 6 of the Preliminaries.

• The figures mentioned in this lemma need not be similar. Any two figures will do, of any size and any shape.

• The bases of these parallelograms need not be equal.

• We only care about the ratio that the corresponding parallelograms have at the limit, not whether they have this ratio before the limit. They may or may not have the same ratio throughout their diminishing.

• Throughout our demonstrations of the lemmas, propositions, and corollaries of *Principia*, it is important to note which assertions are true generally of the relationships in question and which are only true in the ultimate case, at the limit. Careless use of the former for the latter will lead to false conclusions. Every indication of relationship (such as ::, ~, $\propto$, =), when it applies only at the limit, should have expressed with it (that is, written before or over the relationship symbol) the condition "ult." For example, one can write:
" $\overline{AB}$ ult$=$ $\widehat{AB}$ " or " $\overline{AB} \overset{ult}{=} \widehat{AB}$ ".

Expansion of Newton's Sketch of Lemma 4

Given:

"If in two figures $AacE$, $PprT$, there be inscribed two series of parallelograms (as before), and if the number of the two be the same, and if, where the breadths are diminished *in infinitum*, the ultimate ratios of the individual parallelograms in one figure to the [corresponding] individual parallelograms in the other, be the same..."

1. Two figures with the same number of parallelograms inscribed.

2. Ratios of corresponding parallelograms are *ultimately* the same.

To Prove:

"...I say that the two figures $AacE$, $PprT$, are to one another in that same ratio."

The ratio of the figures is the same as the ultimate ratio of the parallelograms.

Proof:

"For as individual parallelogram is to individual parallelogram, so, *componendo*, would the sum of them all be to the sum of them all, ..."

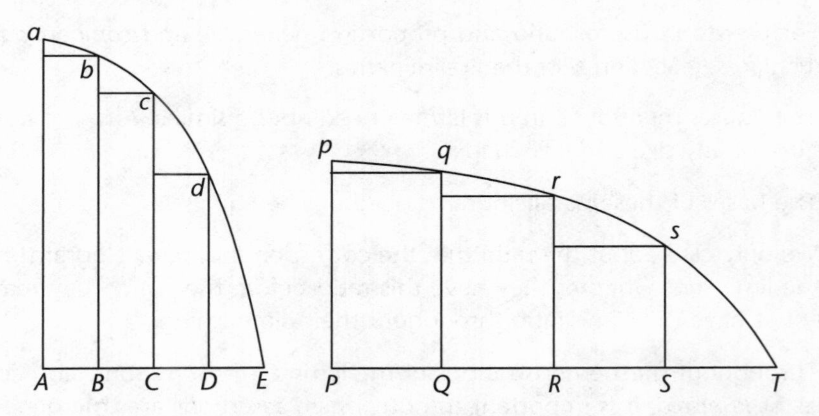

Let the magnitudes M, N be taken such that they are in the same ratio as the ultimate parallelograms Ab and Pq.

We are given that:

$Bc : Qr \overset{ult}{::} M : N$,

$Cd : Rs \overset{ult}{::} M : N$,

etc.

Then by Euclid V.18,

$(Ab + Bc + Cd + \ldots) : (Pq + Qr + Rs + \ldots) \overset{ult}{::} M : N.$

> **Exercise:** Euclid V.18 applies to ratios that are always the same. We apply it here to ratios that only approach and are *ultimately* the same.

This must be justified. How would you argue the applicability of Euclid V.18 to ultimate ratios?

"and thus the figure to the figure; because, by Lemma 3, the former figure stands to the former sum, and the latter figure to the latter sum, in the ratio of equality."

But, by Lemma 2 or 3,

$$(Ab + Bc + Cd + \ldots) \overset{ult}{=} \text{fig}(AacE)$$

and

$$(Pq + Qr + Rs + \ldots) \overset{ult}{=} \text{fig}(PprT)$$

Therefore

$$\text{fig}(AacE) : \text{fig}(PprT) = M : N.$$

(Note that we can drop the "ult" because these are fixed nonvanishing magnitudes.)

Q.E.D.

Lemma 4 Corollary

Hence if two quantities of any kind whatever be divided into the same number of parts in any manner, and those parts, as their number is increased and their size is decreased in infinitum, *attain* a given ratio to each other, the first to the first, the second to the second, and the rest to the rest in order, the wholes will be to each other in that same given ratio. For if in the figures of this lemma the parallelograms be taken to be to each other as the parts, the sums of the parts will always be as the sums of the parallelograms, and consequently, where the number of the parts and of the parallelograms be increased and the size be decreased* in infinitum, *[the sums of the parts will be] in the ultimate ratio of parallelogram to parallelogram, that is (by the hypothesis) in the ultimate ratio of part to part.*

Notes on Lemma 4 Corollary

- We now show that we can compare two magnitudes of any sort using the method of the previous lemmas. These magnitudes need not be areas under

---

* *Obtineant*, third person plural subjunctive of *obtinere*.

curves; they need not even be geometric at all. They could be numbers, velocities, magnetic flux, or anything else someone wanted to explore using this technique.

- **Difference between what is true always and what is true ultimately**

"...as their number is increased and their size is decreased *in infinitum*, attain a given ratio to each other..."

Although the usual classical meaning of *obtinere* would be "to hold" or "maintain," it can also have a meaning closer to its English cognate, i.e. "achieve" or "attain." Here the logic of the proof and Newton's subsequent use of the corollary show that he was taking the latter sense of the word. All other translations have used the standard lexical substitution, without checking the logic of the proof.

To translate the word as "maintain" (Cohen/Whitman) or "have" (Motte) makes the enunciation claim something that Newton's sketch does not prove and his subsequent invocations of the corollary do not use. It implies that the two quantities have and maintain this relationship through the process of increasing the number or parts and decreasing their size. In fact they only have that given ratio at the limit, when the size is decreased *in infinitum.* That is, they *attain,* or *obtain* (the English cognate that Newton evidently had in mind when he was writing his Latin text for this corollary), that given ratio ultimately.

Whether a proposition holds generally or ultimately is something that it is important to know when applying the proposition. When one refers to the enunciation statement, one would like to be able to rely on what it says.

Expansion of Newton's Sketch of the Corollary

Given:

"Hence if two quantities of any kind whatever be divided into the same number of parts in any manner, and those parts, as their number is increased and their size is decreased *in infinitum*, attain a given ratio to each other, the first to the first, the second to the second, and the rest to the rest in order..."

Two magnitudes F and V, of any kind whatever, divided into the same number of parts $ghk...$, $wxy...$, respectively, such that

$$g : w \overset{ult}{::} h : x \overset{ult}{::} k : y \overset{ult}{::} \dots \overset{ult}{::} M : N \qquad (1)$$

where $M : N$ is some given ratio.

To Prove:

"...the wholes will be to each other in that same given ratio."

The whole magnitudes will be to each other in that same ratio:

$F : V :: M : N.$

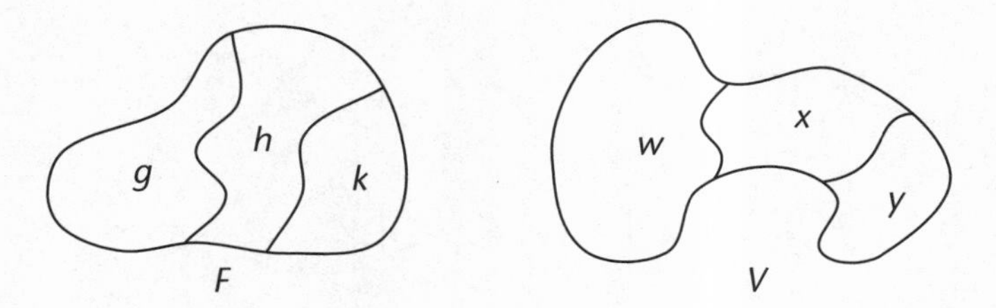

Proof:

"For if in the figures of this lemma the parallelograms be taken to be to each other as the parts..."

We are going to invoke the conclusion of the lemma. We set up two sets of parallelograms so that any pair of parallelograms in either set is in the same ratio as the corresponding parts of *F* and *V*. Note that we must also arrange for the cross-proportionality between the two sets of parallelograms and the parts of the two magnitudes "of any kind" *F* and *V* shown in Equations 4 below. (We at least need the first proportion of Equations 4: if we have that we can derive the rest.)

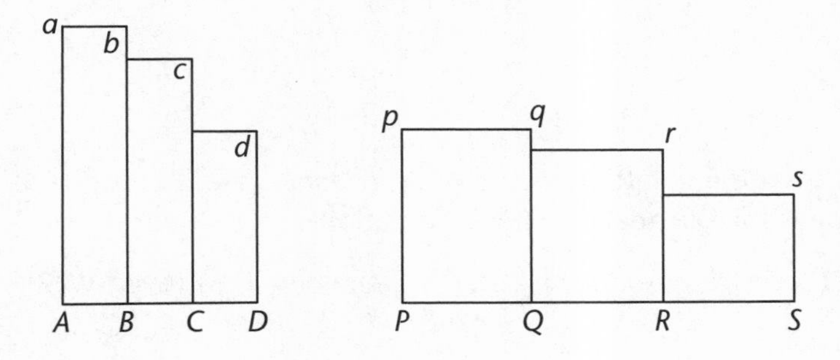

Set up two sets of parallelograms that are contrived *always* to have the same ratio as corresponding parts of *F* and *V*:

$$Ab : Bc :: g : h, \quad Bc : Cd :: h : k, \quad \text{etc.} \tag{2}$$

and

$$Pq : Qr :: w : x, \quad Qr : Rs :: x : y, \quad \text{etc.} \tag{3}$$

And these parallelograms must also be set up so that

$$Ab : Pq :: g : w, \quad Bc : Qr :: h : x, \quad \text{etc.} \tag{4}$$

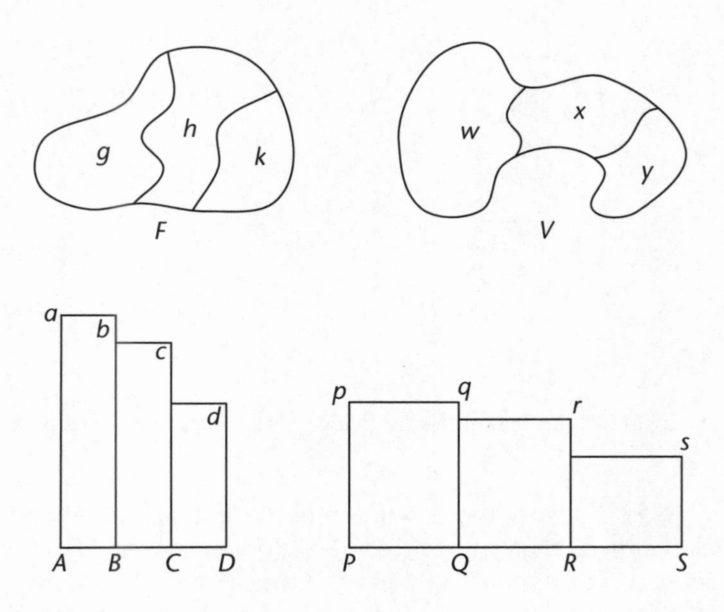

"the sums of the parts will always be as the sums of the parallelograms..."

By *componendo* on Equations 2:

$$Ab+Bc : Bc :: g+h : h \quad \text{[Euclid V.18]}$$

and

$$Bc : Cd :: h : k. \quad \text{[Eq. 2]}$$

Cancelling intermediate terms by *ex aequali* (see Section 6.2 of the Preliminaries for explanation of this operation),

$$Ab+Bc : Cd :: g+h : k. \quad \text{[Euclid V Definition 17; proved V.22]}$$

Componendo,

$$Ab+Bc+Cd : Cd :: g+h+k : k. \tag{5}$$

And similarly, using Equations 3,

$$Pq+Qr+Rs : Rs :: w+x+y : y.$$

Inverting,

$$Rs : Pq+Qr+Rs :: y : w+x+y. \tag{6}$$

By construction $Cd : Rs :: k : y$. (7)

By Equations 5, 7, and 6, we have now established that:

$$Ab+Bc+Cd : Cd :: g+h+k : k$$
$$Cd : Rs :: k : y$$
$$Rs : Pq+Qr+Rs :: y : w+x+y.$$

Cancelling intermediate terms Cd, Rs, k, and y by *ex aequali*, we have:

$$(Ab+Bc+Cd) : (Pq+Qr+Rs) :: (g+h+k) : (w+x+y).$$

This can be continued to any number of parts.

Therefore the sums of the parts will be as the sums of the parallelograms.

"... and consequently, where the number of the parts and of the parallelograms be increased and the size be decreased *in infinitum*, [the sums of the parts will be] in the ultimate ratio of parallelogram to parallelogram, that is (by the hypothesis) in the ultimate ratio of part to part."

Now let the number of parallelograms and parts increase *in infinitum*.

$$(Ab+Bc+Cd+\ldots) : (Pq+Qr+Rs+\ldots) :: (g+h+k+\ldots) : (w+x+y+\ldots). \quad (8)$$

By Equations 4 and 1, by Euclid V.11,

$$Ab : Pq :: g : w \overset{ult}{::} M : N,$$

$$Bc : Qr :: h : x \overset{ult}{::} M : N,$$

etc.

By Lemma 4, if the ultimate ratios of the individual parallelograms inscribed in one figure to the corresponding individual parallelograms in the other be the same, the two curvilinear figures are to each other in the same ratio.

$$M : N :: \text{Area } A\widehat{ad}D : \text{Area } P\widehat{ps}S. \quad \text{[Lemma 4]} \quad (9)$$

$$\text{Area } A\widehat{ad}D : \text{Area } P\widehat{ps}S :: \lim_{n\to\infty} (Ab+Bc+\ldots) : (Pq+Qr+\ldots). \quad \text{[Lemma 3]} \quad (10)$$

$$\lim_{n\to\infty} (Ab+Bc+\ldots) : (Pq+Qr+\ldots) :: \lim_{n\to\infty} (g+h+\ldots) : (w+x+\ldots). \quad \text{[Eq. 8]} \quad (11)$$

$$\lim_{n\to\infty} (g+h+\ldots) : (w+x+\ldots) :: F : V. \quad (12)$$

Note that, since the whole is made up of its parts, $(g+h+\ldots) : (w+x+\ldots)$ is always the same ratio as $F : V$ and therefore is the same at the limit as well.

Combining Equations 9 through 12,

$$F : V :: M : N. \quad \text{[Eu. V.11]}$$

Q.E.D.

Lemma 5

All the sides of similar figures that correspond to each other mutually, curvilinear as well as rectilinear, are proportional, and the areas are in the duplicate ratio of the sides.

Notes on Lemma 5

• **Duplicate Ratio:** Section 6 of Preliminaries explains Newton's uses of ratio and proportion in general and duplicate ratio in particular.

• **Geometry of similar curvilinear figures:** Newton had in mind a geometry of similar curvilinear figures, of which Lemma 5 was, in effect, just one proposition. He was not referring to a text whose theorems we might cite as we cite Euclid; nor, apparently, did he himself write out the propositions he had in mind.

Rather than attempt to reconstruct proofs of the present Lemma and the other propositions that make up this body of curvilinear geometry (an intriguing and worthwhile project, but one that would be its own book), this guidebook will just identify the ones Newton uses and cite them as though he had written them out for us.

• **Definition of Similarity:** Lemma 5 appears either to presuppose, or to implicitly contain, a definition of *similarity of curvilinear figures*. (In the same way, Lemma 6 will raise the question of defining *tangency to a curve*.) Can we discover such a definition? Consider Euclid's treatment.

Euclid defined similarity of *rectilinear* figures this way in Book VI (Definition 1):

> *Similar rectilineal figures* are such as have their angles severally equal and the sides about the equal angles proportional.

Euclid did not give a general definition of similarity of *curvilinear* figures, but he defined similar segments of circles this way in Book III (Definition 11):

> *Similar segments of circles* are those which admit equal angles, or in which the angles are equal to one another.

Newton may have been defining similarity of curvilinear figures this way:

> *Similar curvilinear figures* are such as are always inscribable with an equal number of respectively similar and similarly situated rectilineal figures as that number is increased and the size of the rectilineal figures diminished *in infinitum*.

This definition would be consistent not only with the Euclidean definitions but also with the constructions of Lemmas 2, 3, and 4. The logic follows naturally from the previous lemmas.

Projection:

An alternative and seemingly plausible picture of similarity between curvilinear figures is created by "lines of projection": lines that originate from a single point and intersect corresponding points on the figures in question (see the sketch below). Newton uses this sort of magnification in Lemma 7, relying on Lemma 5 for its justification. But a picture is not a proof. Any proof using projection must build on Euclid's definitions quoted above.

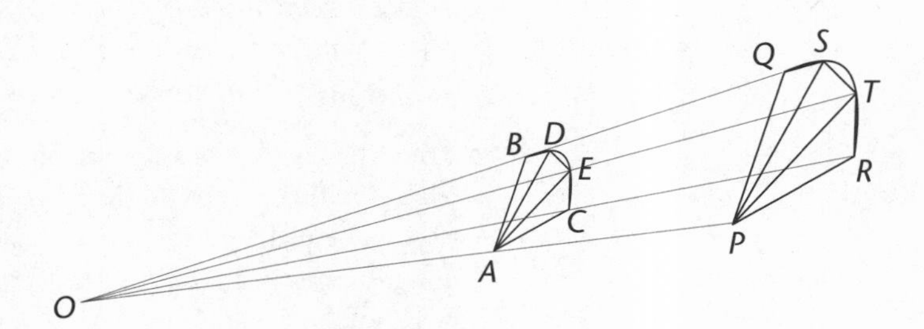

While the method of projection suffices to articulate similarity between *rectilinear polygons* such as *ABDEC* and *PQSTR*, it relies on Euclidean definitions and propositions to do so; it adds nothing to what Euclid demonstrated. Although we may find the projection drawing helpful to the imagination, a legitimate demonstration must still bridge the gulf between *rectilinear* figures (which Euclid did expound) and *curvilinear* figures (which, other than the circle, he did not).

How may we do this? One approach is to imagine the sides of the polygons becoming infinitely small and infinite in number: by this process each polygon indefinitely approaches coincidence with its corresponding curvilinear figure. At every finite stage the polygons are similar—so it is often asserted that the curvilinear figures must be "similar," too.

But can we really assert that a property which holds at every stage up to the limit (if limit there be) must necessarily apply to the curves that the polygons approach? The Pause After Lemma 3 gave as an example the lengths of the perimeters of rectilinear figures holding constant up to their limit, the curve, whose length might be radically different. Furthermore, some quantities vanish at the limit.

Evidently, then, what we need is a proof that the chords are ultimately the same length as the arcs. This proof is provided in Lemma 7—but that, unfortunately, makes use of Lemma 5 in its proof. We are thus left with a puzzle as to how Newton would have filled in the proof of this lemma, and how it would relate to the other lemmas.

Lemma 6

If any arc ACB given in position be subtended by a chord AB, and at some point A, in the middle of continuous curvature, it be touched by a straight line AD produced both ways, and then if points A, B approach each other and come together, I say that the angle BAD, contained by the chord and the tangent, is diminished in infinitum *and ultimately vanishes.*

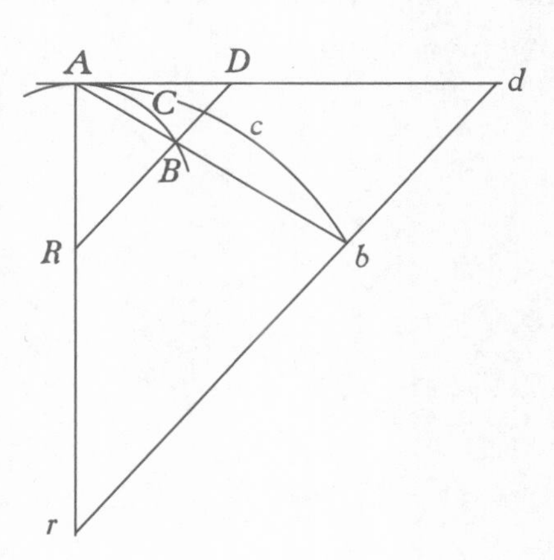

For if that angle does not vanish, the arc *ACB* will contain with the tangent *AD* an angle equal to a rectilinear angle, and for that reason the curvature at point *A* will not be continuous, contrary to the hypothesis.

Note on Newton's Diagram: Newton gave one diagram for Lemmas 6 through 8. Most of the points and lines are not required for Lemma 6 and should be ignored. As you prove the lemma you will want to make your own drawing using only the points and lines called upon by Newton's demonstration here.

Notes on Lemma 6

• Euclid defines a tangent to a circle. "Tangent" means "touching" (sharing one point with) as opposed to "cutting" or intersecting (one line crossing the other). Euclid defines tangent in Book III Definition 2: "A straight line is said to *touch a circle* which, meeting the circle and being produced, does not cut the circle."

Euclid Book III Proposition 16 proves that a line perpendicular to a diameter of a circle at its extremity falls outside the circle (that is, it will not cut the circle—will not extend inside the circle—on either side of the point of tangency). Proposition III.16 tells us how to construct the tangent to a circle, thus establishing that the geometrical entity defined earlier can exist. III.16 goes on to prove two other properties of the tangent: another straight line cannot be interposed between the tangent and the circumference through the point of tangency; and the angle between those two lines (the tangent and the circumference) is less than any rectilinear angle.

The writers on conic sections define the tangent to a conic: it is a line that touches the curve at one point but lies outside the curve at all other points. (Apollonius proves in I.17 of *On Conic Sections* that a line through the vertex of a conic section parallel to an ordinate will lie outside the section and says that therefore it will be a tangent.)

But will this work as a definition for a tangent to curves other than conics? And a merely descriptive definition like "touches the curve at one point but remains outside at all others" doesn't tell us how to construct a tangent, or how to know whether there might even *be* any tangent to a given curve at a given point, or whether a tangent would be unique.

This lemma gives us Newton's general definition and method of construction* of a tangent for any curve. We will need to be able to construct tangents to various sorts of curve in *Principia*, so this is something important to have established.

The definition implicit in this lemma is this:

> A tangent to a curve at a given point is the limiting position of a chord from that point to another as the arc diminishes *in infinitum* towards that point.

One marvels at Newton's originality and power. Instead of having to come up with a definition and method of construction of tangents for each new type of curve, we now have a method that gives us a tangent at any point of continuous curvature on any curve.

Newton breaks through the old difficulty of needing to determine the tangent for each sort of curve by using his method of limits, his calculus, as the technique powerful enough to generate a universal construction.

• **Assumptions about Tangency:** Note that the definition of the tangent given in the previous note is not presented at the beginning of the lemma but emerges as its conclusion. Newton assumes a tangent and shows that it will have to be a limiting position of the chord. Working backwards, using this lemma, we now have a construction for a tangent.

This means that we cannot use the "definition" to construct the tangent at the beginning of the proof of the lemma. We must suppose a tangent to be given without knowing how it was constructed.

Further, we must suppose some things about it in the course of the proof. We should look very closely at what he does assume about the characteristics of a tangent to any curve, and compare that to the characteristics of the tangent for which he has given us a construction by the end of the lemma. Since he is proving something about the tangent, one should be very careful that he, and we, do not assume something possibly true only about circles or other conics, and apply that to this new entity, "tangent to any continuous curve."

---

* This is not a construction in the Euclidean sense: it cannot be done with a straightedge and compass. It does show us how the tangent can be found, and satisfies us that the geometrical entity exists. It is like the construction of conic sections in Apollonius.

It seems to me that he makes only two assumptions about the tangent we are "given." The first is that the tangent, line *AD*, does not cut the curve between the point of tangency *A* and the point *B* that he is moving in towards *A*. The second is that the tangent line *AD* makes an angle with arc $\widehat{AB}$ which is less than any rectilinear angle. Both these conditions are included in Euclid's characteristics for tangent to a circle, as specified in his definition and in III.16 (see previous bulleted note).

- **Continuous curvature:** We are given that our curve has continuous curvature at point *A*. What does this mean?

Visually and intuitively, we could say that it means that there are no sharp corners, no *abrupt* changes of direction.

Here are some pictures of curves that have continuous curvature at all points:

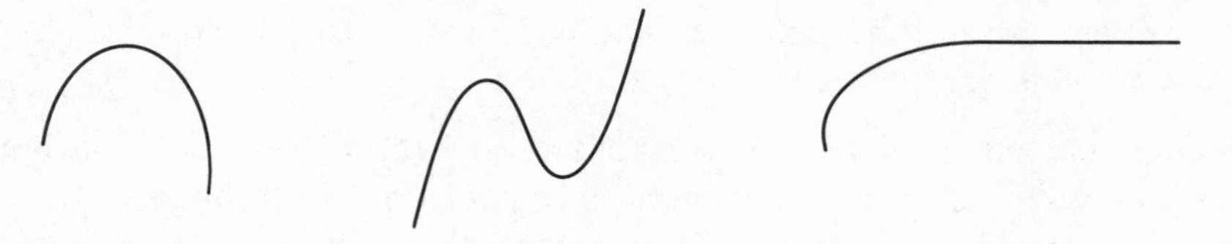

Here are some pictures of curves that do not have continuous curvature everywhere; the noncontinuous points are shown:

Mathematically, what Newton appears to mean by continuous curvature is the following:

1. There exists a tangent for each point throughout the continuous curvature.

2. The arc contains with the tangent an angle not equal to any rectilinear angle; that is, no line can fall between the arc and the tangent. This proved in Euclid *Elements* III.16 to hold for circles, in Apollonius *On Conic Sections* I.32 for other conics.

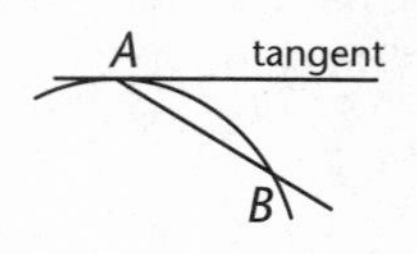

What Euclid and Apollonius meant was that if you construct a rectilinear angle, with one leg being the tangent, on the same side of the tangent as the arc, then no matter how small the angle is, it will always *cut* the arc. It follows from this that there can be only one tangent at a point.

There will always be some finite chord *AB* "within" the arc, and the arc segment $\widehat{AB}$ will fall between the tangent and the line *AB*.

• Euclid III.16 states that no line can be interposed between a tangent and the circumference of a circle. This says, in effect, that any line inside a tangent will cut a circle. How is this lemma different from that proposition, other than being more general?

Here is one difference that strikes me. Euclid begins with the construction of the tangent (perpendicular to a diameter) and then shows that a line inside (which will turn out to be a chord) is sharply different in kind—it cuts. Newton, by contrast, begins with a tangent whose construction is unknown and shows that this tangent is the *limiting position* of the chord—that is, in a certain sense, it is *not* different in kind.

Expansion of Newton's Sketch of Lemma 6

Given:

"If any arc *ACB* given in position be subtended by a chord *AB*, and at some point *A*, in the middle of continuous curvature, it be touched by a straight line *AD* produced both ways, and then if points *A*, *B* approach each other and come together…"

1. Arc *ACB*;

2. Chord *AB* subtending that arc;

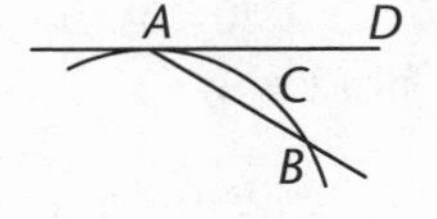

3. Point *A* is "in the middle of continuous curvature," that is, the curvature is continuous at *A* and on both sides of *A*;

4. Suppose a tangent *AD* constructed touching the curve at *A*;

5. Point *B* is imagined moving towards *A* until it coincides.

To Prove:

"…I say that the angle *BAD*, contained by the chord and the tangent, is diminished *in infinitum* and ultimately vanishes."

The angle *BAD* between the tangent and the chord will be diminished *in infinitum* as the arc diminishes. Ultimately there will be no angle between them: that is, the tangent is be the limiting position of the chord.

Proof:

"For if that angle does not vanish, the arc *ACB* will contain with the tangent *AD* an angle equal to a rectilinear angle, …"

This curvilinear angle between the arc and the tangent is the so-called "horned angle" between a circle (or another curved line) and its tangent. Euclid and Apollonius (see the notes above) had proved that if you try to fit a third line between the two, no matter how close you bring it to the tangent it will always cut the curved line in two places. Thus the horned angle is not equal to any rectilinear angle. If another line could be fitted in between some imagined curve and a tangent, then the angle between the curve and the tangent would be equal to some rectilinear angle.

We demonstrate the lemma by *reductio ad absurdum*. Suppose that the angle between the chord and the tangent does not, after all, vanish. Suppose that no matter how close *B* gets to *A*, the chord *AB* never gets closer to *AD* than some line (such as *AE* here), which contains with *AD* some small rectilinear angle *EAD*. It is evident that the arc *ACB* also never gets closer to *AD* than the line *AE*, for then some of the chords would also be closer to *AD* than *AE*. Therefore, the arc will contain with the tangent *AD* an angle equal to a rectilinear angle.

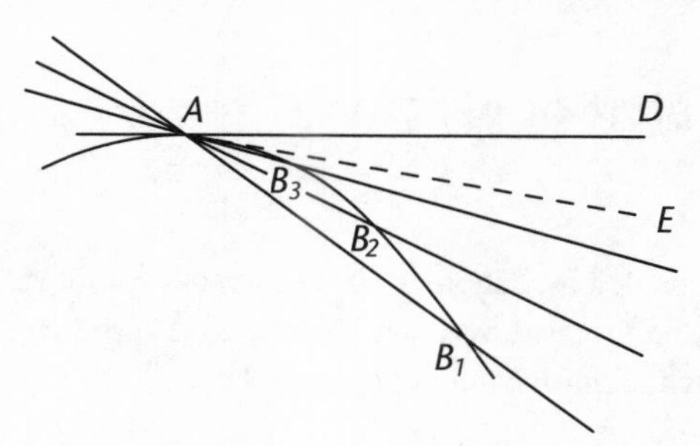

"...and for that reason the curvature at point *A* will not be continuous, contrary to the hypothesis."

Now it is one of the characteristics of a continuous curvature (see bulleted note Continuous Curvature above) that no line may be fitted between the arc and the tangent; that is, that the angle between the arc and the tangent is not equal to any rectilinear angle.

Therefore the curvature of the arc must not be continuous. But that is contrary to our requirement that the curvature be continuous. Thus it is absurd to suppose that there is an angle remaining between the chord and tangent as the arc vanishes. More precisely, it is absurd to suppose that there can be an angle between the ultimate position of the chord, *AE*, and the tangent *AD*.

Therefore there is no angle between them, and they are the same line at the limit. Thus the tangent is the limiting or ultimate position of the chord as the arc vanishes. The angle between the chord and the tangent ultimately vanishes.

Q.E.D.

Note on Sidedness of Tangency

This lemma, and the generality of the curve to which we have now defined a tangent, leave open the possibility of a new characteristic of a tangent. In circles and other conics, tangents were taken to a curve which is convex on both sides of the point of tangency, and tangents did not cut the curve at any point.

Newton has defined a tangent here with "sidedness." The tangent is the limit of a chord as the arc diminishes to the point of tangency from a particular side. Thus the tangent is tangent on that side, and will not cut the curve on that side.

The specification of the curve, that it be continuous, includes curves that have "inflection points," such as the one illustrated in the diagram. An inflection point is where, if one were traveling along the curve as a path, one stops turning right and starts turning left or vice versa. The curve goes from being convex to concave (or the reverse) at an inflection point. In the diagram, as one travels from B' to B, the change of direction occurs at A.

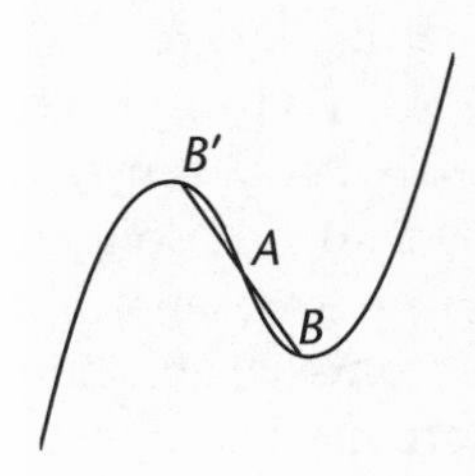

Newton's definition would allow us to have a tangent at inflection points, a tangent that is the limiting position of chords on either side of that point as the arc vanishes. If the curvature is continuous, the limiting position of both chords would fall in a straight line. But each tangent, extended, would cut the curve. This is unlike tangents defined previously for circles and other conics.

For most of the curves Newton will be concerned with, including conics and spirals and cycloids, the tangents will not cut the curve at any point. However, his definition is general enough to include curves that have points of continuous curvature with unique tangents that cross the curve at the point of tangency and yet are genuine tangents and not ordinary intersecting lines.

Pause Before Lemma 7

Notes on Ratios Between Vanishing Quantities: Lemma 1 Revisited

We noted in Lemma 1 that although the lemma could be applied in a simple and straightforward way for quantities that are not themselves vanishing, the application would be a little trickier when we must compare quantities which themselves are "diminishing *in infinitum*."

We also noted that Newton's use of the phrase "ratio of quantities" will be the key to being able to deal with vanishing quantities. Any quantities that approach the same nonzero limit will be moving towards a ratio of equality. But it is not true that every pair of magnitudes that approach a limit of zero will vanish with a ratio of equality.

For vanishing quantities, we cannot compare them directly in the simple terms of Lemma 1; we must follow Newton in his reliance on a comparison of ratios. This section explores why that is so, and how to negotiate this dangerous water safely and in a way that will keep us with Newton.

• In struggling to understand what happens to vanishing quantities, the temptation is to conclude that "at the limit everything is equal." And indeed, there is a trivial sense in which this is true. All vanishing quantities are ultimately zero, and so the limit towards which any two vanishing quantities are tending is the same in both cases. But Newton has a deeper view of the matter, one which, though more of a struggle to grasp initially, turns out to be more useful.

He is not just looking at the limit which *each* quantity approaches, and seeing whether those limits are the same. He looks at the *ratio* of the two quantities, and sees whether that ratio is approaching a ratio of equality. He might say, "I don't ask whether they are equal *after* they vanish: of course they are equal when they are both nothing. I ask whether they are in a ratio of equality *as* they vanish." What does this mean, and how does it affect our understanding of this lemma?

Let's explore this by taking some examples of vanishing quantities and ratios of vanishing quantities that do and *don't* ultimately reach (that is, vanish with) a ratio of equality.

• **First Example: Arc of Circle and Chord of That Arc:** Let's take as our first example one that does fulfill the conditions of Lemma 1 (as Newton will later prove in Lemma 7). Picture a short arc of a circle and the chord of that arc as smaller and smaller arcs are taken over time. As the arc gets smaller, its chord gets closer to being equal in length to the arc.

Lemma 1's first condition: "Quantities, as well as the ratios of quantities, which in any finite time you please constantly tend towards equality..."

The chord is "constantly" tending in that direction, because the ratio between the two doesn't fluctuate between getting closer to equality and further from it as the arc gets smaller. Rather, through the whole range of diminution of the arc, for every bit the arc gets smaller, the ratio gets closer to one of equality.

Lemma 1's second condition: "...and before the end of that time approach nearer to each other than by any given difference you please, [become ultimately equal]."

No matter how small a difference one might posit between the arc and the chord, as the arc keeps getting smaller the chord will, within a finite length of time, get closer in length to the arc than that posited difference. But further, this works with the *ratio* too. We can posit a ratio with however small a difference from equality we please, and the ratio between the arc and chord will get closer to equality than that. This is not the same as the difference beween the two quantities becoming less, as we shall see in the next example.

As the ratio of lengths between arc and chord changes, it approaches a limiting ratio. In this example, that limit is a ratio of equality. Of course, they can never actually reach that limit or *be* equal, since the limit lies where the arc (and therefore also the chord) has vanished. But the lemma asserts that if these conditions are met we will say that they are "ultimately equal."

Suppose we began with an arc that was half a circle. The ratio between arc and chord at that point would be $\pi r : 2r$, or $3.14\ldots : 2$. As the arc is taken continuously smaller, the ratio of its length to that of its chord would change. It will get closer and closer to a ratio of equality, and never pass it (that is, the chord will never become longer than the arc).

It is true that the ratio will never actually reach one of equality (they will never be identical as long as the arc has some finite length). The ratio with which the two magnitudes vanish is not a ratio of actual arcs and chords, but a limit towards which the changing ratio is constantly tending.

Newton calls this limit an "ultimate ratio": the limit is a definite, finite, and existing thing, in this case equality, not something strange and mysterious like "vanishing quantities." The limit is what Newton means by "ultimate." Here the quantities are in an ultimate ratio of equality. For ease of expression, they are said to be "ultimately equal."

• **Second Example: Hypotenuse and Side of a Right Triangle:** In contrast, we could look at an example of a ratio of quantities which are *not* "ultimately equal."

Suppose we were given a right triangle *ABC*, with equal sides *BC* and *BA* about the right angle *B*. Let there be lines parallel to the hypotenuse *AC* moving closer and closer to *B*; that is, let the hypotenuse itself move towards *B*, maintaining the equal angles at *A* and *C*.

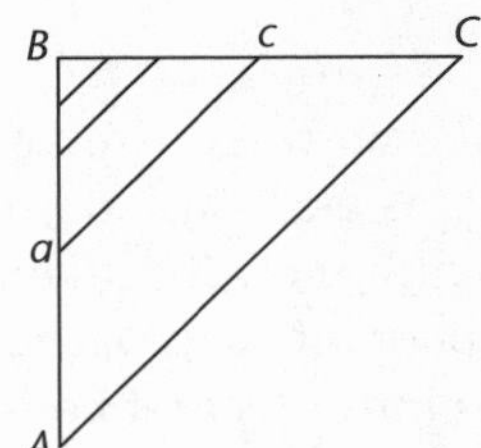

Lemma 1's first condition: "Quantities, as well as the ratios of quantities, which in any finite time you please constantly tend towards equality…"

Our two quantitites are the ever-decreasing *BC* and the ever-decreasing *AC*. The difference between those two quantities becomes continuously less; in a finite time the difference will become less than any given quantity. Thus the conditions of Lemma 1 are satisfied by the quantities directly.

But what happens to the ratio between the two? Does that ratio "tend constantly towards equality"? No, the ratio stays the same because the triangles are all similar. So even though the two *quantities* are approaching equality in some sense, the *ratio* does not. These two quantities *BC* and *AC* vanish with the same ratio they had throughout their diminishing, namely the ratio of $1 : \sqrt{2}$. Thus the ratios do not meet the first condition of

Lemma 1, and we know from Newton's use of Lemma 1 that he would not say that BC and AC are ultimately equal.

• **Third Example: Right Triangle with Rotating Hypotenuse:** The previous example showed that the first condition might appear to be satisfied if we look only at the fact that the quantities are continuously approaching equality, but that if we look at the ratio of the quantities we see that the ratio does not approach equality at all.

But the ratios must meet not only both parts of the first condition but the second condition as well. Consider this third example.

Let there be a triangle with a right angle at B and hypotenuse AC defined by a line projected from and rotating around point P outside the triangle. P is positioned such that angle $PBA = 45°$. Thus as AC rotates it ultimately coincides with PB, and therefore angle BAC will be ultimately equal to 45°.

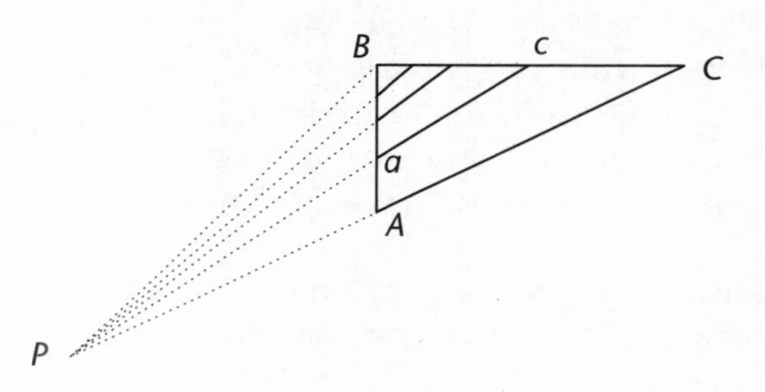

Let our two quantities be sides AC and AB which at some starting point are in some starting ratio, say $\sqrt{5} : 1$. As AC pivots towards PB, the quantities diminish, and the triangle approaches an isosceles right triangle; the ratio continuously approaches a limit of $\sqrt{2} : 1$. [Euclid I.47]

Lemma 1's first condition: "Quantities, as well as the ratios of quantities, which in any finite time you please constantly tend towards equality..."

As AC departs from its initial position, the ratio $AC{:}AB$ decreases from its initial $\sqrt{5} : 1$ value. With every new position, the ratio decreases further. Now the ratio of equality (1 : 1) is less than $\sqrt{2} : 1$; so as AC continues to rotate, the ratio $AC{:}AB$ constantly tends toward the ratio of equality, thus fulfilling the first condition.

Lemma 1's second condition: "...and before the end of that time approach nearer to each other than by any given difference you please, [become ultimately equal]."

If we look only at quantities, the second part appears to be fulfilled as well: the two quantities AC and AB approach nearer to each other than any given difference however small. But the *ratio* $AC{:}AB$ does not: it never approaches nearer to a ratio of equality than $\sqrt{2} : 1$. And, as we know from his uses of Lemma 1, Newton would not say these quantities are "ultimately equal."

Therefore, if we are dealing with vanishing quantities, we need the parenthetical "as well as the ratios of quantities" to fulfill the second condition. The ratio itself, not just the vanishing quantities, must approach closer to equality than any ratio you please, however close that ratio might be to equality. We now understand why Newton needed both formulations in Lemma 1.

• Newton gives a commentary on what he means by "ultimate ratio" in the Scholium that follows the lemmas. The full text of the scholium appears in its place after Lemma 11. But since it will help us now to consider some of what he says there, let us consider these excerpts:

Excerpts from Scholium to the Lemmas

There is the objection that there is no ultimate ratio of evanescent quantities, for before they vanish it is not the ultimate, and when they vanish there is no ratio....

[B]y the ultimate ratio of evanescent quantities is to be understood the ratio of quantities, not before they vanish, not afterward, but [that] with which they vanish. Likewise also, the first ratio of nascent [quantities] is the ratio with which they come into being. And the first and ultimate sum is that with which they begin and cease to be (or to increase or to decrease)....

And the account of the limit of all quantities and proportions that are beginning and ceasing to be, is the same. And since this limit is certain and definite, it is really a geometrical problem to determine it....

It can also be urged that if the ultimate ratios of evanescent quantities be given, the ultimate magnitudes will also be given, and thus every quantity will consist of indivisibles, contrary to what Euclid has demonstrated concerning incommensurables in the tenth book of the Elements. However, this objection rests upon a false hypothesis. Those ultimate ratios with which quantities vanish are not truly ratios of ultimate quantities, but limits to which the ratios of quantities decreasing without limit always approach, and which they can attain to more nearly than by any given difference you please, but never go beyond, nor arrive at before the quantities are diminished *in infinitum*....

Accordingly, in what follows, whenever out of concern for the easy conception of the subject I shall speak of quantities [that are] least, or evanescent, or ultimate, beware of understanding [them to be] quantities determinate in magnitude, but think of them as always to be diminished without limit.

Notes on Excerpts from Scholium to the Lemmas

• **Ratios:** It will help to remember, as we work through the propositions of *Principia*, certain of the tips he gives us here about his definition of "ultimate" quantitites, because, as he warns us in the last sentence quoted here, he will not always be this careful in his language when he is sketching his proofs.

On the contrary, he will use language that has a "concern for the easy conception of the subject" and speak directly of "least," "evanescent," or "ultimate" quantities. We need to remember that he is not talking about what we might call "infinitesimals."

One tip he gives us is to remind us that he is dealing with ratios. This is important because, while infinitely small magnitudes introduce logical problems, and might be objected to on several grounds, the limit towards which a ratio is tending, if there is a demonstrable one, may be a definite thing such as "equality" or "two" or something else specific and finite, which, as Newton says, we can determine geometrically.

"[B]y the ultimate ratio of evanescent quantities is to be understood the ratio of quantities, not before they vanish, not afterward, but [that] with which they vanish."

Newton is not claiming that the ratio is reached before the quantities vanish; after the quantities vanish there is no ratio; but perhaps it will help us, Newton is suggesting, to imagine what the ratio would be *as* they vanish, at the moment of arriving at the limit.

But we might still go astray: we might think that these "ultimate" ratios are ratios of "ultimate" quantities, that is, that the magnitudes imagined in ratio to each other are "indivisibles" or infinitesimals. This again would open him to objections and his science to logical problems. He sees us about to go astray and reminds us that, whatever language he may use later, he means the limiting ratio.

"Those ultimate ratios with which quantities vanish are not truly ratios of ultimate quantities, but limits to which the ratios of quantities decreasing without limit always approach, and which they can attain to more nearly than by any given difference you please, but never go beyond, nor arrive at before the quantities are diminished *in infinitum*."

- **Indivisibles:**

"It can also be urged that if the ultimate ratios of evanescent quantities be given, the ultimate magnitudes will also be given, and thus every quantity will consist of indivisibles, contrary to what Euclid has demonstrated concerning incommensurables* in the tenth book of the *Elements*."

Indivisibles would be bits of magnitude so small as not to be able to be divided further. The concept thus contradicts the traditional Euclidean understanding of magnitude, which views magnitudes as continuous and able to be divided without limit.

In the tenth book of Euclid's *Elements*, the first proposition invokes this process to establish an important principle which lies behind Newton's

---

* Book X of Euclid is about incommensurables, but the first proposition, with which we are concerned, is general; we need not worry about how the difficulties of incommensurability fit in here.

Lemma 1. Euclid's Proposition X.1 says,

> Two unequal magnitudes being set out, if from the greater there be subtracted a magnitude greater than its half, and from that which is left a magnitude greater than its half, and if this process be repeated continually, there will be left some magnitude which will be less than the lesser magnitude set out.

If we could reach some magnitude that could no longer be divided, we could not "repeat this process continually," nor could we suppose with Lemma 1 that the quantities could approach nearer to each other than by any given difference we please.

The idea of indivisibles, like our modern idea of infinitesimals, refers to a bit of magnitude in a special realm between being finite (and thus able to be divided as per Euclid X.1) and being nothing. Newton is trying to avoid the conceptual and logical difficulties with this, using mathematical rigor in his formulation of the limiting ratio, and calling on our intuition with his image of the ratio *as the quantities vanish*.

Lemma 7

The same things being supposed, I say that the ultimate ratio of the arc, chord, and tangent to each other is the ratio of equality.

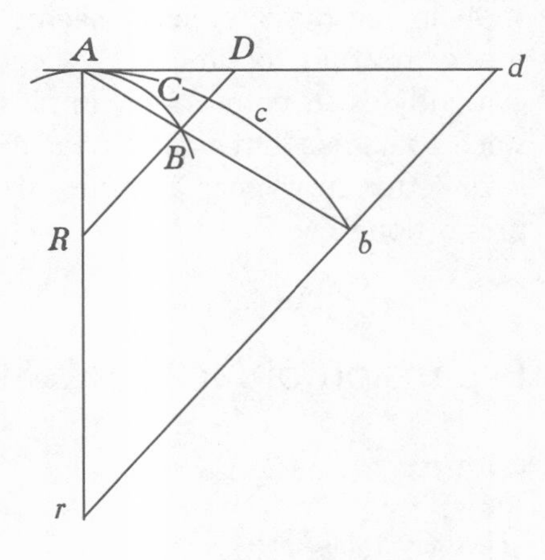

For as the point B approaches the point A, let AB and AD be understood always to be produced to the distant points b and d, and let bd be drawn parallel to the intersecting line BD. And let the arc Acb be always similar to the arc ACB. And, the points A, B, coming together, the angle dAb, by the previous lemma, vanishes; and therefore the straight lines Ab, Ad, always finite, and the arc Acb lying between them, coincide, and therefore will be equal. Therefore, both the straight lines AB, AD always proportional to these, and the arc ACB lying between, will vanish, and will have the ultimate ratio of equality.

Q.E.D.

Note on Newton's Diagram: This diagram applies to Lemma 8; it was given with Lemma 6 and evidently meant to serve for all three. Points and lines not mentioned or used must be ignored.

Notes on Lemma 7

• **Finite Similar Triangles:** This technique is often called Newton's "microscope." It is a mistake to take it simply as a visual aid. Rather, it is a way of construction that avoids the logical difficulties of trying to place vanishing quantities into ratio with one another. It allows us to make ratios of finite quantities that we know to be in the same ratio as the vanishing ones. This is the same technique Leibniz used with his "assignables" in his approach to calculus. Through the use of similar triangles, Leibniz kept lines of finite length in the same ratio as the lines that were vanishing, thus avoiding the logical and conceptual problems of working with ratios of quantities that were themselves vanishing.

• As *B* moves towards *A* whole new sets of triangles and projected triangles appear. The proof is based on similarities within one set of triangles for a given *B* and doesn't depend on parallels or similarities from one position of *B* to the next.

• Some means of selection of *D* must be given or devised such that *D* must move towards *A* as *B* moves towards *A* (this can be written "$D \to A$ as $B \to A$"), and *D* must reach *A* at the same time *B* does. Line *BD* need not remain parallel to itself as $B \to A$; it might be projected from some outside point through *B* to the tangent line.

• As in Lemma 5, Newton seems to be assuming a body of mathematical theory relating to similarity of curvilinear figures—a body of theory that evidently exists comfortably in his mind. Specifically, he supposes here that since the small-letter curvilinear triangles are simply magnifications of the large-letter curvilinear triangles, they will be similar and their areas will be proportional.

Expansion of Newton's Sketch of Lemma 7

Given:

"The same things being supposed..."

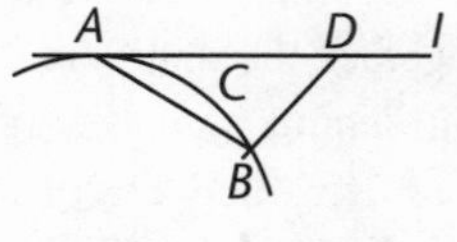

1. Arc *ACB*;

2. Chord *AB* subtending that arc;

3. Point *A* is in the middle of continuous curvature;

4. Tangent *AI* touching the curve at *A*;

5. Point *B* is imagined moving towards *A* until it coincides: this will mean that the arc and the chord both become smaller and smaller until they both vanish.

To Prove:

"...I say that the ultimate ratio of the arc, chord, and tangent to each other is the ratio of equality."

Arc, chord, and the segment of the tangent associated with the chord will vanish with a ratio of equality each to each.

Proof:

Let D be any point on the tangent Al on the same side of A as B. Connect line BD.

"For as the point B approaches the point A, let AB and AD be understood always to be produced to the distant points b and d, and let bd be drawn parallel to the intersecting line BD."

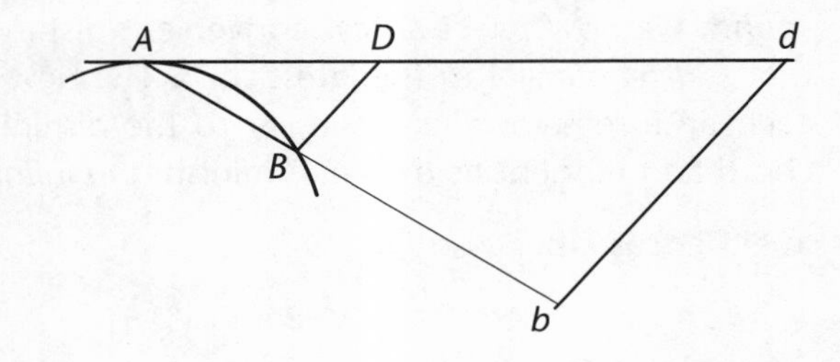

The purpose of using the "distant" points is to create a triangle and arc that will remain finite while similar to the triangle and arc that will be vanishing (see bulleted note on Finite Similar Triangles).

Choose a convenient point b on the other side of B from A or at least such that Ab is a comfortable finite length. Through b draw a line parallel to BD intersecting the tangent at d. Ad should now also be a comfortable finite length, and the projected triangle Abd easy to see and work with.

Because BD is parallel to bd, triangle ABD is similar to triangle Abd by Euclid VI.2 and VI Definition 1.

We now imagine B moving towards A. We will create a new projected reference triangle Abd at each point, since our new chord AB' will no longer pass through our previous distant b. We must pick a new D' each time, and it must be according to some consistent plan, but it need not be selected such that $B'D'$ remains parallel to the previous (or original) BD. We only require D' to remain on the same side of A and to move continuously towards A as B does, not arriving at A before B does.

If $B'D'$ does remain parallel to the original BD, then we may use our original line bd for each new projection. In this case we would find new points b', b'', etc., moving up the original bd towards d.

Case 1: *BD* stays parallel

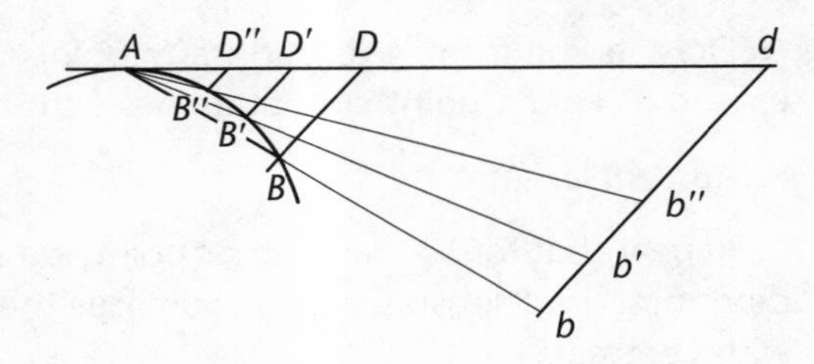

Case 2: BD does not stay parallel

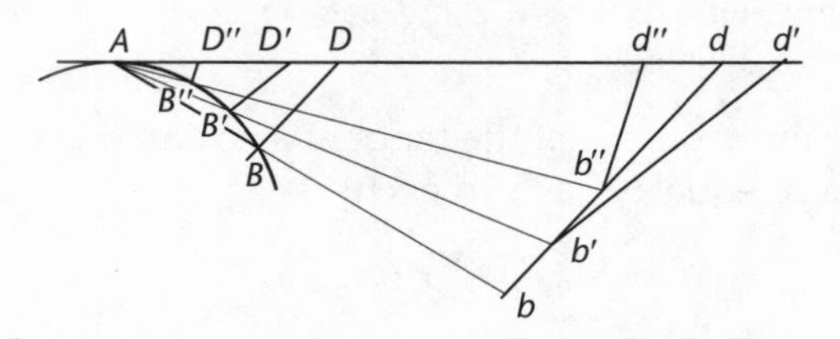

If line $B'D'$ does *not* stay parallel to the original BD, we have to do a little more construction work. Having moved to a new B, with a new D to correspond, we now select a new convenient distant b' and find a new d' such that $b'd'$ is parallel to the latest $B'D'$. The new reference triangle we make each time remains always similar to the diminishing triangle made by the chord and tangent as the arc diminishes *in infinitum*.

In either case, because $db \parallel DB$,

$AB : Ab :: AD : Ad.$ [Eu. VI.2 and V.16] (1)

$\triangle ABD \sim \triangle Abd.$ [Eu. VI.5 and VI Definition 1] (2)

"And let the arc Acb be always similar to the arc ACB."

We must now draw to our (relatively) distant point b an arc Acb similar to the arc ACB on the original curve. We can describe this similar arc point-by-point in this way.

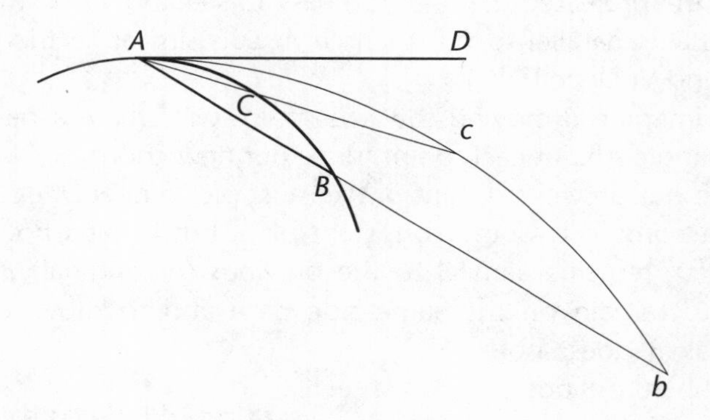

Draw a line from A through any point C on the original curve ACB and extend it. Find a point c on that line such that:

$AB : Ab :: AC : Ac.$ (3)

Repeat that for as many other points on arc AB as desired. Every point so determined will lie on the similar arc [see the definition of similarity discussed with Lemma 5].

Lemmita:* Common Tangent to Similar Curves

This method reveals another useful characteristic of similar arcs sharing an endpoint *A*. We will use this property in later propositions and will refer back to its proof, placed here because it uses the construction and relationships that we have just drawn out.

For the purpose of drawing out this characteristic, let's let the arcs be fixed, not vanishing. *B* and *b* are both fixed, with $\widehat{AB}$ and $\widehat{ab}$ finite arcs.

Consider the line *AC* when *C* arrives at *A*. By Lemma 6 line *CA* will lie on the tangent *AD*. It does not cut the original arc *ACB* in two places, but only touches it in one, *A*. Thus *AC* has a length of zero.

Therefore, by Equation 3, *Ac* must also have a length of zero, whatever the ratio *AB* : *Ab*. This means that line *AD* does not intersect the projected similar arc in two places, but only touches it in one, *A*. Therefore *AD* is also tangent to the projected similar arc *Acb*.

Thus the two similar curves have the same tangent, and both are caught between tangent *ADd* and straight line *ABb*.

Q.E.D.

The construction of the similar arc depends an the implicit definition of similarity of curvilinear figures, as we noted in the discussion with Lemma 5. In Lemma 7 here we construct similar inscribed rectilinear figures (ones with sides *Ac* and *AC*, for example) and move as close as we please to the points on the similar curvilinear figures.

If $b'd'$ remains parallel to the original *BD*, our diagram gives the easiest visual intuition of the final step of the proof. Suppose our line *bd* were to remain given in position, with the points *b*, b', b'', etc. moving up towards *d*.

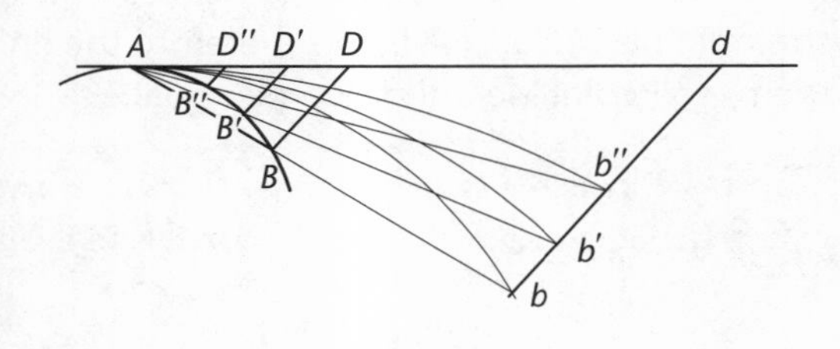

Here we can see that the line Ab' moves towards coincidence and therefore equality with the line *Ad* with the projected arc $\widehat{Acb}$ trapped in between.

However, Newton doesn't require that *BD* remain parallel. He asks us to see that with any projected triangle the vanishing angle *BAD* will lead to coincidence and therefore equality of *Ab* and *Ad* with the arc trapped between. He depends on similarity of the projected triangle with the evanescent triangle *at each point B*, not on any similarity between one point and the next.

---

* To distinguish lemmas implicit in Newton's proofs from those he makes explicit as numbered Lemmas, I will call the former "lemmitas," small lemmas.

As B moves towards A, the arc will change to a new $AC'B'$. We must construct a new similar arc $Ac'b'$ reaching to a new distant b. Each of these successive projected reference arcs remains finite and correlates to the respective reference triangle, always lying between the tangent Ad and the base of the triangle Ab.

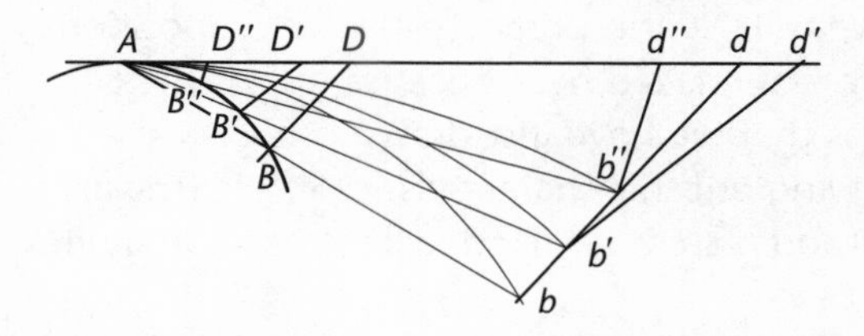

By Equation 1, $AB : Ab :: AD : Ad$.

As B moves towards A, we maintain this relationship to a finite projected reference triangle:

$$AB' : Ab' :: AD' : Ad'. \tag{4}$$

"And, the points A, B, coming together, the angle dAb, by the previous lemma, vanishes..."

Equation 4 will hold until, as B reaches A, the angle dAb (or $d'Ab'$, where the primes would increase *in infinitum*) vanishes, as it will do by Lemma 6.

"...and therefore the straight lines Ab, Ad, always finite, and the arc Acb lying between them, coincide, and therefore will be equal."

When the angle between the chord and the tangent vanishes, the finite straight lines Ab and Ad, and therefore the finite arc $\overset{\frown}{Acb}$ which lies between them, will coincide and become equal.

"Therefore, both the straight lines AB, AD always proportional to these, and the arc ACB lying between, will vanish, and will have the ultimate ratio of equality.

Q.E.D."

But by Equation 4, the vanishing chord and tangent will always be in the same proportion, and must remain so as they vanish as well. Since when they vanish the convenient projected lines are in a ratio of equality, the vanishing lines must be in a ratio of equality as well. Their ultimate ratio is a ratio of equality. Since the arc is trapped between them, it must ultimately be in a ratio of equality with both the chord and the tangent.

Q.E.D.

Lemma 7 Corollary 1

Hence, if through B there be drawn BF parallel to the tangent, always

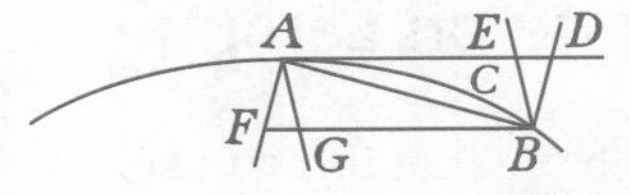

intersecting at F any straight line you please AF passing through A, this [line] BF will ultimately have the ratio of equality to the evanescent arc ACB, because, when the parallelogram AFBD is completed, [BF] will always have the ratio of equality to AD.

Note on Lemma 7 Corollary 1

Newton will later refer to line segment *BF* as a "sine." He calls it this because in a circle it would be half the chord of twice the arc, which in trigonometry is the sine of the angle subtended by that (original undoubled) arc. This line segment will be used in later propositions, so note that it will vanish in a ratio of equality with the arc, chord, and tangent.

Expansion of Newton's Sketch of Corollary 1

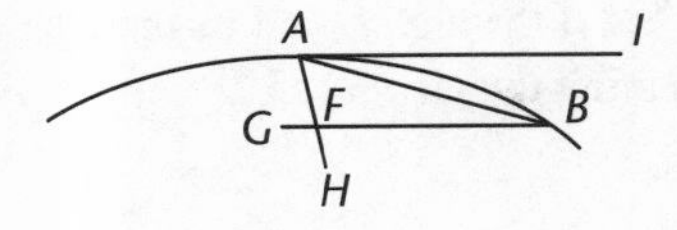

"Hence, if through *B* there be drawn *BF* parallel to the tangent, always intersecting at *F* any straight line you please *AF* passing through *A*..."

Through point *B* construct *BG* ∥ *AI*. [Eu. I.31]

Let there be any line *AH* passing through *A*. Let the point in which it intersects *BG* be called *F*.

"...this [line] *BF* will ultimately have the ratio of equality to the evanescent arc *ACB*, because, when the parallelogram *AFBD* is completed, [*BF*] will always have the ratio of equality to *AD*."

Through point *B* construct *BK* ∥ *AF* intersecting *AI* in point *D*.

BF = *AD*. [Eu. I.34]

Since *AD* vanishes with a ratio of equality to the arc, so will *BF*.

Lemma 7 Corollary 2

And if through B and A there be drawn many straight lines BE, BD, AF, AG, intersecting the tangent AD and its parallel BF, the ultimate ratio of all the lines AD, AE, BF, BG, cut off, and of the chord and the arc AB, to each other, will be the ratio of equality.

Note on Lemma 7 Corollary 2

Note that this corollary is *not* saying that *BD* or any of the other cutting lines vanishes with a ratio of equality with the arc, chord, and tangent. It is only the segment cut off on the tangent or its parallel that does so. A careless student might suppose that *everything* vanishes with a ratio of equality; Newton does not think this (see the Pause Before Lemma 7). Only certain things can be shown by geometric demonstration to be ultimately equal to certain other things, even though many more things may vanish at the same moment.

We must keep in mind, for the later extensive use we will make of these lemmas, just which lines have been proved to vanish in a ratio of equality with chord, arc, and tangent, and which have not.

Expansion of Newton's Sketch of Corollary 2

"And if through *B* and *A* there be drawn many straight lines *BE*, *BD*, *AF*, *AG*, intersecting the tangent *AD* and its parallel *BF*..."

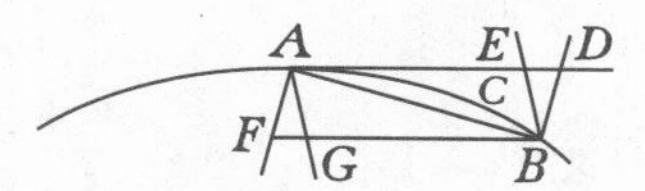

"... the ultimate ratio of all the lines *AD*, *AE*, *BF*, *BG*, cut off, and of the chord and the arc *AB*, to each other, will be the ratio of equality."

We are identifying some more lines which vanish in a ratio of equality with the chord, arc, and tangent.

Lines such as *AE*, which are made by running lines from *B* to intersect the tangent line *AD*, could themselves have been tangents in this lemma: remember that we did not require that *D* be selected by any particular rule. Thus what we proved of *AD* there will also apply to any line *AE* cut off by a line from *B*.

Lines such as *BG*, parallel to the tangent, could be lines parallel and equal to the tangent line, had *D* been appropriately selected. Thus by Corollary 1, they would also attain a ratio of equality with the evanescent arc.

Question: Are there any restrictions on where *BE* can be drawn?

Answer: Point *E* can be anywhere *D* might have been selected. This means that *E* is not permitted to get to *A* before *B* does.

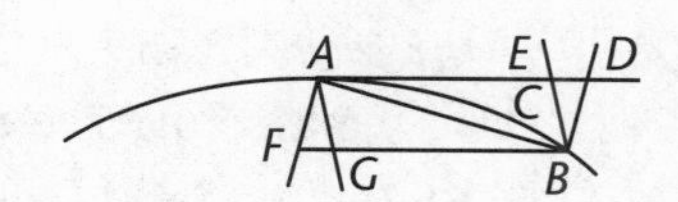

Lemma 7 Corollary 3

And therefore all these lines can be used in place of each other in all considerations of ultimate ratios.

Lemma 8

If given straight lines AR, BR, together with the arc ACB, the chord AB, and the tangent AD, form the three triangles RAB, RACB, RAD, and then the points A and B approach each other, I say that the ultimate form of the evanescent triangles is similarity, and the ultimate ratio equality.

For while point *B* approaches point *A*, let *AB*, *AD*, *AR*, be understood to be extended to the distant points *b*, *d*, and *r*, and *rbd* to be drawn parallel to *RD*, and let the arc *Acb* be always similar to the arc *ACB*. And when the points *A*, *B* come together, the angle *bAd* will vanish, and therefore the three triangles *rAb*, *rAcb*, *rAd*, always finite, will coincide, and for that reason are similar and equal. Hence, *RAB*, *RACB*, *RAD*, always similar and proportional to these, will ultimately become similar and equal to each other.

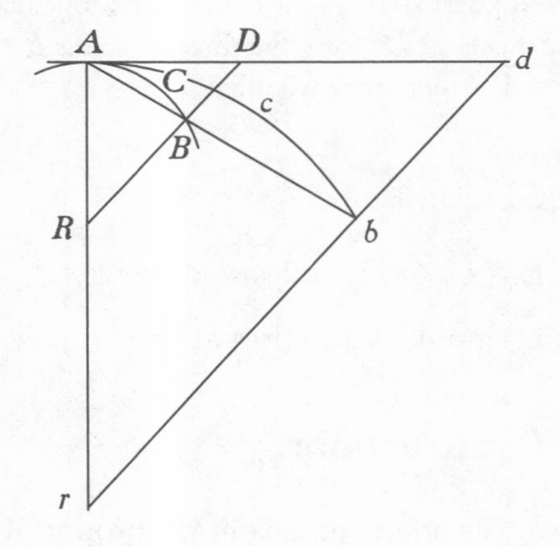

Q.E.D.

Notes on Lemma 8

- The term "evanescent" is explained in the Scholium to the Lemmas. The explanation is excerpted in the Pause Before Lemma 7. Newton says: "[B]y the ultimate ratio of evanescent quantities is to be understood the ratio of quantities, not before they vanish, not afterward, but [that] with which they vanish."

It is crucial, however, that we understand not evanescent quantities placed into ratio, but a ratio of finite quantities looked at as they decrease to the point of vanishing. (The Pause Before Lemma 7 ponders this fine distinction and gives Newton's explanation of it.)

• The same conditions set out in Lemma 7 are necessary here since Newton uses Lemma 7 to prove Lemma 8.

As in the case of Lemma 7, as B moves to A, a whole new set of triangles appears, including, perhaps, a new means of selecting R or D and a new projected line rd.

The proof is based on similarities within one set of triangles for a given B and doesn't depend on parallels or similarities from one position of B to the next.

• As in Lemma 7, D must move towards A as B moves towards A. But R need not move towards A. It could be fixed in position. Perhaps D is found by projecting always through R and the changing B to the tangent.

Expansion of Newton's Sketch of Lemma 8

"If given straight lines AR, BR, together with the arc ACB, the chord AB, and the tangent AD, form the three triangles RAB, $RACB$, RAD, and then the points A and B approach each other..."

Given:

1. Arc $\widehat{ACB}$, chord AB, tangent AI;

2. Line AX through A at any angle within the tangent.

Construction:

We are given a line AX through A at any angle on the same side of the tangent as point B. Then D can be chosen anywhere on tangent AI. A line from D through B will define R on AX. Or else choose R on line AX and from R through B construct a line defining D on the tangent.

We have now formed three triangles: $\triangle RAD$, $\triangle RAB$, $\triangle R\widehat{AC}B$.

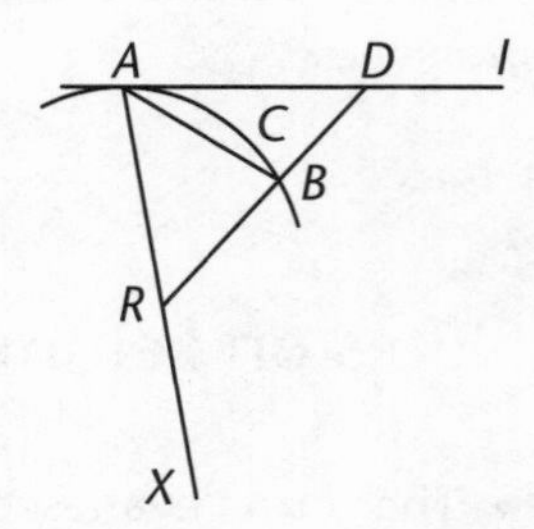

To Prove:

"I say that the ultimate form of the evanescent triangles is similarity, and the ultimate ratio equality."

As $B \to A$, the vanishing triangles will ultimately be similar, and the ratio of any two ultimately one of equality.

Proof:

"For while point B approaches point A, let AB, AD, AR, be understood to be extended to the distant points b, d, and r, and rbd to be drawn parallel to RD, and let the arc Acb be always similar to the arc ACB."

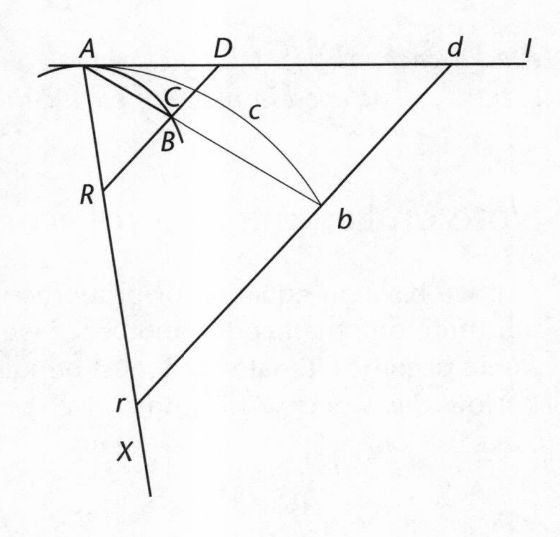

Produce AB or AD or AR to some distant point (b or d or r), and through this point draw rd parallel to RD, the intersections of AB, AD, and AR with rd being b, d, and r respectively.

$$AR : Ar :: AB : Ab :: RB : rb. \quad \text{[Eu. VI.2]} \qquad (1)$$

Therefore $\triangle RAB \sim \triangle rAb$. [Eu. VI.5]

$$AR : Ar :: AD : Ad :: RD : rd. \quad \text{[Eu. VI.2]}$$

Therefore $\triangle RAD \sim \triangle rAd$. [Eu. VI.5]

Construct an arc $\widehat{Ab}$ similar to arc $\widehat{AB}$. [see method described on page 86 in the expanded proof for Lemma 7]

$$\text{Arc } \widehat{ACB} : \text{Arc } \widehat{Acb} :: AB : Ab. \quad \text{[construction]}$$

By Equation 1, the remaining sides AR, Ar and RB, rb of $\triangle R\widehat{ACB}$ and $\triangle r\widehat{Acb}$ are in the same ratio.

Therefore $\triangle R\widehat{ACB} \sim \triangle r\widehat{Acb}$.

"And when the points A, B come together, the angle bAd will vanish, and therefore the three triangles rAb, $rAcb$, rAd, always finite, will coincide, and for that reason are similar and equal. Hence, RAB, $RACB$, RAD, always similar and proportional to these, will ultimately become similar and equal to each other.

Q.E.D."

But by Lemma 7, as $B \rightarrow A$, straight lines Ab, Ad, and arc $\widehat{Ab}$, will ultimately coincide and be equal. Thus triangles $\triangle rAb$, $\triangle rAd$, and $\triangle r\widehat{Ab}$ will ultimately coincide, and thus be similar and equal.

And since $\triangle RAB$, $\triangle RAD$, and $\triangle R\widehat{AB}$ are similar to them, as we have just shown, they will become themselves ultimately similar and equal to one another.

Q.E.D.

Lemma 8 Corollary

And hence, those triangles, in all considerations involving ultimate ratios, can be used in place of each other.

Note on Lemma 8 Corollary

If we have an equation or proportion involving ultimate ratios, we may substitute one triangle for another. If we have a nonultimate proportion, it would become ultimate (and must be so noted) if we make this substitution.

How this works will become clear as we delve into the propositions to come.

Lemma 9

If the straight line AE and the curve ABC, both given in position, mutually intersect in a given angle A, and to that straight line, BD and CE be applied ordinatewise in another given angle, meeting the curve at B and C, and then the points B and C simultaneously approach the point A, I say that the areas of the triangles ABD, ACE, will be ultimately to each other in the duplicate ratio of the sides.

For while the points B, C approach the point A, let AD always be understood to be produced to the distant points d, e, so that Ad, Ae are proportional to AD, AE, and let the ordinates db, ec, be set up parallel to the ordinates DB, EC, meeting AB, AC extended at b and c. Let the curve Abc, similar to ABC, be understood to be described, and likewise the straight line Ag, touching both curves at A and intersecting DB, EC, db, ec, applied ordinatewise, at F, G, f, g. Then, with the length Ae remaining constant, let the points B, C, come together with the point A, and, since angle cAg vanishes, the curvilinear areas Abd, Ace, will coincide with the rectilinear areas Afd, Age; and therefore (by Lemma 5) they will be in the duplicate ratio of the sides Ad, Ae. But to those areas the areas ABD, ACE, are always proportional, and to those sides the sides AD, AE [are always proportional]. Therefore, the areas ABD, ACE, also, are ultimately in the duplicate ratio of the sides AD, AE.

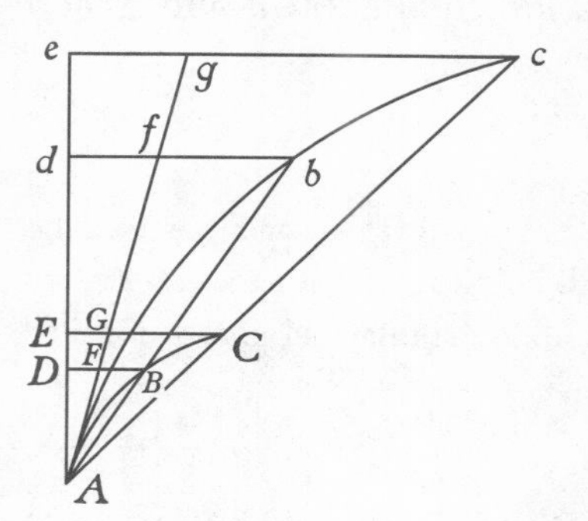

Q.E.D.

Notes on Lemma 9

- **"Given angle *A*"** is an angle between straight line *AE* and curve *ABC*. Line *AE* is not a tangent but a line that cuts the curve at *A*. (Line *Ag* is drawn as the tangent at *A* in the course of the proof.)

- **"Ordinatewise"** in this case just means that lines *BD* and *CE* are drawn from the curve to another line at a given angle. We will later encounter more specific meanings of this term for particular conic sections.

- **"Duplicate ratio"**: See Section 6.3 of Preliminaries.

$\triangle \widehat{ABD} : \triangle \widehat{ACE} \overset{ult}{::} dup(AD : AE)$ can be expressed in more algebraic form as:

$\triangle \widehat{ABD} : \triangle \widehat{ACE} \overset{ult}{::} \overline{AD}^2 : \overline{AE}^2.$

- **Similarity of curvilinear figures.** As discussed in notes to Lemma 5 and Lemma 7, Newton is assuming a body of mathematical theory (possibly of his own devising) relating to similarity of curvilinear figures.

Specifically, he supposes here that since the small-letter curvilinear triangles are simply magnifications of the large-letter curvilinear triangles, they will be similar and their areas will be proportional. (This takes place at Equation 10 in the expanded proof.) We must either accept this property of curvilinear figures as a given mathematical tool or construct our own proof. Notes to Lemma 5 discuss what the definition of similarity for curvilinear figures could be and how one might go about proving that curvilinear figures exhibit properties of similarity parallel to those exhibited by similar rectilinear figures.

Expansion of Newton's Sketch of Lemma 9

Given:

"If the straight line *AE* and the curve *ABC*, both given in position, mutually intersect in a given angle *A*, and to that straight line, *BD* and *CE* be applied ordinatewise in another given angle, meeting the curve at *B* and *C*, and then the points *B* and *C* simultaneously approach the point *A* ..."

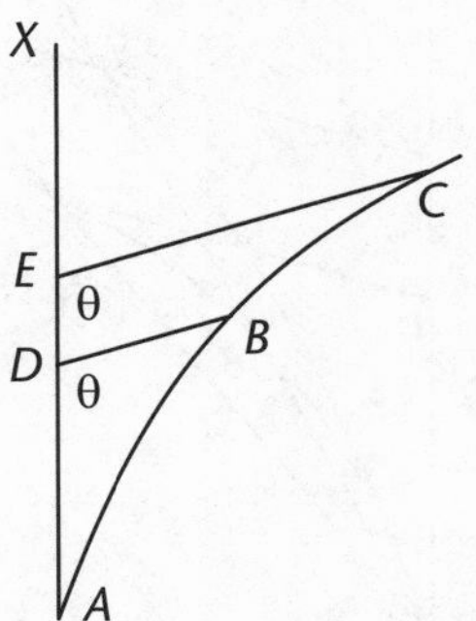

1. Straight line *AX* on which *E* will be found;
2. Curved line $\widehat{ABC}$;
3. Given curvilinear angle *EABC* between *AX* and curve $\widehat{ABC}$;
4. Angle θ at which lines *BD* and *CE* meet *AX*;
5. $B \to A$ and simultaneously $C \to A$.

Construction:

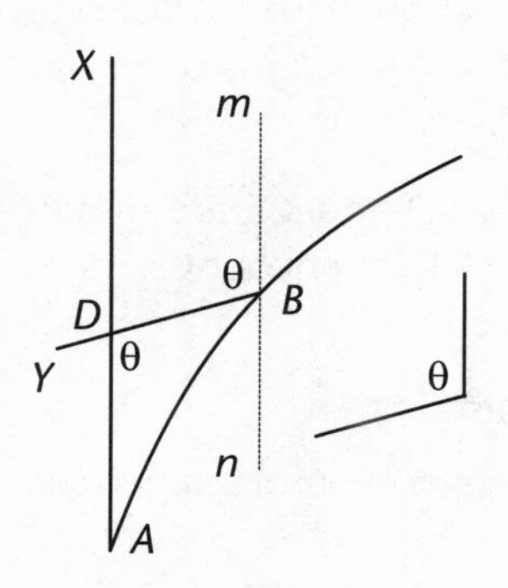

Through B construct $mn \parallel AX$. [Eu. I.31]

On line mn at point B construct angle mBY equal to the given angle θ. [Eu. I.23]

Let BY intersect line AX in point D. Angle ADB = angle θ. [Eu. I.29]

Repeat procedure to construct angle AEC = angle θ.

We now have constructed the triangles whose properties this lemma is going to demonstrate.

To Prove:

"I say that the areas of the triangles ABD, ACE, will be ultimately to each other in the duplicate ratio of the sides."

$\triangle A\widehat{BD} : \triangle A\widehat{CE} \overset{ult}{::} dup(AD : AE)$.

Proof:

"For while the points B, C approach the point A, let AD always be understood to be produced to the distant points d, e, so that Ad, Ae are proportional to AD, AE, and let the ordinates db, ec, be set up parallel to the ordinates DB, EC, meeting AB, AC extended at b and c. Let the curve Abc, similar to ABC, be understood to be described…"

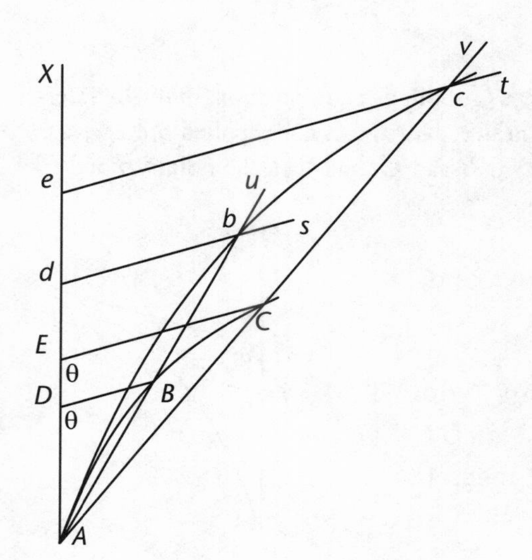

Let given lines AX and $\widehat{ABC}$ meet in given curvilinear angle $XABC$.

Find points d and e beyond D on AX such that

$$AD : AE :: Ad : Ae. \qquad (1)$$

Let ds and et be drawn parallel to BD and EC.

Extend line AB to u and AC to v.

Let ABu intersect ds at b; let ACv intersect et at c.

Because $DB \parallel db$ in $\triangle Adb$,

$AD : Ad :: AB : Ab.$ [Eu. VI.2] (2)

Because EC ∥ *ec* in △*Aec*,

$AE : Ae :: AC : Ac.$ [Eu. VI.2] (3)

Alternating Equation 1,

$AD : Ad :: AE : Ae.$ (4)

Using Equation 4 to combine Equations 2 and 3,

$AB : Ab :: AC : Ac.$ [Eu. V.11] (5)

Therefore the point *b* will lie on a curve *Abc* similar to *ABC*, constructed according to the method described in Lemma 7. Let this curve "be understood to be described" by repeating this operation through other points (in thought, through every point) on arc *AC*. The result will be a complete curve *Abc*.

"... and likewise the straight line *Ag*, touching both curves at *A* and intersecting *DB*, *EC*, *db*, *ec*, applied ordinatewise, at *F*, *G*, *f*, *g*."

Now draw straight line *Aw* to touch both curves at *A*. [See lemmita entitled Common Tangent to Similar Curves in the expanded proof for Lemma 7.]

Let tangent *Aw* intersect lines *DB* in *F*, *EC* in *G*, *db* in *f*, and *ec* in *g*.

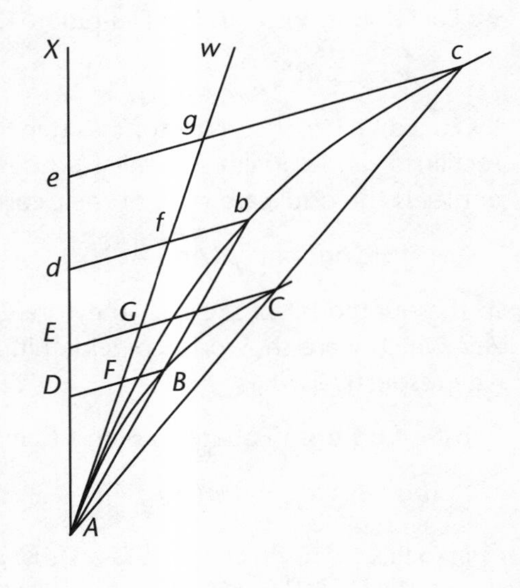

"Then, with the length *Ae* remaining constant, let the points *B*, *C*, come together with the point *A*, and, since angle *cAg* vanishes, the curvilinear areas *Abd*, *Ace*, will coincide with the rectilinear areas *Afd*, *Age*; ..."

Let $B \rightarrow A$ and $C \rightarrow A$. As this happens *D* and *E* may change, and the ratio $AD : AE$ may also change. However, we may choose to hold length *Ae* constant, which will determine the fourth proportional *Ad*. Let *Ae* remain its original length.

By Lemma 6, angle *CAg*, the angle between the chord and the tangent, will vanish at the limit. Therefore chord *AC*, and thus its extension *Ac*, comes to coincide with tangent *Ag*, with arc *Ac* trapped in the middle.

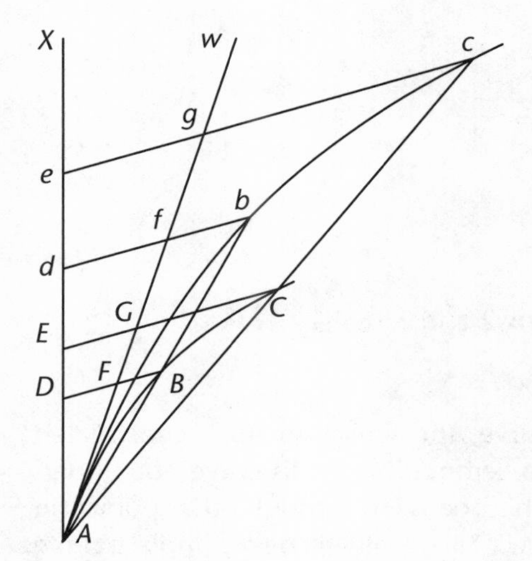

As in Lemma 7, we can conclude that the arc Ac will ultimately coincide with straight line Ag, and therefore curvilinear area Ace will ultimately coincide with the rectilinear area Age.

$$\triangle \widehat{Ace} \overset{ult}{=} \triangle Age. \qquad (6)$$

Similarly, as angle bAf vanishes and line ABb coincides with line Af, curvilinear area $\widehat{Abd}$ will ultimately coincide with rectilinear area Afd.

$$\triangle \widehat{Abd} \overset{ult}{=} \triangle Afd. \qquad (6')$$

"...and therefore (by Lemma 5) they will be in the duplicate ratio of the sides Ad, Ae."

In $\triangle Age$, line $df \parallel$ base eg.

By Euclid VI.2, VI.5, and VI Definition 1, this creates similar triangles.

$$\triangle Age \sim \triangle Afd.$$

By Euclid VI.19, or, as Newton has generalized it in Lemma 5 to include both rectilinear and curvilinear similar figures, the ratio of the areas of similar triangles is the duplicate ratio of their corresponding sides. Therefore:

$$\triangle Afd : \triangle Age :: dup(Ad : Ae). \qquad (7)$$

In the quoted bit of sketch, "they" refers back to curvilinear areas Abd and Ace which were shown to coincide ultimately with rectilinear areas Afd and Age respectively.

Substituting from Equations 6 and 6′ into Equation 7,

$$\triangle \widehat{Abd} : \triangle \widehat{Ace} \overset{ult}{::} dup(Ad : Ae) \qquad (8)$$

"But to those areas the areas ABD, ACE, are always proportional, and to those sides the sides AD, AE [are always proportional]. Therefore, the areas ABD, ACE, also, are ultimately in the duplicate ratio of the sides AD, AE.

Q.E.D."

("Those areas" is still referring back to curvilinear areas Abd and Ace.)

In $\triangle Abd$, line $BD \parallel$ base bd.

By Euclid VI.2, VI.5, and VI Definition 1,

$$\triangle ABD \sim \triangle Abd;$$

$$AD : Ad :: DB : db. \quad (9)$$

Now arc $\widehat{AB}$ is similar to arc $\widehat{Ab}$, and they are in the same ratio as lines AD and Ad (see Equation 2 and the construction of the similar curve); that is,

$$\widehat{AB} : \widehat{Ab} :: AD : Ad.$$

And sides DB and db are in the same ratio by Equation 9.

Thus in curvilinear triangles $\triangle ABD$ and $\triangle Abd$ we have three corresponding sides respectively proportional, with the curvilinear sides similar. This gives us a strong plausibility that the triangles are similar, with the areas of the triangles in the duplicate ratio of those sides. Unfortunately, we don't have a Euclidean proposition to prove this for us, nor does *Principia* give us a Newtonian one. Accepting the plausibility in lieu of explicit proof, we conclude:

$$\triangle \widehat{AB}D \sim \triangle \widehat{Ab}d.$$

Again, by Euclid VI.2, VI.5, and VI Definition 1,

$$\triangle ACE \sim \triangle Ace,$$

$$AE : Ae :: EC : ec.$$

We conclude, as before,

$$\triangle \widehat{AC}E \sim \triangle \widehat{Ac}e.$$

By the proportionality of the areas, we conclude:

$$\triangle \widehat{Ab}d : \triangle \widehat{Ac}e :: \triangle \widehat{AB}D : \triangle \widehat{AC}E. \quad (10)$$

Combining Equations 8 and 10,

$$\triangle \widehat{AB}D : \triangle \widehat{AC}E \overset{ult}{::} dup(Ad : Ae).$$

Substituting from Equation 1,

$$\triangle \widehat{AB}D : \triangle \widehat{AC}E \overset{ult}{::} dup(AD : AE).$$

Q.E.D.

Lemma 10

The spaces which a body describes when any finite force whatever urges [it], whether that force be determinate and immutable, or whether it be continually increased or continually decreased, are at the very beginning of the motion in the duplicate ratio of the times.

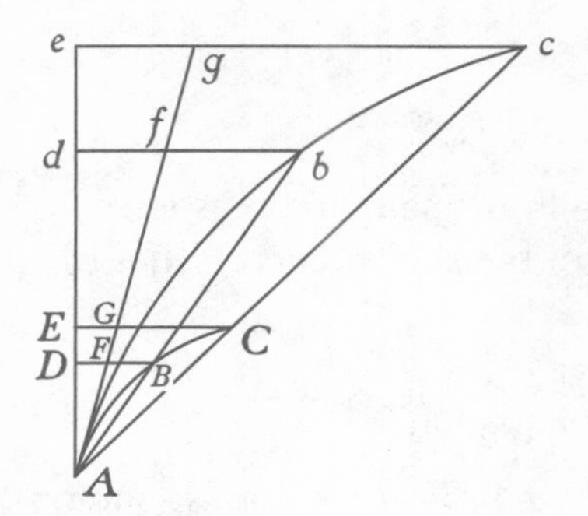

Let the times be displayed by the lines AD, AE, and the generated velocities by the ordinates DB, EC; and the spaces described by these velocities will be as the areas ABD, ACE described by these ordinates, that is, at the very beginning of the motion (by Lemma 9), in the duplicate ratio of the times AD, AE.

Q.E.D.

[**Note on Newton's diagram:** This diagram was given with Lemma 9 and is evidently intended to serve for Lemma 10 as well.]

Notes on Lemma 10

- **Physical Application:** Lemma 10 applies Lemma 9 to bodies and forces.

- The curves associated with the two times are $\widehat{AB}$ and $\widehat{AC}$. A too-hasty glance at the enunciation and diagram might suggest that the curves *ABC* and *Abc* are the paths of the body over the two times. $\widehat{Abc}$ is the magnification given in the proof of Lemma 9; it not used in this proof.

- **A graph:** Nor does curve *ABC* represent the path of the body over the two times; the lengths of the arcs *AB* and *AC* are not the distances traveled. Rather, this diagram is a graph, plotting velocity against time.

  "Ordinates" *DB* and *EC*, whose lengths represent velocity, are drawn, parallel to each other, from points on the time axis *ADE* to the curve. They define curvilinear areas $\widehat{ABD}$, $\widehat{ACE}$ between the curve on one side and the length of time from the origin *A* to the points *B* and *C* where the ordinates intersect the time axis on the other. We are taking the area "under" the curve (in this diagram, it is to the left of the curve). These areas represent the distances the body travels in those times.

- **"Spaces":** In this lemma and its corollaries, distances (which Newton sometimes calls "spaces" as usual in his time) are represented geometrically by areas. It may take some extra attention to remember that we are not talking about areas but line distances, especially if the word "spaces" itself makes us think of areas.

- Lemma 10 applies to one body moving under a single force and compares the distances traveled in two different times from the same beginning point. The single force could be changing, but we look at the motion just as it comes into being, "at the very beginning of motion." At the limit, the point from which both motions emerge, the force would be the same. But we are not interested in the point at which distance, force, and time are equal, but the ratio of the distances and the times just after they have come into being.

• **Nascents: Questions for discussion:** Quantities are nascent when movement is just beginning. The earlier lemmas derived some mathematical tools for ratios that approach a limit as the quantities vanish or "evanesce." Here, finally, are the nascents mentioned in the excerpts from the Scholium.

How are nascents and evanescents different and the same? Mathematically, they appear to be the same. Physically, they are the same only if we imagine time running backwards.

If we are to use the tools developed for evanescents for real-world cases when movement is just beginning and the quantities are "nascent," we must feel satisfied that that physical application is legitimate. How would you justify this?

We looked at several examples of processes that approach a limiting ratio as the quantities change over time. But what does it mean for quantities to approach a limiting ratio as they *come into being*? One might want to talk about something "departing," but we aren't *approaching* a limit but at best moving away from one. And perhaps we remember all the difficulties Galileo's friends get into trying to think about naturally accelerated motion starting from rest.*

Expansion of Newton's Sketch of Lemma 10

Given:

"The spaces which a body describes when any finite force whatever urges [it], whether that force be determinate and immutable, or whether it be continually increased or continually decreased ..."

1. A body describes distances on a path along which a finite force acts;

2. The body is urged by this force, which may be constant or increasing or decreasing.

To Prove:

"[These spaces] are at the very beginning of the motion in the duplicate ratio of the times."

The distances these bodies describe along the path vary in the duplicate ratio of the times if we look at the very beginning of motion.

$d \overset{ult}{\propto} t^2$.

---

* Galileo, *Two New Sciences,* Third Day.

Proof:

"Let the times be displayed by the lines AD, AE, and the generated velocities by the ordinates DB, EC; and the spaces described by these velocities will be as the areas ABD, ACE described by these ordinates…"

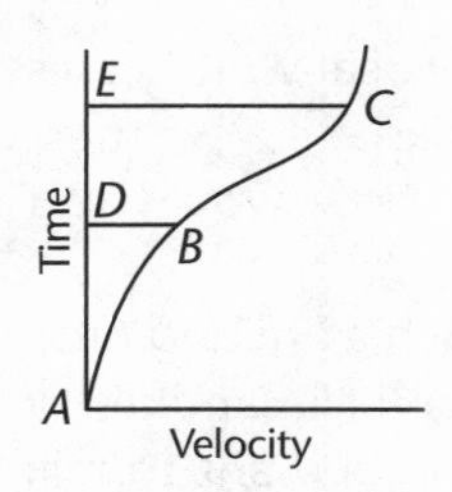

Suppose the velocity of a body to be a function of time.

The area under the curve will be the distance traveled ("the spaces which a body describes"). In equable motion, distances vary as the speeds compounded with the times.* If velocity is changing, we may use the method of Lemma 2 to find the area under the curve by summing inscribed rectangles, each of which will have as its area speed times time.

Thus the curvilinear area $\widehat{ABD}$ is the distance covered in time AD, and the curvilinear area $\widehat{ACE}$ is the distance covered in time AE.

"…that is, at the very beginning of the motion (by Lemma 9), in the duplicate ratio of the times AD, AE.

Q.E.D."

We will now invoke Lemma 9, which was about evanescent quantities, in this lemma, which is about nascent quantities. Did you satisfy yourself, in pondering the questions in the bulleted note entitled Nascents, that we may do that?

By Lemma 9, as $B \rightarrow A$ and $C \rightarrow A$, the areas $\widehat{ABD}$ and $\widehat{ACE}$ will be to each other in the duplicate ratio of sides AD and AE.

Therefore the spaces described at the beginning of motion will be in the duplicate ratio of the times.

Q.E.D.

Notes on Lemma 10 Corollaries

• The lemma derives something about one body moving under a single force. It compares the distances traveled in two different times from the same beginning point. The corollaries expand and rearrange the conclusions of the lemma (and the previous corollaries) to cover other situations. They allow us to compare distances traveled by different bodies under the same, proportional, and different forces, first in similar parts of similar figures and then without restriction. They then rearrange to give forces in terms of the distances and times, and times in terms of distances and forces.

---

* Galileo, *Two New Sciences,* fourth proposition of equable motion.

The first two corollaries deal with a special case, namely where two bodies are moving under the same or equal forces (Cor. 1) or proportional forces (Cor. 2) and describe similar parts of similar figures. This case is not used in Newton's central argument and where the corollaries are used they are applicable only by making an approximation. For these reasons the first two corollaries will be stated without expanded proof.

The remaining three corollaries are general and Cor. 4 is used explicitly in I.6 and implicitly in III.4. Both of these propositions are elements of Newton's central argument. The applications in the central argument are general and do not involve similar parts of similar figures. Because of this generality, Newton could not have used the first two corollaries there.

The need for generality in the application has implications for how we construct the proofs of the later corollaries. If we make their proofs depend on the first two corollaries, they will carry the same condition of similar parts of similar figures.

As it happens, we can prove the first two corollaries without the condition of similar parts of similar figures. Such a proof was included in the previous editions of this guidebook. It covers the special case of similar parts of similar figures, but is not limited to that case. This approach simplified the proofs of the later corollaries by allowing us to derive them from the earlier ones without having to remove the pesky condition or to be limited in application. However, I have taken a different approach in this edition because my previous approach required us to ignore the fact that Newton specified the condition in the enunciations of the corollaries.

Because the special cases of the first two corollaries aren't used in the central argument, and because the way they build into the general case is problematic and any expanded proof conjectural, it seems best to prove the later corollaries directly, without reference to the earlier ones. This may indeed have been Newton's intention: to remove the condition from the general later corollaries by a proof that doesn't rely on the earlier ones.

- **The symbol** $\propto$: See Section 6.5 of the Preliminaries.

$d \propto t^2$ is a shorthand expression for the proportion that distances (at least two, put in ratio) are in the duplicate ratio of the corresponding times.

$d_1 : d_2 :: dup(t_1 : t_2)$ or, more algebraically,

$d_1 : d_2 :: t_1^2 : t_2^2$

Lemma 10 Corollary 1

And hence it is easily deduced that when bodies describe similar parts of similar figures in proportional times, the deviations that are generated by any equal forces you please similarly applied to the bodies, and that are measured by the distances of the bodies from those places of the similar figures to which the same bodies would have arrived in the same proportional times without those forces, are as the squares of the times in which they are generated, very nearly.

Lemma 10 Corollary 2

Further, the deviations which are generated by proportional forces similarly applied to similar parts of similar figures, are as the forces and the squares of the times conjointly.

Lemma 10 Corollary 3

The same is to be understood of any spaces you please which the bodies describe when different forces urge them. At the very beginning of the motion, these are as the forces and the squares of the times conjointly.

Expansion of Newton's Sketch of Corollary 3

Given:

"[A]ny spaces you please which the bodies describe when different forces urge them."

Bodies travel over some distances under the influence of different forces.

To Prove:

"At the very beginning of the motion, these are as the forces and the squares of the times conjointly."

$d \overset{ult}{\propto} t^2 \times F.$

Proof:

In the following diagram, PR represents time of travel along any distance d_1, RQ represents the velocity generated by the force, and thus the curvilinear area $\widehat{PQR}$ represents the distance d_1.

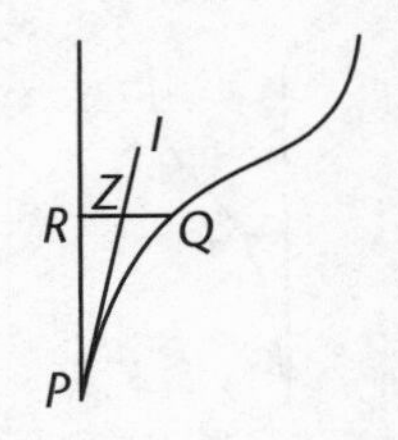

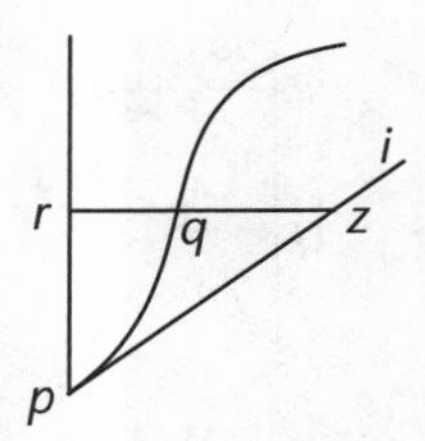

Likewise pr represents the time of travel along any other distance d_2, rq represents velocity, and curvilinear area $\widehat{pqr}$ represents the distance d_2.

Let us first consider equal times; let $PR = pr$.

Draw tangents PI, pi intersecting QR, qr (extended, if necessary) at Z, z.

By the demonstration of Lemma 9, the rectilinear areas and the curvilinear areas ultimately coincide. Therefore,

$$\text{area } \widehat{PQR} \overset{\text{ult}}{=} \triangle PRZ \quad \text{and} \quad \text{area } \widehat{pqr} \overset{\text{ult}}{=} \triangle prz. \tag{1}$$

and

$$QR \overset{\text{ult}}{=} RZ \text{ and } qr \overset{\text{ult}}{=} rz \tag{2}$$

But $\triangle PRZ$ and $\triangle prz$ are on equal heights (PR and pr, respectively); therefore their areas will be as their bases RZ and rz.

$$\triangle PRZ : \triangle prz :: RZ : rz. \quad \text{[Eu. VI.1]} \tag{3}$$

Substituting from Equations 1 and 2 into Equation 3:

$$\text{area } \widehat{PQR} : \text{area } \widehat{pqr} \overset{\text{ult}}{::} QR : qr.$$

But for equal times PR and pr, the velocities will be as the forces.

$$F_1 : F_2 :: QR : qr. \quad \text{[Def. 7]}$$

Therefore

$$\text{area } \widehat{PQR} : \text{area } \widehat{pqr} \overset{\text{ult}}{::} F_1 : F_2. \tag{4}$$

Now let the times be unequal, and suppose $PR > pr$. Let $PU = pr$ and draw $UV \parallel RQ$. Let the new area $\widehat{PVU}$ = old area $\widehat{PQR}$.

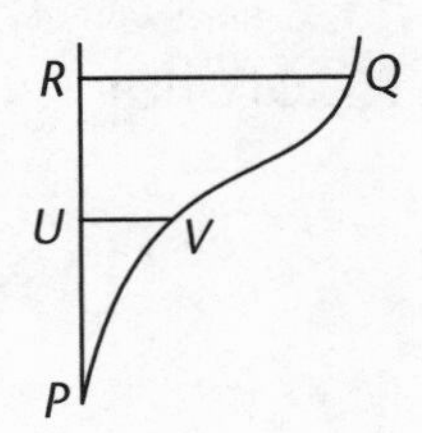

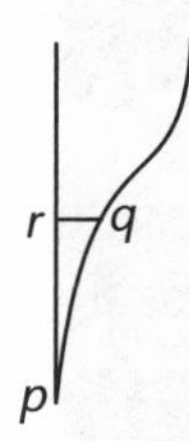

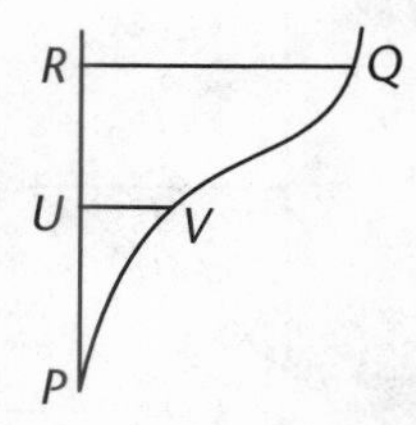

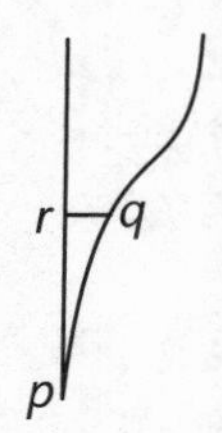

Rewriting Equation 4, area $\widehat{PVU}$: area $\widehat{pqr}$ :: $F_1 : F_2$. (5)

But by Lemma 9,

area $\widehat{PQR}$: area $\widehat{PVU}$ $\overset{ult}{::}$ $PR^2 : PU^2$.

But $PU = pr$, so

area $\widehat{PQR}$: area $\widehat{PVU}$ $\overset{ult}{::}$ $PR^2 : pr^2$. (6)

Compounding Equations 5 and 6 *ex aequali*,

area $\widehat{PQR}$: area $\widehat{pqr}$ $\overset{ult}{::}$ $PR^2 : pr^2$ comp $F_1 : F_2$.

But area $\widehat{PQR} = d_1$ and area $\widehat{pqr} = d_2$;

and $PR = t_1$ and $pr = t_2$.

So $d_1 : d_2 \overset{ult}{::} {t_1}^2 : {t_2}^2$ comp $F_1 : F_2$

or $d \overset{ult}{\propto} t^2 \times F$.

Thus the distances that the bodies describe when different forces urge them are at the beginning of motion as the forces and the squares of the times conjointly.

Q.E.D.

Lemma 10 Corollary 4

And for that reason, at the very beginning of the motion, the forces are as the spaces described directly and the squares of the times inversely.

Expansion of Newton's Sketch of Corollary 4

By the previous corollary,

$$d \overset{ult}{\propto} t^2 \times F;$$

$$F \overset{ult}{\propto} \frac{d}{t^2}.$$

Q.E.D.

Lemma 10 Corollary 5

And the squares of the times are as the spaces described directly and the forces inversely.

Expansion of Newton's Sketch of Corollary 5

By the previous corollary,

$$F \overset{ult}{\propto} \frac{d}{t^2}.$$

Therefore

$$t^2 \overset{ult}{\propto} \frac{d}{F}.$$

Q.E.D.

Scholium

If indeterminate quantities of different kinds are compared among themselves, and one of them is said to be as any other you please directly or inversely, the meaning is that the former is increased or decreased in the same ratio as the latter, or as its reciprocal. And if one of them is said to be as two or more others directly or inversely, the meaning is that the first is increased or decreased in the ratio compounded of the ratios in which the others, or the others' reciprocals, are increased or decreased. Thus, if A be said to be as B directly and C directly and D inversely, the meaning is that A is increased or decreased in the same ratio as $B \times C \times \frac{1}{D}$; that is, that A and $\frac{BC}{D}$ are to each other in a given ratio.

Notes on Scholium After Lemma 10

- **Compounding ratios:** This methodology of operating on proportions is discussed in Section 6.3 of the Preliminaries.

"And if one of them is said to be as two or more others directly or inversely, the meaning is that the first is increased or decreased in the ratio compounded of the ratios in which the others, or the others' reciprocals, are increased or decreased. Thus, if A be said to be as B directly and C directly and D inversely, the meaning is that A is increased or decreased in the same ratio as $B \times C \times \frac{1}{D}$ ..."

Newton states here that what he means by compounding ratios is multiplication.

The way Newton would expand this example into a proper proportion would be $A : a :: (B \times C \times {}^{1}/_{D}) : (b \times c \times {}^{1}/_{d})$, where the capital letters are the magnitudes at one time and the small letters are the magnitudes at another time. This is valid whether the quantities A and $(B \times C \times {}^{1}/_{D})$ are both finite, or both evanescent, or if one is finite and the other evanescent.

- **Placing finite and evanescent quantities into ratio:**

"...that is, that A and $\frac{BC}{D}$ are to each other in a given ratio."

This is valid if the quantities A and $\frac{BC}{D}$ are both finite or (with care, see Pause Before Lemma 7) if both are evanescent. But if they are mixed, they cannot be placed in ratio. It won't be a proper "relationship in respect of size" (see definition of ratio in Preliminaries Section 6.1) if one quantity is infinitely larger than the other.

Newton understands perfectly well how these quantities can be used and would never set mixed types into relationship. Some of his readers over the centuries have, however, made this mistake, including some scholars who should know better. Take heed, and you won't be one of them.

Lemma 11

The evanescent subtense of the angle of contact, in all curves having a finite curvature at the point of contact, is ultimately in the duplicate ratio of the subtense of the conterminous arc.

Case 1. Let that arc be AB, its tangent AD, the subtense of the angle of contact perpendicular to the tangent BD, subtense of the arc AB. Let AG, BG be erected perpendicular to the latter subtense AB and to the tangent AD, meeting at G, and then let the points D, B, G, come to the points d, b, g, and let J be the intersection of the lines BG, AG, occurring ultimately, where the points D, B, come all the way to A. It is obvious that the distance GJ can be less than any assigned [magnitude] you please. Now, by the nature of circles passing through the points ABG, Abg, square AB is equal to $AG \times BD$, and square Ab is equal to $Ag \times bd$, and therefore the ratio of square AB to square Ab is compounded of the ratios AG to Ag and BD to bd. But seeing that GJ can be taken to be less than any assigned length you please, it can happen that the ratio AG to Ag differ from the ratio of equality by less than any assigned difference you please, and therefore, that the ratio square AB to square Ab differ from the ratio BD to bd by less than any assigned difference you please.

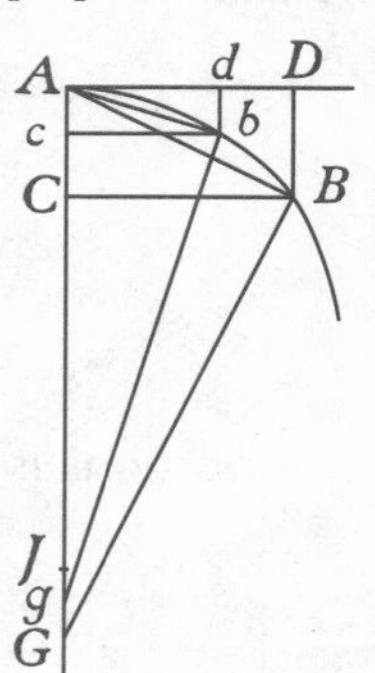

Therefore, by Lemma 1, the ultimate ratio of square AB to square Ab is the same as the ultimate ratio of BD to bd.

Q.E.D.

Case 2. Now let BD be inclined to AD at any given angle you please, and the ratio BD to bd will always be the same as before, and therefore will be the same as square AB to square Ab.

Q.E.D.

Case 3. And even if the angle D not be given, but that the straight line BD converge toward some given point, or be established by any other law whatever, nevertheless, the angles D, d, being established by the law in common, will always converge to equality and will approach nearer to each other than by any assigned difference you please, and therefore will ultimately be equal, by Lemma 1; and on that account the lines BD, bd are in the same ratio to each other as before.

Q.E.D.

Notes on Lemma 11

• **Finite Continuous Curvature:** Continuous curvature is discussed in the notes to Lemma 6.

Finite curvature means that the curvature at the point in question (call it point *A*) is the same as that of a circle with a finite radius. This circle is the "circle of curvature" at point *A*.

If a circle can be constructed through *A* that has the same curvature at *A* as the given curve and that shares with the given curve a common tangent *AD*, we will say that the given curve has finite continuous curvature at *A*. The construction of this circle of curvature is given in this lemma.

• **Point of Contact:** *A* is a "point of contact" between the curve and tangent *AD*.

• **Angle of Contact:** The curvilinear angle $D\widehat{AB}$ between the tangent and the arc is the "angle of contact" of the tangent.

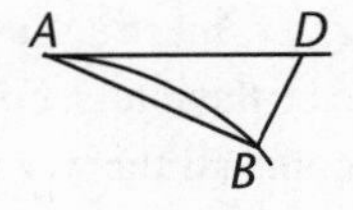

• **Subtense of the Angle of Contact:** *BD* is the "subtense of the angle of contact." It extends from point *B* on the curve to meet the tangent at some point *D*, making some angle *BDA*. As $B \rightarrow A$, *D* is also supposed to move towards *A* according to some consistent method of selection of *D*. Thus as $B \rightarrow A$, subtense *BD* evanesces.

• **Subtense of the Conterminous Arc:** Chord *AB* is the "subtense of the conterminous arc," that is, the arc having the same terminal points *A* and *B*.

Expansion of Newton's Sketch of Lemma 11

Given:

"[I]n all curves having a finite curvature at the point of contact..."

Finite continuous curvature at the point of tangency A is given.

To Prove:

"The evanescent subtense of the angle of contact ... is ultimately in the duplicate ratio of the subtense of the conterminous arc."

$BD \overset{ult}{\propto} dup\ AB$ or, more algebraically,

$BD \overset{ult}{\propto} \overline{AB}^2$.

Proof:

Case 1

"*Case 1.* Let that arc be AB, its tangent AD, the subtense of the angle of contact perpendicular to the tangent BD, subtense of the arc AB."

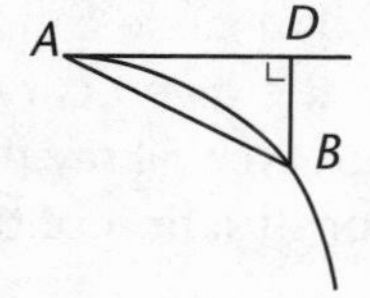

For this first case, let the subtense of the angle of contact be perpendicular to the tangent.

$BD \perp AD$

AB is the subtense of the arc, or the chord.

"Let AG, BG be erected perpendicular to the latter subtense AB and to the tangent AD, meeting at G, and then let the points D, B, G, come to the points d, b, g, and let J be the intersection of the lines BG, AG, occurring ultimately, where the points D, B, come all the way to A."

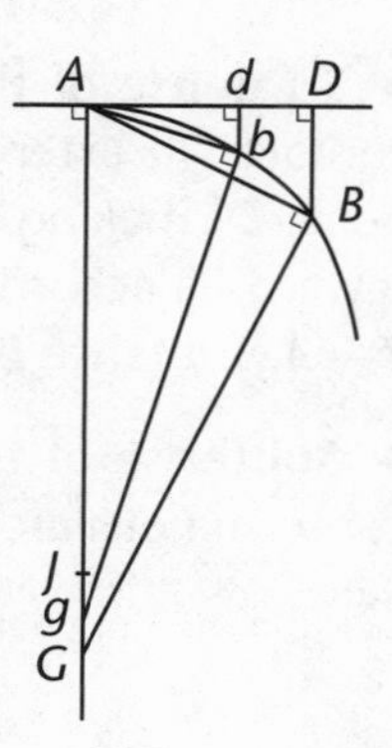

Draw line $BG \perp AB$.

Draw line $AG \perp AD$.

Let points D and B move towards A. We will have new points d, b, g.

Bisect AG at Q [Eu. I.10]. Construct a circle with center Q and radius AQ [Eu. Post. 3]. This circle will pass through B because of the right angle at B [Eu. III.31].

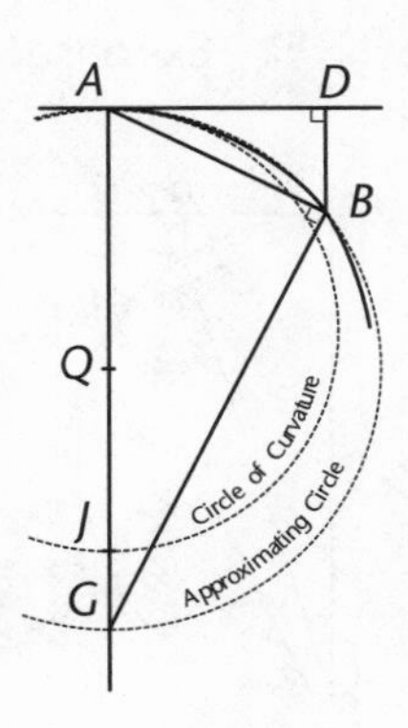

Let's call this circle which passes through *A*, *B*, and *G* "the approximating circle." At the limit, as *B* arrives at *A*, it will be the "circle of curvature" at *A*. We have already made the assumption, or hypothesis, that there is such a circle for *A* in our given condition that the curve has finite continuous curvature in the immediate vicinity of *A* (see bulleted note above entitled Finite Continuous Curvature).

The circle of curvature at *A* will have a finite diameter, by our given condition that the curvature at *A* is finite. Let that finite diameter be *AJ*.

J will be the ultimate intersection of *BG* and *AG*, as *B* and *D* arrive at *A*.

That is, we assume that, as $B \rightarrow A$, *G* approaches a limit *J*.

"It is obvious that the distance GJ can be less than any assigned [magnitude] you please."

As *B* starts moving towards *A*, some busy curves might have approximating circles through *A* and *B* whose diameters *AG* get larger and smaller. However, as *B* gets very close to *A*, either *G* comes smoothly to *J* or it must jump to it. Since this is a well-behaved curve of continuous curvature at *A*, we may assume that it does not jump, but as *B* gets very close to *A* the curvature comes to change in a continuous way moving towards the curvature at *A*.

This would mean that, near *A*, *G* moves continuously towards *J*, and thus as *B* gets as close as you please to *A*, then *G* will get as close as you please to *J*.

"Now, by the nature of circles passing through the points ABG, Abg, square AB is equal to $AG \times BD$, and square Ab is equal to $Ag \times bd$, and therefore the ratio of square AB to square Ab is compounded of the ratios AG to Ag and BD to bd."

Lemmita: We need an extra little lemma here, not explicitly given by Euclid but easily proved.

In any circle the chord is a mean proportional between the diameter and the perpendicular from one end of the chord to the tangent from the other end of the chord.

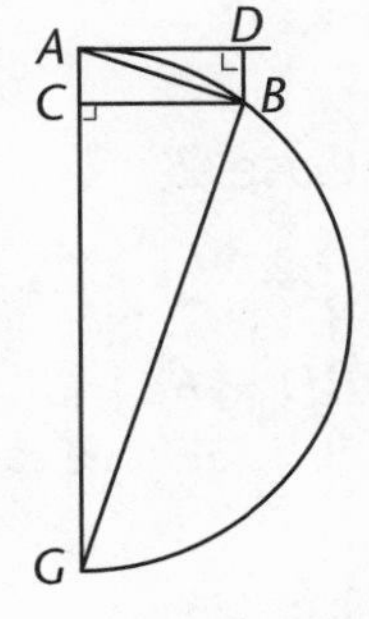

Let the chord be *AB*, the diameter *AG*. We are given that angle *ABG* is an angle in a semicircle; *AD* is a tangent at *A*; $BD \perp AD$.

We must prove that

$AG : AB :: AB : DB.$

Exercise: Can you write out the proof?

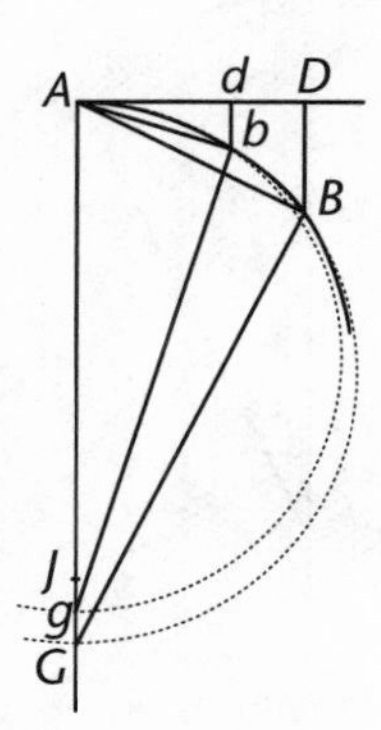

Consider the approximating circles ABG passing through A and B and Abg passing through A and b.

Applying the conclusion of the lemmita you just proved to both approximating circles,

$$AG : AB :: AB : DB,$$
$$Ag : Ab :: Ab : db.$$

Since these three straight lines are proportional, we can use Euclid VI.17 to say:

$$AB^2 = AG \times DB, \quad \text{and} \quad Ab^2 = Ag \times db.$$

Dividing equals by equals,

$$\frac{AB^2}{Ab^2} = \frac{AG \times DB}{Ag \times db}.$$

"But seeing that GJ can be taken to be less than any assigned length you please, it can happen that the ratio AG to Ag differ from the ratio of equality by less than any assigned difference you please, and therefore, that the ratio square AB to square Ab differ from the ratio BD to bd by less than any assigned difference you please. Therefore, by Lemma 1, the ultimate ratio of square AB to square Ab is the same as the ultimate ratio of BD to bd.

Q.E.D."

But since G and g will get as close as we please to J (see expansion two steps back), the ratio $AG : Ag$ will get as close as we please to unity, and will, in Newton's terms, ultimately be a ratio of equality.

Therefore as $(AG : Ag) \to 1$,

$$AB^2 : Ab^2 \overset{ult}{::} DB : db.$$

Q.E.D.

Case 2

"*Case 2*. Now let BD be inclined to AD at any given angle you please, and the ratio BD to bd will always be the same as before, and therefore will be the same as square AB to square Ab.

Q.E.D."

Let BD be inclined to AD in any given angle; that is, angle Adb = angle ADB, even though they need not be right angles.

From b and B drop perpendiculars to line AD. Let them intesect at d' and D'. Thus d' and D' will correspond to d and D in Case 1.

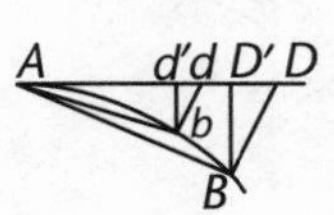

Because they have two angles respectively equal,

$$\triangle dbd' \sim \triangle DBD'.$$

Therefore by Euclid VI.4,

$$db : DB :: d'b : D'B. \tag{1}$$

By case 1, using the lettering of this case,

$$AB^2 : Ab^2 \overset{ult}{::} D'B : d'b.$$

Substituting from Equation 1,

$$AB^2 : Ab^2 \overset{ult}{::} DB : db.$$

Q.E.D.

Case 3

"*Case 3.* And even if the angle D not be given, but that the straight line BD converge toward some given point, or be established by any other law whatever, nevertheless, the angles D, d, being established by the law in common, will always converge to equality and will approach nearer to each other than by any assigned difference you please, and therefore will ultimately be equal, by Lemma 1; and on that account the lines BD, bd are in the same ratio to each other as before.

Q.E.D."

Suppose that angle θ is not given; it could change as $B \rightarrow A$. We might want it to be determined by a line projected through B from some external point S.

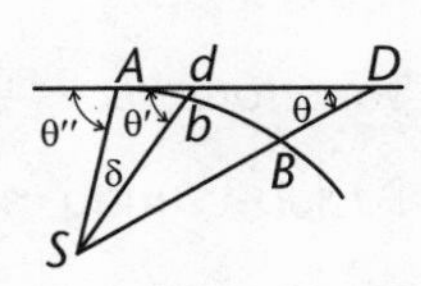

The ratio $db : DB$ is not constant, but the changing angles θ approach equality in the terms of Lemma 1.

$$\theta'' = \theta' + \delta \quad \text{[Eu. I.32]}$$

As $d \rightarrow A$, $\delta \rightarrow 0$, and therefore

$$\theta' \rightarrow \theta''.$$

Similarly, $\theta \rightarrow \theta''$.

The difference between θ and θ' gets as small as you please as D and d move to A, satisfying the conditions of Lemma 1.

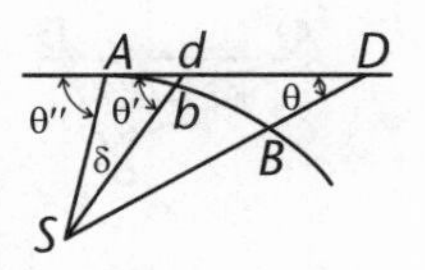

Since at the limit the angles are equal, DB ult|| db by Euclid I.28. Thus Case 3 reduces to Case 2. Therefore, at the limit, we may follow the reasoning of Case 2 to the conclusion

$$AB^2 : Ab^2 \overset{ult}{::} DB : db.$$

Q.E.D

Lemma 11 Corollary 1

Hence, since the tangents AD, Ad, the arcs AB, Ab, and their sines BC, bc, become ultimately equal to the chords AB, Ab, their squares will also be ultimately as the subtenses BD, bd.

Notes on Lemma 11 Corollary 1

- Line BC would be a trigonometric sine if the curve were a circle. See the note to Lemma 7 Corollary 1, where Newton introduces this line and proves it is ultimately equal to the chord, arc, and tangent.

- The conclusion of this corollary suggests that every curve, at any point of continuous finite curvature, acts like a parabola. Here's how.

This corollary proves that $BD \propto CB^2$.

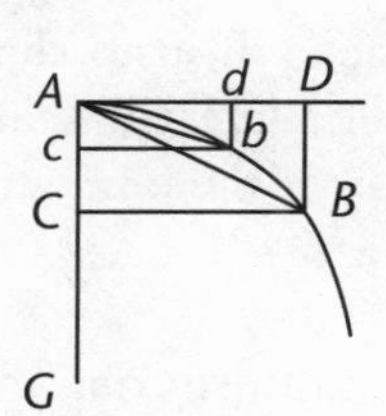

By the construction of Lemma 7 Corollary 1,

$BC \parallel AD$,

$BD \parallel AC$.

Therefore $BD = AC$. [Eu. I.34]

So substituting into the conclusion of this corollary,

$AC \propto CB^2$.

By Apollonius I.20, it is characteristic of a parabola that $AC \propto CB^2$.

Expansion of Newton's Sketch of Corollary 1

Lemma 7 showed that the tangents, arcs, and chords vanish with a ratio of equality. Lemma 7 Corollary 1 added the sines to that list by showing that they are ultimately in a ratio of equality with the tangent.

Since Lemma 11 has proved that the subtenses will vary as the squares of the chords, we can conclude that the subtenses will vary as the squares of the tangents, arcs, and sines as well. Therefore,

$$BD \propto AD^2 \tag{1}$$

$$BD \propto \widehat{AB}^2 \tag{2}$$

$$BD \propto BC^2. \tag{3}$$

Lemma 11 Corollary 2

The squares of the same are also ultimately as are the sagittae of the arcs, which bisect the chords and converge to a given point. For these sagittae are as the subtenses **BD**, *bd.*

Note on Lemma 11 Corollary 2

Sagittae (singular, *sagitta*) of arcs lie on lines which originate from a point on the concave side of the curve, and, passing through the points of bisection of the chords to those arcs, extend to reach the curve. The actual sagitta is the segment of this line lying between the chord and the arc. (The word *sagitta* is Latin for "arrow.")

Note that, unless the curve is a circle, with the point of origin the center of the circle, the sagitta will probably not be perpendicular to the chord. If it is perpendicular it has a special name: it is then called a *versed sine*.

When we use this corollary, the point of origin, or the "point towards which the sagittae converge," will be the center of centripetal force.

Expansion of Newton's Sketch of Corollary 2

We will prove that the sagittae are as the subtenses of the tangents.

By Lemma 11 Corollary 1, the tangent subtense varies ultimately as the square of the tangent:

$$EH : BD \overset{ult}{::} AH^2 : AD^2. \tag{1}$$

By the reasoning used in Case 3 of the lemma, as $D \rightarrow A$, angle AHS and angle ADS both approach angle ω. Lines SH and SD are therefore ultimately parallel to each other by Euclid I.28.

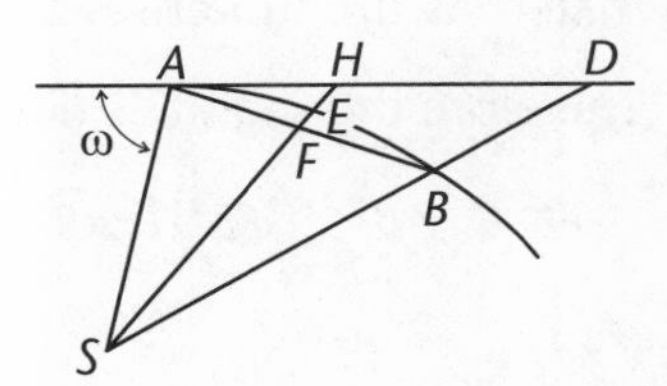

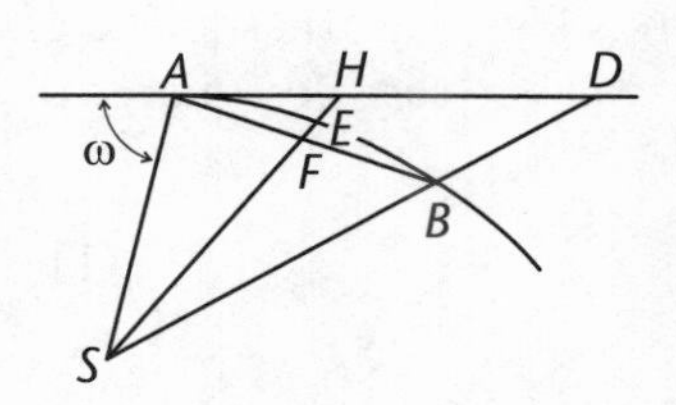

$FH \overset{ult}{\parallel} DB$.

$AF : AB \overset{ult}{::} AH : AD$. [Eu. VI.2]

Squaring, $AF^2 : AB^2 \overset{ult}{::} AH^2 : AD^2$.

Therefore, substituting into Equation 1,

$EH : BD \overset{ult}{::} AF^2 : AB^2$.

But the sagitta by definition bisects the chord, so $AF : AB :: 1 : 2$.

$$\frac{EH}{BD} \overset{ult}{=} \frac{AF^2}{AB^2} = \frac{1^2}{2^2} = \frac{1}{4}.$$

$$BD \overset{ult}{=} 4EH. \qquad (2)$$

By the same reasoning on other sides of the triangles,

$FH : BD \overset{ult}{::} AF : AB :: 1 : 2$

$FH : BD \overset{ult}{::} 1 : 2$

$$BD \overset{ult}{=} 2FH. \qquad (3)$$

Combining Equations 2 and 3:

$4EH \overset{ult}{=} 2FH$; so $FH \overset{ult}{=} 2EH$; therefore $FE \overset{ult}{=} EH$.

Rewriting Equation 2, $EH \overset{ult}{=} \frac{1}{4}BD$.

Therefore $FE \overset{ult}{=} \frac{1}{4}BD$.

Therefore the sagitta FE is ultimately a quarter of subtense BD. And the same will hold for all pairs of sagittae and subtenses of the angles of contact.

Therefore the sagitta ultimately varies as the subtense of the angle of contact.

$FE \overset{ult}{\propto} DB$.

Note: This conclusion, although given by Newton as a step on the way to the results of this corollary, is an important one and is called upon in its own right in later propositions.

By Corollary 1, the squares of the chords, arcs, tangents, and sines are all ultimately as the subtenses BD, bd.

Therefore the sagitta is ultimately as the square of any of those lines.

$FE \overset{ult}{\propto} AB^2$; $FE \overset{ult}{\propto} \widehat{AB}^2$; $FE \overset{ult}{\propto} AD^2$; $FE \overset{ult}{\propto} BC^2$.

Q.E.D.

Lemma 11 Corollary 3

Therefore, the sagitta is in the duplicate ratio of the time in which a given body describes the arc with a given velocity.

Expansion of Newton's Sketch of Corollary 3

Given:

Body travels along an arc with a constant velocity.

To Prove:

sagitta $\overset{ult}{\propto}$ time^2.

Proof:

By Corollary 2, the sagitta $\overset{ult}{\propto}$ arc^2.

By Galileo's first proposition of equable motion, when speeds are equal, the times are as the distances.

So the times the body takes to describe the arcs will vary as the lengths of the arcs.

Therefore sagitta $\overset{ult}{\propto}$ time^2.

Q.E.D.

[**Corollaries 4 and 5** are not invoked in any proposition of Newton's central sequence. They are therefore omitted from the selections in this guidebook.

Scholium after Lemma 11

For the rest, in all these [cases] we suppose that the angle of contact is neither infinitely greater nor infinitely less than the angles of contact which circles contain with their tangents; that is, that the curvature at point A is neither infinitely small nor infinitely great, or that the interval AJ is of finite magnitude. For DB can be taken as AD^3, in which case no circle can be drawn through the point A

between the tangent AD and the curve AB, and thus the angle of contact will be infinitely less than circular ones. And by a like argument, if DB be made successively as AD^4, AD^5, AD^6, AD^7, and so on, there will be obtained a series of angles of contact continuing to infinity, any later of which you please is infinitely less then a previous one. And if DB be made successively as AD^2, $AD^{3/2}$, $AD^{4/3}$, $AD^{5/4}$, $AD^{6/5}$, $AD^{7/6}$, and so on, there will be obtained another infinite series of angles of contact, of which the first is of the same kind as circular ones, the second infinitely greater, and any later one you please is infinitely greater than a previous one. But further, between any two of these angles you please, a series continuing to infinity on both sides can be inserted, of which any later one you please will be infinitely greater or less then a previous one. For example, between the terms AD^2 and AD^3 there may be inserted the series $AD^{13/6}$, $AD^{11/5}$, $AD^{9/4}$, $AD^{7/3}$, $AD^{5/2}$, $AD^{8/3}$, $AD^{11/4}$, $AD^{14/5}$, $AD^{17/6}$, and so on. And again, between any two angles you please of this latter series, there can be inserted a new series of intermediate angles differing from each other by infinite intervals. Nor does nature acknowledge a limit.

Notes on Scholium after Lemma 11

• This first paragraph of the scholium following the lemmas applies to Lemma 11. The remainder of the scholium applies to all the lemmas, and is presented following these notes.

• In Lemma 11 we made an assumption, or hypothesis, that the curve in question, at the point *A* in question, could have its curvature matched by a circle of finite radius.

Newton knows that this hypothesis will not be true of all curves at all points. But by calling upon this hypothesis in his given conditions, he explicitly limits the result of the lemma to points that meet this criterion. This paragraph of the scholium further draws attention to the particular condition, and reminds us that there are some points on some curves to which it will not apply.

In particular, he notes the following. First, all second degree curves, including all conic sections, will meet the condition. Conic sections are examples of curves for which we may freely invoke Lemma 11, and Newton does so in the propositions which follow. Second, we may not freely invoke Lemma 11 for curves whose equations contain variables raised to powers greater or less than two.

In the first example Newton gives, where DB varies as AD^3 (the so-called "cubic parabola"), the curve has continuous curvature everywhere,

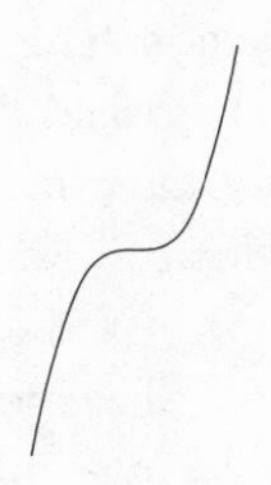

including at the origin (where its inflection point falls), and thus may have a tangent there by Lemma 6. But that curve does not have *finite* continuous curvature at the inflection point. The curvature there is what Newton calls "infinitely less" than the curvature of a circle, because at an inflection point the curve has zero curvature (like a straight line); it is less than that of a circle of any finite radius.

Simple power function curves which are not second degree (*DB* varying as *AD* to a power greater or less than two) will contain one point that does not meet the criterion: the point at the origin. All other points on these simple power function curves will have a finite curvature. Should we wish to apply this lemma to them, we would need only to make sure that our point *A* did not fall at the origin.

There are many other curves that are not conic sections or second degree equations but that are well-behaved and continuous, with finite continuous curvature all the way around. These curves also meet the criterion of Lemma 11, namely that the curvature can be matched by a circle with finite radius at any point on the curve.

The equiangular spiral is an example of such a curve. Its equation ($r = k^{-\theta}$) is not the sort of power function being referred to here; rather than a variable raised to a fixed power other than two, it is a constant raised to a variable power. This equation does not create a curvature infinitely less or greater than a circle with finite radius. Newton will apply Lemma 11 to this spiral in Book I Proposition 9.

Scholium to the Lemmas (continued)

Those things that have been demonstrated concerning curved lines and enclosed surfaces are easily applied to the curved surfaces and contents of solids. I have, moreover, put these lemmas at the beginning in order to avoid the tedium of drawing long demonstrations out to an absurdity, in the manner of the ancient geometers. For the demonstrations are made more concise by the method of indivisibles. But because the hypothesis of indivisibles is harder, and that method is accordingly thought to be less geometrical, I have preferred to lead the demonstrations of the following material back to the ultimate sums and ratios of evanescent quantities, and to the first sums and ratios of nascent quantities; that is, to the limits of the sums and ratios; and therefore to place at the beginning the demonstrations of those limits as briefly as I could. For the same things are established thereby as are established by the method of indivisibles, and we shall use the principles more safely now that they have been demonstrated. Further, in what follows, if at any time I should consider quantities as if they consisted of small fixed parts, or if I should take small curved lines for straight ones, I do not wish these to be understood as indivisibles, but always as divisible evanescents; not as the sums and ratios of determinate parts, but always as the limits of sums and ratios; and I wish the force of such demonstrations always to be referred back to the method of the preceding lemmas.

There is the objection that there is no ultimate ratio of evanescent quantities, for before they vanish it is not the ultimate, and when they vanish there is no ratio. But it could also likewise be urged by the same argument that there is no ultimate velocity of a body arriving at a certain place where the motion ceases, for the velocity before the body reaches the place is not the ultimate, and where it reaches the place, it has no velocity. And the reply is easy: by the ultimate velocity I understand that by which a body is moved, neither before it reaches the ultimate place and the motion stops, nor afterward, but just then when it reaches it; that is, that very velocity with which the body reaches the ultimate place and with which the motion ceases. And likewise, by the ultimate ratio of evanescent quantities is to be understood the ratio of quantities, not before they vanish, not afterward, but [that] with which they vanish. Likewise also, the first ratio of nascent [quantities] is the ratio with which they come into being. And the first and ultimate sum is that with which they begin and cease to be (or to increase or to decrease). There exists a limit which the velocity can reach, but not go beyond, at the end of the motion. This is the ultimate velocity. And the account of the limit of all quantities and proportions that are beginning and ceasing to be, is the same. And since this limit is certain and definite, it is really a geometrical problem to determine it. But it is legitimate to make use of every geometrical [means] to determine and demonstrate other geometrical [matters].

It can also be urged that if the ultimate ratios of evanescent quantities be given, the ultimate magnitudes will also be given, and thus every quantity will consist of indivisibles, contrary to what Euclid has demonstrated concerning incommensurables in the tenth book of the *Elements*. However, this objection rests upon a false hypothesis. Those ultimate ratios with which quantities vanish are not truly ratios of ultimate quantities, but limits to which the ratios of quantities decreasing without limit always approach, and which they can attain to more nearly than by any given difference you please, but never go beyond, nor arrive at before the quantities are diminished *in infinitum*. The point will be more clearly understood in infinitely great [magnitudes]. If two quantities whose difference is given be increased *in infinitum*, their ultimate ratio will be given, namely, the ratio of equality, but this does not mean that the ultimate or maximum quantities of which that is the ratio will be given. Accordingly, in what follows, whenever out of concern for the easy conception of the subject I shall speak of quantities [that are] least, or evanescent, or ultimate, beware of understanding [them to be] quantities determinate in magnitude, but think of them as always to be diminished without limit.

Note on Remainder of Scholium to the Lemmas

The section Pause Before Lemma 7 contains notes on this latter part of the scholium.

SECTION 2

On the Finding of Centripetal Forces

Proposition 1

The areas which bodies driven in orbits [gyros] *describe by radii drawn to an immobile center of forces, are contained in immobile planes and are proportional to the times.*

Let the time be divided into equal parts, and in the first part of the time let the body, by its inherent force, describe the straight line AB. In the second part of the time, the same body, if nothing were to impede it, would pass on by means of a straight line to c (by Law 1), describing the line Bc equal to AB, with the result that, radii AS, BS, cS being drawn to the center, the areas ASB, BSc would come out equal. But when the body comes to B, let the centripetal force act with an impulse that is single but great, and let it have the effect of making the body depart from the straight line Bc and - continue in the straight line BC. Let cC be drawn parallel to BS, meeting BC at C; and, the second part of the time being completed, the body (by Corollary 1 of the Laws) will be located at C, in the same plane as the triangle ASB.

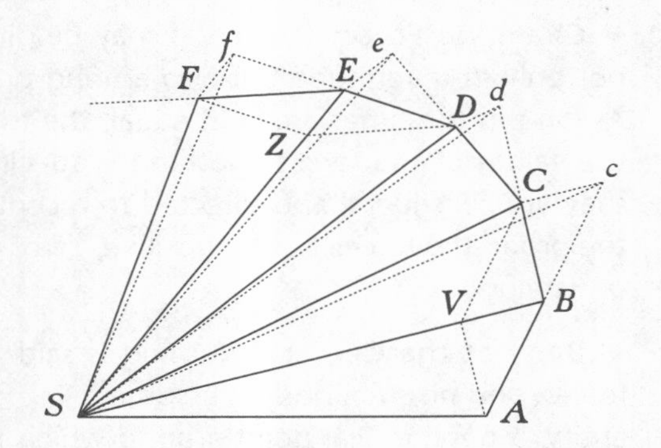

Connect SC, and, because of the parallels SB, Cc, triangle SBC will be equal to triangle SBc, and therefore also to triangle SAB. By a similar argument if the centripetal force should act successively at C, D, E, and so on, making the body describe the individual straight lines CD, DE, EF, and so on, in the individual particles of time, all these will lie in the same plane, and the triangle SCD will be equal to the triangle SBC, and SDE to SCD, and SEF to SDE. Therefore, in equal times equal areas are described in a motionless plane; and, *componendo*, any sums whatever of areas $SADS$, $SAFS$ are to one another as are the times of description. Now let the number of the triangles be increased and their breadth decreased *in infinitum*, and their ultimate perimeter ADF (by Corollary Four of Lemma Three) will be a curved line: and therefore the centripetal force, by which the body is perpetually drawn back from the tangent of this curve, will act without ceasing, while any described areas whatever $SADS$, $SAFS$, always proportional to the times of description, will be proportional to those same times in this case.

Q.E.D.

Notes on Book One Proposition 1

One stands in awe of the amazing power of this simple proposition, on which rests all of Newton's celestial mechanics, and indirectly his proofs for universal gravitation. Consider the following aspects of the range of this proposition's applicability.

• **Any Force Law:** This proposition will apply with any force law. Beware of assuming now or later that this somehow depends on an inverse square force law. The forces could be completely independent of the distance of the body (as well as of any other property of the body). Or, if they depend on distance, they could be directly as the distance to some power. They could also vary with time since the creation of the universe or depend on the movement of sunspots or the population of tent caterpillars.

• **Changing Force:** The forces may be different from moment to moment not only in magnitude but also among positive, negative, and zero values. As long as the times are held equal, the triangles will be equal regardless of the values of the forces. Nothing is specified about the forces except that they are centripetal and directed to a center that is immobile. As shown in the proof, the forces can be positive, zero, or negative; they can be constant or varying.

• **Body at the Center:** Nothing is said about any body at the center of forces, and no such body is assumed. The center of forces is a mathematical entity, a geometrical point around which the equal areas may be found. It will be a key discovery in the development of Newton's theory of universal gravitation that in certain circumstances there *are* bodies at this geometrical center of forces.

We must be cautious here, as we always must as we work through this development, not to be assuming what we "know" as a consequence of this book's conclusions having become part of our current world view. Remember that as we go into *Principia* we understand gravity only as terrestrial heaviness, and what makes the heavenly bodies orbit is everybody's guess. (Section 3 of the Preliminaries gave what some of everybody's guesses were—it might be a good idea to review that to help you create or remain in the appropriate pre-Newtonian frame of mind.)

For more motivation on this, remember (or refer back to) the encouragements and warnings in Section 3 of the Preliminaries. Encourage yourself by thinking about the importance and excitement of discovering the true foundations for all our physical intuitions, and warn yourself about the frustrations to be encountered if we bring in our assumptions and then must deal with a Newton who seems to have been too stupid to see how obvious his proposition is, and so diabolically perverse as to insist on a long difficult proof for what we learned in third grade.

• **Center of Force, Force at the Center:** We might be inclined to think of the center as a center of *force* (singular) because we are imagining something there exerting that force. But Newton is presenting it as the point towards which all the forces around the path, the forces at each of the infinite numbers of points on the path, are directed, and is consistent in this enunciation, and in all the corollaries, in calling it "center of forces." This is not a consequence of his speaking of *bodies* (plural) in the enunciation, since in Corollaries 1–3 the body is singular.

We might even catch ourselves talking about "the force at the center." But Newton has not given us a force at the center, only a center towards which the forces are directed. On the other hand, during the course of the sketch, he does allow himself to speak about "the centripetal force" (singular) acting with an impulse, acting successively, and (at the limit) perpetually drawing back the body from the tangent to the curve.

The important thing is that none of Newton's proofs in any way assumes or depends upon the force being singular or being exerted from the center. This generality of the proofs has an important significance in the larger system of *Principia*. Newton says he does not know the cause of gravity, and that he contrives no hypotheses about it (see the General Scholium, page 488). Thus we don't know whether the force that turns the planets out of their tangential inertial path is a single force somehow pulling from the center, or many forces pushing from behind the planets, or something different yet. Because we don't know, we need our demonstrations to be general enough to allow for the different possibilities.

• **Orbit:** Although the word "orbit" may suggest a closed path to you, there is no assumption in the proof of this proposition that the path be closed.

• **No "ghost curve":** The bases of triangles in the proposition—the distances the body travels between impulses of force—do not circumscribe, and are not inscribed in, this ultimate curve, nor do they connect to it or follow it in any other way as a kind of "ghost curve." They are more like tangents to the curve-to-be than like chords; but they are not tangents, either, because the curve does not exist as long as the force is impulsive: it exists only at the limit. The points *A, B, C,*... of the finite case polygon do not necessarily fall anywhere on the ultimate curve, nor do any of the points on the sides such as *AB, BC, CD*.

• **Accelerative force.** Here and in general throughout Books I and II, when Newton says "force" he means accelerative quantity of force (Definition 7). It is a measure of force proportional to the velocity that is generated by that force in a given time. It is only in Book III that he will begin to work with motive quantity of force (Definition 8), which is measured by the motion generated in a given time, and thus involves mass.

Newton has a complex and difficult structure to build. He begins with the simpler case, one in which mass is ignored. Only when he has established what he can about accelerative force does he take the next step, which is bringing in the mass of the attracted body. This gives us another level of complication. Then, finally, he brings in the third level of complication, the mass of the attracting body. We may be grateful to him for building gradually in this way, since, had he not, what is already difficult would have been much more so. These second and third levels of complication are introduced in Book III.

Expansion of Newton's Sketch of I.1

"The areas which bodies driven in orbits [*gyros*] describe by radii drawn to an immobile center of forces, are contained in immobile planes and are proportional to the times."

Given:

Immobile center of forces.

To Prove:

1. The areas that bodies driven in orbits describe by radii drawn to that center of forces are proportional to the times;
2. the path of the body and the center of forces remain in the same plane.

Proof:

Part 1

Step 1: Equal Areas in Equal Times

"Let the time be divided into equal parts, and in the first part of the time let the body, by its inherent force, describe the straight line AB. In the second part of the time, the same body, if nothing were to impede it, would pass on by means of a straight line to c (by Law 1), describing the line Bc equal to AB, with the result that, radii AS, BS, cS being drawn to the center, the areas ASB, BSc would come out equal."

The First Law of Motion stated:

> Every body continues in its state of resting or of moving uniformly in a straight line, except insofar as it is driven by impressed forces to alter its state.

Suppose that in the first part of the time the body moves from A to B by its inherent force. That is, it has some velocity at A and continues with that same velocity in a straight line unless some added force changes it,

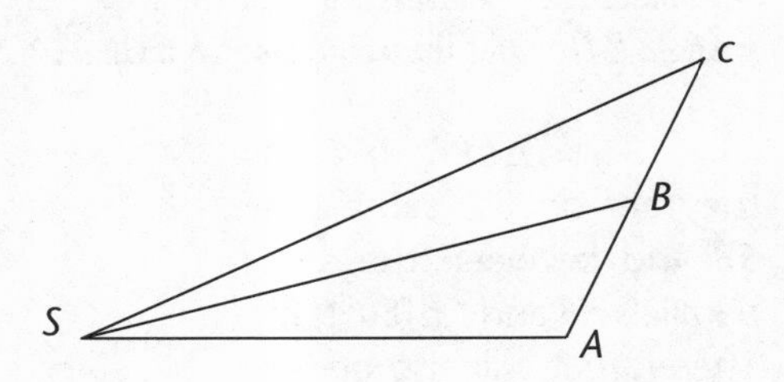

according to the First Law of Motion. From A to B we are assuming no external force is operating on the body.

In the second part of the time it would, again by Law 1, if not hindered, move directly in a straight line to c, where $Bc = AB$ (since the times are equal and no force has been impressed to change the velocity).

By Euclid I.38 (same height, equal bases), $\triangle SAB = \triangle SBc$. Therefore in this case, the situation or instance of a zero force, equal areas will be described by the radii in equal times.

"But when the body comes to B, let the centripetal force act with an impulse that is single but great, and let it have the effect of making the body depart from the straight line Bc and continue in the straight line BC. Let cC be drawn parallel to BS, meeting BC at C; and, the second part of the time being completed, the body (by Corollary 1 of the Laws) will be located at C, in the same plane as the triangle ASB."

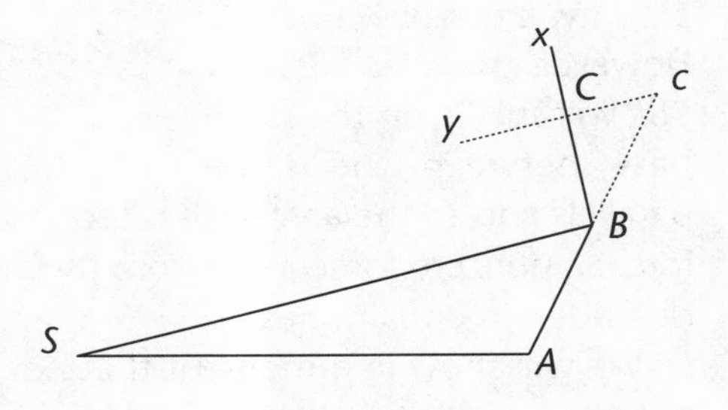

Now suppose a centripetal force acts on the body at B, turning it aside into line Bx. This force is to be understood as a single impulse of significant magnitude operating at that moment. That magnitude is measurable in the amount by which Bx is deflected from its default line Bc. To find that magnitude, draw cy from c parallel to SB, meeting Bx at C.

The centripetal force is impelling the body from B towards S. BS is its direction of force. Its magnitude is found by the actual deflection cC that brings the body, in a line parallel to BS, to the actual point C. C is reached in the same time that would have taken the body to c.

The times here for all these motions AB, Bc, BC, and so on, are equal; but the distances traveled AB, BC, and so on, are not necessarily equal.

We will call cC the effect of the force exerted as an impulse on the body at B for purposes of resolving the two forces.

BC is the resolution of the first force Bc (where its innate force is impelling it) and the second force cC (what centripetal force has accomplished in moving it to C instead of c). BC is the body's actual motion in the given time.

Note that although Bc is necessarily equal to AB, BC is not necessarily the same length as Bc. One must find C using the parallel to line BS.

"Connect SC, and, because of the parallels SB, Cc, triangle SBC will be equal to triangle SBc, and therefore also to triangle SAB."

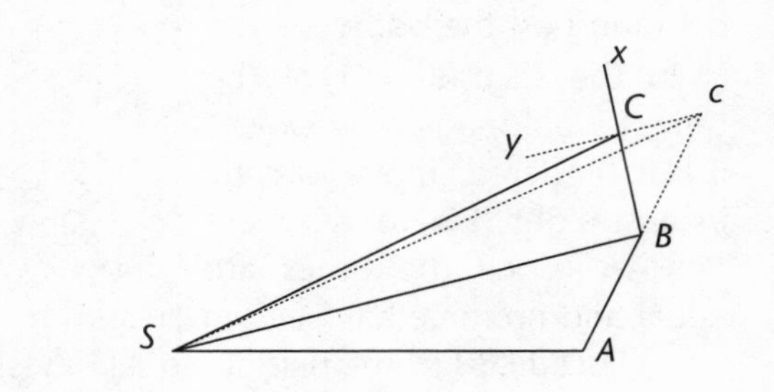

$\triangle SBc = \triangle SBC$ because they are on the same base SB and between the same parallels SB and Cc [Eu. I.37]. Therefore it will also be true that when we have a positive centripetal force, a nonzero force directed towards the center, equal areas will be described by the radii in equal times.

Suppose finally that we have a repulsive force directed away from the center of forces, a negative centripetal force. Bx will now lie on the other side of Bc, as will our new point C. However, triangles SBC and SBc will still lie on the same base between the same parallels and so the areas will be equal. Thus if we had a negative centripetal force, a nonzero force away from the center, we would also have equal areas described in equal times.

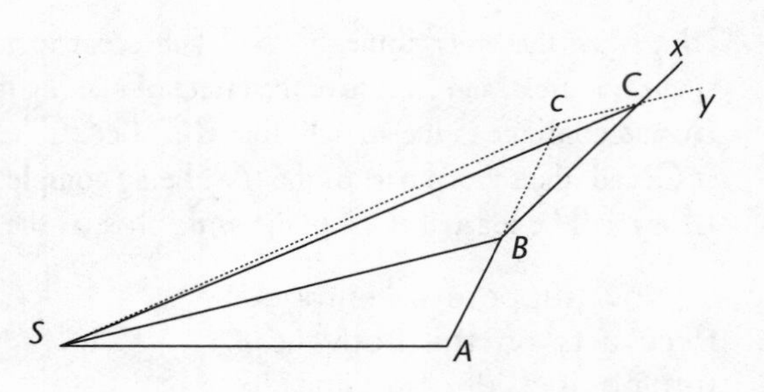

We will show in a moment that not only does C lie in the plane of $\triangle SAB$ but also the center of forces and all the triangles lie in one plane, as asserted in the enunciation.

"By a similar argument if the centripetal force should act successively at C, D, E, and so on, making the body describe the individual straight lines CD, DE, EF, and so on, in the individual particles of time, all these will lie in the same plane, and the triangle SCD will be equal to the triangle SBC, and SDE to SCD, and SEF to SDE. Therefore, in equal times equal areas are described in a motionless plane;..."

Now suppose this process to be repeated with always the same increments of time. In the equal-time increment (the "individual particle of time") the body will travel the resultant between the motion due to its inherent force and the motion it would have had under the influence of the centripetal force, in a way exactly analogous to our demonstration above. Thus the area of every triangle will remain equal, regardless of how the forces may vary.

Step 2: The Center of Forces and All the Triangles Lie in One Plane

Triangle *SAB* defines a plane, so those three points are in one plane [Eu. XI.2].

A body travelling in a straight line *ABc* will continue in the same plane, so line *Bc* is also in the "immobile plane" [Eu. XI.1].

Since *c* and *S* are both in the plane, line *cS* will be, and so will triangle *SBc*.

Since *B* is impelled toward *S* by the centripetal force, the motion from that force *cC* will be in the plane of *B* and *S*, namely the given plane. So *C* is in that plane, and triangle *SBC* also.

And so on for all successive changes of motion.

Step 3: Proportionality

"...and, *componendo*, any sums whatever of areas *SADS*, *SAFS* are to one another as are the times of description."

Having demonstrated the equal areas for equal times, we can add together any number of consecutive triangles (having proved in Step 2 that they all lie in one plane) to form polygonal areas; these areas will be proportional to the times of their description.

Part 2

"Now let the number of the triangles be increased and their breadth decreased *in infinitum*, and their ultimate perimeter *ADF* (by Corollary Four of Lemma Three) will be a curved line: and therefore the centripetal force, by which the body is perpetually drawn back from the tangent of this curve, will act without ceasing, while any described areas whatever *SADS*, *SAFS*, always proportional to the times of description, will be proportional to those same times in this case.

Q.E.D."

Since any described area (sum of triangles) is proportional to the time, this will also hold for evanescent triangles, the limiting case as the equal times approach zero. This limit of a sum of triangles, as in Lemma 3 Corollary 4, will be a curvilinear area. In this case as well, the areas will be proportional to the times. Q.E.D.

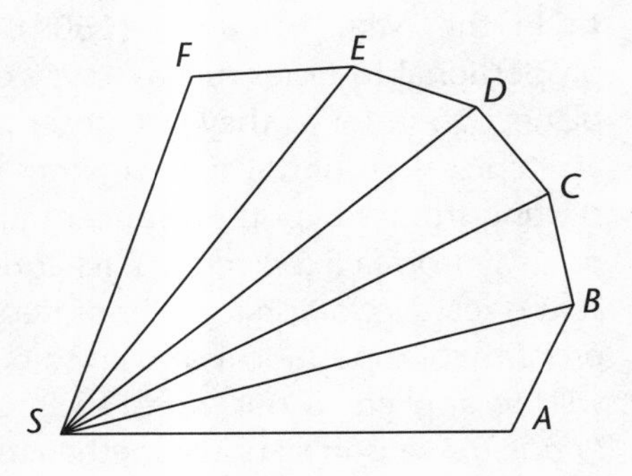

Note that as the time approaches zero, the forces, while in theory remaining impulsive, approach closer and closer to acting at every point. The limiting case, corresponding to an actual curvilinear path, may be thought of as a continuous force, although that might misleadingly suggest a single force. Perhaps we should say "forces acting continuously turning the body out of its tangential motion."

Pause After Proposition 1

• When we first encountered Definition 3, the definition of inherent force, we asked ourselves in what sense this was a force (or perhaps the question was what force meant to Newton if he included inertia as a force). One might be struck, therefore, to see him using both inherent force and impressed force to calculate the path of the body in this proposition; this seems to be an application in which their homologous effects would justify seeing them as the same sort of entity.

But do note that Newton is not actually using the two sorts of force to yield a resultant force; rather, he is letting each individual sort of force result in a motion and resolving the two motions into a resultant.

• Before going into the proposition we noted the many ways it was amazingly open. But there is one respect in which we might want to note that application of the proposition is limited: it does not claim that this method of reducing time intervals will lead to any curve you please. It is explicitly applied to a certain class of curves: those generated when a body is moving in response to forces applied along radii to a fixed center of forces.

• Note that Newton has not offered an argument that the limiting case in the proposition is a unique curve. This was not his goal. He has undertaken to prove that, *whatever* curve results, there will be equable description of areas and that the body and the center of forces remain coplanar. This is what he needs, and what he does in fact prove.

• In the *New Astronomy* (1609), Johannes Kepler invented the areas-proportional-to-times rule as a way of mathematizing the observed fact that planets go faster as they get closer to the sun. This later came to be known as Kepler's Second Law. Newton is here deriving the same conclusion starting from a hypothesis of central forces (forces directed radially towards or away from a fixed center). Its agreement with what had been recognized as the real-world applicability of Kepler's rule does not, however, consititute proof that Proposition 1 applies to our world. It is still only hypothetical, and will be applied to our world in Book III. At this point in the unfolding of *Principia,* we don't know whether there are such things as central forces in our world. What this proposition tells us is that if there are such forces operating, the result will be equable description of areas.

Notes on I.1 Corollaries

• The following are corollaries to the proposition; the arcs being invoked should be understood as arcs of the curve that is the limiting case of the proposition.

• Remember that in all these propositions and corollaries, it is important to note which assertions are of relationships true generally, and which are only true of the ultimate case as the arcs evanesce. Careless use of the latter for the former will lead to false conclusions. Every indication of relationship (such as $::$, $\sim$, $\propto$, $=$), when it applies only at the limit, needs to have expressed with it (that is, written before or over the relationship symbol) the condition "ult."

I.1 Corollary 1

The velocity of a body attracted to an immobile center in nonresisting spaces is inversely as the perpendicular dropped from that center to the rectilinear tangent of the orbit. For in those places A, B, C, D, E, the velocity is as the bases of the equal triangles AB, BC, CD, DE, EF, and these bases are inversely as the perpendiculars dropped to them.

[Note on translation: Though the wording of the last sentence may seem awkward, we have chosen to leave it as Newton wrote it, with velocity in the singular. To change this to a plural presumes to edit Newton and deprives us of what may be an insight into the way he was thinking about these matters.]

[Note on diagram: Newton gives no separate diagram for the corollaries. For Corollary 1, it is safe to use the diagram he gives with the proposition.]

Notes on I.1 Corollary 1

• This first corollary has not entirely left the impulse model. What the corollary proves is true for finite times as well as at the limit where the forces are exerted continuously, so we could use the diagram of the proposition.

• Kepler said the elapsed times over equal arcs vary directly as the distances *SA*, *SB*, etc. It would follow from this that the speeds were inversely as the distances *SA*, *SB*, etc. Newton is saying that Kepler's formulation isn't quite right, that the speeds are inversely as the perpendiculars to *AB*, *BC*, etc.

Does it strike you as odd that the speeds should depend, not on the distance to the body, but on the distance to some place where nothing is?

• In the proposition Newton spoke of bodies being driven in orbits and of centripetal forces acting with the effect that the body departs from the straight line of the tangent. These are neutral images. Here, however, he speaks of a body being "attracted" to an immobile center.

This language was the subject of a warning in Section 4.2.1 of the Preliminaries; review Newton's disclaimers quoted in that section if you feel you may be about to make some assumption about the cause or mechanism of these centripetal forces.

Expansion of Newton's Sketch of I.1 Corollary 1

Given:

1. Immobile center of forces;
2. curvilinear orbit formed by impulses as in proposition.

To Prove:

Velocities will be inversely as the perpendiculars to the tangents.

Proof:

We begin here with an orbit—a curvilinear path. We saw in the proposition that we can get to a curved path by letting the equal-time increment approach zero. This final figure of the proposition is the curved path made up of evanescent bases *AB*, *BC*, *CD*, etc.

Now let's approach the curved path another way. Suppose we are given a curved orbit, say the final limiting case curve of the proposition or maybe one given by observation of a body moving under forces directed to a center. Take equal-time arcs along this orbit, and construct the chords of those arcs.

Let's look at three of the equal-time arcs on the given curved path—arcs *AB*, *LM*, and *PQ*—and at the triangles on bases *AB*, *LM*, and *PQ*.

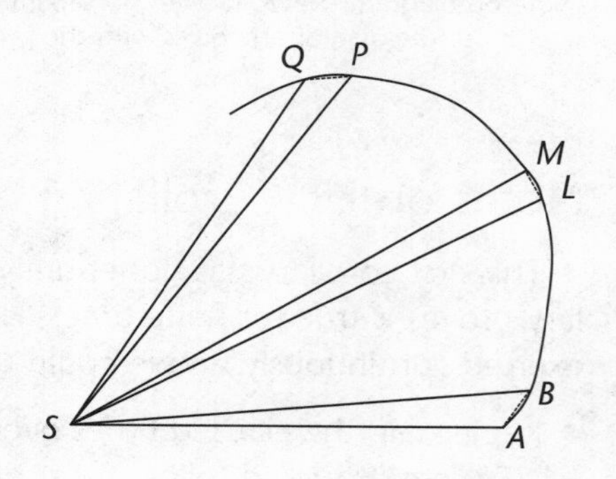

Since the arcs are described in equal times, they will be to one another in length as the average velocities along them. We can write this as:

arcs $\propto$ average velocities.

> **Note:** Remember that this "$\propto$" notation is shorthand for a proportion. Whenever we see it, we know that we are actually talking about at least four terms.
> So here, for example, we could expand this to:
>
> $$\frac{\overset{\frown}{AB}}{\overset{\frown}{LM}} = \frac{\text{av. vel. over arc } AB}{\text{av. vel. over arc } LM}, \quad \frac{\overset{\frown}{LM}}{\overset{\frown}{PQ}} = \frac{\text{av. vel. over arc } LM}{\text{av. vel. over arc } PQ}, \text{ etc.}$$

But we're looking now at the limiting case as $\Delta t \to 0$. Here $B \to A$, $M \to L$, and $Q \to P$. Therefore the arcs AB, LM, etc. will be to one another as the instantaneous velocities at A, L, and P,

$$\frac{\overset{\frown}{AB}}{\overset{\frown}{LM}} \overset{ult}{=} \frac{\text{vel}_A}{\text{vel}_L}, \qquad \text{etc.}$$

But by Lemma 7, the chords are ultimately as the arcs.

$$\frac{\overline{AB}}{\overline{LM}} \overset{ult}{=} \frac{\text{vel}_A}{\text{vel}_L}, \quad \text{etc.}$$

The arcs, and therefore ultimately the chords, are the equal-time distances traveled.

It can be proved starting from Euclid I.39 and Euclid VI.1 that bases of equal triangles are inversely as the heights of the triangles. (Find the heights of the triangles by dropping perpendiculars from S to the bases extended as necessary.)

So bases $\propto$ 1 / perpendiculars to chords.

Therefore velocities $\overset{ult}{\propto}$ 1 / perpendiculars to chords.

> **Note:** Remember that this "$\propto$" notation is shorthand for a proportion. Here, however, it is an inverse proportion—the second ratio is the inverse of the first. That is, for example:
>
> $$\frac{\text{vel}_A}{\text{vel}_L} \overset{ult}{=} \frac{SY}{SX} \qquad \text{or} \qquad \frac{\text{vel}_P}{\text{vel}_A} \overset{ult}{=} \frac{SX}{SZ}.$$

By Lemma 6, as the arcs evanesce so that $Q \to P$, $M \to L$, and $B \to A$, the limiting positions of the chords are the tangents.

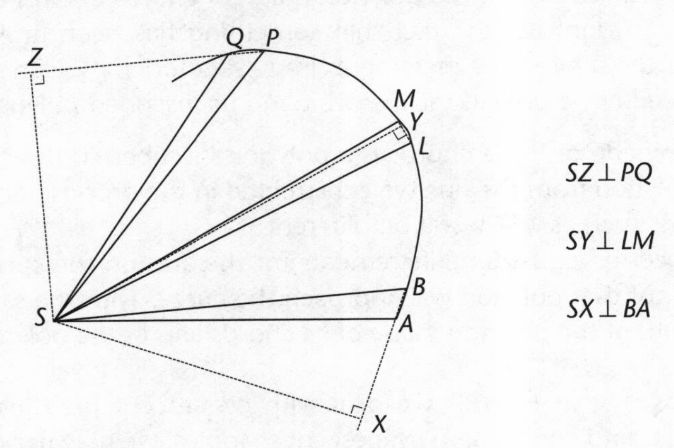

$SZ \perp PQ$

$SY \perp LM$

$SX \perp BA$

Therefore the limiting positions of the perpendiculars to the chords are the perpendiculars to the tangents.

Thus the velocities at these points P, L, A will ultimately be inversely as the perpendiculars to the tangents to the evanescent arcs.

velocities $\overset{ult}{\propto}$ 1 / perpendiculars to tangents. Q.E.D.

I.1 Corollary 2

If the chords AB, BC of two arcs described successively in equal times by the same body in nonresisting spaces be completed into the parallelogram $ABCV$, and its diagonal, in that position which it ultimately has when those arcs are diminished in infinitum, *be produced in both directions, it will pass through the center of forces.*

Notes on I.1 Corollary 2

• The claims of this and the following corollaries can be recognized as true in the case of the polygonal areas set up in the main part of the proof of the proposition. But the enunciations will not talk about that case. They will assert certain things about geometric entities built in curved paths. We will look at perpendiculars to tangents to curved paths and parallelograms built on chords of arcs of such curves.

We will have to be very careful in these corollaries, and in Proposition 2, not to confuse the polygonal area enclosed by chords of the curved orbit with the polygonal area made up of the triangles of impulsive forces. They are not the same. Remember that the polygon of the impulsive model was not following a "ghost curve." (See the last bullet in the notes to the proposition.)

Most of what is true of the finite case of the impulsive-model polygon is not true of the finite case of the polygon made of chords on arcs of a given curved path. So don't let the fact that something has been proved (or is obvious) for the finite-time-increment impulse model mislead you into thinking it needn't be proved for the orbit and its inscribed polygon.

• The figure made of these chords is a polygon inscribed in the curve. It is a different polygon from the one we constructed in the proposition, and the parallelograms such as *ABCV* will be different.

If we reduce the equal-time increments for the arcs on the curved path, the perimeter of that polygon will approach the curve. Thus the same curve is the limit both of the polygon made of its chords and of the polygon of the proposition.

This means that as long as we deal with evanescent time increments, evanescent arcs and chords, and evanescent triangles, we may use either the figure of the proposition or a figure made of chords of equal-time arcs. Since very different things will be true of the two figures for finite time increments, we will have to proceed with care and make sure all our steps are justified. But this will be the key to our strategy in this and the following corollaries.

Expansion of Newton's Sketch of I.1 Corollary 2

Given:

1. Immobile center;
2. two arcs of a curved orbit described successively in equal times by a body moving under the influence of forces directed towards a center.

To Prove:

Diagonal of parallelogram completed out of chords of those two arcs will ultimately pass through the center of forces when the arcs evanesce.

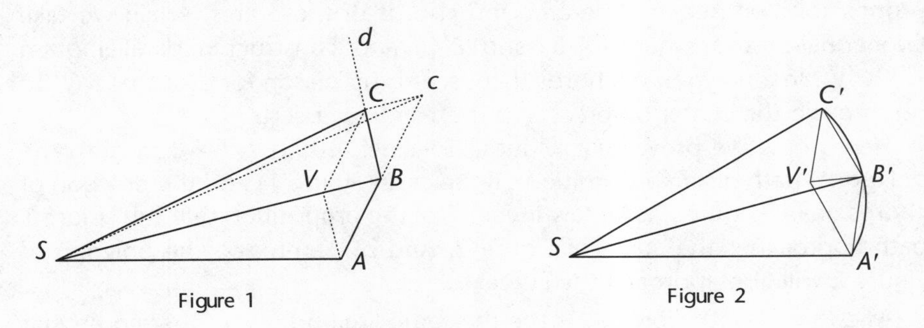

Figure 1 Figure 2

Proof:

Let's look at two situations now. The first, shown in Figure 1, is the polygon of the finite case in the proposition and shows two equal-time polygonal bases with a parallelogram *ABCV* constructed on them.

The second, shown in Figure 2, is the ultimate case of the proposition, where the triangles have evanesced and the path is a curve. This figure shows equal-time arcs with chords drawn across those arcs.

In Figure 1, $Bc = AB$ by the construction of the proposition; and therefore, by Euclid I.34, $Bc = CV$. Therefore diagonal *BV* of the parallelogram *ABCV* is the side of the parallelogram *BcCV* in Newton's diagram for the proposition, and $BV \parallel cC$.

By the proposition, a line from *B* parallel to *cC* will go through the center of forces. Therefore diagonal *BV* of the parallelogram *ABCV* will go through the center of forces. This is true by the construction of the proposition and applies always, not just in the limiting case.

Thus the diagonal of the parallelogram here will converge towards the center of forces, as asserted in the enunciation of this corollary. But this corollary is not talking about the finite case polygon of the proposition, but of the figure made from chords of equal-time arcs taken on the curved orbit of the ultimate case of the proposition.

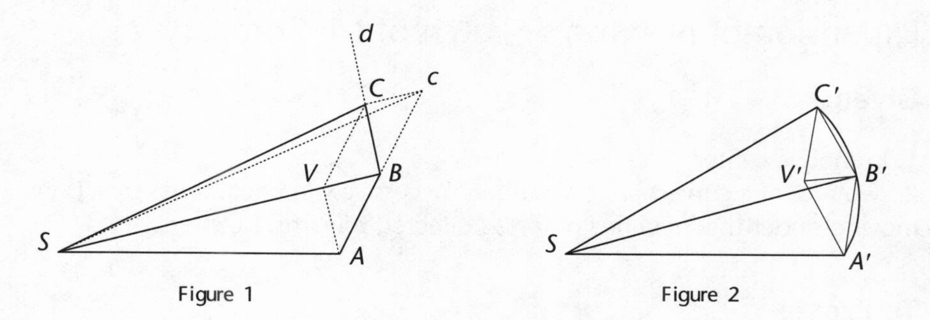

Figure 1

Figure 2

So let's look at the curved orbit in Figure 2. This is the ultimate case of the proposition, with equal-time arcs and chords of those arcs. Again we take two equal-time segments $A'B'$ and $B'C'$ and construct a parallelogram $A'B'C'V'$. In a polygon of chords there seems no reason for diagonal $B'V'$ to go through the center of forces S as the corollary asserts.

Here's how we prove that it does. Consider letting $\triangle t \to 0$ in both the polygonal path of the proposition's finite case (Figure 1) and the polygon of chords of arc (Figure 2). We saw in Part 2 of the proposition that the Figure 1 path approaches the curve of Figure 2. And by Lemma 7, the polygon of Figure 2 will also approach the curve.

As $\triangle t \to 0$, both polygons have the same limiting curve; therefore what is always true of the proposition's finite polygonal path will be *ultimately* true of the figure made by the chords of equal-time arcs on the orbit.

We showed that with the polygonal path of the proposition (Figure 1), the diagonal of the parallelogram BV, which is parallel and equal to cC, will always lie on line BS. Therefore it will do so in the limiting case of the curved orbit and its chords used in this corollary (Figure 2).

Q.E.D.

I.1 Corollary 3

If the chords AB, BC, and DE, EF, of arcs described in equal times in nonresisting spaces be completed into parallelograms ABCV, DEFZ, the forces at B and E are to each other in the ultimate ratio of the diagonals BV, EZ, where these arcs are diminished in infinitum. *For the motions BC and EF of the body are (by Corollary 1 of the Laws) composed of the motions Bc, BV, and Ef, EZ; and BV and EZ, equal to Cc and Ff, were in the demonstration of this proposition generated by the impulses of the centripetal force at B and E, and therefore are proportional to these impulses.*

Notes on I.1 Corollary 3

• This is not, properly speaking, a corollary of Proposition 1, and should, perhaps, have been presented as a separate proposition. It does not depend on Proposition 1. Rather, it is a simple application of Corollary 1 of the Laws of Motion to bodies and forces.

• Proposition 1 applies to a single orbit. Newton's use in Corollary 3 of sequential alphabetical letters for his orbital points suggests that he also has in mind a single orbit here (or maybe he just wanted to avoid having to commission another diagram).

However, nothing is assumed or asserted in this corollary about centers of forces or centripetal forces, and nothing in the logic assumes a single orbit. Furthermore, Newton uses this corollary in the proof of Corollary 4, which explicitly refers to different orbits.

We thus make a shift in Corollary 3 and now consider what may be different orbits and forces.

• This corollary, whose immediate applicability may seem obscure, lays a foundation that will be drawn upon first in Proposition 4, then further generalized in Proposition 6, and finally, in Proposition 6 Corollary 1, gathered into a powerful expression that Newton will use to determine the relationship of forces under many different force laws.

Expansion of Newton's Sketch of I.1 Corollary 3

"If the chords AB, BC, and DE, EF, of arcs described in equal times in nonresisting spaces be completed into parallelograms $ABCV, DEFZ, \ldots$"

Given:

1. Immobile center of forces;

2. pairs of successive chords, all of which are of equal-time arcs;

3. nonresisting spaces.

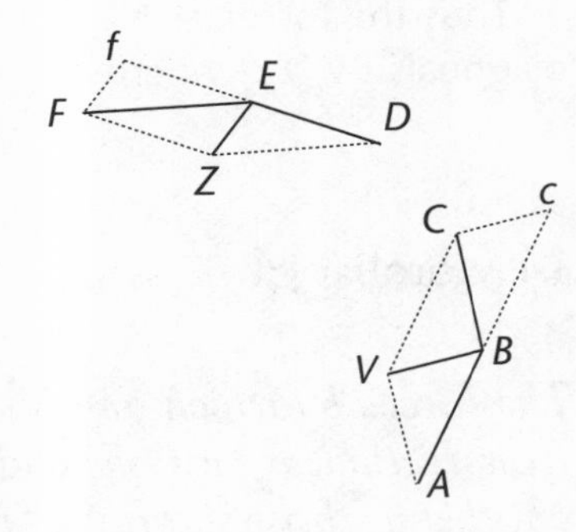

To Prove:

"... the forces at B and E are to each other in the ultimate ratio of the diagonals BV, EZ, where these arcs are diminished *in infinitum*."

The diagonals of parallelograms made from each pair of successive chords will be to one another ultimately as the forces at the intersection of the two chords.

Proof:

We are supposing evanescent parallelograms with impulsive forces being added at *B* and *E*. We suppose further that in the same evanescent time that the body actually goes to *C* and *F*, it would have gone to *c* and *f*.

"For the motions *BC* and *EF* of the body are (by Corollary 1 of the Laws) composed of the motions *Bc*, *BV*, and *Ef*, *EZ*;..."

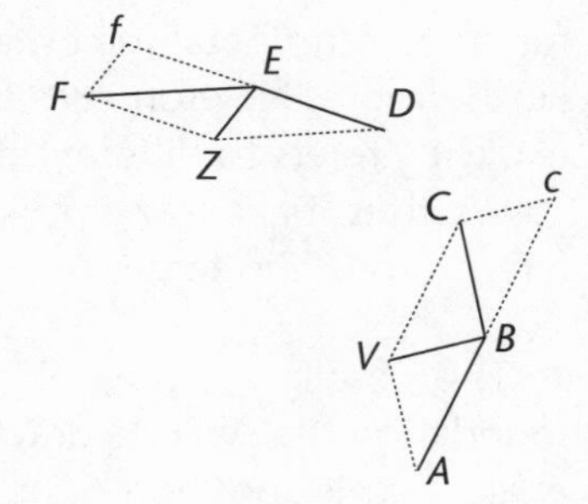

By Corollary 1 of the Laws of Motion, the actual paths *BC* and *EF* are composed of the sides of the parallelogram *Bc*, *BV*, and *Ef*, *EZ*.

"... and *BV* and *EZ*, equal to *Cc* and *Ff*, were in the demonstration of this proposition generated by the impulses of the centripetal force at *B* and *E*, and therefore are proportional to these impulses."

As in the construction for Proposition 1, the magnitudes of force generated by impulses at *B* and *E* are represented by *BV* and *EZ*. The length of the lines representing these forces can be found by the lengths of *cC* and *fF*, which measure the actual deflection. By Euclid I.34, in parallelogram *FfEZ*, $fF = EZ$ and in parallelogram *CcBV*, $cC = BV$.

Now we consider the path *AB* taken by the body in the same evanescent time period before the impulse was added at *B*. By Law 1, $Bc = AB$. But by Euclid I.34, $Bc = VC$. Therefore $AB = VC$. Because *BC* continues in straight line *AB*, $AB \parallel VC$. Thus by Euclid I.33, area *ABCV* is a parallelogram, with *BV* its diagonal.

Similarly, *EZ* is the diagonal of parallelogram *DEFZ*.

Thus the forces at *B* and *E* are to each other in the ultimate ratio of the diagonals *BV*, *EZ*, where arcs $\widehat{ABC}$ and $\widehat{DEF}$ are diminished *in infinitum*.

Q.E.D.

I.1 Corollary 4

The forces by which any bodies in nonresisting spaces are drawn back from rectilinear motions and are deflected into curved orbits are to one another as those sagittae of arcs described in equal times which converge to the center of forces and bisect the chords, when those arcs are diminished in infinitum. *For these sagittae are the halves of the diagonals with which we have been concerned in Corollary 3.*

Notes on I.1 Corollary 4

• Corollary 4, which extends Corollary 3, is also applicable to different orbits, as Newton makes explicit by referring to "any bodies" (plural). It requires only that each orbit has a center towards which its own forces are directed. In each individual case, Proposition 1 Corollary 2 is called upon to note that the crucial diagonals go through the respective centers of forces.

For future applications, we should note that the logic of the argument in no way assumes or depends upon there being a jointly shared center. In fact, Newton himself invokes this corollary in Proposition 4 to apply to a case where there are different orbits around different centers of forces.

However, because Newton's wording makes "center" singular, the diagram provided here shows a shared center for both orbits. This case, which is also a possible application of the corollary, is the one found when several planets orbit the sun as a shared center of forces.

• As in Lemma 11, the sagittae are lines that converge to a point, here the center of forces, and bisect the chords. The segment that is the actual sagitta is the part between the chord and the curve.

Expansion of Newton's Sketch of I.1 Corollary 4

Given:

1. Immobile center of forces;
2. nonresisting spaces;
3. equal-time arcs of curved orbits.

To Prove:

Centripetal forces ultimately vary as the sagittae of the equal-time arcs.

That is, $f_1 : f_2 \overset{ult}{::} \text{sagitta}_1 : \text{sagitta}_2$

or $f \overset{ult}{\propto} \text{sagittae}$.

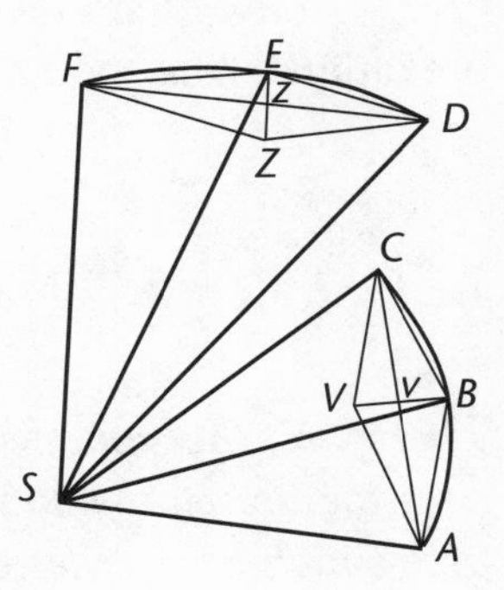

Proof:

To keep the same conditions as Corollary 3, let arcs *AB*, *BC*, *DE*, and *EF* be arcs described in equal times. Arcs *AC* and *DF* are also equal-time arcs, with times doubled. Draw chords *AC*, *DF*.

Note: We can draw lines from point S, bisecting the chords AC and DF, but they might not cut the curve at B and E. Alternatively we can connect SB and SE, but those lines may not bisect the chords AC, DF. In fact, for curves other than circles with the center of forces at the center, lines SB and SE would generally not bisect the chords in the finite case.

Complete the parallelograms $FEDZ$ and $CBAV$, in each of which two of the original equal-time chords form two sides. Diagonals AC, BV and DF, EZ bisect each other (prove using Euclid I.34, I.29, I.15, I.26). Therefore Bv, Ez are the lines that actually bisect the chords AC and DF.

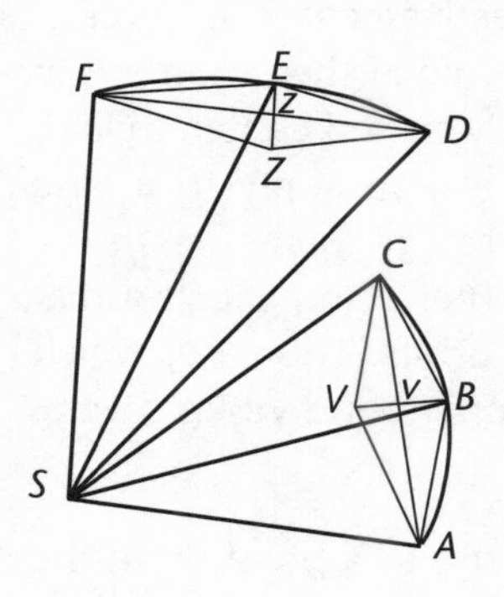

We set up our equal-time arcs of the curved path, and, as in the previous two corollaries, we made parallelograms on chords of two pairs of successive equal-time arcs such as $\overset{\frown}{AB}, \overset{\frown}{BC}$ and $\overset{\frown}{DE}, \overset{\frown}{EF}$. One diagonal of each of those parallelograms constitutes the chord of the two successive equal-time arcs; they will be chords of new double-length equal-time arcs such as $\overset{\frown}{AC}$, $\overset{\frown}{DF}$. These are the equal-time arcs to which the sagittae mentioned in the corollary apply.

Since by Corollary 2 the diagonals BV, EZ converge towards the center of forces only at the limit, consider now the ultimate case of the orbit and chords as $\triangle t \to 0$.

Bv ultimately lies on SB, and Ez ultimately lies on ES, by Corollary 2.

And $BV : EZ \overset{ult}{::} f_B : f_E$ by Corollary 3.

Since Bv, Ez (the ultimate sagittae) are half BV, EZ,

$Bv : Ez \overset{ult}{::} f_B : f_E$.

That is, forces that draw bodies out of rectilinear motion into a curved orbit are ultimately as the sagittae of arcs traversed in equal times.

Q.E.D.

I.1 Corollary 5

Therefore, the same forces are to the force of gravity [gravitas] *as these sagittae are to the sagittae, perpendicular to the horizon, of parabolic arcs which projectiles describe in the same time.*

Notes on I.1 Corollary 5

This corollary is not used in this basic sequence of propositions; nevertheless some comments will be helpful in placing it in its context.

• Corollary 5 (not present in the first edition) was evidently included because it struck Newton that the groundwork for some of what he would develop later had already been laid here. It is a look ahead at how this proposition (I.1) is going to be applied in I.4 Corollary 9 and in III.4.

• The corollary is saying that the deflections under the respective forces, geometrically the sagittae, are a measure of the force, both in the case of a body in orbit and a cannonball. The term *gravitas*, it must be remembered, means only terrestrial heaviness, and so far may only be applied to the cannonball.

But in the case of both curves, the curve of the body in orbit and curve of the cannonball in its parabola, there is something geometrically common. The deflection that the centripetal force produces and that which the force of gravity produces are both deflections from the tangential inertial path; and this deflection can be measured in both cases by the sagittae.

• However, nothing is being said here about the nature of the forces; it is only a geometrical observation. Thus it would be premature to conclude the cannonball and the heavenly body are being deflected by *the same* force; it would certainly be false to think Newton believed he had proved any such thing by this point.

I.1 Corollary 6

All the same things pertain, by Corollary 5 of the Laws, where the planes in which the bodies move, along with the centers of forces located in them, are not at rest, but move uniformly in a straight line.

Note on I.1 Corollary 6

Proposition 1 specifies that the center of forces is immobile; this corollary extends the proposition to apply to the situation when the whole plane is moving with uniform rectilinear motion. By Corollary 5 of the Laws,

> The motions of bodies contained in a given space are the same among themselves whether that space be at rest or move uniformly in a straight line without circular motion.

Proposition 2

Every body that moves in some curved line described in a plane, and, by a radius drawn to a point that either is immobile or proceeds uniformly in a straight line, describes areas about that point proportional to the times, is urged by a centripetal force tending to the same point.

Case 1. For every body that moves in a curved line is deflected from a rectilinear course by some force acting upon it (by Law 1). And that force by which a body is deflected from a rectilinear course, and is driven to describe the least triangles *SAB*, *SBC*, *SCD*, and so on (all equal), about an immobile point *S* in equal times, acts at the place *B* along a line parallel to *cC* (by I.40 of the *Elements* and Law 2), that is, along line *BS*; and at the place *C* along a line parallel to *dD*, that is, along the line *SC*, and so on. Therefore, it acts always along lines tending to that immobile point *S*.

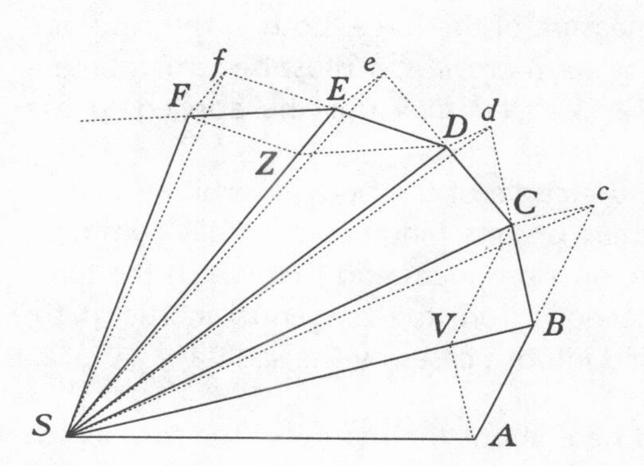

Q.E.D.

Case 2. And, by the fifth corollary of the Laws, it is just the same, whether the surface in which the body describes the curvilinear figure be at rest, or move along with the body, the figure described, and its point *S*, uniformly in a straight line.

Q.E.D.

[Newton gives no diagram for this proposition, evidently intending the diagram he gives with Proposition 1 to serve for both. The lines will have a different meaning here, however; see the second bulleted note below.]

Notes on I.2

• This is the converse of Proposition 1, and will be used in Book III (extended by the following proposition when necessary) for actually applying the tools of Book I to the celestial mechanics of our universe. In Book III we will find bodies moving around a point so as to describe areas proportional to the times and conclude that they are urged by a centripetal force toward that point. This will be the first step to a series of dazzling conclusions.

• Although we appear here to be showing the converse by working with the same diagram we used in Proposition 1 (and indeed Newton gives us only the diagram for Proposition 1 to serve for both propositions), there is an important difference to be noted.

In Proposition 2 we begin with a curved orbit, so we must keep in mind that we may use this diagram only in its ultimate case, with continuous force and evanescent bases *AB*, *BC*, and so on.

• It might have been easier if Newton had proved Proposition 2 as a converse of Proposition 1 by a *reductio ad absurdum* proof or by a proof involving the inscribing of finite triangles in the curve and then diminishing them *in infinitum*. The proof he has given instead is tricky, demands alert attention, and includes some difficult steps among the infinitesimals, a notoriously slippery region.

In Proposition 1, and in his calculus generally, Newton takes finite triangles or rectangles or other lines or figures, and looks at what happens as a limit is approached. Here in the proof he gives for Proposition 2 we must work with a more challenging situation: lines and figures that are always at the point of vanishing. To visualize and refer to the relationships we are working with, we use a diagram showing them as finite—a diagram that, most confusingly, looks just like the diagram for Proposition 1, where we had different finite figures that were then taken to a limit.

From shortly after Newton's time to the present, physicists have annoyed mathematicians by treating the vanishing quantity as if it were something that could be operated on as a magnitude. For example, a *dx* in an expression of integral or differential calculus is, for a mathematician, an operator; the physicist, on the other hand, often treats it as a little magnitude that can be divided or multiplied using algebra. (When treated as if it were the tiniest of magnitudes, the term *infinitesimal* is often used for this vanishing quantity.)

Expansion of Newton's Sketch of I.2

"Every body that moves in some curved line described in a plane, and, by a radius drawn to a point that either is immobile or proceeds uniformly in a straight line, describes areas about that point proportional to the times, is urged by a centripetal force tending to the same point."

Given:

1. A body moves in a curved line in a plane;

2. areas proportional to the times are described about a point that is immobile or moving with uniform rectilinear motion.

To Prove:

That body is urged by a centripetal force directed to that point.

Proof:

"*Case 1*. For every body that moves in a curved line is deflected from a rectilinear course by some force acting upon it (by Law 1)."

Since by the First Law of Motion "every body continues in its state of resting or of moving uniformly in a straight line, except insofar as it is driven by impressed forces to alter its state," if we observe a body moving in a curved path, we may conclude that some force has been impressed upon it to change its motion from a straight line.

"And that force by which a body is deflected from a rectilinear course, and is driven to describe the least triangles SAB, SBC, SCD, and so on (all equal), about an immobile point S in equal times,..."

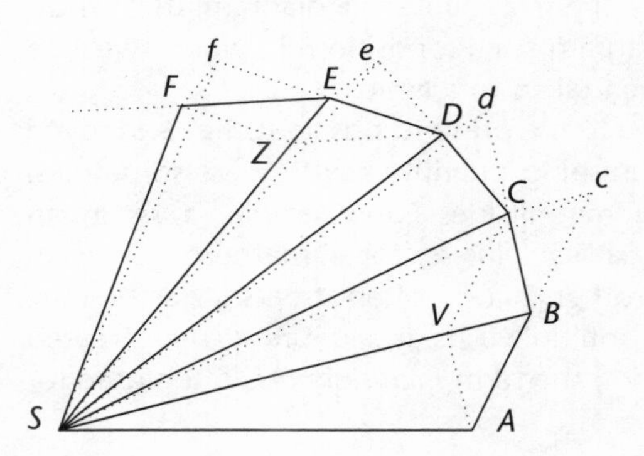

From our second given, above, evanescent $\triangle SBA$ = evanescent $\triangle SBC$. Note that we are not given finite triangles that we will later diminish *in infinitum*. In these triangles the arc, chord and tangent must be understood as coinciding.

Equal evanescent triangles $\triangle SBA$, $\triangle SBC$, etc., are given to be equal areas around an immobile point S and within the curved path. We must prove that S is also the center of forces.

"...acts at the place B..."

How do we know that the force acts at point B? We're considering a curved path in Proposition 2. Let us take any point on it, say point B. At that point a body will, by Law 1, continue in a straight line unless a force acts upon it. Since we are given a curved path, we know that the body is not continuing in a straight line. Therefore a force must be acting upon it at point B.

We can think of the curve we are given as made up of infinitely small segments of straight lines, with impulses of force added at the beginning of each infinitely small segment, with an infinitely short coasting phase in between. Lines *AB, BC,* etc., are the infinitely short coasting phases. Impulses are added at sequential points *A, B, C,* etc. (A mathematician, of course, would balk at the idea of sequential points, since points have no length. But we must think like physicists here to visualize this proof.)

"... along a line parallel to *cC* (by I.40 of the *Elements* and Law 2),... "

Step 1: Determining Point c

We are not explicitly told in this proposition, but for the proof to work *Bc* must be the distance the body would have traveled from *B* in the given time if no force were applied at *B*.

If no force is exerted at *B, Bc* would be a tangential path. But in these evanescent triangles *Bc* is also the extension of line *AB*, since chord and tangent coincide in the limit. Thus *Bc* is the extension of *AB*, and the path that would be traced in the same time the body took to traverse *AB* if no force were exerted at *B*. It is also the path that would be followed in the same time the body actually takes in going from *B* to *C*.

Step 2: That the Force Is Parallel to Line *cC*

Now we come to our next challenge: how do we know that the force acts along a line parallel to *cC*?

By Law 2, a change of motion takes place following the straight line in which that force is impressed. The force is impressed at *B*, so we know that the line along which the force is impressed begins there. But at what angle? Towards what point as a center of that force?

We are not given the force; all we are given is the result of the force: the curve. A measure of how much the motion has changed is found in *cC*, the line between where it would have been without the force and where it is with the force.

It is not the magnitude of *cC* that interests us here, but what the position of line *cC* tells us about the direction of the line of force from *B*. It appears to be Newton's idea that the angle *BcC* gives us that direction; that the line of the force must be parallel to that line *cC*.

> **Question for Discussion:** This certainly seems plausible. Does it seem to you proved? That is, does it seem to you to follow from what we have been given here that the direction of the force must be parallel to *cC*? Can you construct a tighter proof?

Thus the direction of the force is from *B* in a line parallel to *cC*.

Step 3: That *BS* Is Parallel to *cC*

By Law 1, every body continues in its state of rest or of moving uniformly in a straight line unless driven by impressed forces to alter that state. During the infinitely short coasting phases, like *AB* and its continuation *Bc*, no force is being impressed. Therefore the body continues in uniform motion. Uniform motion is the same as equable motion, motion at a constant speed.

By Galileo's second proposition of equable motion, if a body passes through two spaces in equal times, these spaces will be to each other as the speeds. In the case of equal-time paths *AB* and *Bc*, the speeds are the same (since no force is supposed to have intervened that could have changed the speed); therefore the distances will be the same.

Therefore $Bc = AB$.

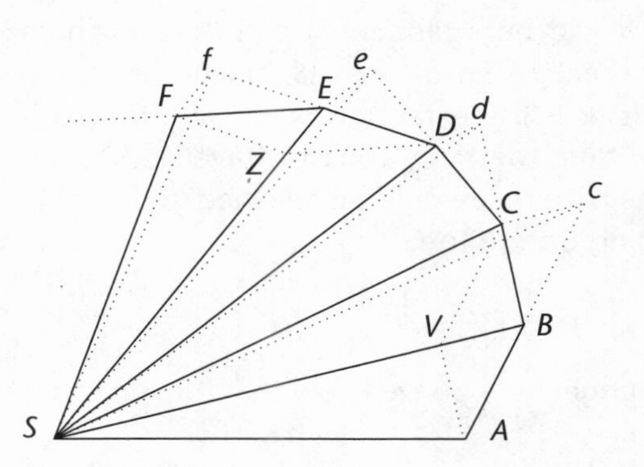

By Euclid I.38 (equal bases, same height), $\triangle SAB = \triangle SBc$.

Therefore $\triangle SBc = \triangle SBC$.

By Euclid I.39, equal triangles on the same base are within the same parallels.

Therefore $cC \parallel SB$.

(Euclid I.40, which Newton cites, says that equal triangles on *equal* bases are within the same parallels. This is equivalent.)

"... that is, along line *BS*; and at the place *C* along a line parallel to *dD*, that is, along the line *SC*, and so on. Therefore, it acts always along lines tending to that immobile point *S*."

Therefore *BS* is the line from *B* that is parallel to *cC*, that is, at the angle indicating the direction of the change of motion.

"*Case 2*. And, by the fifth corollary of the Laws, it is just the same, whether the surface in which the body describes the curvilinear figure be at rest, or move along with the body, the figure described, and its point *S*, uniformly in a straight line.

Q.E.D."

The conclusion of Case 1 is extended by using Corollary 5 of the Laws to apply also when the body and its center are moving with uniform rectilinear motion. Q.E.D.

[Corollaries 1 and 2 are omitted because they are not used in Newton's basic sequence of propositions.]

Scholium

A body can be urged by a centripetal force compounded of many forces. In this case, the sense of the proposition is that that force which is compounded of all of them tends to the point *S*. Further, if some force act always along a line perpendicular to the surface described, this will make the body deflect from the plane of its motion, but it will neither increase nor decrease the quantity of the surface described, and so may be neglected in the compounding of forces.

Notes on Scholium after I.2

• The case presented in this scholium of many different forces acting on a body, forces that may be compounded into a single resultant force toward a particular center, is another reason not to assume that there is a body at point *S*. (See note after Proposition 1 for the first reason.)

Even though we may sometimes find a body at the center of forces, when the situation becomes more complex and the force on a body is compounded of several forces, the resolved center of forces is more obviously a geometrical, rather than a physical, point. This is explored further in Book III when we find the center of forces for the planets, and discover that it is not at the center of the sun.

• In the second part of this scholium, center of forces *S* must be regarded as moving too; the whole plane is accelerated in a direction perpendicular to its original position, and within that moving plane the areas of the equal-time triangles remain equal. This perpendicular force will nevertheless make the body deviate from the original plane and describe a spiral in absolute space. In this instance, the areas described in absolute space will not be proportional to the times of description: this proportionality holds only for areas in the accelerated plane.

This is meant to correspond to the situation described in Corollary 6 of the Laws of Motion.

However, nothing is said here about this perpendicular force acting on a *system* of two or more bodies. At this point we may be expected to assume that this force is operating continuously on the one body of the previous propositions along a line perpendicular to the plane of its movement, as specified in the previous propositions.

If one pictures a body at the center of centripetal force, it might not be inconsistent with the scholium to imagine that the outside force, which Newton says acts on the "surface" of that plane, would act also on that body at the center of forces and thus on the system of two bodies. This would take a step beyond what is stated, however.

Proposition 3

Every body that, by a radius drawn to the center of another body moved in any way whatever, describes areas about that center proportional to the time, is urged by a force compounded of a centripetal force tending towards that other body, and of all the accelerative force by which that other body is urged.

Let the first body be *L*, and the other body *T*: and (by Corollary 6 of the Laws) if both bodies be urged in parallel lines by a new force equal and opposite to that by which the other body *T* is urged, the first body *L* will continue to describe the same areas as before about the other body *T*. However, the force by which the other body *T* was being urged is now annulled by a force equal and opposite to it, and therefore (by Law 1) that other body *T*, now left entirely to itself, will either be at rest or move uniformly in a straight line. And the first body *L*, being urged by the difference of the forces (that is, by the remaining force) continues to describe areas proportional to the time about the other body *T*. Therefore, by Proposition 2, the difference of the forces tends towards that other body *T* as center.

Q.E.D.

[Newton gives no diagram for this proposition.]

Notes on I.3

• This proposition considers an actual body at the center of forces. It is somewhat anomalous for Book I to do so, but it is needed in this proposition because the center of forces *T* of the movement of one body, *L*, is itself being urged towards another center of forces. Forces can operate only on bodies, not on geometrical points; so the center of forces *T* around which body *L* is driven must be at the center of another physical body.

• Although we have a body at the center of forces, Newton makes no reference to any *mass* which the body at *T* may possess. He cannot, therefore, be intending to consider *motive force* here, because mass (quantity of matter) enters into the motive quantity of force. Rather, as he does almost exclusively throughout Books I and II Newton is using *accelerative* quantity of force. See the last bulleted note in the Notes on Book I Proposition 1 for an explanation of his strategy and tactics on this usage.

• The outside accelerative forces of this proposition could be impelling the body at the center in some curved and/or nonuniform motion. One thinks immediately of a situation in which, for example, a planet is being impelled in an orbit around a center of forces while a moon orbits around it.

Note, however, that this proposition corresponds only approximately to that situation. Given equable description of areas, it shows that any disturbing forces must act identically on *L* and *T*—that is, in parallel lines. But the disturbing forces on a planet and its moon, being directed towards the sun *S*, are not truly parallel.

Nevertheless, they will be nearly so under two conditions, which may be found together reinforcing each other. First, when the planet and satellite are sufficiently close together, the angle from the outside center of forces is small, and *TS* and *LS* are effectively collinear. Second, when the center of forces is sufficiently distant, the angle again becomes small, even if the satellite's orbital radius is somewhat larger, and the lines of force to the two bodies are still effectively parallel. Because the various planet-satellite systems satisfy these conditions, the proposition will apply approximately. Corollary 2 is provided to make the approximation explicit.

• Proposition 1 Corollary 6, and Proposition 2 Case 2 considered uniform motion of the plane of the orbit in a straight line. They applied Corollary 5 of the Laws.

The second part of the scholium following Proposition 2 considered an accelerative force acting on a body in a direction always perpendicular to the plane of that body's motion, but did not ask us to consider a body at the center of centripetal force also being acted on by the outside force.

This proposition extends the scholium by considering a body at the center of forces being acted on by an outside force. Furthermore, it allows the force to be in any direction, not just perpendicular to the plane of the single body's motion.

• Note that what is specified as "given" in Proposition 3 is accelerative forces operating on a body at the center of equable description of areas. We prove that they must be operating on the whole system. Then when we use Proposition 3 we are using an extension of Proposition 2 to the situation where the whole system is being operated on by other accelerative forces.

• Note that, following the structure of Proposition 2 here, we are given equable description of areas and conclude that the body is being urged by a particular compounding of forces. In using this proposition, then, we look for the equable description of areas and then know something about the force environment. It will be a crucial step in the application of the hypothetical propositions to our actual universe when we find we can use this with Jupiter and its moons.

Expansion of Newton's Sketch of I.3

"Every body that, by a radius drawn to the center of another body moved in any way whatever, describes areas about that center proportional to the time, is urged by a force compounded of a centripetal force tending toward that other body, and of all the accelerative force by which that other body is urged."

Given:

1. Equable description of areas by one body L about another body T;
2. T, the body at the center of forces, is urged by some accelerative force.

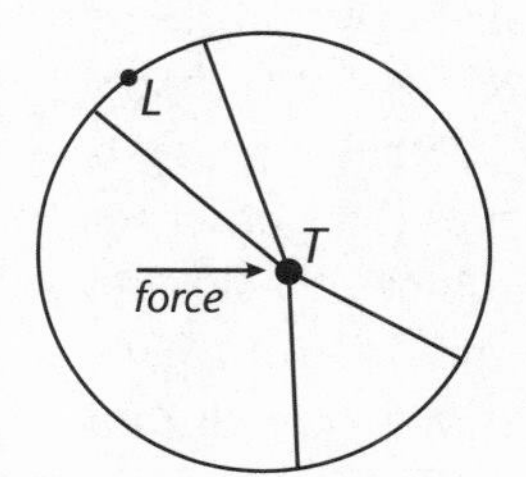

To Prove:

L is urged by a force compounded of the centripetal force impelling it toward T and of all the accelerative forces by which T is urged.

Proof:

Because we are "given" equable description of areas around a point, we might want to say that Proposition 2 tells us that the center towards which the body is impelled is that point. But remember that Proposition 2 requires that the point around which we have equable description of areas is at rest or moving in uniform rectilinear motion, and here we have accelerative outside forces.

Therefore, to apply Proposition 2, we first have to bring the situation back to the one used in that proposition: we must get the center of forces at rest or moving uniformly. Newton does this by cancelling out the accelerative force on it by a hypothetical equal and opposite force. We'll call this force F_H.

"Let the first body be L, and the other body T; and (by Cor. 6 of the Laws) if both bodies be urged in parallel lines by a new force equal and opposite to that by which the other body T is urged, the first body L will continue to describe the same areas as before about the other body T."

We were given that L describes areas proportional to the times around T. Suppose that the system of L and T (Figure 1) be urged by equal accelerative forces F_H in parallel lines and that the magnitude and direction be equal and opposite to the force(s) F_C by which body T is being urged.

Consider the body L describing equal areas around body T. By Corollary 6 of the Laws, it will continue to describe equal areas if the plane of L and T and the orbit (in one plane by Proposition 1) is urged in the direction of parallel lines by equal accelerative forces.

Thus the equable description of areas by *L* around *T* as its center will be unaffected.

"However, the force by which the other body *T* was being urged is now annulled by a force equal and opposite to it, and therefore (by Law 1) that other body *T*, now left entirely to itself, will either be at rest or move uniformly in a straight line. And the first body *L*, being urged by the difference of the forces (that is, by the remaining force) continues to describe areas proportional to the time about the other body *T*. Therefore, by Proposition 2, the difference of the forces tends towards that other body *T* as center."

Consider body *T*. The hypothetical forces equal and opposite to the accelerative forces on it annul the latter and leave it at rest or moving with uniform rectilinear motion.

Consequently we have reduced this case to Proposition 2 and can say that because *L* is describing areas proportional to the times around *T*, *T* is the center of forces for *L*.

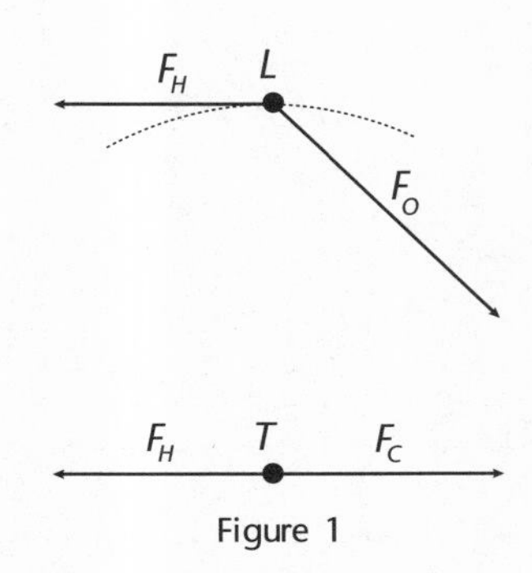

Figure 1

Conclusion:

By the enunciation,

"Every body that, by a radius drawn to the center of another body moved in any way whatever, describes areas about that center proportional to the time, is urged by a force compounded of a centripetal force tending toward that other body, and of all the accelerative force by which that other body is urged."

We need another step to show that the total force on *L* is compounded of that force and all the forces on *T* in those cases when the latter have not been annulled by a hypothetical force. We will now take the hypothetical force away and see what the original force on *L* must have been.

Let's look at the forces on *L* alone. The resultant of those forces is a force towards *T*. One of the forces we know: it's the hypothetical force that is equal and opposite to the forces on *T*. The others, if any, we don't know.

But because we know the resultant between them and the hypothetical force, we can find a second component that, with the hypothetical force, will yield that resultant. This second component will be the total original force on *L*. It will itself be a resultant of the force towards *T* and any other original forces on *L*.

Figure 2 shows the situation at the end of Newton's sketch giving resultants only.

Figure 3 shows the components that are going into those resultants. In the case of T, we have an original force F_C (we've used the subscript C knowing that it will turn out to be the common force on L and T) exactly balanced by the hypothetical force F_H. In the case of L we have the hypothetical force F_H and the other side of the parallelogram whose first side is F_H and diagonal is the centripetal force F_L. Call this force F_O, the total original force on L.

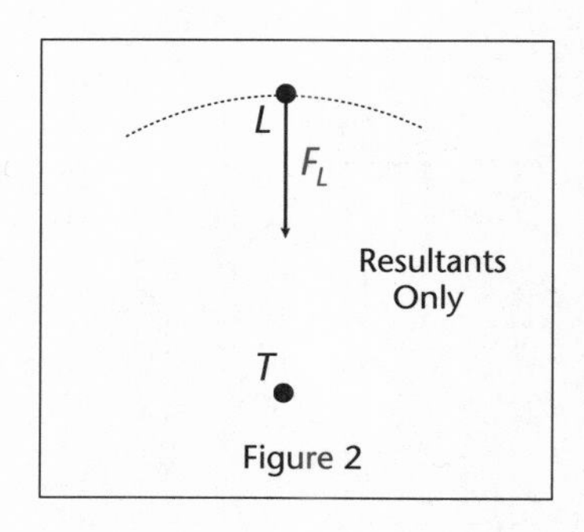

Figure 2

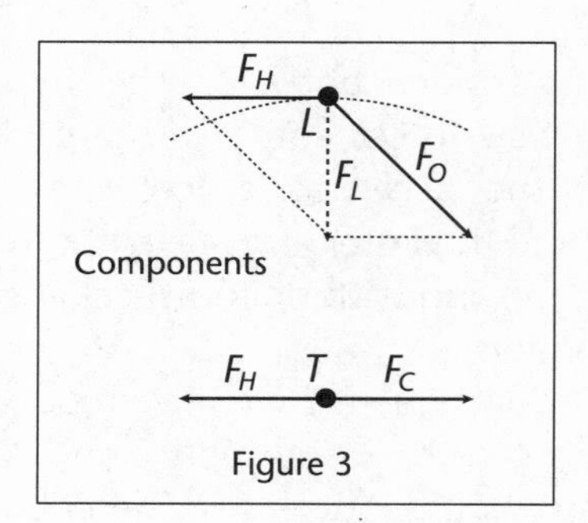

Figure 3

Now, as in Figure 4, construct $LM = LH$ on the opposite side. Since $IKLH$ is a parallelogram, $IK = LH$ and therefore $LM = IK$ [Eu. I.34].

Therefore $IKML$ is a parallelogram and $\overrightarrow{LK}$ represents the resultant of $\overrightarrow{LI}$ and $\overrightarrow{LM}$ [Cor 1 of the Laws].

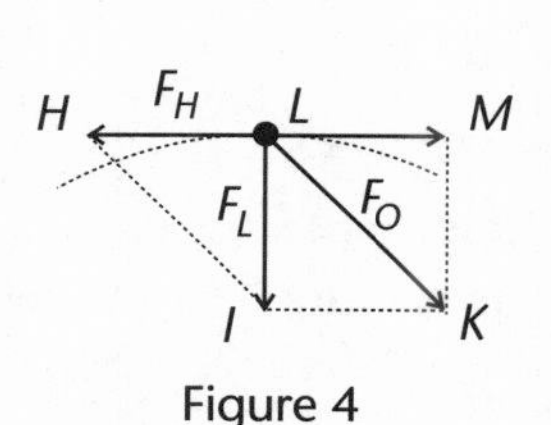

Figure 4

But $\overrightarrow{LK} = F_O =$ the total original force on L

and $\overrightarrow{LI} = F_L =$ the force on L towards T

and $\overrightarrow{LM} = -F_H = F_C$.

Therefore the original force on L is compounded of the force on L towards T and the common force on L and T.

Q.E.D.

I.3 Corollary 1

Hence if one body L by a radius drawn to another T describes areas proportional to the times, and from the whole force by which the former body L is urged (be it simple or compounded of several forces as in the second corollary of the Laws) there be subtracted (by the same corollary of the Laws) the whole accelerative force by which the other body is urged, all the remaining force by which the former body is urged tends towards the other body T as center.

Note on I.3 Corollary 1

This is the form of Proposition 3 that will be used in Book III to conclude from equable description of areas of moons that the moons are urged towards the point that is the geometric center of the equable areas as a center of centripetal force.

I.3 Corollary 2

And, if the areas are nearly proportional to the times, the remaining force nearly tends towards the other body T.

Note on I.3 Corollary 2

As noted above, this covers the case in which the accelerative force on the whole system is centripetal and not perfectly parallel.

I.3 Corollary 3

And vice versa, if the remaining force tends nearly towards the other body T, those areas will be approximately proportional to the times.

[Corollary 4 is omitted because it is not used in Newton's central argument.]

Scholium

Since the uniform description of areas is the indicator of the center regarded by that force by which the body is chiefly affected, and by which it is drawn back from rectilinear motion and kept in its orbit, why may we not, in what follows, take the uniform description of areas as the indicator of the center about which all circular motion in free spaces is performed?

Notes on Scholium After I.3

It is indeed true that the uniform description of areas indicates a center. However, we need to be careful to note that this may be the center of only a part of the force on a body, perhaps only a small part.

For example, a satellite or moon may describe areas uniformly about a planet, but may still be urged much more forcefully towards the sun than towards the planet. (This happens to be the situation with the earth and its moon.) In such a case, one component of the total force on the satellite is directed towards the planet, and the other, greater component is directed towards the sun.

Newton considers the relative size and interaction of these components in Proposition I.66 (see Appendix B).

Proposition 4

The centripetal forces of bodies that describe different circles with uniform motion tend towards the centers of the same circles, and are to one another as the squares of arcs described in the same times applied to [i.e., divided by] the radii of the circles.

These forces tend to the centers of the circles (by Proposition 2 and Proposition 1 Corollary 2), and are to one another as the versed sines of arcs described in least equal times (by Proposition 1 Corollary 4); that is, as the squares of the same arcs applied to [i. e., divided by] the diameters of the circles (by Lemma 7), and therefore, since these arcs are as the arcs described in any equal times whatever, and the diameters are as the radii of the same, the forces will be as the squares of any arcs whatever described in the same times, applied to [i. e., divided by] the radii of the circles.

Q.E.D.

[Newton gives no diagram for this proposition.]

Notes on I.4

• In this proposition and its first eight corollaries, Newton is moving from given orbits to a corresponding force law governing the bodies in those given orbits. The force law he finds may not govern the movements of any other bodies at other distances. He is not, in short, assuming or deducing a field.

• Consider also that nothing in this proposition requires that the orbits have the same centers. One might be comparing Jupiter's moons and Venus's. This would indeed give us a ratio of forces, but not anything generalizable to other orbits.

• This proposition and its corollaries give us a great collection of useful relationships that will be very handy later. Although we must be careful not to use relationships that we proved for equable concentric circular motion as if we had proved them for other sorts of paths, the proposition and its first seven corollaries prove more useful than they might seem for two reasons.

First, the extension given in Corollary 8 allows us to use the relationships for any similar parts of similar figures where we have equable description of areas. Second, in applying our mathematical demonstrations to our world, we may use the relationships that result from the conditions of the proposition as an approximation if our observations show us motion very close to concentric circular and equable.

Corollary 9 is particularly important in its implications for applying our mathematical demonstrations to our world. It allows us to relate motion of a body in orbit over a given time, which we can determine from observation, to distance the same body would have fallen in the same time, which will turn out to be a measure of gravity. Thus it plays a key role in the proposition that I find the most beautiful and thrilling in *Principia*—Book III Proposition 4.

• Note that Proposition 4 and Corollaries 1–7 and 9 all deal with circles and equable motion. It turns out (see proof) that this means that the conclusions arrived at are all true generally, not just at the limit.

Corollary 8 re-derives all the relationships of the proposition and the first seven corollaries for similar parts of similar figures rather than circles with equable motion. Although the corollary does not say so explicitly, it turns out (see proof) that Corollary 8 relationships hold only for the ultimate ratios.

• When Newton speaks of a force (or anything else) varying as one thing directly and something else inversely, we are to understand a simultaneous dependence. He does not mean to make two independent statements. To get the ratio of forces in this example, we must compound the ratio of the first with the inverse ratio of the second.

To compound ratios to see how force varies with two variables, we must set up the conditions of Euclid VI.23, which defines compound ratio. We will have set up those conditions when we have established relationships between force and each of the variables individually, with the other held constant. (See Section 6 of Preliminaries.)

• Versed sines are sagittae at right angles to the chords. Since this proposition concerns circles, with the center of forces at the center of the circle, the sagittae of Proposition 1 Corollary 4 will be versed sines.

• Some students are tempted to try to use Lemma 11 in this proof. If you are, slow down and look more carefully. Lemma 11 concerns the relationships between various lines from a single point along a single curve. It cannot be applied to Proposition 4, which makes comparisons between two different orbits.

Expansion of Newton's Sketch of I.4

"The centripetal forces of bodies that describe different circles with uniform motion tend towards the centers of the same circles, and are to one another as the squares of arcs described in the same times applied to [i.e., divided by] the radii of the circles."

Given:

1. Bodies moving around the circumferences of circles;

2. the bodies move with equable motion.

To Prove:

1. Centripetal forces on the bodies will be directed to the centers of the circles.

2. The centripetal forces on bodies in any two circles will be to each other as squares of equal-time arcs in the two different circles divided by the circles' respective radii, even for finite arcs.

That is, the ratio of forces will be as the ratio of the squares of the arcs compounded with the inverse of the radii:

$$f_1 : f_2 :: (\text{arc}_1{}^2 : \text{arc}_2{}^2) \text{ comp } (\text{radius}_2 : \text{radius}_1).$$

Or, $f \propto \text{arc}^2 / \text{radius}$.

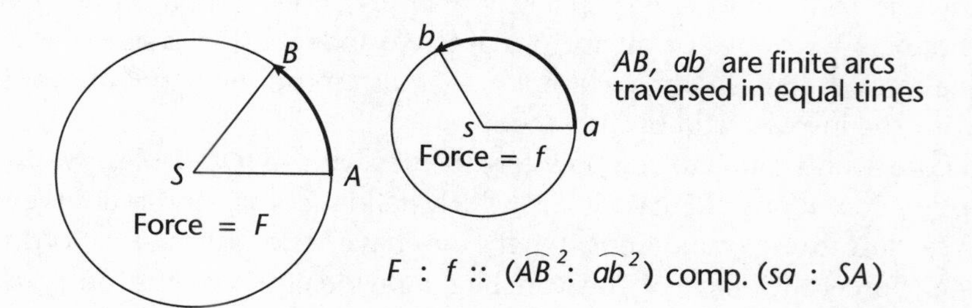

Proof:

Part 1: Force Directed to Centers of Circles

"These forces tend to the centers of the circles (by Proposition 2 and Proposition 1 Corollary 2),..."

The bodies are given as moving uniformly around circles. In equal times we will have equal arcs because of the uniform motion. The radii within one circle will be equal by the nature of circles. Therefore the sectors will be equal round the center of the circle. Thus the areas described in equal times will be equal for each circle and the bodies describe areas proportional to the times around the center of the respective circles.

By Proposition 2, if a body describes areas proportional to the times about a point, it is urged by a centripetal force directed to that point.

Part 2: The Ultimate Case: Force $\overset{ult}{\propto}$ Arc^2 / Diameter

The proposition and sketch state the relationship generally, not just at the limit. Newton first demonstrates it for the limiting case, invoking Proposition 1 Corollary 4 and Lemma 7, both true only ultimately. But what is true at the limit may not be true generally. In Part 3 we will demonstrate the relationship for the finite case.

"...and are to one another as the versed sines of arcs described in least equal times (by Proposition 1 Corollary 4);..."

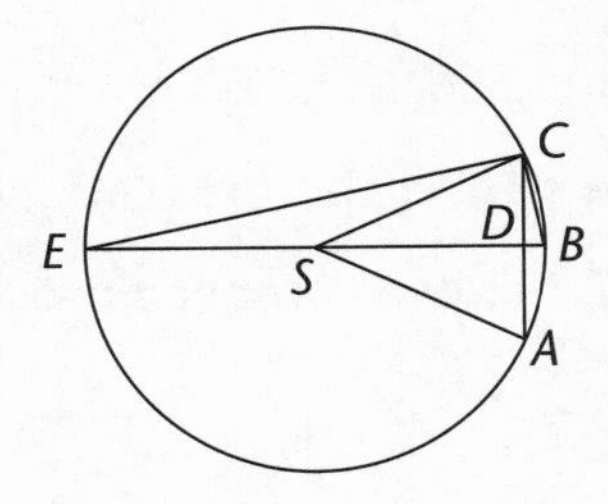

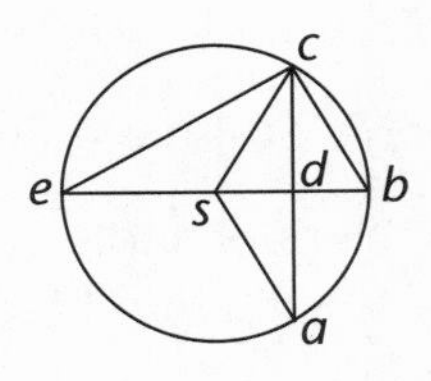

Arcs $\overset{\frown}{AB}$ and $\overset{\frown}{BC}$ are traversed in equal times in the first circle, and arcs $\overset{\frown}{ab}$ and $\overset{\frown}{bc}$ are traversed in the same equal times in the second.

By Galileo, *Two New Sciences*, Proposition 2 of Equable Motion, "If a moveable passes through two spaces in equal times, these spaces will be to one another as the speeds." The speed is constant in each circle; therefore the equal-time arcs will be equal within each circle.

$$\overset{\frown}{AB} = \overset{\frown}{BC} \quad \text{and} \quad \overset{\frown}{ab} = \overset{\frown}{bc}.$$

By Proposition 1 Corollary 4, $f \overset{ult}{\propto}$ sagittae for two bodies each in the middle of equal evanescent times. Arcs AB, BC, ab, and bc are all described in equal times. DB, db are the sagittae of the double arcs AC, ac.

Thus in this case,

$$F \overset{ult}{\propto} DB. \tag{1}$$

"...that is, as the squares of the same arcs applied to [i. e., divided by] the diameters of the circles (by Lemma 7),..."

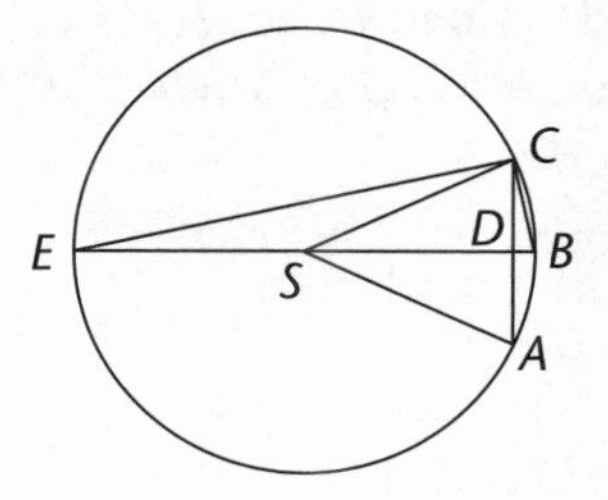

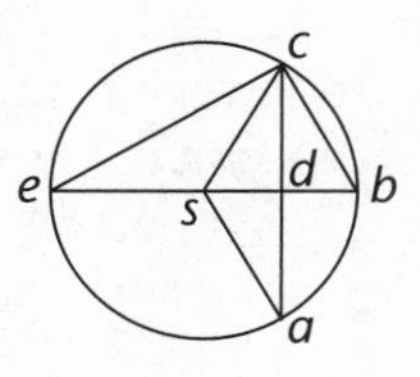

$SC = SA$. [Radii of the same circle are equal.]

Angle CSB = angle BSA. [By Euclid VI.33, equal arcs are subtended by equal angles.]

$SD = SD$. [identity]

Therefore $\triangle CSD \cong \triangle ASD$ [Eu. I.4]

Therefore angle SDC = angle SDA.

Therefore angle SDC is a right angle. [Prove using Euclid I.13]

But angle ECB is a right angle because it lies in a semi-circle. [Eu. III.31]

From the right angle of $\triangle ECB$ a perpendicular CD has been dropped making $\triangle CDB$. By Euclid VI.8,

$\triangle CDB \sim \triangle ECB$.

Therefore $EB : CB :: CB : DB$, [Eu. VI.4]

and $CB^2 = EB \times DB$. [Eu. VI.17]

$$DB = CB^2 \,/\, EB. \tag{2}$$

Following the same reasoning for the second body,

$$db = cb^2 \,/\, eb. \tag{2a}$$

Substituting from Equation 1:

$F \overset{ult}{\propto} CB^2 \,/\, EB$.

$\overline{CB} \overset{ult}{=} \widehat{CB}$, by Lemma 7.

Therefore $F \overset{ult}{\propto} \widehat{CB}^2 \,/\, EB$. Or,

$$\frac{F}{f} \overset{ult}{=} \frac{\widehat{CB}^2}{\widehat{cb}^2} \times \frac{eb}{EB}.$$

Generally,

$$\text{Force} \overset{ult}{\propto} \text{arc}^2/\,\text{diameter}. \qquad (3)$$

Part 3: Proof of Finite Case: Force $\propto$ Arc^2/ Radius

"... and therefore, since these arcs are as the arcs described in any equal times whatever, and the diameters are as the radii of the same, the forces will be as the squares of any arcs whatever described in the same times, applied to [i. e., divided by] the radii of the circles.

Q.E.D."

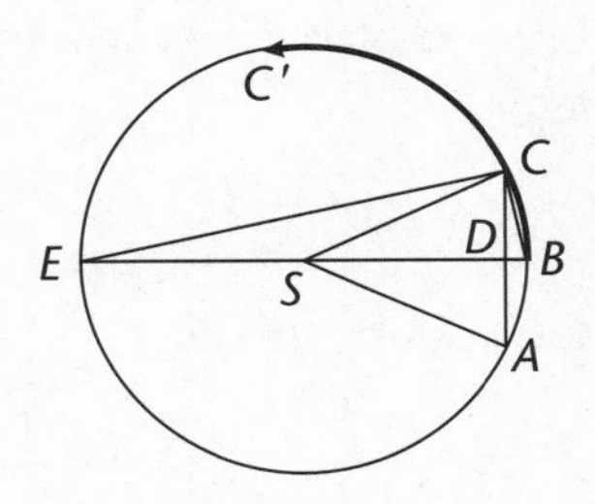

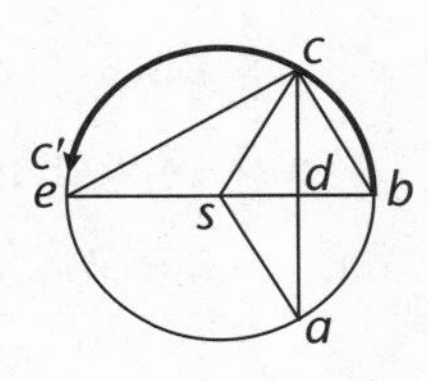

Now take two finite arcs BC' and bc' traversed in any equal finite time with the same respective speeds at which the evanescent arcs BC, bc are traversed.

By Galileo, *Two New Sciences*, Proposition 2 of Equable Motion, "If a moveable passes through two spaces in equal times, these spaces will be to one another as the speeds." This can easily be extended to two different bodies each in equable motion.

Therefore

$$\widehat{BC}' : \widehat{bc}' :: \text{speed on circle } S : \text{speed on circle } s,$$

and

$$\widehat{BC} : \widehat{bc} :: \text{speed on circle } S : \text{speed on circle } s.$$

And thus, by Euclid V.11,

$$\widehat{BC}' : \widehat{bc}' :: \widehat{BC} : \widehat{bc}.$$

Under these conditions, then, we may substitute the squares of the finite arcs for the squares of the evanescent ones.

By Equation 3 in Part 2, $f \overset{ult}{\propto} \text{arc}^2/\,\text{diameter}$.

So also, since radius $\propto$ diameter,

$$f \overset{ult}{\propto} \text{arc}^2/\,\text{radius}.$$

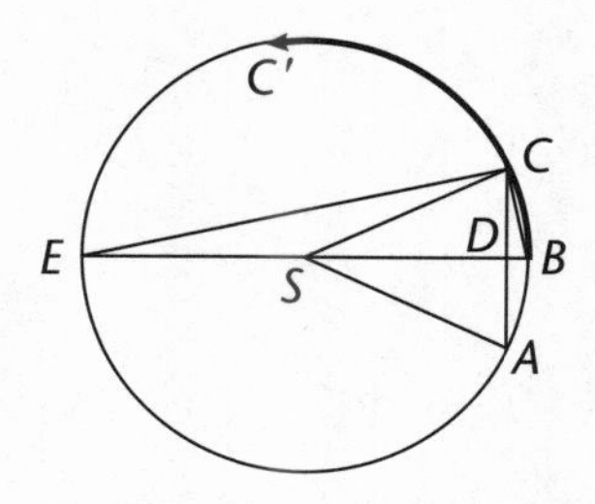

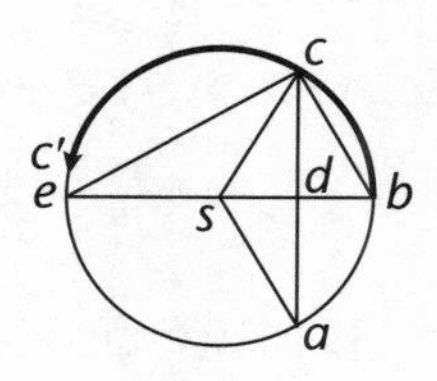

In terms of proportions, this is:

$F_1 : f_2 \overset{ult}{::} (\text{arc}_1{}^2/R_1) : (\text{arc}_2{}^2/r_2)$.

Or, $F_1 : f_2 \overset{ult}{::} (\text{arc}_1{}^2 : \text{arc}_2{}^2)$ comp $(r_2 : R_1)$.

Substituting the constant ratio of squares of finite arcs for the evanescent:

$F_1 : f_2 :: (\text{arc}_1{}^2 : \text{arc}_2{}^2)$ comp $(r_2 : R_1)$.

Or, $f \propto \text{arc}^2/\text{radius}$,

for finite equal-time arcs of circles with bodies in equable motion.

Q.E.D.

I.4 Corollary 1

Since those arcs are as the velocities of the bodies, the centripetal forces will be in the ratio compounded of the duplicate ratio of the velocities directly, and the simple ratio of the radii inversely.

Expansion of Newton's Sketch of I.4 Corollary 1

Given:

1. circles;
2. equable motion;
3. a given time t.

To Prove:

The forces will be as the squares of the velocities divided by the radii.

That is, $f \propto v^2/r$.

In terms of proportions: $f_1 : f_2 :: (v_1^2/r_1) : (v_2^2/r_2)$.

Or, $f_1 : f_2 :: (v_1^2 : v_2^2) \text{ comp } (r_2 : r_1)$.

Proof:

By definition, velocity = distance / time.

$v \propto \text{arc}/t$.

For a given constant time t, $v \propto \text{arc}$.

Squaring both sides, $v^2 \propto \text{arc}^2$.

By the proposition, $f \propto \text{arc}^2/\text{radius}$.

Substituting for arc^2,

$f \propto v^2/r$.

Q.E.D.

I.4 Corollary 2

And, since the periodical times are in the ratio compounded of the ratio of the radii directly and the ratio of the velocities inversely, the centripetal forces are in the ratio compounded of the ratio of the radii directly and the duplicate ratio of the periodic times inversely.

Expansion of Newton's Sketch of I.4 Corollary 2

Given:

1. motion in circles with center of forces at center of circle;
2. equable motion;
3. given time t.

To Prove:

Forces are in the ratio compounded of the ratios of the radii directly and the square of the periodic times inversely:

$f \propto r/P^2$.

Proof:

By definition, velocity = distance / time. Therefore time = distance / velocity.

$P = 2\pi r/v.$

$P \propto r/v.$

$v \propto r/P. \quad (1)$

$f \propto v^2/r.$ [Cor. 1]

Substituting for v from Equation 1:

$$f \propto \frac{(r/P)^2}{r}.$$

$f \propto r^2/(P^2 \times r).$

$f \propto r/P^2.$

Q.E.D.

I.4 Corollaries 3–7

Notes on I.4 Corollaries 3–7

• We are given in these corollaries that periodic times are proportional to the *n*th power of the distance from the center (or radius). In Corollary 3, $n = 0$; in Corollary 4, $n = 1/2$; in Corollary 5, $n = 1$; and in Corollary 6, $n = 3/2$. (Corollary 7 gives the generalized relationship.) That is, we are given two bodies whose periodic times are in the same ratio as the *n*th power of their radii. From that we prove that the centripetal forces are proportional to the radii in particular ways: directly (Corollary 3), in the subduplicate ratio (Corollary 4), inversely (Corollary 5), inversely as the squares (Corollary 6), or generally, inversely as the R^{2n-1} power (Corollary 7).

• We are so accustomed to thinking of gravity's inverse square law as continuous, of gravitation as a continuous field, that it may not occur to us that it must be proved. We have a conditional statement here, but even if its premise is established in the phenomena, its conclusion applies only to the two bodies at the two distances for which we are given the particular periodic times and radii.

In thought, or in mathematics, one may place a body at any distance and thereby calculate a periodic time for an inverse square force; but that fact has no relevance to conclusions about the physical world.

Corallary 6, which deduces the inverse square force law, and more generally all of Corollaries 3–7, may seem to say more than they do; one should not read into them generalizations that later investigations may or may not show to hold for our world. What we must note is that even when we have a particular force law deduced from a particular relationship of periodic times and distances, this relationship is argued only for *those two bodies.* It does not follow from these corollaries that the force on a third body placed at any other distance will vary in the stated ways in relation to either of the two originally given or to yet a fourth at some other distance.

For example, suppose Russia has a satellite in an orbit at 80,000 miles from the center of the earth. France places one at 20,000 miles out. Suppose further that their periodic times are observed to be in a ratio of 2:1. The Grand Duchy of Fenwick is readying a launch for an orbit at 5,000 miles out. Do these corollaries allow Fenwick to count on its satellite's periodic time being half that of France's satellite? They do nothing of the sort. The only thing Corollary 6 allows us to conclude is the relationship of the forces on Russia's satellite and France's. Should we in fact observe the relationship between the orbits of existing satellites given above, we can calculate that the periodic times are in the subduplicate ratio of the radii and so we can apply Corollary 4 and say that the forces on those two satellites will be equal.

But at this point in *Principia* we don't know whether the Grand Duchy of Fenwick will even be able to place its satellite 5,000 miles out. There might be some distances for which no orbits can exist, just as there are some orbitals at which electrons could not be placed. And it might be that for some orbits that do exist there is no discernable relationship between distance and periodic time that applies to all of them.

The discoverer of the 3/2 relationship of the periodic times of the six planets—Kepler—did not assume that another body could be dropped in and be expected to have the same periodic time relationship; he didn't imagine that there was anything universal here. On the contrary, the 3/2 ratio resulted from a delicate balancing of factors.

In Kepler's theory (*Epitome* I.1.4 and I.2.4) the force (a transverse, not a centripetal, force) varied inversely as the planet's distance from the sun. Its effect upon a planet was proportional to the planet's volume, which he supposed to be directly proportional to the planet's distance from the sun. Thus these two cancelled, and if there were no other factor, the periodic time would be directly proportional to the path length. But Kepler also thought that the speed would be influenced by the planet's weight or quantity of matter, which in his cosmology was proportional to the square root of the distance. Combining this with the effect of the path length turns out to produce a 3/2 power relation for the primary planets at their particular distances, volumes, and quantities of matter. In this view, another body placed in orbit around the sun would not follow this rule unless its volume and mass were adjusted appropriately.

If looking through modern eyes misleads us into assuming continuity, looking with modern eyes should also suggest caution. Modern physics teaches that the fact that we find a phenomenon in nature that exhibits certain relationships at certain distances does not entitle us to universalize the relationship to all distances. Some phenomena are limited by quantum states, or wavelengths and their multiples—interference or resonance phenomena, for example. The fact that an electron has a certain energy level at a certain size orbit (a certain distance from the nucleus) doesn't mean that one can extrapolate to any energy level at any corresponding orbit size. Some things are governed by physical laws that determine what is possible, regardless of how it might be imagined in thought or modeled in geometry.

• Corollaries 3–7 are easily proved using the first two corollaries. The algebraic manipulations are left to the student.

I.4 Corollary 3

Whence if the periodic times be equal, and therefore the velocities be as the radii, the centripetal forces will also be as the radii, and conversely.

I.4 Corollary 4

If both the periodic times and the velocities be in the subduplicate ratio of the radii, the centripetal forces will be equal among themselves, and conversely.

I.4 Corollary 5

If the periodic times be as the radii, and therefore the velocities equal, the centripetal forces will be inversely as the radii, and conversely.

I.4 Corollary 6

If the periodic times be in the sesquiplicate [i. e., the subduplicate of the triplicate] ratio of the radii, and therefore the velocities inversely in the subduplicate ratio of the radii, the centripetal forces will be inversely as the squares of the radii, and conversely.

Notes on I.4 Corollary 6

• See Section 6.3 of the Preliminaries for discussion of the powers of ratios. More algebraically, the triplicate ratio is a ratio of cubes; the subduplicate ratio is a ratio of square roots. The sesquiplicate ratio is a ratio of two quantities each taken to the 3/2 power. In this case, it is

$$P_1 : P_2 :: r_1^{3/2} : r_2^{3/2}.$$

• Kepler discovered that the six planets known to him moved such that their periodic times were in the sesquiplicate ratio of the radii (or, more prosaically, that their periodic times were as the $^3/_2$ powers of their radii) from the center of forces. This is what we now call his Third Law.

Newton is not here asserting Kepler's Third Law, but rather exploring what consequences such a discovery would have in a particular case, namely circles with equable motion and the center of forces at the center. This proposition and its corollaries, like all in Books I and II, are hypothetical and make no assertions about our physical universe. (That is reserved for Book III.) Instead, they make different suppositions and note the results: these if-thens form a toolchest of templates that he can match against the phenomena.

I.4 Corollary 7

And universally, if the periodic time be as any power R^n of the radii R, and therefore, the velocity be inversely as the R^{n-1} power of the radii, the centripetal forces will be inversely as the R^{2n-1} power of the radii, and conversely.

I.4 Corollary 8

All the same things concerning the times, velocities, and forces, by which bodies describe similar parts of any similar figures whatever, having the centers similarly situated in those figures, follow from the demonstration of the preceding applied to these particular cases. It is applied, however, by substituting the uniform description of areas for uniform motion, and by taking the distances of bodies from the centers for the radii.

Notes on I.4 Corollary 8

• This corollary extends all the relationships of Proposition 4 and Corollaries 1–7, which there applied to equal-time parts of equable motion in circles, to apply to equal-time similar parts of similar figures without equable motion.

• Despite the claim that "all the same things" apply, there is this significant difference: that for equable motion around circles the relationships are generally true, but for similar parts of similar figures we aren't given equable motion and it turns out (see proof) that the relationships are only true ultimately, that is, at the limit as $\Delta t \to 0$ and the corresponding arcs evanesce.

• **Geometry of Similar Curvilinear Figures.** As discussed in the Notes on Lemma 5, Newton has in mind a geometry of similar curvilinear figures with a concept of points "similarly situated" in the two similar figures. This geometry evidently includes the following theorems that Newton tacitly employs in Corollary 8.

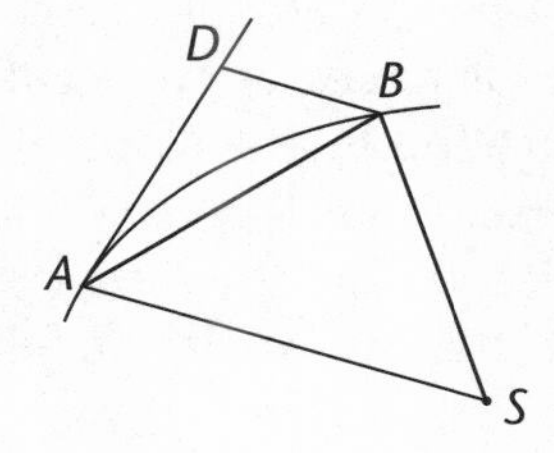

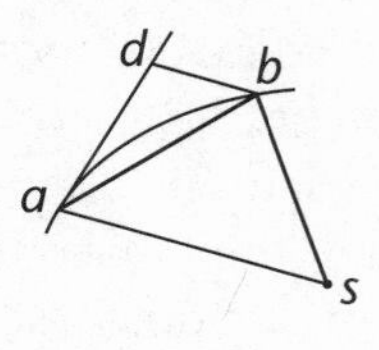

Suppose $\widehat{AB}$, $\widehat{ab}$ to be similar parts of similar figures, with S, s in similar positions. AD, ad are tangent at A and a.

Theorem 1. The angles between the two ends of similar arcs from a similarly situated point are equal. Angle ASB = angle asb.

Theorem 2. The ratios between the lines from the two ends of similar arcs to a point similarly situated are the same. $AS : SB :: as : sb$.

Theorem 3. The angles between the tangents at points on the ends of similar arcs and lines from those points of tangency to points similarly situated with respect to those arcs are equal. Angle DAS = angle das.

Theorem 4. In similar arcs, the chords are in the same ratio as the arcs. $\widehat{AB} : \widehat{ab} :: AB : ab$.

Theorem 5. The figures made up of similar arcs and another pair of points similarly situated are also similar. Sector $ABS \sim$ sector abs.

• The "demonstration" here is a good example of what I have called Newton's "hints and sketches." The unwary student might imagine that in some automatic way "uniform descriptions of areas" allows the substitution of similar parts of similar figures for the arcs of circles, and that no actual proof is required. That unwary student faces a rude awakening when he or she gets up to demonstrate the corollary and realizes that saying how uniform description of areas allows that substitution is not obvious to everyone. Suddenly, in fact, the formerly trusting soul at the board is being expected to supply that explanation.

This unpleasant experience can be avoided by remembering that Newton is not giving us the proof; rather, we are coming up with the proofs ourselves with help from Newton. In this case the help consists of a single succinct hint (slightly paraphrased): "Uniform description of areas can be used to derive these same relationships for similar parts of similar figures."

So, in a spirit of fun and adventure, take on Newton's challenge. All the pieces you will need have been given. And if the going isn't getting anywhere, and you've had all the fun to be had from it, a proof follows.

Expansion of Newton's Sketch of I.4 Corollary 8

Given:

1. similar figures;
2. centers of forces in similar positions in these figures;
3. equable description of areas.

To Prove:

We must re-prove Proposition 4 and Corollaries 1–7 using distances of bodies for radii (but note, if you're referring back here later to see what you proved, that they turn out to apply only ultimately).

Setting Out and Construction:

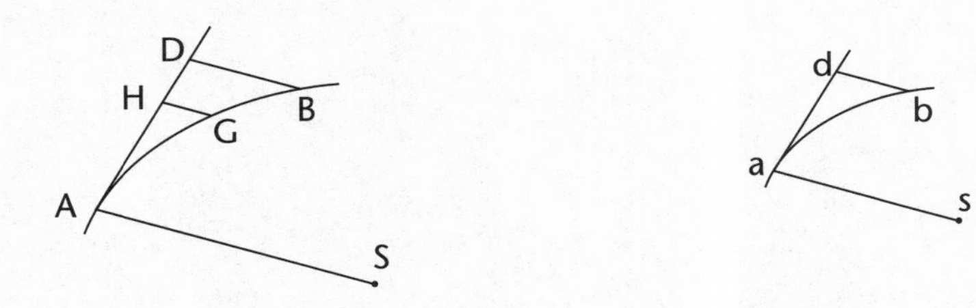

Suppose two orbits whose paths make similar figures. The centers of forces, *S* and *s*, around which areas are swept out in proportional times are similarly situated.

Arc $\widehat{AB}$ is taken similar to $\widehat{ab}$.

Suppose that in some time *t*, *A* goes to *G* while *a* goes to *b*.

Draw *AD*, *ad* tangents at *A*, *a*.

Draw *SA* and *sa* from the start of the similar arcs to the similarly-situated centers of forces for the orbits.

Draw *GH*, *bd* subtenses parallel to *SA*, *sa*.

Draw $BD \parallel AS$. Remember that in the lemmas we could find point D to construct the subtense BD any way we wanted. Now we make use of that freedom to make the subtense BD parallel to the line to the similarly situated point S. This will be needed in our proof.

Note: Remember that $\widehat{AG}$, $\widehat{ab}$ are arcs described in *equal times*; $\widehat{AB}$, $\widehat{ab}$ are the *similar* arcs.

Part 1: Proof for Proposition 4 Equivalence

To Prove for Part 1:

1. Centripetal forces on the bodies in any two similar figures will be directed to the centers of the figures.

2. The centripetal forces will be to each other as squares of equal-time arcs in the two similar figures divided respectively by the respective distances from the centers.

Proof for Part 1:

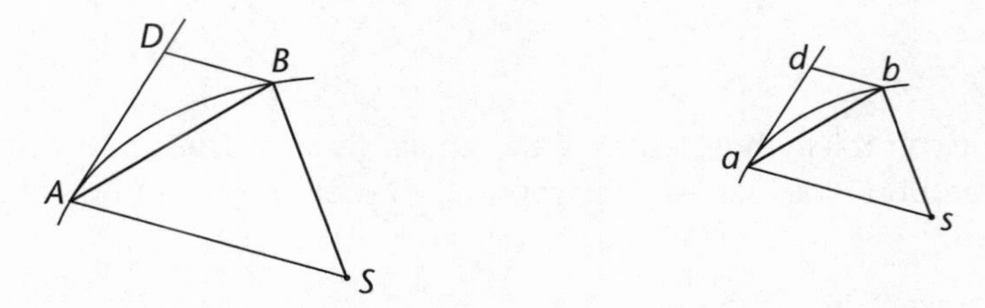

Because $DB \parallel AS$ and $db \parallel as$, by Euclid I.29,

$$\text{Angle } DBA = \text{angle } BAS \quad \text{and} \quad \text{angle } dba = \text{angle } bas. \tag{1}$$

(Theorems cited below are in the third bulleted note, Geometry of Similar Curvilinear Figures, in Notes on I.4 Corollary 8, above.)

Because S and s are "similarly situated,"

Angle ASB = angle asb. [Theorem 1 in note]

$AS : SB :: as : sb$. [Theorem 2 in note]

Therefore $\triangle ABS \sim \triangle abs$. [Eu. VI.6]

Therefore angle BAS = angle bas. (2)

By Equations 1 and 2,

angle DBA = angle dba. (3)

Angle DAS = angle das. [Theorem 3 in note] (4)

Therefore, subtracting Equation 2 from Equation 4,

angle DAB = angle dab. (5)

By Equations 3 and 5, and by Euclid I.32, VI Def. 1, and VI.4,

$\triangle DBA \sim \triangle dba$, and

$BD : bd :: AB : ab$.

But $AB : ab :: \widehat{AB} : \widehat{ab}$ [Theorem 4 in note]

Therefore $BD : bd :: \widehat{AB} : \widehat{ab}$. (6)

By Lemma 11 Corollary 2,

sagitta of arc $ab \overset{ult}{\propto}$ subtense bd;

sagitta of arc $AG \overset{ult}{\propto}$ subtense GH.

By Proposition 1 Corollary 4,

$f \overset{ult}{\propto}$ sagittae of arcs in equal times.

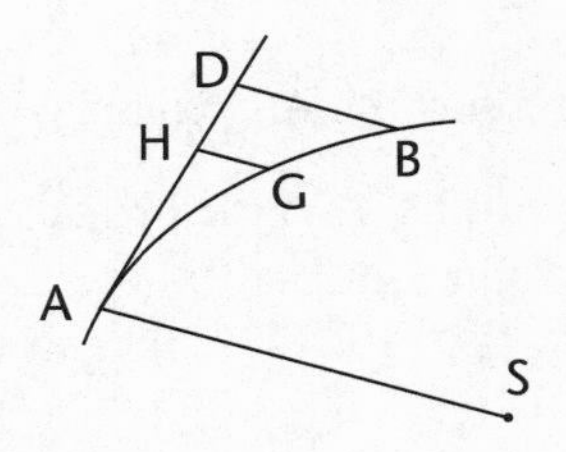

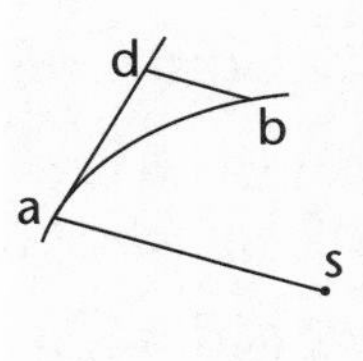

Arcs AG and ab are the equal-time arcs. GH and bd are the subtenses of the equal-time arcs.

Therefore

$f_A : f_a \overset{ult}{::} GH : bd$. (7)

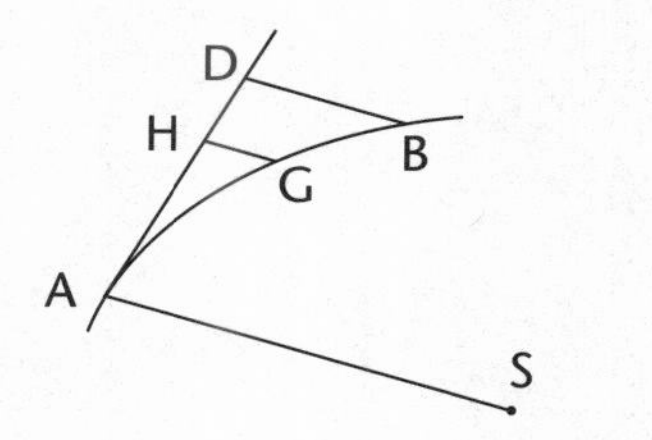

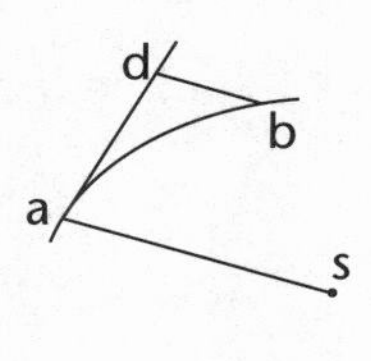

Now let's see what the ratio $GH : bd$ turns out to be.

By Lemma 11 Corollary 1, in a single curve, subtenses $\overset{ult}{\propto}$ arcs^2.

Therefore,

$$GH : BD \overset{ult}{::} \widehat{AG}^2 : \widehat{AB}^2. \tag{8}$$

By definition of compound ratio (see Euclid VI.23 or Sec. 6 of Preliminaries):

$$\widehat{AG}^2 : \widehat{AB}^2 :: (\widehat{AG}^2 : \widehat{ab}^2) \text{ comp. } (\widehat{ab}^2 : \widehat{AB}^2),$$

so Equation 8 becomes

$$GH : BD \overset{ult}{::} (\widehat{AG}^2 : \widehat{ab}^2) \text{ comp. } (\widehat{ab}^2 : \widehat{AB}^2).$$

But the ratio $\widehat{ab}^2 : \widehat{AB}^2$ is the *duplicate* of $\widehat{ab} : \widehat{AB}$; that is, it is the ratio $\widehat{ab} : \widehat{AB}$ compounded with itself (see section 6.3 of Preliminaries). Thus

$$GH : BD \overset{ult}{::} (\widehat{AG}^2 : \widehat{ab}^2) \text{ comp. } (\widehat{ab} : \widehat{AB}) \text{ comp. } (\widehat{ab} : \widehat{AB}). \tag{9}$$

Recall, though, that what we set out to find is the ratio $GH : bd$, not $GH : BD$. So, again by the definition of compound ratio, write

$$GH : bd :: (GH : BD) \text{ comp } (BD : bd).$$

Substituting for $(BD : bd)$ from Equation 6 and for $(GH : BD)$ from Equation 9 above,

$$GH : bd \overset{ult}{::} (\widehat{AG}^2 : \widehat{ab}^2) \text{ comp. } (\widehat{ab} : \widehat{AB}) \text{ comp. } (\widehat{ab} : \widehat{AB}) \text{ comp. } (\widehat{AB} : \widehat{ab}).$$

But the compound of a ratio such as $\widehat{ab} : \widehat{AB}$ with its own reciprocal, $\widehat{AB} : \widehat{ab}$, yields a ratio of *unity*. Thus the previous equation reduces to

$$GH : bd \overset{ult}{::} (\widehat{AG}^2 : \widehat{ab}^2) \text{ comp. } (\widehat{ab} : \widehat{AB}). \tag{10}$$

Now, since $\widehat{ab}$, $\widehat{AB}$ are similar parts of similar figures, with s, S in similar positions, we may invoke Theorem 5 in the note above to infer

$$\text{sector } abs \sim \text{sector } ABS;$$

therefore $\widehat{ab} : \widehat{AB} :: sa : SA$. [Lem. 5]

Substituting into Equation 10,

$$GH : bd \overset{ult}{::} (\widehat{AG}^2 : \widehat{ab}^2) \text{ comp. } (sa : SA).$$

Now we may substitute for $GH : bd$ from Equation 7, and obtain

$$F : f \overset{ult}{::} (\widehat{AG}^2 : \widehat{ab}^2) \text{ comp. } (sa : SA).$$

The force is proportional to the ratio of the squares of the arcs compounded with the inverse ratio of the distances. Compounding, then,

$$F : f \overset{ult}{::} \widehat{AG}^2 \times sa : \widehat{ab}^2 \times SA.$$

In the right-hand ratio, divide both antecedent and consequent by $sa \times SA$ to obtain:

$$F : f \overset{ult}{::} \frac{\widehat{AG}^2}{SA} : \frac{\widehat{ab}^2}{sa}.$$

Forces are ultimately proportional to the squares of the arcs described *in equal times* divided by their distance from the center, for similar parts of similar figures.

Q.E.D.

Proposition 4 concluded that $f \propto \text{arc}^2/r$. But since in Corollary 8 we don't have uniform motion, our conclusion can't be extended to the finite case.

Part 2: For Corollary 1

To Prove:

The forces in similar points of similar figures will be directly as the squares of the velocities at those points and inversely as the distances from the centers.

$$f \overset{ult}{\propto} v^2/r.$$

Proof:

By Galileo, *Two New Sciences*, Proposition 2 of Equable Motion, "If a moveable passes through two spaces in equal times, these spaces will be to one another as the speeds." Extending this to two different movables, the equal-time distances will be as the speeds.

These bodies are not in equable motion, but as $\Delta t \to 0$, the average velocities over $\widehat{AG}$ and $\widehat{ab}$ approach the velocities at A and a, so it would be true that:

$$V_A : v_a \overset{ult}{::} \widehat{AG} : \widehat{ab}.$$

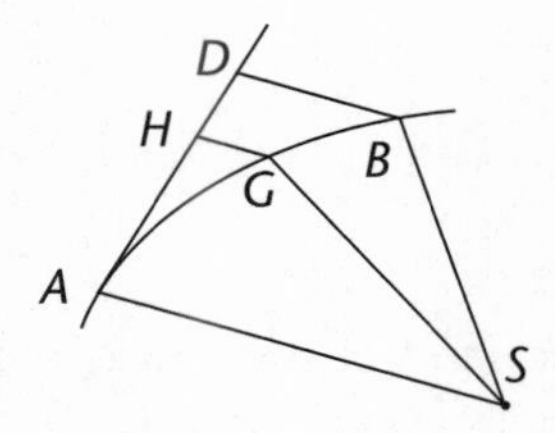

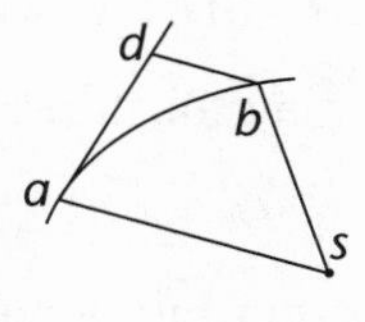

Squaring both sides, $V_A{}^2 : v_a{}^2 \overset{ult}{::} \widehat{AG}^2 : \widehat{ab}^2$.

Or, $v^2 \overset{ult}{\propto} \text{arc}^2$.

By the previous proof, $f \overset{ult}{\propto} \text{arc}^2/r$.

Substituting, $f \overset{ult}{\propto} v^2/r$.

Q.E.D.

Part 3: For Corollary 2

To Prove:

The forces in similar points of similar figures will be as the distances from similar centers of forces directly and as the squares of the periodic times inversely.

Proof:

We are given that the areas described are proportional to the times for each orbit. Letting P and p be the periodic times and taking the same time t for both,

$p : t$ :: orbit a : sector abs,

and $P : t$ :: orbit A : sector AGS.

Inverting the ratios of the second proportion and compounding the ratios of the two proportions, or, what is equivalent, treating them as equations and dividing the first by the second, we get

$p : P$ :: (orbit a : sector abs) comp (sector AGS : orbit A).

$p : P$ :: (orbit a : orbit A) comp (sector AGS : sector abs). (1)

Since the figures are similar, and $\widehat{AB} \sim \widehat{ab}$,

orbit a : orbit A :: sector abs : sector ABS.

Substituting into Equation 1,

$$p : P :: (\text{sector } abs : \text{sector } ABS) \text{ comp } (\text{sector } AGS : \text{sector } abs);$$

$$p : P :: \text{sector } AGS : \text{sector } ABS. \qquad (2)$$

But we have given that the areas described will be as the times; therefore

$$\text{sector } AGS : \text{sector } ABS :: t_{\widehat{AG}} : t_{\widehat{AB}}. \qquad (3)$$

Remember that we're not supposing equable motion, only equable description of areas. So in the finite case, the ratio of the times will not be as the arcs. The times over arcs will be proportional to the distances divided by the average velocities:

$$t_{\widehat{AG}} : t_{\widehat{AB}} :: (\widehat{AG} / V_{\widehat{AG}}) : (\widehat{AB} / V_{\widehat{AB}}).$$

But as arcs AG, AB evanesce, the average velocity along each approaches the instantaneous velocity at A, and therefore the ratio of velocities approaches a ratio of equality.

$$V_{\widehat{AG}} \overset{ult}{=} V_{\widehat{AB}}.$$

Therefore,

$$t_{\widehat{AG}} : t_{\widehat{AB}} \overset{ult}{::} \widehat{AG} : \widehat{AB}.$$

Substituting from Equation 3,

$$\text{sector } AGS : \text{sector } ABS \overset{ult}{::} \widehat{AG} : \widehat{AB}.$$

Substituting from Equation 2,

$$p : P \overset{ult}{::} \widehat{AG} : \widehat{AB}.$$

Squaring,

$$p^2 : P^2 \overset{ult}{::} \widehat{AG}^2 : \widehat{AB}^2.$$

By Part 1 of this corollary,

$$F : f \overset{ult}{::} (\widehat{AG}^2 : \widehat{AB}^2) \text{ comp } (R : r).$$

Substituting from Equation 4,

$$F : f \overset{ult}{::} (p^2 : P^2) \text{ comp } (R : r).$$

Thus the forces are ultimately proportional to the radii directly and the periodic times inversely.

$$f \overset{ult}{\propto} r / P^2.$$

Q.E.D.

Parts 4–8

The conclusions analogous to Corollaries 3–7 follow from the relationships derived above in Parts 1–3 just as they did for circles; but they will all be only ultimately true.

I.4 Corollary 9

From the same demonstration it also follows that the arc, which a body describes by revolving uniformly in a circle by a given centripetal force in any time whatever, is the mean proportional between the diameter of the circle and the descent traversed by the body falling under the same given force and in the same time.

Notes on I.4 Corollary 9

- Now we're back to circular equable motion with the center of forces at the center of the circle.

- You may be tempted, when you see the phrase "distance fallen," to mentally substitute "$d = \frac{1}{2}at^2$" from your post-Newtonian knowledge of gravitational acceleration. But that formula is not Newtonian style; Newton stays with proportions, not setting anything absolutely equal to anything else or specifying constants of proportionality. Instead of pulling in a modern formulation, try to put yourself in Newton's mind. How would he be thinking of these things? How did Galileo think of them? (No, Galileo nowhere speaks of "$d = \frac{1}{2}at^2$"; that isn't his style or his thinking either.) By going back to the most basic understanding of naturally accelerated motion, you will have what you need for this proposition.

Expansion of Newton's Sketch of I.4 Corollary 9

Given:

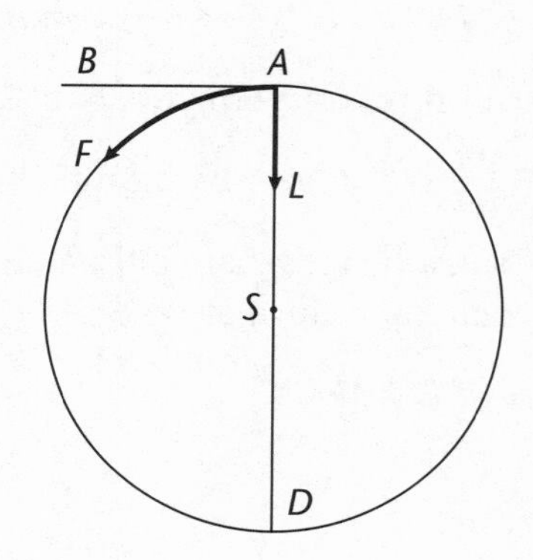

1. Circle *AFD* with diameter *AD* and center *S*, with the center of forces at the center of the circle.

2. A body describes arc *AF* in given time *t* in equable motion around center of forces *S*.

3. The same body falls from the same initial point *A* towards the same center of forces *S* covering distance *AL* in the same time *t*.

To Prove:

The arc traveled by the body in circular motion is a mean proportional between the circle's diameter and the distance fallen in the same time.

That is, $AD : \widehat{AF} :: \widehat{AF} : AL$.

Proof:

Step 1: The Ratio in the Finite Case

The distance traveled around the circle in equable motion is a function of time. According to Galileo, in *Two New Sciences*, Proposition 4 of Equable Motion, distances traveled in equable motion vary as the speeds times the times.

$$\text{arc} \propto v \times t.$$

So since v is constant, arc $\propto t$, and

$$\text{arc}^2 \propto t^2. \quad (1)$$

By Galileo, *Two New Sciences*, Proposition 2 of Naturally Accelerated Motion,

> If a moveable descends from rest in uniformly accelerated motion, the spaces run through in any times whatever are to each other as the duplicate ratio of their times; that is, are as the squares of those times.

$$\text{distance fallen} \propto t^2. \quad (2)$$

Combining Equations 1 and 2,

$$\text{arc}^2 \propto \text{distance fallen}. \quad (3)$$

This proportionality will hold for any two arcs and corresponding distances fallen, even if one is evanescent.

Let's see what the relationship between those two magnitudes is in the evanescent case, that is, at the limit as the given time approaches zero. We will then know the relationship in all cases.

Step 2: Finding the Relationship in the Ultimate Case

Everything we deduce in Step 2 must be understood as applying only to the limiting case at the very beginning of motion.

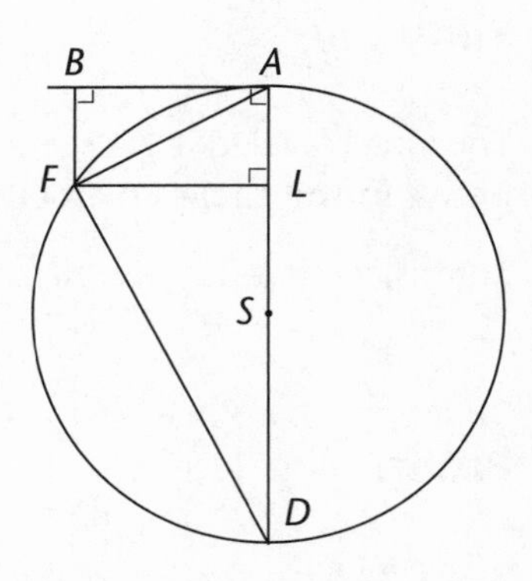

We are given that a body at A, with centripetal force directed to S, will follow path $\widehat{AF}$ in the time it would have fallen AL if it had no tangential velocity. [Given]

Let's take a look at least arc $\widehat{AF}$.

We are going to be treating this as Newton treated the force in Proposition 1 and pretend that the force over $\widehat{AF}$ acts all at once at point A. But we must remember that we are dealing with a nascent time interval and least arc.

The tangential velocity the body has at A would have taken it to B in the least time t that it takes to traverse arc $\widehat{AF}$.

Under the impulse towards S alone it would have gotten to L in the nascent time. Observe that $AL \perp AB$. By Corollary 1 to the Laws, the resultant is the diagonal AF of the parallelogram $ABFL$ (since we're at the limit, chord and arc coincide). And by Euclid I.28, the parallelogram will be a rectangle, and consequently $FL \perp AD$.

But since $\triangle AFD$ is in a semicircle it is a right triangle [Eu. III.31] and FL is dropped from the right angle to the base,

$\triangle ALF \sim \triangle AFD$. [Eu. VI.8]

$AD : AF :: AF : AL$. [Eu. VI.4]

But $\overline{AF} \overset{\text{ult}}{=} \widehat{AF}$. [Lem. 7]

So $AD : \widehat{AF} \overset{\text{ult}}{::} \widehat{AF} : AL$.

$$\widehat{AF}^2 \overset{\text{ult}}{=} \text{rect } AD, AL. \quad \text{[Eu. VI.17]} \tag{4}$$

Step 3: Conclusion

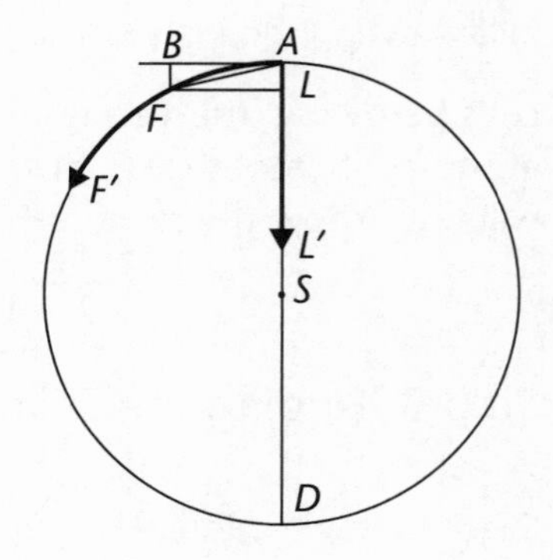

Now we put what we have discovered about the ultimate case together with what we could say generally.

Let's take two time intervals, one evanescent and one finite. The evanescent arc we will call AF, and the corresponding distance fallen will be AL. The finite arc will be AF', and the corresponding distance fallen AL'.

Because by Equation 3 in Step 1, for all time intervals, $\text{arc}^2 \propto$ distance fallen,

$$\widehat{AF}^2 : \widehat{AF'}^2 :: AL : AL'.$$

Applying a common side (constant AD, the diameter) to the third and fourth magnitudes of the proportion,

$$\widehat{AF}^2 : \widehat{AF'}^2 :: \text{rect } AL, AD : \text{rect } AL', AD.$$

We now have a proportion made of two ratios of areas.

Alternating, by Euclid V.16,

$$\widehat{AF}^2 : \text{rect } AL, AD :: \widehat{AF'}^2 : \text{rect } AL', AD. \tag{5}$$

Equation 4 in Step 2 concluded: $\widehat{AF}^2 \overset{ult}{=} \text{rect } AD, AL.$

Thus the square of the evanescent arc is equal to the rectangle of the evanescent distance fallen with the diameter.

Therefore, since the evanescent antecedent and consequent on the left hand side of Equation 5 are equal, the finite antecedent and consequent on the right hand side are also equal.

Therefore $\widehat{AF'}^2 = \text{rect } AD, AL';$

$$AD : \widehat{AF'} :: \widehat{AF'} : AL'. \quad \text{[Eu. VI.17]}$$

Therefore the arc, which a body describes by revolving uniformly in a circle by a given centripetal force in any time whatever, is the mean proportional between the diameter of the circle and the descent traversed by the body falling under the same given force and in the same time.

Q.E.D.

Scholium

The case of the sixth corollary applies in the celestial bodies (as our countrymen Wren, Hooke, and Halley have separately inferred), and for that reason I have decided to set forth matters pertaining to a centripetal force decreasing in the duplicate ratio of the distances from the centers more fully in what follows.

Moreover, with the help of the preceding proposition and its corollaries, is also deduced the ratio of a centripetal force to any known force, such as is the force of gravity. For if a body by the force of its gravity revolves in a circle concentric with the earth, this gravity is its centripetal force. Further, from the descent of heavy bodies we are given both the time of one revolution and the arc described in any time you please, by Cor. 9 of this proposition. And through propositions of this kind, Huygens, in his excellent treatise *On the Pendulum Clock*, has compared the force of gravity with the cetrifugal forces of revolving bodies.

[The rest of the scholium, which gives an alternative proof of I.4, is omitted.]

Proposition 5

Proposition I.5 shows how to find a center of forces given velocities at three places on an orbit. It is not used or referred to in any of the propositions in Newton's central argument, and so is omitted.

Proposition 6

If a body revolve in any orbit whatever about an immobile center in a non-resisting space, and describe any arc you please just now nascent, in the least possible time, and the sagitta of the arc be understood to be drawn, which bisects the chord and being produced passes through the center of forces, the centripetal force in the middle of the arc will be as the sagitta directly and the time twice inversely [i.e., inversely in the duplicate ratio of the time].

For the sagitta in a given time is as the force (by Proposition 1 Corollary 4), and by increasing the time in any ratio you please, because the arc increases in the same ratio, the sagitta is increased in the duplicate of that ratio (by Lemma 11 Corollary 2 and 3). Therefore, it is as the force once and the time twice. Let the duplicate ratio of the time be removed on both sides, and the force comes out to be as the sagitta directly and the time twice inversely.

Q.E.D.

The same thing is also easily demonstrated by Lemma 10 Corollary 4.

[Newton gives no diagram for this proposition, nor does he refer to the points of any diagram in the proof. He does provide a diagram for Corollary 1, which is included with the corollary.]

Notes on I.6

• Newton states this proposition for a body in a single orbit; however, nothing in the logic of the argument in the basic proposition precludes it being used for two different orbits. (This will not be the case for the all-important Corollary 1, which follows. As we shall see, a step of the derivation of

Corollary 1's expression for relative force requires something that in general is true only of a single orbit.)

- Newton gives us different tiny arcs on the orbit, traversed in "least" but unequal times. We wish to know the relationship between the forces in the middle of these least arcs in terms of the respective sagittae drawn from the center of forces, and the respective times of traversing the least arcs.

- We must remind ourselves that although in this enunciation and in future ones Newton speaks of "any arc" as if he were dealing with individual arcs, he is still working in proportions made up of ratios of forces and their corresponding arcs and times. We are not being told what the force in the middle of a particular nascent arc is absolutely, but how it compares to the force in the middle of another nascent arc.

Question for discussion: How can there be "least" times that are unequal?

If you would like a hint, read this paragraph. Consider that if unequal "least" arcs sounds confusing it may be because you are picturing the "least arcs" as finite in size: in that case, it is indisputable that if one is larger than another that is "least", the first cannot also be least. But remember that we have a process of vanishing here. Both arcs are diminishing without limit and the technical term *least* means that it has gotten smaller than any assignable magnitude. You might need to revisit the examples in the Pause Before Lemma 6: we saw that lines could vanish in a ratio of, say, $1 : \sqrt{2}$. Thus *as* they vanish one was larger than the other, even though both were "least quantities" at that point.

Question for discussion: Why does Newton insist in this enunciation that we are looking at "any arc you please just now nascent, in the least possible time" rather than just saying that we should take evanescent (or nascent) arcs?

Expansion of Newton's Sketch of I.6

"If a body revolve in any orbit whatever about an immobile center in a nonresisting space, and describe any arc you please just now nascent, in the least possible time, and the sagitta of the arc be understood to be drawn, which bisects the chord and being produced passes through the center of forces, the centripetal force in the middle of the arc will be as the sagitta directly and the time twice inversely [i.e., inversely in the duplicate ratio of the time]."

Given:

1. Nascent arcs and times;
2. sagittae through center of forces.

To Prove:

$f \overset{ult}{\propto} \text{sagitta}/t^2$.

Proof:

"For the sagitta in a given time is as the force (by Proposition 1 Corollary 4), ..."

Let there be given nascent arcs AC and ad traversed with least times T and t. Bodies describing these arcs are urged towards center of forces S. No pair of these arcs is supposed constant or equal.

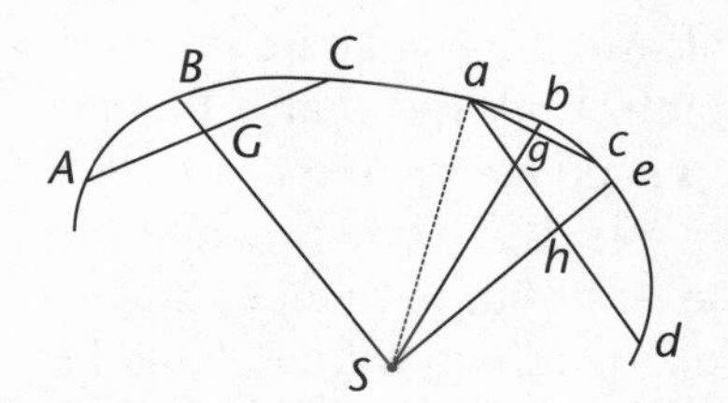

Let GB be the sagitta from S bisecting the chord AC.

Let he be the sagitta from S bisecting the chord ad.

We mean to prove that

$$F : f \overset{ult}{::} (GB : he) \text{ comp } ({t_{ad}}^2 : {T_{AC}}^2),$$

that is, that the force varies as the sagitta directly and the square of the time inversely.

Newton cites Proposition 1 Corollary 4, which requires equal-time arcs.

Let $\widehat{AC}$, $\widehat{ac}$ be arcs described in equal times. (Note that arc ac may be either smaller or larger than arc ad.) Let gb be the sagitta from S bisecting chord ac.

Now GB, gb are the sagittae that bisect the chords AC, ac of equal-time arcs. So we can apply Proposition 1 Corollary 4.

For equal times,

$$GB : gb \overset{ult}{::} F : f. \quad \text{[I.1 Cor. 4]} \tag{1}$$

"... and by increasing the time in any ratio you please, because the arc increases in the same ratio, ..."

Since $\widehat{ad}$ and $\widehat{ac}$ are both least arcs, they are both being considered at the moment of departing from point a. The velocity of the body as it departs from a is the same regardless of which arc it is imagined as travelling over. Thus the velocities over both arcs are ultimately equal. From this it follows that the arcs increase in the same ratio as the times.

"...the sagitta is increased in the duplicate of that ratio (by Lemma 11 Corollary 2 and 3)."

By Lemma 11 Corollary 3, for a given velocity, on the same curve, the sagittae vary ultimately as squares of the times. (Note that this is an ultimate ratio because only as *c* and *d* reach *a* are the velocities equal.)

For equal velocities on the same curve,

$$gb : he \overset{ult}{::} T_{ac}^{2} : t_{ad}^{2}. \qquad (2)$$

"Therefore, it is as the force once and the time twice."

Putting Equations 1 and 2 together following Euclid VI.23:

$$\text{GB} : \text{he} \overset{ult}{::} (F : f) \text{ comp } (T_{ac}^{2} : t_{ad}^{2}).$$

But $T_{AC} = T_{ac}$ because these are equal-time arcs; therefore:

$$\text{GB} : \text{he} \overset{ult}{::} (F : f) \text{ comp } (T_{AC}^{2} : t_{ad}^{2}).$$

"Let the duplicate ratio of the time be removed on both sides, and the force comes out to be as the sagitta directly and the time twice inversely."

Compound both sides with the inverse ratio of squares of times:

$$(GB : he) \text{ comp } (t_{ad}^{2} : T_{AC}^{2}) \overset{ult}{::} F : f.$$

That is, the forces are ultimately as the sagittae of least arcs directly and the squares of the corresponding least times inversely.

Q.E.D.

"The same thing is also easily demonstrated by Lemma 10 Corollary 4."

After giving the preceding proof, it evidently struck Newton that this proposition could also be proved using Lemma 10 Corollary 4, which states that at the very beginning of motion, the forces are as the spaces described directly and the squares of the times inversely.

The "spaces" mentioned in the corollary are the deviations from a path already in progress. The conclusion applies only to those deviations under a force, not to any other motion, including inertial tangential motion. Therefore the spaces are not the arcs traveled, but rather are the lengths of the subtenses. And, by Lemma 11 Corollary 2, the subtenses vary as the sagittae. Thus the proof falls out easily.

I.6 Corollary 1

If a body P in revolving about center S describe the curved line APQ, while the straight line ZPR touch that curve at any point you please P, and from any other point on the curve you please Q there be drawn QR to the tangent parallel to the distance SP, and QT be dropped perpendicular to that distance SP, the centripetal force will be

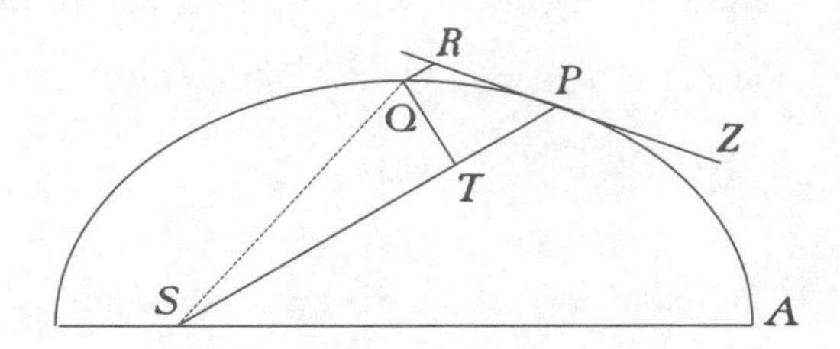

inversely as the solid $\frac{\text{sq.}SP \times \text{sq.}QT}{QR}$, *provided that the quantity of that solid always be taken which it has ultimately, when the points P and Q come together. For QR is equal to the sagitta of double the arc QP, at whose middle P is; and twice the triangle SQP, or* $SP \times QT$, *is proportional to the time in which that double arc is described; and therefore may be used to represent the time.*

Notes on I.6 Corollary 1

• This corollary makes an assumption (see proof) of equable description of areas between the points being compared. By Proposition 1, for centripetal forces, this will always be true for a particular orbit. It will in general *not* be true for different bodies in different orbits, even if they share a center of forces. For example, it would be the purest coincidence if in one hour the earth described the same area around the sun that one of Jupiter's moons described around Jupiter. If in one hour the earth and Jupiter described equal areas around the sun, the periods would be as the areas; that is, periodic times would be as the squares of the radii. By Proposition 4 Corollary 7 it would follow that an inverse cube force law was at work.

Thus I.6 Corollary 1 may be used for different orbits only if it can be proved that there is equable description of areas between the two orbits (not just in each individually).

• We are working with two adjacent positions of a body traveling in an orbit, points *P* and *Q*, which correspond to points *B* and *A* in the Lemmas.

On the orbit, the body is moving from *P* to *Q*. In the finite case, *Q* is a later position of the body. In the ultimate case, *Q* is the "next" position, or the position as the body is just leaving position *P*.

In our thought experiment of taking the limit, we imagine *Q* to be moving towards *P*. This is the direction opposite to the one in which we suppose the body actually to be moving.

Thus, in the taking-the-limit thought exercise, $Q \rightarrow P$. This is happening while the motion of the body remains $P \rightarrow Q$.

This will come up repeatedly in this and the following propositions. It can be confusing to students who are trying to move through the demonstrations too quickly. But if you think it through at the start, it won't snag you.

- The "solid" $\frac{SP^2 \times QT^2}{QR}$, for different times and forces but for the same center of forces, will turn out to vary inversely as the forces. This entity, whether we think of it as geometric or algebraic, will be a useful parameter for us in the following propositions.

- **Why "solid"?** As a geometric entity $\frac{SP^2 \times QT^2}{QR}$ is a "solid," that is, a three-dimensional figure. It has four dimensions—sides or magnitudes—in the numerator and one in the denominator. This may strike the modern mind as a bit fanciful, but it's worth pondering the fact that Newton envisioned it this way, at least enough to allow him to use the terminology. And although we no longer speak of solids (or rectangles to mean a product, as Newton does throughout this book), we are perfectly comfortable talking about "squares" of numbers.

- **Question for Discussion:** We can also think of the inverse, $QR/(SP^2 \times QT^2)$, as varying directly with the force. (This direct form, however, could not be called a "solid" since it no longer has three dimensions.) But Newton doesn't choose to express what the forces are proportional to as direct variation. Why do you think Newton chose a more cumbersome expression? Keep this question in mind as you watch him unfold his system in the upcoming propositions.

- **Let's think about the elements of "the solid."**

Because this entity $\frac{SP^2 \times QT^2}{QR}$ is so important in the upcoming propositions as the determinate of the relative forces, it can be helpful to have an intuitive sense of the elements that make it up.

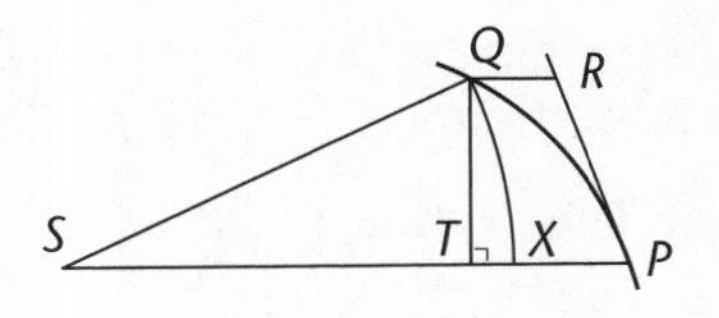

QR is the subtense of tangent at P and arc PQ; by Lemma 11 Corollary 2, the sagittae vary ultimately as QR. By Proposition 1 Corollary 4, for equal times, the forces $F \overset{ult}{\propto}$ sagittae. Therefore the force at P varies ultimately as this QR for equal times. This agrees with what may be our intuitive sense that QR is the measure of how far the body is deflected from the rectilinear course PR in going from P to Q.

QT is how much the body has moved around S. Motion PQ is composed of PX (or RQ) towards S plus motion XQ around S. As $Q \rightarrow P$, SX (which = SQ) $\rightarrow ST$. QT is also the height of $\triangle SQP$, the area described in the proportional time. (Note that QT is not $\overset{ult}{=} RP$ unless $SP \perp$ tangent.)

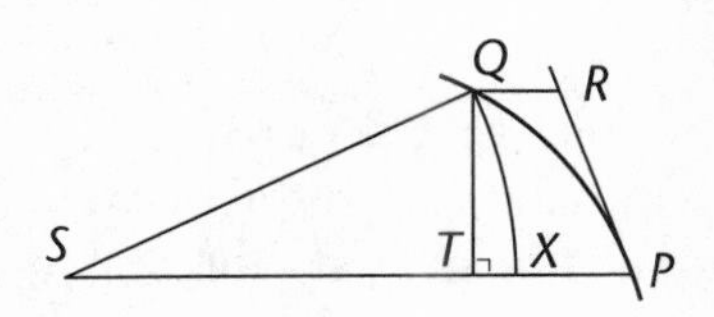

SP is the distance of the body from the center of forces. *SP* is also the base of $\triangle SQP$, the area described in proportional times.

$SP, QT \propto \triangle SQP \propto$ the times, given equable description of areas.

Question for Discussion: What will happen to the ratio of forces as the sizes of the different elements change relative to one another? For example, what does it mean for *SP* and *QT* to stay the same while *QR* changes?

- This corollary restates in a different way what the proposition has already given us, with the added required assumption of a single orbit or at least equable description of areas across the two orbits (see first bulleted note above). The proposition said that the forces are ultimately as the sagittae directly and the squares of the times inversely. This corollary says that the forces are ultimately as *QR* directly and the squares of *SP, QT* inversely.

What Newton needs to prove in this corollary is that *QR* does vary ultimately as the sagittae and furthermore that the product of *SP* and *QT* does vary ultimately as the times.

Expansion of Newton's Sketch of I.6 Cor 1

"If a body *P* in revolving about center *S* describe the curved line *APQ*, while the straight line *ZPR* touch that curve at any point you please *P*, and from any other point on the curve you please *Q* there be drawn *QR* to the tangent parallel to the distance *SP*, and *QT* be dropped perpendicular to that distance *SP*, the centripetal force will be inversely as the solid $\frac{\text{sq.}SP \times \text{sq.}QT}{QR}$, provided that the quantity of that solid always be taken which it has ultimately, when the points *P* and *Q* come together."

Given:

1. Points *P* and *Q* on curve with a body moving from *A* through *P* towards *Q*;

2. line *ZPR*, tangent at point *P*;

3. $QR \parallel SP$ and $QT \perp SP$;

4. let $Q \to P$.

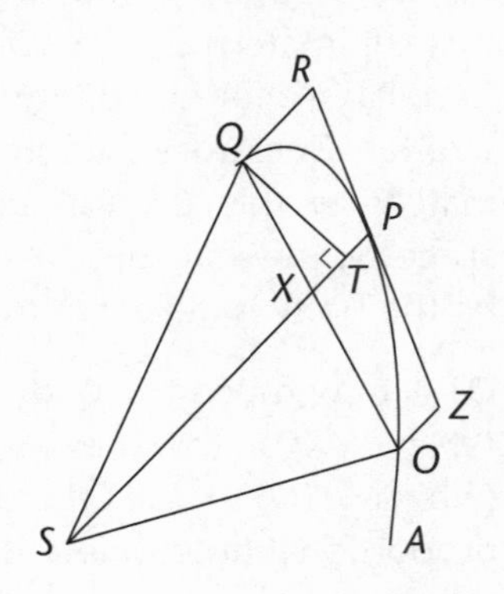

To Prove:

$$F \overset{ult}{\propto} 1/\frac{SP^2 \times QT^2}{QR}.$$

Proof:

Let $\widehat{PO} = \widehat{QP}$.

Draw $OZ \parallel QR$. (Thus OZ, QR, SP are all parallel; this is needed later.)

"For QR is equal to the sagitta of double the arc QP, at whose middle P is; ..."

We must prove that the subtense is ultimately equal to the sagitta of twice the arc; in particular that XP is ultimately the sagitta of the double arc $\widehat{QO}$, and that QR is ultimately equal to that sagitta.

By Lemma 11,

$QR : OZ \overset{ult}{=} PR^2 : PZ^2$

But $PR \overset{ult}{=} \widehat{PQ}$ and $PZ \overset{ult}{=} \widehat{PO}$ [Lem. 7]

and $\widehat{PQ} = \widehat{PO}$; [by construction]

therefore $PR \overset{ult}{=} PZ$.

Therefore $QR \overset{ult}{=} OZ$.

And $QR \parallel OZ$. [construction]

Therefore, by Euclid I.33, area $QRZO$ is ultimately a parallelogram,

as are $QRPX$ and $XPZO$ (this is needed later); (1)

therefore

$QX \overset{ult}{=} PR$. (2)

and by Equation 1,

$XO \overset{ult}{=} PZ$. (3)

But we proved that $PR \overset{ult}{=} PZ$; (4)

Therefore by Equations 2, 3, and 4, $QX \overset{ult}{=} XO$; (5)

and therefore

XP is ultimately the sagitta of the double arc $\widehat{QO}$.

But $XP \overset{ult}{=} QR$, since they are opposite sides of $QXPR$, which by Equation 1 is ultimately a parallelogram.

Therefore subtense $QR \overset{ult}{=} XP$, the sagitta of the double arc $\widehat{QO}$.

Q.E.D.

"... and twice the triangle SQP, or $SP \times QT$, is proportional to the time in which that double arc is described; and therefore may be used to represent the time."

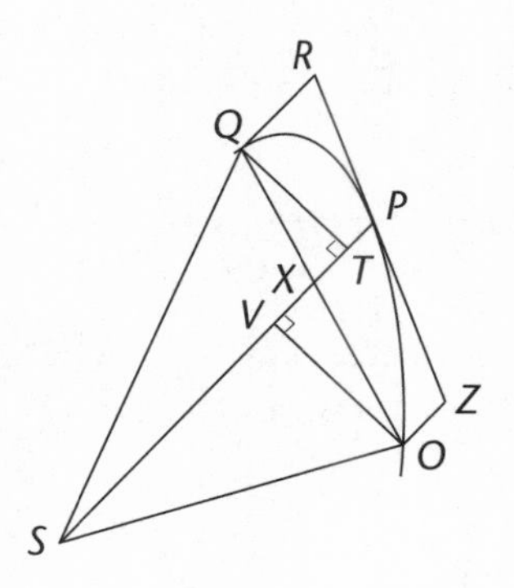

Now we prove that twice the curvilinear area $\widehat{\triangle}QSP$ is proportional to the time in which the double arc is described.

Note that we picked point O such that arcs $\widehat{PO} = \widehat{QP}$; these arcs are not necessarily traversed in equal times.

Newton now makes a step slightly inconsistent with his usual style and sets the area of $\widehat{\triangle}QSP$ actually equal (ultimately) to half its base times its height.

Rectilinear $\triangle QSP = \frac{1}{2}(SP \times QT)$.*

By Lemma 7, arc $QP \overset{ult}{=}$ chord QP.

Therefore $\widehat{\triangle}QSP \overset{ult}{=} \triangle QSP$.

So $\widehat{\triangle}QSP \overset{ult}{=} \frac{1}{2}(SP \times QT)$.

Given this, it follows that twice that curvilinear triangle will ultimately be equal to its base times its height. Thus

$2 \times \widehat{\triangle}QSP \overset{ult}{=} (SP \times QT)$.

But what Newton is claiming is that $\widehat{\triangle}QSO \overset{ult}{=} (SP \times QT)$.

We have picked O such that $\widehat{OQ}$ is twice $\widehat{PQ}$.

But is $\widehat{\triangle}QSO \overset{ult}{=} 2 \times \widehat{\triangle}QSP$?

To answer this, we must draw $OV \perp SP$.

By Equation 5, $QX \overset{ult}{=} XO$.

Angle OVX = angle QTX. [both right angles]

Angle VXO = angle QXT. [Eu. I.15]

Therefore triangles QXT and OXV are ultimately congruent [Eu. I.26], and so

$QT \overset{ult}{=} OV$.

* The area of a rectilinear triangle is half the product of its base and its height. You can derive this from Euclid I.42 (*To construct ... a parallelogram equal to a given triangle*) and VI.23 (*Equiangular parallelograms have to one another the ratio compounded of the ratios of their sides*). Bear in mind that the unit of measurement for area is a square whose side is 1.

Thus the two curvilinear triangles $\widehat{\triangle}QSP$ and $\widehat{\triangle}PSO$ are ultimately equal, because they are on the same base SP and ultimately have equal heights QT and OV.

Therefore, we may conclude that since $\widehat{\triangle}QSO \overset{ult}{=} \widehat{\triangle}QSP + \widehat{\triangle}PSO$,

$$\widehat{\triangle}QSO \overset{ult}{=} 2\,\widehat{\triangle}QSP.$$

Consequently, $\widehat{\triangle}QSO \overset{ult}{=} SP \times QT$, as Newton claimed.

But by Proposition 1, given a center of forces, the areas described will be proportional to the times. Thus the double area $\widehat{\triangle}QSO$ will take twice the time.

By the proposition, $F \overset{ult}{\propto} \text{sagittae}/t^2$.

So now for twice the arc:

XP is ultimately the sagitta of twice the arc; $SP \times QT$ varies as the time of the double arc.

$$F \overset{ult}{\propto} XP\,/\,(SP \times QT)^2.$$

But because $QRPX$ is ultimately a parallelogram, by Equation 1, $XP \overset{ult}{=} QR$.

$$F \overset{ult}{\propto} QR\,/\,(SP \times QT)^2;$$

$$F \overset{ult}{\propto} 1/\frac{SP^2 \times QT^2}{QR}.$$

Q.E.D.

Question for Discussion: Note the odd conclusion we have come to on the way to this result as Newton has proved it. We have seen that, ultimately, successive equal arcs define equal areas. In Proposition 1, we proved that successive equal-*time* arcs define equal areas, and noted that the distances could be different. Now, for the sake of our proof, we seem to have shown that equal-time arcs are equal-distance arcs. What do you make of this conclusion?

If you want a hint, read this paragraph. You might notice that Newton's proof depends on these arcs being two successive evanescent arcs. Does that suggest to you that we're really talking about (essentially) the same point? If it would not work for two equal-time arcs at different places on the orbit, maybe we haven't thrown out the distinction between equal-time and equal-distance arcs after all.

Question for Discussion: Why did Newton use this complicated and difficult way to approach the proof? Wouldn't it have been easier to have simply substituted $\frac{1}{2}(SP \times QT)$ for the time, and QR for the sagitta, on the single arc PQ? (In the course of proving Lemma 11 Corollary 2 it was proved that the subtense QR is ultimately four times the sagitta.) It seems unlikely that such a line of proof would not have occurred to him. What does his more complicated way bring out or allow for that the shorter way does not? Is it more in keeping with the approach he has been taking in the previous propositions? Does the simpler approach make some assumption he would not want to make at this point? What do you think?

[Corollaries 2–4 were introduced in the second edition to support a series of alternative proofs (headed *idem aliter*, "the same thing differently") in the following propositions. This alternative series, though interesting in itself, does not constitute a necessary component of our "toolkit," and so will be omitted. Had Newton wished to replace his original proofs with these apparently simpler alternative proofs, he had two editions after the first in which to do so; he chose to leave the ones we are doing as his basic demonstrations.]

I.6 Corollary 5

Hence if any curvilinear figure you please APQ be given, and in it be given also the point S to which the centripetal force is always directed, the law of centripetal force may be found by which any body you please P, perpetually drawn back from a rectilinear course, will be confined to the perimeter of that figure, and will trace it out by revolving. That is, one computes the solid $\frac{\text{sq.}SP \times \text{sq.}QT}{QR}$ *... [that is] inversely proportional to this force. We shall give examples of this procedure in the following problems.*

Notes on I.6 Corollary 5

- Newton has now laid the groundwork and is starting out with a new sequence of propositions in which the method developed in Proposition 6 Corollary 1 and summarized in this corollary will be applied. He will look at the ratio between versions of this "solid" at different points on the orbit to get the inverse ratio of forces and thereby determine force laws for various situations.

- The ratio of forces given by Corollary 1, and whose usefulness is spelled out in Corollary 5, will apply to any continuous curve and any force law, except for a very few problem points on certain curves that don't satisfy the conditions of Lemma 11.

Proposition 7

If a body orbit on the circumference of a circle, it is required to find the law of centripetal force tending to any given point.

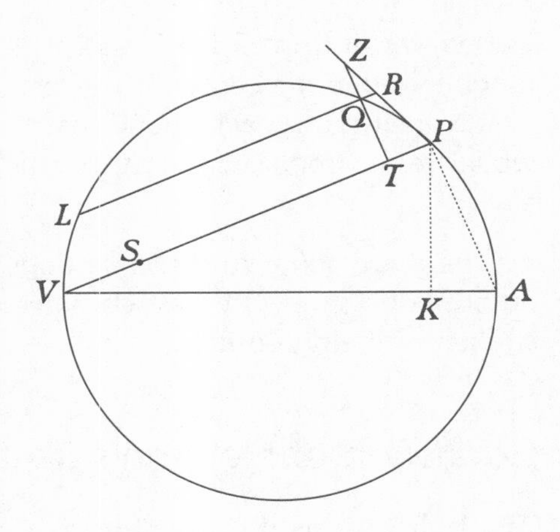

Let the circumference of the circle be $VQPA$, S the given point to which the force tends as to its center, P the body borne on the circumference, Q the next place to which it will move, and PRZ the tangent to the circle at the former place. Through the point S let the chord PV be drawn, and, the circle's diameter VA being drawn, let AP be connected, and QT be drawn perpendicular to SP, and extended to intersect the tangent PR at Z, and finally, let LR be drawn through Q, and let it be parallel to SP and meet the circle at L and the tangent PZ at R. And because of the similar triangles ZQR, ZTP, and VPA, sq.RP (that is, QR, RL) will be to sq.QT as sq.AV to sq.PV. Therefore, $\frac{QR, RL \times \text{sq.}PV}{\text{sq.}AV}$ is equal to sq.QT. Let these equals be multiplied by $\frac{\text{sq.}SP}{QR}$, and, points P and Q coming together, let PV be written for RL. Thus $\frac{\text{sq.}SP \times \text{cub.}PV}{\text{sq.}AV}$ will be made equal to $\frac{\text{sq.}SP \times \text{sq.}QT}{QR}$. Therefore (by Corollaries 1 and 5, Proposition 6), the centripetal force is inversely as $\frac{\text{sq.}SP \times \text{cub.}PV}{\text{sq.}AV}$; that is (because sq.$AV$ is given), inversely as the square of the distance or altitude SP and the cube of the chord PV conjointly.

Q.E.I.

[Q.E.I. is an abbreviation for *quod erat inveniendum*, "that which was to have been found."]

[The diagram shown with Newton's text is the one Newton provided. Point *K* is a remnant of the proof he gave in the first edition, a demonstration replaced in the second edition by the one presented here.]

Notes on I.7

• This proposition deals with eccentric circular motion. The preference from ancient times had been for heavenly bodies to move in equable motion in circles. Ptolemy had given his allegiance to equable motion in circles, even at the cost of having to introduce the equant (which referred the equable motion to a center other than the center of the circle). This preference for circularity goes well into the seventeenth century. Copernicus eschewed eccentric motion and the Ptolemaic equant and made all motions on all circles take place around the centers of the circles, thus adhering to an even simpler principle of equable circular motion than that used by Ptolemy, but at the cost of having to offer a more complicated model. Ptolemy could have done something equivalent to what Copernicus did, but chose the simpler model at the sacrifice of the simpler principle. Kepler abandoned circular orbits for ellipses, but his noncircular orbits were not immediately accepted;

Galileo, for example, didn't accept them, staying with Copernican circular motion around the centers of circles.

It is against this background that Newton here explores the force law that would result from circular motion around a center not the center of the circle.

- There are some further questions and notes given in the Pause After Proposition 7; they are best considered after working through the demonstration of the proposition.

Expansion of Newton's Sketch of I.7

"If a body orbit on the circumference of a circle, it is required to find the law of centripetal force tending to any given point."

Given:

1. Body revolving in the circumference of a circle;
2. center of forces not necessarily at the center of the circle.

To Find:

The force law directed to any given point as center of forces.

Proof:

"Let the circumference of the circle be *VQPA*, *S* the given point to which the force tends as to its center, *P* the body borne on the circumference, *Q* the next place to which it will move, and *PRZ* the tangent to the circle at the former place. Through the point *S* let the chord *PV* be drawn, and, the circle's diameter *VA* being drawn, let *AP* be connected, and *QT* be drawn perpendicular to *SP*, and extended to intersect the tangent *PR* at *Z*, and finally, let *LR* be drawn through *Q*, and let it be parallel to *SP* and meet the circle at *L* and the tangent *PZ* at *R*."

Let *S* be the point towards which the force is directed.

Let *VA* be the diameter of the circle, *PV* a chord, *PZ* tangent at *P*.

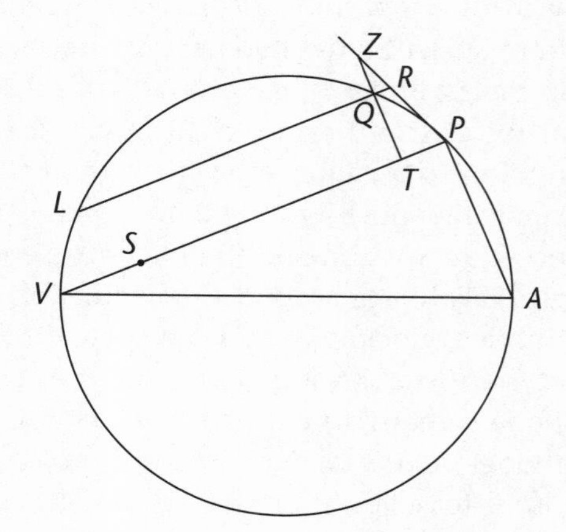

"And because of the similar triangles ZQR, ZTP, and VPA, $sq.RP$ (that is, QR, RL) will be to $sq.QT$ as $sq.AV$ to $sq.PV$. Therefore, $\frac{QR, RL \times sq.PV}{sq.AV}$ is equal to $sq.QT$."

By the construction of Proposition 6 Corollary 1, $QR \parallel SP$.

Therefore in $\triangle ZPT$ a line QR has been drawn parallel to base PT.

$ZR : RP :: ZQ : QT$. [Eu. VI.2]

$ZP : RP :: ZT : QT$. [Eu. VI.17, *componendo*]

Alternating, by Euclid V.16,

$$ZP : ZT :: RP : QT. \quad (1)$$

And by Euclid VI.5, $\triangle ZQR \sim \triangle ZTP$—a fact noted by Newton but, it seems, not strictly needed.

By Euclid III.32, if a straight line PZ touch a circle, and from point of contact P there be drawn across, in the circle, a straight line PV cutting the circle, the angle ZPT it makes with the tangent equals angle PAV in the alternate segment of the circle.

angle ZPT = angle PAV. [Eu. III.32]

By the construction of Proposition 6 Corollary 1, angle ZTP is a right angle.

By Euclid III.31, angle VPA is a right angle.

angle ZTP = angle VPA. [Eu. I Post. 4]

Therefore $\triangle ZPT \sim \triangle VPA$.

$ZP : ZT :: AV : PV$. [Eu. VI.4]

Substituting from Equation 1, $RP : QT :: AV : PV$.

Squaring, $RP^2 : QT^2 :: AV^2 : PV^2$.

$QT^2 \times AV^2 = RP^2 \times PV^2$. [Eu. VI.16]

Rewriting, $QT^2 = RP^2 \times PV^2 / AV^2$.

But by Euclid III.36, if from a point R outside a circle a line RL cuts and another line RP touches, the rectangle made up of the whole RL and part outside the circle QR is equal to the square on RP.

RP^2 = rect RL, QR. [Eu. III.36]

Substituting,

$$QT^2 = \frac{(\text{rect } RL, QR) \times PV^2}{AV^2}. \quad (2)$$

"Let these equals be multiplied by $\frac{\text{sq.}SP}{QR}$, and, points P and Q coming together, let PV be written for RL. Thus $\frac{\text{sq.}SP \times \text{cub.}PV}{\text{sq.}AV}$ will be made equal to $\frac{\text{sq.}SP \times \text{sq.}QT}{QR}$."

Multiply both sides of Equation 2 by SP^2/QR:

$$QT^2 \times \frac{SP^2}{QR} = (RL \times QR)\frac{PV^2}{AV^2} \times \frac{SP^2}{QR}.$$

Cancelling QR and rewriting,

$$\frac{QT^2 \times SP^2}{QR} = \frac{RL \times PV^2 \times SP^2}{AV^2}.$$

At the limit, as $Q \to P$, $RL = PV$; so substituting:

$$\frac{QT^2 \times SP^2}{QR} \overset{ult}{=} \frac{PV^3 \times SP^2}{AV^2}. \tag{3}$$

"Therefore (by Corollaries 1 and 5, Proposition 6), the centripetal force is inversely as $\frac{\text{sq.}SP \times \text{cub.}PV}{\text{sq.}AV}$; that is (because sq. AV is given), inversely as the square of the distance or altitude SP and the cube of the chord PV conjointly.

Q.E.I."

Now by Proposition 6 Corollary 1,

$$f \overset{ult}{\propto} 1 \Big/ \frac{QT^2 \times SP^2}{QR}.$$

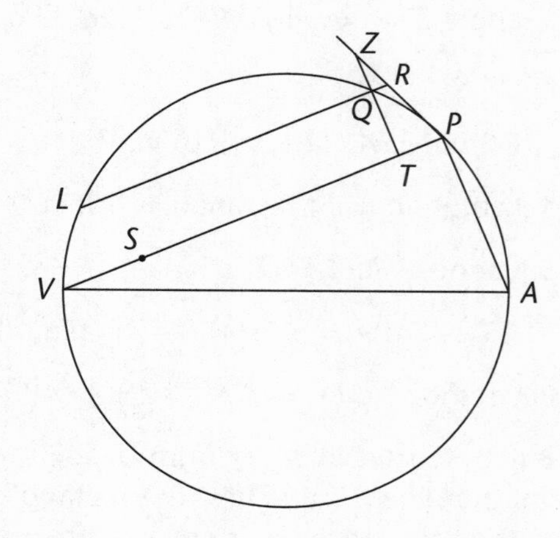

Substituting from Equation 3,

$$f \propto \frac{AV^2}{SP^2 \times PV^3}.$$

But diameter AV is constant; therefore

$$f \propto \frac{1}{SP^2 \times PV^3}.$$

(Note that we do not need an "ult" in this conclusion because the variables are all finite.)

Q.E.I.

Pause After Proposition 7

- **Question for Discussion**: What could this proposition mean? This force law says that the force on a body at *P* depends on a distance *PV* to a point to be reached at some later time some distance around the orbit.

What are the elements of this force law? $f \overset{ult}{\propto} 1/(PV^3 \times SP^2)$.

SP = the distance of the body from the center of forces.

PV = the chord from the body through the center of forces to the circle on the other side.

This seems to mean that the force has to know where the body will be later in order to know how much to pull or push. Or the body must be going in a track. Or either the body or the force must be intelligent. What sense can you make of it?

- It is worth remarking that in Chapter 39 of the *New Astronomy*, Kepler takes up the question of how a body may be made to move in a circle about a center other than the circle's own, using either physical forces or moving intelligences. His conclusion is that circular motion is extraordinarily complicated and cannot be generated by those means. Some sort of track or rigid sphere (which devices Kepler rejected) would be required in addition to the moving forces.

Newton's proof here may be intended as a commentary on Kepler's physical speculations. Using an entirely different physical theory, he nonetheless comes to the same conclusion: there seems to be something *unnatural* about eccentric circular motion.

- In the second edition of *Principia* Newton added some alternate proofs to this proposition and several of the the following ones (I.9–I.12). These alternative proofs were introduced by the words "the same thing differently" (*idem aliter*). These addenda used a circle of curvature for their proof. As discussed in the Preliminaries, this book omits them. Since working with the circle of curvature seems an interesting approach, why drop them?

As I explained there, this edition is a tracing of what Newton gave us, not what he might have done or what might have been more elegant, or more interesting to a modern reader. The *idem aliter* was not how Newton worked out the proofs of these propositions: it is a quite different approach to the proofs. If Newton had preferred this alternative approach, he could have replaced his original proofs. He had two editions (and at least thirteen years) in which to do so, but he chose to leave the original approach as the primary presentation, with the *idem aliter* proof tacked on afterwards as just another way to look at the problem.

Moreover, *Principia* is full of interesting side paths. It would take a lifetime to explore them all. This book avoids most of these side paths to follow the central argument Newton himself recommended. The goal is to allow readers to get a clear picture of his structure and method, and to enable classes to encompass the whole central argument in a semester.

[We omit **Corollary 1**, which is not used in Newton's central argument. It gives the force law for a rather fanciful hypothetical case in which the center of forces is on the circumference of the circle. In this case as soon as *P* got to *S* there would be no more circle. It would be hard to say what the body would do from there, since at that point, under an inverse power force law (this corollary concludes that it would be an inverse fifth power force law), there would be an infinite force. If the force at that point remained finite, the body would just yo-yo back and forth in a straight line tangent to the former circle and centered on the center of forces. What a body would do under the influence of an infinite force is beyond geometry to determine.]

I.7 Corollary 2

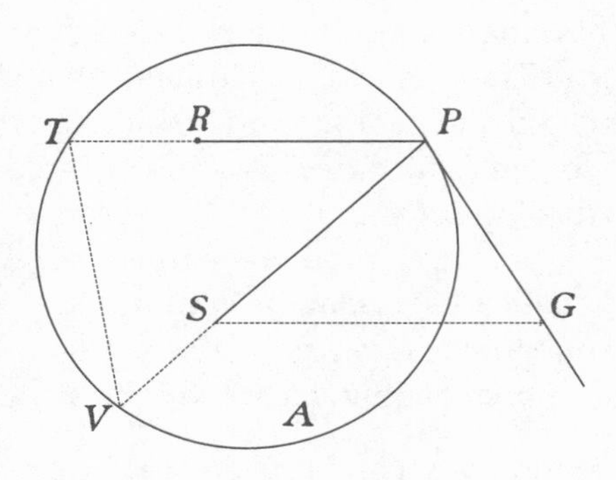

The force by which a body P is revolved in a circle APTV about the center of forces S, is to the force by which the same body P can be revolved in the same circle and in the same periodic time about any other center of forces you please R, as sq. RP × SP is to the cube of the straight line SG which is drawn from the first center of forces S to the tangent of the orbit PG and is parallel to the distance of the body from the second center of forces. For by the construction of this proposition, the former force is to the latter force as sq. RP × cub. PT is to sq. SP × cub. PV; that is, as SP × sq. RP is to $\frac{\text{cub.}\,SP \times \text{cub.}\,PV}{\text{cub.}\,PT}$*; or (because of the similar triangles PSG, TPV), to cub. SG.*

Expansion of Newton's Sketch of I.7 Corollary 2

"The force by which a body *P* is revolved in a circle *APTV* about the center of forces *S*, is to the force by which the same body *P* can be revolved in the same circle and in the same periodic time about any other center of forces you please *R*, as *sq. RP* × *SP* is to the cube of the straight line *SG* which is drawn from the first center of forces *S* to the tangent of the orbit *PG* and is parallel to the distance of the body from the second center of forces."

Given:

1. Two possible centers of forces, R and S, in the same circle;
2. periodic times around these two possible centers of forces are the same.

Construction:

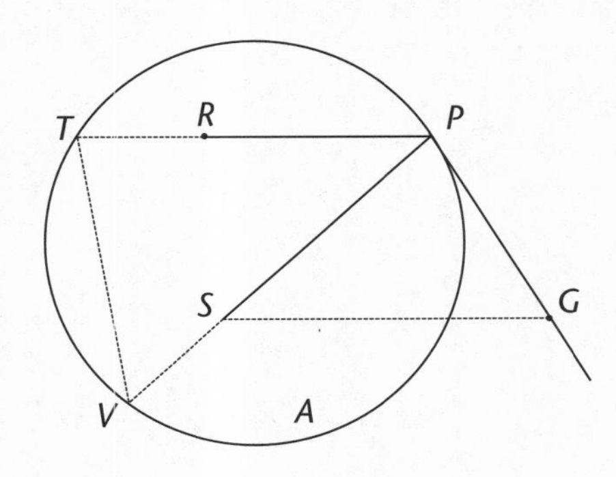

1. PG tangent at P;
2. $SG \parallel RP$.

To Prove:

$F_S : F_R :: (RP^2 \times SP) : SG^3.$

Proof:

"For by the construction of this proposition, the former force is to the latter force as sq. $RP \times$ cub. PT is to sq. $SP \times$ cub. PV; ..."

By the proposition, force varies inversely as the square of the distance from the center of forces times the cube of the length of the chord:

Around center S, $\quad F_S \propto 1/(SP^2 \times PV^3)$. (1)

Around center R, $\quad F_R \propto 1/(RP^2 \times PT^3)$. (2)

The sketch asserts that:

$$F_S : F_R :: (RP^2 \times PT^3) : (SP^2 \times PV^3).$$

Although this may look plausible, it does not follow directly, and would not be true unless the constants of proportionality were the same in each case.

What Equations 1 and 2 are actually saying, after we unpack the shorthand of the $\propto$ symbol, is this:

Around center S,

$$F_S \text{ at } P : F_S \text{ at } p :: \frac{1}{SP^2 \times PV^3} : \frac{1}{Sp^2 \times pv^3}. \tag{1a}$$

Around center R,

$$F_R \text{ at } P : F_R \text{ at } p :: \frac{1}{RP^2 \times PT^3} : \frac{1}{Rp^2 \times pt^3}. \tag{2a}$$

What we must prove is this:

$$F_S \text{ at } P : F_R \text{ at } P :: (RP^2 \times PT^3) : (SP^2 \times PV^3).$$

This makes it clear that the assertion of the sketch is not obvious, and must be proved.

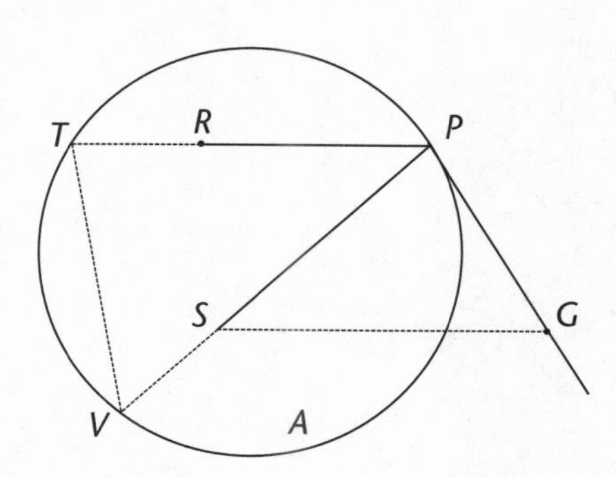

To effect this proof, we must use the conclusions of Proposition 6 Corollary 1 and Proposition 7, which depend on a single center of forces, or at least on the same areas described in the same times around both centers (see the first note to I.6 Corollary 1). Before we can invoke them we must prove that the motions around the two centers describe the same proportional areas.

Lemmita: Equable Areas Swept Out around Either Center

Let the equal periodic times around S and R be called T. Let this time be divided into n equal intervals, and in time $\frac{T}{n}$ let distance P_1Q_1 be traveled around center S and distance P_2Q_2 be traveled around center R.

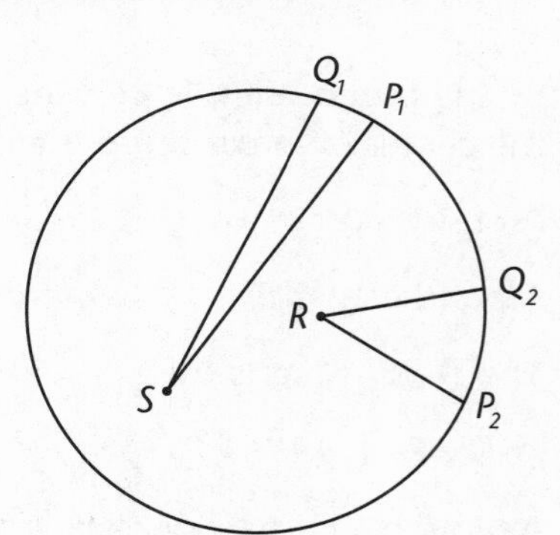

By Proposition 1:

For circle around center of forces S,

$$\frac{\text{Area of Circle}}{SP_1Q_1} = \frac{\text{Periodic Time } T}{(T/n)}$$

and for the circle around center of forces R,

$$\frac{\text{Area of Circle}}{RP_2Q_2} = \frac{\text{Periodic Time } T}{(T/n)}.$$

Since we are given that periodic time T is the same,

$$\frac{\text{Area of Circle}}{SP_1Q_1} = \frac{\text{Area of Circle}}{RP_2Q_2}.$$

And since the total area is the same,

$$SP_1Q_1 = RP_2Q_2 .$$

But SP_1Q_1 and RP_2Q_2 are the equal-time areas.

Therefore areas around S and R are in the same ratio to the times and we can apply I.6 Corollary 1 and I.7 across the two centers of forces.

By Proposition 1 Corollaries 3 and 4, forces on *equal time* arcs are as the subtenses or sagittae, regardless of centers.

By Proposition 6, forces on *any* arcs are as subtenses divided by the squares of the times, regardless of centers. Or, by I.6 Corollary 1, the forces are as $QR/(SP^2 \times QT^2)$, if $(SP^2 \times QT^2)$ represents the time. But we proved in the lemmita just above that areas swept out for equal time arcs are equal from either center; therefore $(SP^2 \times QT^2)$ represents the time in comparisons across both cases.

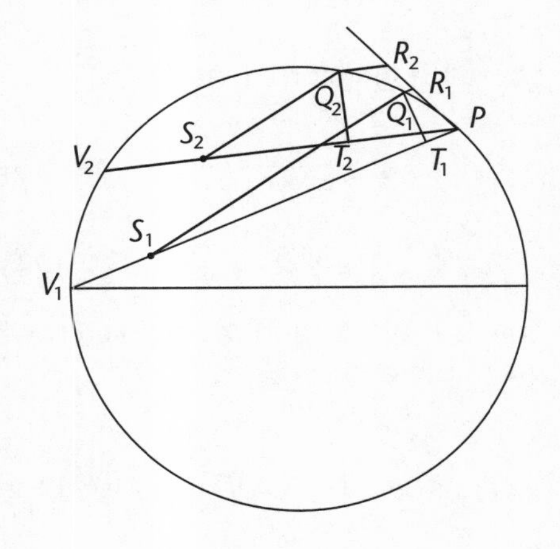

Now let's make a new diagram, with a single point P and equal time arcs starting at P around our two centers. Let's (carefully) use the letters of the last two propositions to follow the next couple of steps. In the time the body is understood to move to Q_1 around center S (which we'll temporarily call S_1), it moves to point Q_2 around center R (which we'll temporarily call S_2). We'll call the subtenses Q_1R_1 and Q_2R_2 and the perpendiculars Q_1T_1 and Q_2T_2.

By Proposition 6 Corollary 1,

$$F \text{ at } P \text{ towards } S_1 : F \text{ at } P \text{ towards } S_2 :: \frac{Q_1R_1}{(S_1P \times Q_1T_1)^2} : \frac{Q_2R_2}{(S_2P \times Q_2T_2)^2}.$$

By Equation 3 of Proposition 7,

$$\frac{Q_1R_1}{Q_1T_1{}^2 \times S_1P^2} \overset{ult}{=} \frac{A_1V_1{}^2}{PV_1{}^3 \times S_1P^2}$$

and

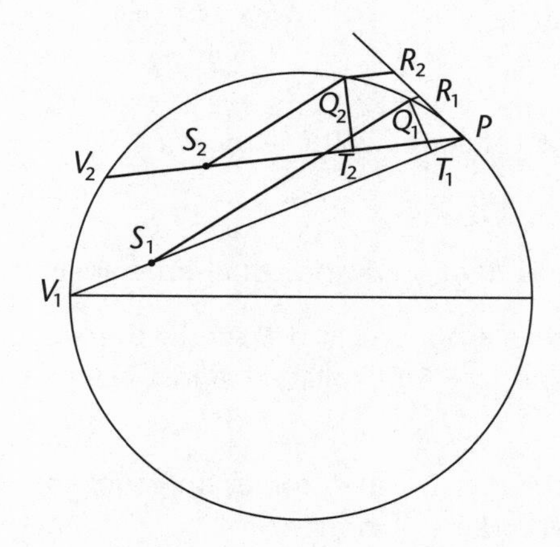

$$\frac{Q_2R_2}{Q_2T_2{}^2 \times S_2P^2} \overset{ult}{=} \frac{A_2V_2{}^2}{PV_2{}^3 \times S_2P^2}.$$

But $A_1V_1 = A_2V_2$ since both are diameters of the circle; therefore:

$$F_{S_1} : F_{S_2} :: \frac{1}{PV_1{}^3 \times S_1P^2} : \frac{1}{PV_2{}^3 \times S_2P^2}.$$

Or, using Newton's letters,

$$F_S : F_R :: \frac{1}{PV^3 \times SP^2} : \frac{1}{PT^3 \times RP^2};$$

$$F_S : F_R :: (RP^2 \times PT^3) : (SP^2 \times PV^3).$$

This is the assertion in the sketch which it took some work to unpack; we may now proceed.

"... that is, as $SP \times$ sq. RP is to $\frac{\text{cub.}\, SP \times \text{cub.}\, PV}{\text{cub.}\, PT}$; ..."

Multiply antecedent and consequent by (SP/PT^3)

$$\frac{F_S}{F_R} = \frac{RP^2 \times PT^3 \times \frac{SP}{PT^3}}{SP^2 \times PV^3 \times \frac{SP}{PT^3}};$$

$$F_S : F_R :: (RP^2 \times SP) : \left(SP^3 \times \frac{PV^3}{PT^3}\right). \tag{3}$$

"... or (because of the similar triangles PSG, TPV), to SG^3."

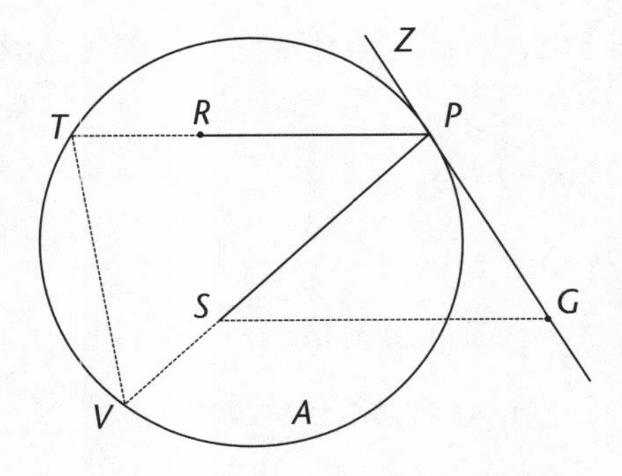

By Euclid III.32, if a straight line PG touch a circle, and from point of contact P there be drawn across, in the circle, a straight line PT cutting the circle, the angle ZPT it makes with the tangent equals angle TVP in the alternate segment of the circle.

So angle ZPT = angle TVP.

And angle ZPT = angle PGS. [Eu. I.28]

Therefore angle TVP = angle PGS.

angle TPS = angle PSG. [Eu. I.29]

Therefore $\triangle TVP \sim \triangle PSG$.

Therefore $PV : PT :: SG : SP$. [Eu. VI.4]

Rect SG,PT = rect PV,SP. [Eu. VI.16]

Rewriting, $SG = (PV/PT) \times SP$.

Cube each side: $SG^3 = (PV^3/PT^3) \times SP^3$.

Substitute from this into Equation 3:

$$F_S : F_R :: (RP^2 \times SP) : SG^3 .$$

Q.E.D.

I.7 Corollary 3

The force by which a body P is revolved in any orbit whatever about a center of forces S, is to the force by which the same body P can be revolved in the same orbit and in the same periodic time about any other center of forces you please R, as SP × sq. RP (contained, clearly, beneath the distance of the body from the first center of forces S and the square of its distance from the second center of forces R) is to the cube of the straight line SG, which is drawn from the first center of forces S to the tangent of the orbit PG and is parallel to the distance RP of the body from the second center of forces. For the forces in this orbit at any point P you please on it are the same as in the circle of the same curvature.

Expansion of Newton's Sketch of I.7 Corollary 3

Given:

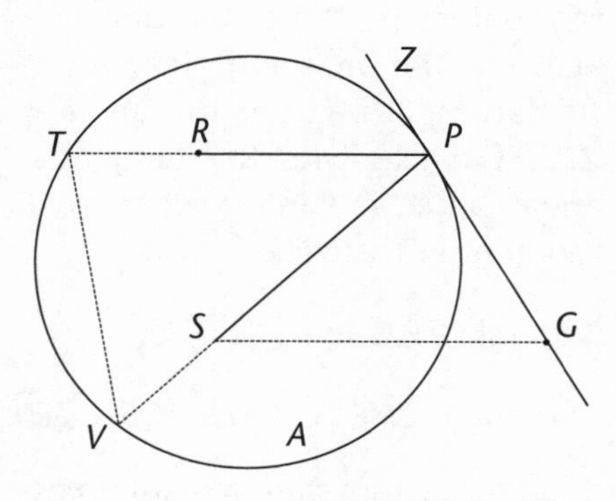

1. Any orbit;
2. ratio of forces in the same orbit;
3. periodic time constant;
4. different centers of forces.

To Prove:

$F_S : F_R :: (RP^2 \times SP) : SG^3$ even though it's not a circle.

"For the forces in this orbit at any point P you please on it are the same as in the circle of the same curvature."

Proof:

We established in Lemma 11 that the paths we are dealing with can be approximated by circles at any given point. Therefore, with this curve too we can draw the circle of curvature at that point and apply the argument of Corollary 2 to two centers of forces. That is, if you know the law of force to one center you can find the law of force to the other. Thus you can find F_R or F_S.

Q.E.I.

Question for Discussion: Suppose you wanted to use this corollary to compare the force law for a center of forces at the center of a conic orbit with the force law for a center of forces at the focus of the conic. Suppose S were the focus, and R the center of the conic. Our demonstration assumed R to be inside the circle of curvature. But for an hyperbola center R would be outside. Could we still use this corollary?

Proposition 8

Proposition 8, like Proposition 7 Corollary 1, is a purely hypothetical case: a body is moving on the circumference of a semicircle attracted to an infinitely distant center of forces. The proposition is never again referred to; perhaps it is included to emphasize the hypothetical and theoretical nature of Book I. Since it is not used in Newton's central argument, it is not included here.

Proposition 9

If a body orbit in a spiral PQS intersecting all radii SP, SQ, and so on, in a given angle, it is required to find the law of centripetal force tending to the center of the spiral.

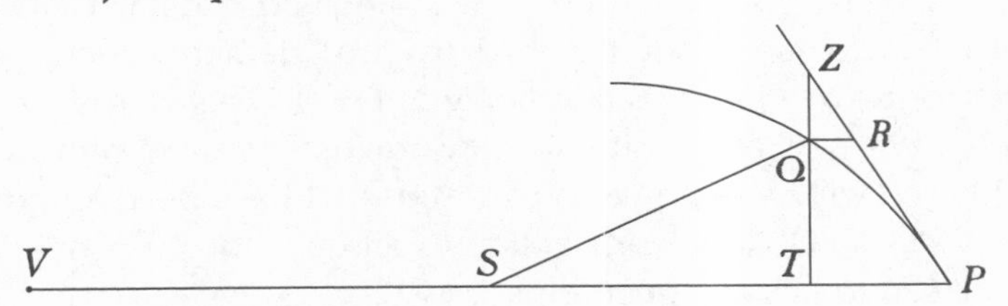

Let the indefinitely small angle PSQ be given, and because all the angles are given the figure $SPRQT$ will be given in shape. Therefore, the ratio QT / QR is given, and sq. QT / QR is as QT, that is (because that figure is given in shape), as SP. Now let the angle PSQ be changed in any way whatever, and the straight line QR subtending the angle of contact QPR will (by Lemma 11) be changed in the duplicate ratio of PR or QT. Therefore, sq. QT / QR will remain the same as before; that is, as SP. Consequently, $\frac{\text{sq.}SP \times \text{sq.}QT}{QR}$ is as cub. SP. And so (by Proposition 6 Corollaries 1 and 5) the centripetal force is inversely as the cube of the distance SP.

Q.E.I.

Notes on I.9

- The sort of spiral referred to here is one such that the tangent at any point makes a constant angle with the radius, as specified in the enunciation. It is the sort in which you can find the Golden Ratio expressed, the ratio into which the line is cut in Euclid II.11: $a : b :: b : a + b$.

  The proposition is not looking for the force law for the spiral defined by Archimedes in his treatise *On Spirals*. In Archimedes' spiral the radius grows in length uniformly as it revolves uniformly around the center. This turns out to be a different geometrical entity.

- When Newton says in his sketch to let angle *PSQ* be "changed in any way whatever," he means us to keep that in mind that he has told us that it is an "indefinitely small" angle. We take a small angle and look at what happens as it vanishes. It couldn't be changed to a large angle and have the other ultimate relations still hold.

- Newton invokes Lemma 11 in his sketch for this proposition.

  We would want to know that this spiral is the sort of curve that Lemma 11 applies to: that is, that its curvature can be approximated at any point by a circle with finite radius. It turns out that the equiangular spiral is one of the curves for which Lemma 11 may legitimately be invoked.

Exercise: The curvature can be shown by calculus and analytic geometry to be approximated by a circle with finite radius. If you enjoy these modern mathematical tools, you might want to make a proof of this for yourself using them.

But here's an exercise even more challenging and original: Can you come up with a Newtonian proof for this? In your demonstration, you must remember to use relationships that apply as $Q \rightarrow P$. That is, they must apply where we have a fixed *P* with *Q* moving in, just as *B* moved to *A* in Lemma 11. You will be tempted to use ratios that have been found to apply to different *P*'s and *Q*'s in the similar figures set up in the proof of Proposition 9. This will lead to fallacious confirmations.

- **Proportional Radii:** In the equiangular spiral, the ratio of lengths of any consecutive radii making equal angles from the center will be constant. In this diagram, that means $SP : SQ :: Sp : Sq$ where angle PSQ = angle pSq. Newton knows this to be true of the equiangular spiral, and had proved it elsewhere by his method of fluxions.*

Here, for the purposes of this proof, he foregoes demonstrating it and takes it as a known property.

- **Construction.** Newton gives a construction of the equiangular spiral in I.41 Corollary 3 (see note on this section of *Principia* on page 294). Newton uses lines cut off by a moving tangent to a conic section to construct the spiral that would be generated by an inverse-cube force law. He states that the proof of this follows from "the quadrature of a certain curve;" that is, by the use of his method of fluxions. This point is particularly relevant to I.9, in that it shows that Newton had been working with the equiangular spiral primarily by means of fluxions, but chose not to interrupt the flow of the *Principia* at that early point with an excursion into those techniques.

Expansion of Newton's Sketch of I.9

"If a body orbit in a spiral *PQS* intersecting all radii *SP*, *SQ*, and so on, in a given angle, it is required to find the law of centripetal force tending to the center of the spiral."

---

* In "Methods of Series and Fluxions," completed in 1670–1671. In *The Mathematical Papers of Isaac Newton*, edited by D.T. Whiteside, Vol. III. The proof is found on pp. 169–173.

Given:

1. A spiral in which the angle between the radius and the tangent is constant.
2. center of forces at the center of the spiral.

To Find:

The force law tending towards the center of the spiral.

Proof:

"Let the indefinitely small angle PSQ be given, and because all the angles are given the figure $SPRQT$ will be given in shape. Therefore, the ratio QT / QR is given, and sq. QT / QR is as QT, that is (because that figure is given in shape), as SP."

Construct figure $SPRQT$ using the standard construction for "the solid" of Proposition 6 Corollary 1 and using equal angles at S for the different instances of it.

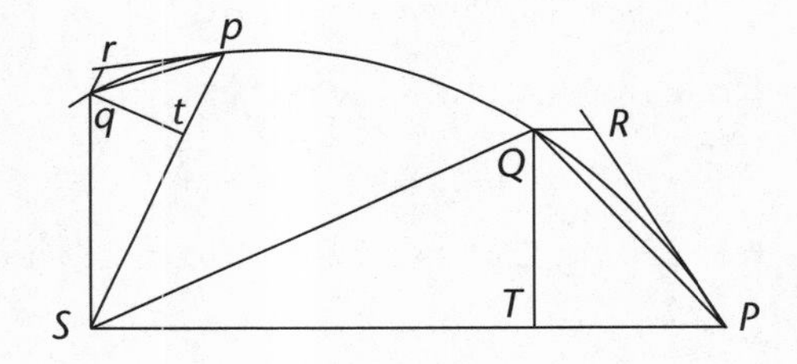

We will show that the figures $SPRQT$ and $Sprqt$ are made up of triangles that are similar and similarly situated. Therefore it is itself given in shape, that is, always similar, with proportional sides as shown.

Connect P, Q and p, q.

Angle QSP = angle qSp. [by construction]

$SP : SQ :: Sp : Sq$. [see bulleted note above entitled Proportional Radii]

Therefore by Euclid VI.4 and Euclid VI Def 1,

$\triangle SPQ \sim \triangle Spq$.

Therefore

$PQ : SP :: pq : Sp$ [Eu. VI.4] (1)

and angle SPQ = angle Spq. (2)

Angle SPR = angle Spr. [given]

Subtracting equals from equals,

angle QPR = angle qpr. (3)

By the construction of the figure of "the solid,"

$$qr \parallel Sp \quad \text{and} \quad QR \parallel SP.$$

Therefore, by Euclid I.29,

$$\text{angle } Spq = \text{angle } pqr$$

and

$$\text{angle } SPQ = \text{angle } PQR.$$

From this and Equation 2 it follows that

$$\text{angle } PQR = \text{angle } pqr. \tag{4}$$

By Equations 3 and 4, $\triangle PQR$ and $\triangle pqr$ are equiangular, so by Euclid VI.4

$$QR : PQ :: qr : pq.$$

$$QR : qr :: PQ : pq. \quad \text{[Eu. V.16]} \tag{5}$$

Similarly it can be shown that $\triangle QTP \sim \triangle qtp$ and that

$$QT : qt :: PQ : pq. \tag{6}$$

Combining Equations 5 and 6, using Euclid V.16,

$$QT : QR :: qt : qr. \tag{7}$$

Writing the ratios as fractions,

$$QT / QR = qt / qr, \quad \text{or}$$

$$\frac{QT/QR}{qt/qr} = \frac{1}{1}.$$

Multiply both sides by QT / qt:

$$\frac{QT^2/QR}{qt^2/qr} = \frac{QT}{qt}.$$

That is,

$$(QT^2/ QR) : (qt^2/ qr) :: QT : qt. \tag{8}$$

Or, $(QT^2/ QR) \propto QT$.

Note that this is true of any pair of equal angles, finite or evanescent, at different places in the spiral.

Combining Equations 1 and 6, using Euclid V.16,

$$QT : qt :: SP : Sp.$$

Substituting into Equation 8:

$$(QT^2/QR) : (qt^2/qr) :: SP : Sp;$$

or, $(QT^2/QR) \propto SP$ (9)

Alternate Conclusion:

Now it looks as if all we have to do is to multiply both sides of Equation 9 by SP^2:

$$\frac{SP^2 \times QT^2}{QR} \propto SP^3.$$

By Proposition 6 Corollaries 1 and 5: $F \overset{ult}{\propto} 1 / \frac{SP^2 \times QT^2}{QR}$.

Therefore $F \propto 1/SP^3$.

Q.E.I.

Question for Discussion: This alternate conclusion still assumes the given angle at S. Newton does it a more complicated way, considering the angle at S changing in both figures after the figures are found to be given in shape. Why does he do this?

"Now let the angle PSQ be changed in any way whatever, and the straight line QR subtending the angle of contact QPR will (by Lemma 11) be changed in the duplicate ratio of PR or QT. Therefore, sq. QT/QR will remain the same as before; that is, as SP. Consequently, $\frac{\text{sq.}SP \times \text{sq.}QT}{QR}$ is as cub. SP. And so (by Proposition 6 Corollaries 1 and 5) the centripetal force is inversely as the cube of the distance SP."

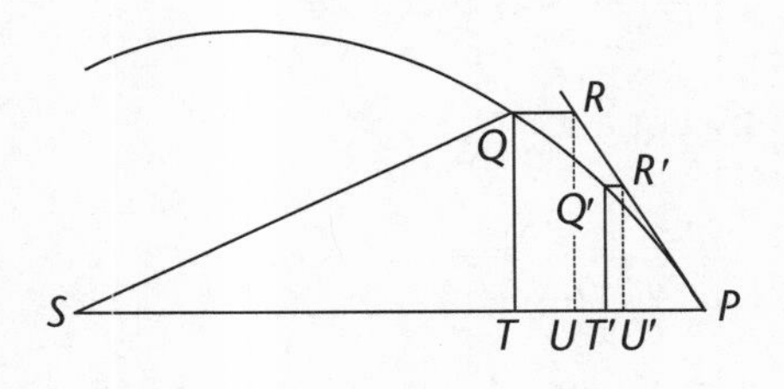

Newton has said the angle is "indefinitely small," so he can call on Lemma 11, which deals with limiting cases. The angle is changing from one "indefinitely small" value to another, and as it does so, by Lemma 11 Corollary 1, subtense QR in the original angle will be to subtense $Q'R'$ in the changed angle as the square on tangent PR is to the square on tangent PR'.

$$QR : Q'R' :: PR^2 : PR'^2 \quad \text{[Lem. 11 Cor. 1]}$$

or, $QR \overset{ult}{\propto} PR^2$. (10)

Now consider lines RU and $R'U'$, both perpendicular to SP.

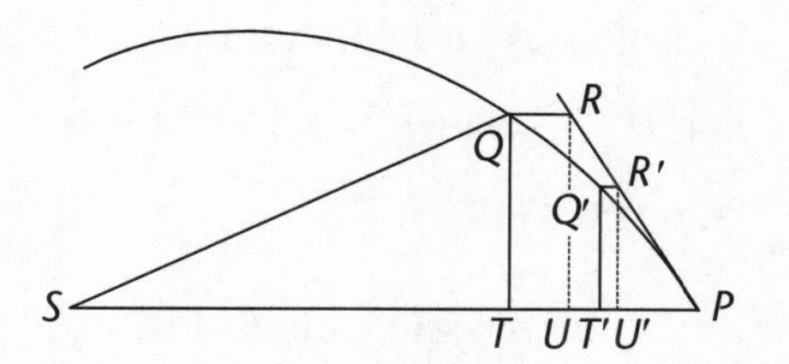

Line $R'U'$ is parallel to base RU in $\triangle RUP$.

By Euclid VI.2 and VI.5,

$$\triangle RUP \sim \triangle R'U'P.$$

Therefore $R'P : R'U' :: RP : RU$

or, *alternando*: $R'P : RP :: R'U' : RU$.

That is, $RP \propto RU$.

But $RU = QT$ and $R'U' = Q'T'$. [Eu. I.34]

So $RP \propto QT$

and $RP^2 \propto QT^2$.

Substituting into Equation 10,

$$QR \overset{ult}{\propto} QT^2.$$

Note: This is variation of the angle within one figure, not between the similar figures we were using before, in which, by Equation 7, $QT : QR :: qt : qr$, that is, where $QT \propto QR$.

This one expands to: $QR : Q'R' \overset{ult}{::} QT^2 : Q'T'^2$.

Alternating and inverting, $QT^2/QR \overset{ult}{=} Q'T'^2/Q'R'$. (11)

"Therefore, sq. QT / QR will remain the same as before; that is, as SP. Consequently, $\frac{\text{sq.}SP \times \text{sq.}QT}{QR}$ is as cub. SP."

By Equation 9,

$$\frac{QT^2/QR}{qt^2/qr} = \frac{SP}{Sp}.$$

Substituting from Equation 11,

$$\frac{Q'T'^2/Q'R'}{qt^2/qr} \overset{ult}{=} \frac{SP}{Sp}.$$

Compound with (SP^2/Sp^2):

$$\frac{Q'T'^2 \times SP^2/Q'R'}{qt^2 \times Sp^2/qr} \overset{ult}{=} \frac{SP^3}{Sp^3}. \tag{12}$$

Or,

$$\frac{Q'T'^2 \times SP^2}{Q'R'} \overset{ult}{\propto} SP^3.$$

"And so (by Proposition 6 Corollaries 1 and 5) the centripetal force is inversely as the cube of the distance SP."

Now we must take it to the limit and look at the ultimate ratios. We must be very careful here. Both pSq and PSQ shrink. In the first part we kept the same angle at S; now we do not assume that. Thus we pull in elements from both lines of reasoning—the one with the similar figures and constant angle at S and the one with the diminishing angle at S.

The numerator of the expression on the left side of Equation 12 is the inverse of this expression for the "changed" angle at S (but still, of course, the constant angle at the tangent as specified in the enunciation), using figure $SPR'Q'T'$.

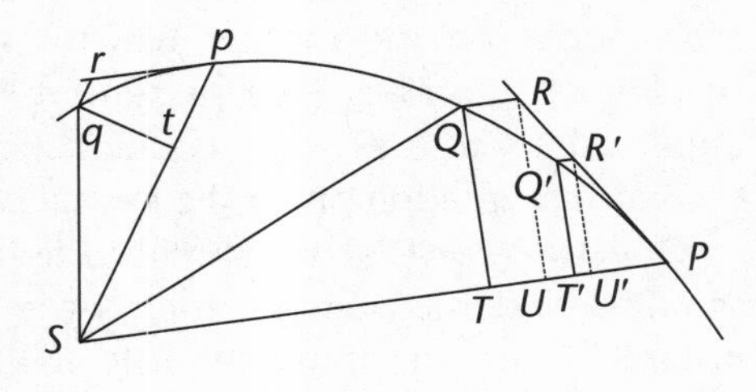

The denominator of the expression on the left side of Equation 12 is the inverse of this expression for the original angle, using figure $Sprqt$.

By Proposition 6 Corollaries 1 and 5:

$$F \overset{ult}{\propto} 1/\frac{SP^2 \times QT^2}{QR}.$$

Thus for any indefinitely small angle at the center S for a spiral with constant angle between radius and tangent,

$$F_P : F_p :: Sp^3 : SP^3 \quad \text{or} \quad F \propto 1/SP^3.$$

Note that we no longer have any evanescent magnitudes in our expression; therefore it is not an ultimate proportion.

Thus, given a body moving in a spiral with constant angle between the radius and the tangent, with the center of forces at the center of the spiral, the body will be attracted to that center inversely as the cube of the distance from the center.

Q.E.I.

Pause After Proposition 9

The converse of this proposition would say that an inverse cube law would lead to a spiral.

Note that Newton has not proved this converse; to be used it would need to be proved. At this point, we have no idea what weird curves might result from an inverse cube force law. It is easy to prove at least that degenerate cases of the spiral like a circle (angle at the tangent = 90°) and a straight line into S (angle at the tangent = 0°) could result from an inverse cube law.

But it turns out that most inverse cube examples would lead to a spiral into the center or out away from it—and result in an unstable universe.

In Proposition 43–45 of Book I, Newton shows that when the line of apsides is moving we get an inverse cube component added to the inverse square component. (See the glossary for *apsides* and pages 295–296 for discussion of I.45.) In Propositions I.64–I.66, considering the effect of three or more bodies on one another, Newton shows that there will be movement of the lines of apsides. (See Appendix B for discussion of I.65–I.66.) If one applied I.45 to this, one could conclude that there is an inverse cube component in addition to the inverse square component.

Question for Discussion: Since an inverse cube law is associated with a spiral, would this suggest that the effect of additional bodies introduces instability into the universe? And if the laws of planetary motion lead to instability, would that suggest that the Creator might be required to intervene periodically to put things back on track?

Lemma 12

All parallelograms circumscribed about any conjugate diameters of a given ellipse or hyperbola are equal among themselves.

This is proved in Conics.

[Newton gives no diagram for this lemma.]

Note on Lemma 12:

This lemma, along with Lemma 13 and Lemma 14 (which appear a little later), cites some relationships among lines in conic sections that Newton will use in the following propositions. These lemmas happen not to be among those proved by Apollonius, and thus cannot be cited as we do the others from that work. But the proofs of these three lemmas are simple manipulations of the standard conic relationships and propositions, and, unlike Lemmas 1–11, there is nothing new or characteristically Newtonian in them. For this reason we will not take the time here to work through their arguments. However, full proofs are given in Appendix A, to be consulted if you need or wish to see a diagram or satisfy yourself about what is being proved.

Proposition 10

If a body orbit in an ellipse, it is required to find the law of centripetal force tending to the center of the ellipse.

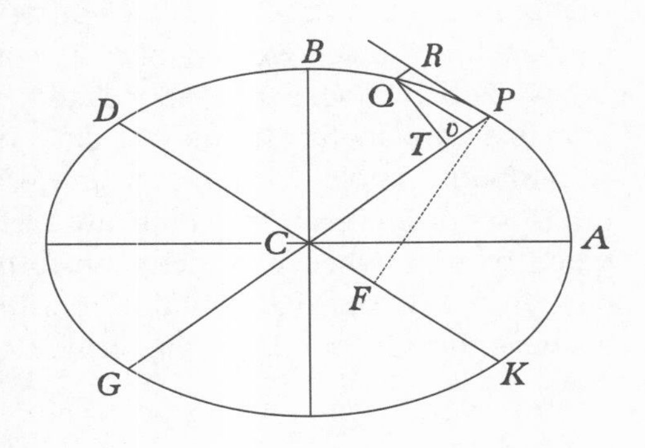

Let CA, CB be semiaxes of the ellipse; GP, DK other conjugate diameters; PF, QT perpendiculars to the diameters; Qv applied ordinatewise to the diameter GP; and (from the conics) if the parallelogram $QvPR$ be completed, the rectangle Pv,vG will be to sq. Qv as sq. PC is to sq. CD; and (because of the similar triangles QvT, PCF) sq. Qv is to sq. QT as sq. PC is to sq. PF; and, compounding ratios, rectangle Pv,vG is to sq. QT as sq. PC to sq. CD and [i.e., compounded with] sq. PC to sq. PF; that is, $vG : \frac{\text{sq.}QT}{Pv} :: \text{sq.}PC : \frac{\text{sq.}CD \times \text{sq.}PF}{\text{sq.}PC}$. Write QR for Pv, and (by Lemma 12) $BC \times CA$ for $CD \times PF$, as well as (points P and Q coming together) $2PC$ for vG, and, multiplying extremes and means by each other, this becomes $\frac{\text{sq.}QT \times \text{sq.}PC}{QR} = \frac{2\,\text{sq.}BC \times \text{sq.}CA}{PC}$. Therefore (by Proposition 6 Corollary 5) the centripetal force is inversely as $\frac{2\,\text{sq.}BC \times \text{sq.}CA}{PC}$; that is (because 2 sq. $BC \times$ sq. CA is given), inversely as $\frac{1}{PC}$; that is, directly as the distance PC.

Q.E.I.

Notes on I.10

- **Question for Discussion:** What sort of universe would this be, where force varies as distance from the center of forces? The farther away the body gets, the more it's impelled in. It sounds as if everything would collapse to one center.

• **The force law of simple harmonic motion.** The force is found to vary directly as the distance from the center, that is, directly as the displacement from the point of equilibrium. This makes the resulting motion *simple harmonic*, like that shown by a pendulum or spring. Hook one end of a bungee cord around a nail driven into a table; affix a body to the other end, and pull the body away from the nail (thus stretching out the cord). If you then release the body, while at the same time giving it a sharp perpendicular push, it will describe an ellipse about the nail as center.

• This proposition, and its converse given in Corollary 1, are of interest because they allow us to consider what the relationships of simple harmonic motion would turn out to look like in terms of planetary motion. One might have looked for those relationships to be played out in the heavens, and indeed some thinkers did. (Thomas Lydiat speculated that the planets moved by buoyancy, bobbing up and down, a form of simple harmonic motion: from this he arrived at elliptical orbits just as Newton does here in I.10—before Kepler worked out his conclusion about elliptical planetary orbits. Robert Hooke also speculated on simple harmonic motion as a principle for planetary motion, as did a number of Cartesians.)

Although simple harmonic motion will turn out not to govern planetary orbits in our universe, this force law and orbit relationship must be investigated so that, when it comes time to match the phenomena against the relationships in our mathematical toolchest, we will be able to say whether we have this sort of relationship or not.

Expansion of Newton's Sketch of I.10

"If a body orbit in an ellipse, it is required to find the law of centripetal force tending to the center of the ellipse."

Given:

1. Body moves in an ellipse;
2. center of forces is at the center of the ellipse.

To Find:

Force law tending to that center.

Proof:

"Let CA, CB be semiaxes of the ellipse; GP, DK other conjugate diameters; PF, QT perpendiculars to the diameters; Qv applied ordinatewise to the diameter GP; ..."

Draw PR tangent to the ellipse at P, the position of the body.

Draw diameter PG through center C.

Draw conjugate diameter $DK \parallel PR$. [Ap. I.32 converse]

From Q, a later position of the body on the ellipse, drop $Qv \parallel PR$ intersecting PC at v.

Qv and DC are both parallel to PR; therefore they are parallel to each other and both are ordinates to diameter GP.

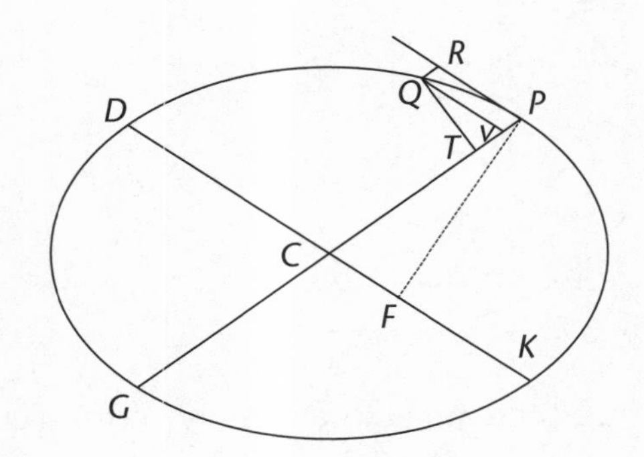

Step 1

"... and (from the conics) if the parallelogram $QvPR$ be completed, the rectangle $Pv.vG$ will be to sq. Qv as sq. PC is to sq. CD; ..."

Apollonius I.21, second part, applied here: If, in an ellipse, straight lines DC, Qv are dropped ordinatewise to diameter GP, the squares on the ordinates DC^2, Qv^2 will be to each other as the areas contained by the straight lines cut off (abscissas) GC,CP and Gv,vP.

That is, $DC^2 : Qv^2 = \text{rect } GC,CP : \text{rect } Gv,vP$.

But by Apollonius I Definition 9, "The midpoint of the diameter is the center of the ellipse." So

$GC = CP$. [Ap. I Def. 9]

Substituting, $DC^2 : Qv^2 :: PC^2 : Gv,vP$.

Alternating, $DC^2 : PC^2 :: Qv^2 : (Gv,vP)$.

Inverting, $PC^2 : CD^2 :: (Pv,vG) : Qv^2$.

$$(Pv,vG) : Qv^2 :: PC^2 : CD^2. \qquad (1)$$

(This is the first equation that will be compounded in Step 3.)

Step 2

"... and (because of the similar triangles QvT, PCF) sq. Qv is to sq. QT as sq. PC is to sq. PF; ..."

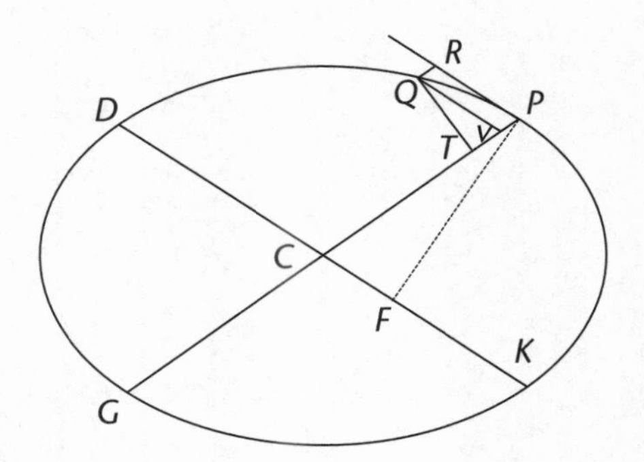

Drop $QT \perp PC$.

Drop $PF \perp DK$.

Draw $QR \parallel PC$ intersecting tangent PR in R.

Tangent $PR \parallel$ conjugate diameter DK.

Therefore $QvPR$ is a parallelogram.

Since both are right angles by construction,

Angle QTP = angle PFC. [Eu. I Post. 4]

Since $Qv \parallel DK$ and they are cut by transversal vC,

Angle PCF = angle QvT. [Eu. I.29]

Therefore $\triangle QTv \sim \triangle PFC$.

$Qv : QT :: PC : PF$. [Eu. VI.4]

Squaring both sides,

$$Qv^2 : QT^2 :: PC^2 : PF^2. \tag{2}$$

(This is the second proportion to be compounded.)

Step 3

"and, compounding ratios, rectangle Pv,vG is to sq. QT as sq. PC to sq. CD and [i. e., compounded with] sq. PC to sq. PF; that is, $vG : \frac{\text{sq. }QT}{Pv} :: \text{sq. }PC : \frac{\text{sq. }CD \times \text{sq. }PF}{\text{sq. }PC}$."

Compounding Equation 1 with Equation 2, cancelling Qv^2:

$$\frac{Pv, vG}{QT^2} = \frac{PC^2}{CD^2} \times \frac{PC^2}{PF^2}.$$

Divide both sides by $(Pv \times PC^2)$:

$$\frac{vG}{PC^2 \times QT^2} = \frac{PC^2}{CD^2 \times Pv \times PF^2}.$$

Rearranging,

$$\frac{vG}{QT^2/Pv} = \frac{PC^2}{CD^2 \times PF^2/PC^2}. \tag{3}$$

"Write QR for Pv, and (by Lemma 12) $BC \times CA$ for $CD \times$ PF, as well as (points P and Q coming together) $2PC$ for vG, and, multiplying extremes and means by each other, this becomes $\frac{\text{sq.}QT \times \text{sq.}PC}{QR} = \frac{2\,\text{sq.}BC \times \text{sq.}CA}{PC}$."

Because figure *QvPR* is a parallelogram, by Euclid I.34,

$$QR = Pv. \tag{4}$$

Construct major and minor axes and the parallelograms whose sides are tangent to the ellipse at the ends of these diameters and at the ends of the diameters first constructed according to Lemma 12.

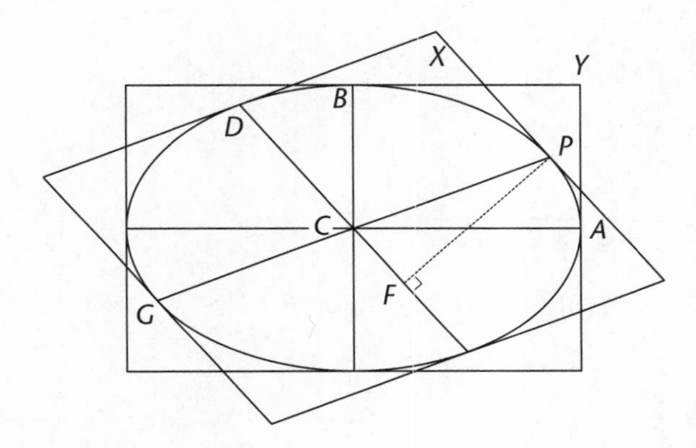

By Lemma 12,

$$\text{parallelogram } BCAY = \text{parallelogram } DCPX.$$

But since *BC* and *CA* are semi-axes and at right angles to each other,

$$\text{parallelogram } BCAY = BC \times CA.$$

Therefore $\quad \text{parallelogram } DCPX = BC \times CA.$

By extension of Euclid VI.1, parallelograms on the same base are to one another as their heights. The height of parallelogram *DCPX* is the perpendicular *PF* to the base *DC* (extended). Although it could without much difficulty be shown in a strictly Euclidean manner by constructing another parallelogram on base *DC*, let's follow Newton, who here jumps out of the language of proportions to assert:

$$\text{parallelogram } DCPX = DC \times PF.$$

Therefore $\quad DC \times PF = BC \times CA.$

Squaring both sides, $\quad DC^2 \times PF^2 = BC^2 \times CA^2.$

Substituting into Equation 3:

$$\frac{vG}{QT^2/Pv} = \frac{PC^2}{BC^2 \times CA^2/PC^2}. \tag{5}$$

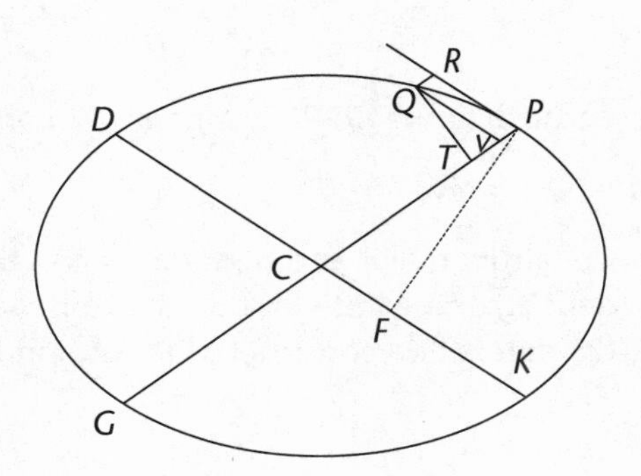

Now let $\Delta t \to 0$. As successively shorter times are taken, the body will have moved less and less from P in going to Q. That is, as $\Delta t \to 0$, $Q \to P$.

As $Q \to P$, $v \to P$ and $vG \overset{ult}{=} PG$.

But by Apollonius I Definition 9,

$$PG = 2 \times PC.$$

So $\quad vG \overset{ult}{=} 2 \times PC.$

Substituting into Equation 5:

$$\frac{2 \times PC}{QT^2/Pv} \overset{ult}{=} \frac{PC^2}{BC^2 \times CA^2/PC^2}.$$

Multiplying means and extremes, [Eu. VI.16],

$$\frac{2 \times PC \times BC^2 \times CA^2}{PC^2} \overset{ult}{=} \frac{PC^2 \times QT^2}{Pv}.$$

By Equation 4, $\quad QR = Pv.$

Substituting and cancelling,

$$\frac{2 \times BC^2 \times CA^2}{PC} \overset{ult}{=} \frac{PC^2 \times QT^2}{QR}.$$

Or, inverting,

$$\frac{PC}{2 \times BC^2 \times CA^2} \overset{ult}{=} \frac{QR}{PC^2 \times QT^2}.$$

That is (using limit notation):

$$\lim_{Q \to P} \frac{QR}{PC^2 \times QT^2} = \frac{PC}{2 \times BC^2 \times CA^2}. \tag{6}$$

"Therefore (by Proposition 6 Corollary 5) the centripetal force is inversely as $\frac{2\,\text{sq.}BC \times \text{sq.}CA}{PC}$; that is (because 2 sq. BC × sq. CA is given), inversely as $\frac{1}{PC}$: that is, directly as the distance PC.

Q.E.I."

By Proposition 6 Corollaries 1 and 5, with radius PC for SP:

$$F \overset{ult}{\propto} \frac{QR}{PC^2 \times QT^2}.$$

Or, using limit notation:

$$F \propto \lim_{Q \to P} \frac{QR}{PC^2 \times QT^2}. \tag{7}$$

Therefore, combining Equations 6 and 7,

$$F \propto \frac{PC}{2 \times BC^2 \times CA^2}.$$

But BC and CA are semi-axes and therefore constant; so the whole magnitude ($2 \times BC^2 \times CA^2$) is constant.

Therefore $F \propto PC$.

If a body revolve in an ellipse, urged by a force directed towards the center of the ellipse, that force will vary directly as the distance of the body from that center.

Q.E.I.

I.10 Corollary 1

Therefore, the force is as the distance of the body from the center of the ellipse; and in turn, if the force be as the distance, the body will move in an ellipse having its center at the center of forces, or possibly in a circle into which an ellipse of course might change.

Notes on I.10 Corollary 1

- Although he didn't do so for Proposition 9, Newton gives us a converse for Proposition 10. Remember that converses of propositions must be proved; they do not necessarily follow from the proposition, and in many cases turn out to be false. (Proposition: If cat, then animal. Converse: If animal, then cat.)

Newton does not give us a proof for this converse. The proof would be rather more difficult than the proposition in this case. We can say from our experience with pendulums, which obey the same force law, that the converse does seem to be true. Newton, who was even more familiar with pendulum phenomena than we are, no doubt felt certain that it could be proved; perhaps elsewhere he had even sketched such a proof for himself. But he chose not to spell it all out here, preferring to get on with the propositions that turn out to bear more directly on his foundations of gravitational theory and celestial mechanics.

As with a number of his so-called corollaries, which would more properly be called scholia, Newton is here just mentioning a relationship for completeness. If and when we want to go more deeply into the matter, we can make our own proof.

I.10 Corollary 2

And in the revolutions made in all ellipses about the same center, the periodic times will be equal. For in similar ellipses, those times will be equal (by Proposition 4 Corollary 3 and 8), and in ellipses having a common major axis they are to each other as the whole areas of the ellipses directly and the small parts of the areas described simultaneously inversely; that is, as the minor axes directly and the velocities of the bodies at the principal vertices inversely; that is, as those minor axes directly and the lines applied ordinatewise to the same point of the common axis inversely; and therefore (because of the equality of the direct and inverse ratios) in a ratio of equality.

Notes on I.10 Corollary 2

• **Questions for discussion.** This corollary is claiming that, no matter how the sides of the ellipse are squished in or stretched out, the periodic times will remain the same. Does this seem surprising to you? Would you expect a body to take as long to go around a little bitty orbit as around a huge one? Are you curious how Newton will prove such a thing? Can we see how the force law of I.10 would have that result? Can we get an intuitive sense of this corollary before working through the proof?

• The second part of this corollary can be illustrated by an analogous phenomenon in the world of our experience.

We know that in our world we get equal periods with a given string length pendulum no matter how far we pull it back. This distance of displacement corresponds to the semi-major axis. Imagine giving the pendulum a push to the side as it is released. Now we have an ellipse with a nonzero semi-minor axis. Look at it in terms of a coordinate system. There is amplitude along the x-axis and amplitude along the y-axis. Looking at either independently we know we would have equal periods regardless of the amplitude. Therefore it is plausible that when we combine them the periods for any pendulums moving in ellipses with the same center will be equal. Now superimpose the two motions. At any given time, the bob is at some position in each motion, that is, with respect to each axis. Since the period of each motion is the same, the bob will return to the same place after each period, traversing a fixed curve—an ellipse, in fact.

• The first part of this corollary requires an extension of I.10. The proposition proves that the accelerative force varies as the distance for one body. In this corollary we use a universalized statement that the accelerative forces vary as the distances for any two or more different bodies compared to one another.

By I.10 the force on a particular body varies as its distance from the center. But that doesn't mean that two bodies at the same distance must have the same accelerative force on them.

An example from our world illustrates this. Two bobs on strings of different lengths could be set moving on the same center. If they are given the same displacement the ellipses in which they move would have the same major axis. They could also be given the same minor axis, either of zero length, which would result in a simple side-to-side swing, or of a positive length, which would result in swings enclosing space. Despite having the same major and minor axes, these pendulums will have different periodic times because of the different string lengths. The different periodic times mean that they have different accelerative forces on them at the same displacement.

Thus it is not necessarily true that two bodies at the same distance must have the same accelerative force on them.

What is not necessarily true must either be proved or assumed. Since it is has not yet been proved, and is not proved in this corollary, it is assumed and should really appear as another "given." This is entirely legitimate, since these are hypothetical propositions and Newton is entitled to assume anything he pleases. We must be alert to what is assumed, however, so that we can assess whether we may apply the propositions to real-world situations.

In his commentary to Definition 7, Newton anticipated what he would prove later when he told us there that the accelerative force of gravity is equal at equal distances from the center of forces. This equality will indeed be proved for our world in III.6, but much groundwork must be laid before we will be in a position to make that proof.

Expansion of Newton's Sketch of I.10 Corollary 2

"And in the revolutions made in all ellipses about the same center, the periodic times will be equal."

Given:

Ellipses with the same center.
[Accelerative forces equal at equal distances.]

To Prove:

Their periodic times will be equal.

Proof:

Part 1: Similar Ellipses

"For in similar ellipses, those times will be equal (by Proposition 4 Corollaries 3 and 8)..."

Suppose we have similar ellipses around the same center.

By the proposition, $F \propto r$.

By Proposition 4 Corollary 3 (converse) and Corollary 8: For similar figures, where $F \propto r$, periodic times will be equal.

Therefore in this case periodic times are equal.

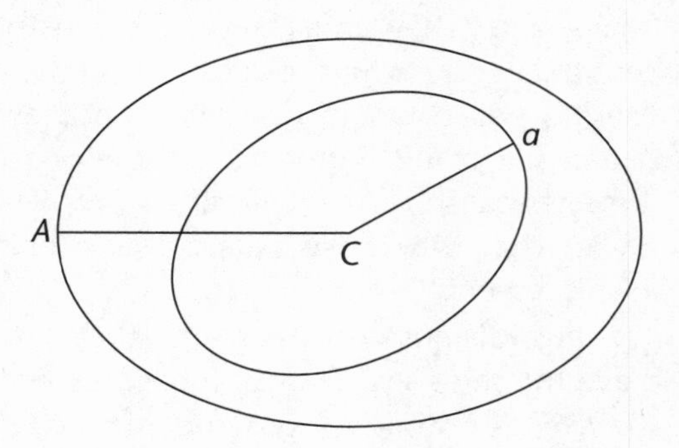

Part 2: Ellipses with the Greater Axis Common

Step 1:

"... and in ellipses having a common major axis, [periodic times] are to each other as the whole areas of the ellipses directly and the small parts of the areas described simultaneously inversely."

By Proposition 1, Δarea$/\Delta$time is constant for one particular orbit. Therefore, letting Δt represent the elapsed time,

$$\Delta\text{area}/\Delta t = \text{Whole area}/\text{Periodic time } P$$

Therefore, for one particular orbit,

$$P = \frac{\text{whole area}}{(\Delta\text{area}/\Delta t)} \qquad (1)$$

For two orbits,

$$\frac{P}{p} = \frac{\text{whole area}_1/(\Delta\,\text{area}_1/\Delta t)}{\text{whole area}_2/(\Delta\,\text{area}_2/\Delta t)}.$$

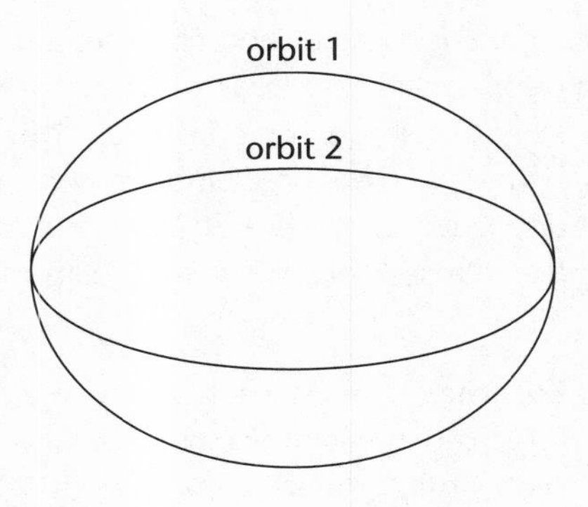

That is,

$$\text{Period} \propto \frac{\text{whole area of ellipse}}{\text{parts of areas in same time}}.$$

Thus Step 1 of Part 2 has been proved.

Step 2:

"... that is, as the minor axes directly and the velocities of the bodies at the principal vertices inversely."

Step 2a:

"... minor axes directly... "

By Apollonius I.21 (second part) for the thinner ellipse: If in an ellipse straight lines *bC*, *DE* are dropped ordinatewise to the diameter, the squares on them bC^2, DE^2 will be to each other as the areas contained by the straight lines cut off (abscissas) *AC*, *CG* and *AE*, *EG*.

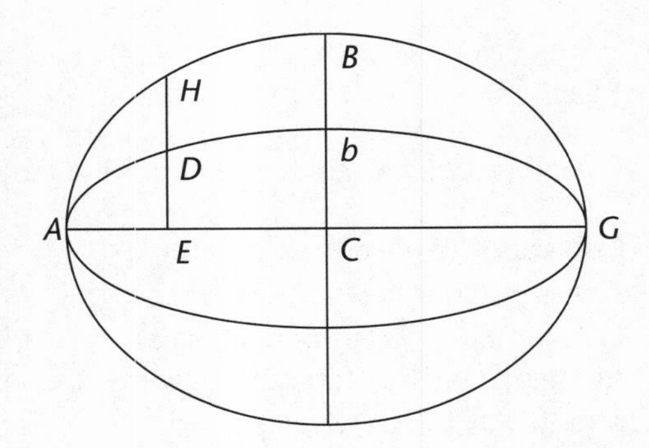

$$bC^2 : DE^2 :: \text{rect } AC,CG : \text{rect } AE,EG.$$

But since $AC = CG$, [Ap. I Def. 9]

$$bC^2 : DE^2 :: AC^2 : AE \times EG. \qquad (2)$$

By Apollonius I.21 for the thicker ellipse,

$$BC^2 : HE^2 :: AC^2 : \text{rect } AE,EG.$$

Combining with Equation 2,

$$BC^2 : HE^2 :: bC^2 : DE^2. \quad \text{[Eu. V.11]}$$

$$BC^2 : bC^2 :: HE^2 : DE^2. \quad \text{[Eu. V.16, } alternando\text{]}$$

Taking square roots,

$$BC : bC :: HE : DE. \qquad (3)$$

Any pair of heights will be in the same ratio.

Take equal bases and make rectangles. The areas of rectangles will be as the heights. Let n increase without limit, the length of the bases getting as small as you please. The area of the sum of the rectangles will approach as a limit the area of the ellipse by Lemma 3 Corollary 4.

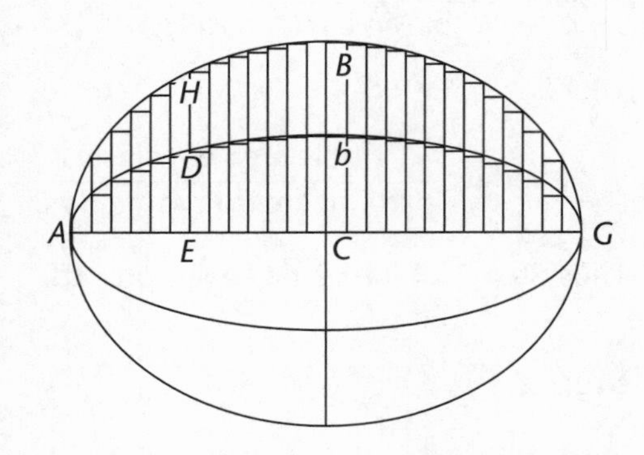

But the area of each rectangle (and thus that of the sum of the rectangles) will vary as this constant ratio of the heights, that is, as $BC : bC$, which is the ratio of the semi-minor axes of the ellipses.

Thus the areas of the ellipses will vary as their semi-minor axes, and therefore as their full minor axes:

Areas of whole ellipses $\propto$ minor axes. (4)

Step 2b

"... and [as] the velocities of the bodies at the principal vertices inversely."

Consider the motion of a body at A, a principal vertex (either end of the major, or "principal" axis). At this point the tangent is perpendicular to the principal axis.

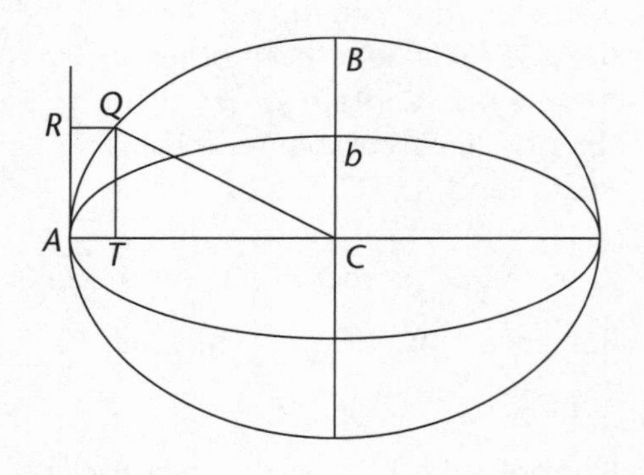

$$\triangle QAC \propto CA \times QT$$

By Lemma 7,

As $Q \to A$, arc $QA \overset{ult}{=}$ chord QA.

$$\widehat{\triangle} QAC \overset{ult}{\propto} CA \times QT. \qquad (5)$$

Since A is a principal vertex, $QT \parallel RA$ and thus is the "sine" mentioned in Lemma 11 Corollary 1. Therefore,

$$QT \overset{ult}{=} \widehat{QA}. \quad \text{[Lem. 11 Cor. 1].}$$

Substituting into Equation 5,

$$\widehat{\triangle} QAC \overset{ult}{\propto} CA \times \widehat{QA}.$$

The semi-major axis CA is shared between the two ellipses, and so CA is constant. Therefore,

$$\text{Area } \widehat{\triangle QAC} \overset{ult}{\propto} \text{arc } \widehat{AQ}\,.$$

Letting Δt represent the time to traverse AQ, and dividing each side by that same time:

$$(\Delta\text{area}/\Delta t) \overset{ult}{\propto} (\Delta\widehat{AQ}/\Delta t).$$

Or, to express it in terms of a process approaching a limit:

$$(\Delta\text{area}/\Delta t) \propto \lim_{\Delta t \to 0}(\Delta\widehat{AQ}/\Delta t). \qquad (6)$$

But $\lim_{\Delta t \to 0}(\Delta\widehat{AQ}/\Delta t) = V_A,$ by definition. (7)

In Newton's language, $\Delta\widehat{AQ}/\Delta t \overset{ult}{\propto} V_A$. (8)

Combining Equations 6 and 7:

$$\Delta\text{area}/\Delta t \propto V_A \text{ or}$$

$$\frac{\Delta\,\text{area}_1/\Delta t}{\Delta\,\text{area}_2/\Delta t} = \frac{V_{A(1)}}{V_{A(2)}}\,. \qquad (9)$$

Note: In Proposition 1 we saw that $\Delta\text{area}/\Delta t$ was constant for a given orbit. Equation 9 compares different orbits for ellipses with common major axis.

Now we can take the conclusion of Step 1:

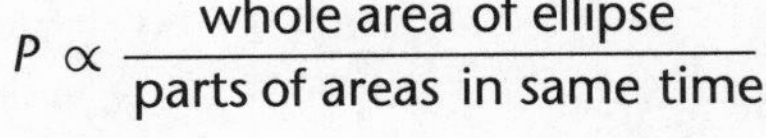

$$P \propto \frac{\text{whole area of ellipse}}{\text{parts of areas in same time}}$$

and substitute from Equation 9:

$$P \propto \frac{\text{whole area of ellipse}}{V_A}.$$

Then substituting from Equation 4:

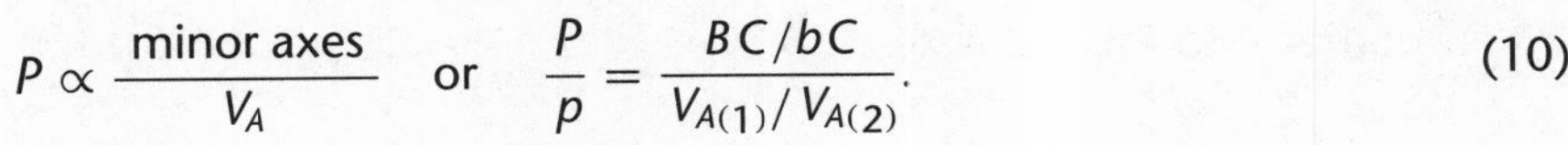

$$P \propto \frac{\text{minor axes}}{V_A} \quad \text{or} \quad \frac{P}{p} = \frac{BC/bC}{V_{A(1)}/V_{A(2)}}. \qquad (10)$$

Step 3

"... that is, as those minor axes directly and the lines applied ordinatewise to the same point of the common axis inversely."

By Equation 8,

$$V_A \overset{ult}{\propto} \frac{\Delta \widehat{AQ}}{\Delta t}.$$

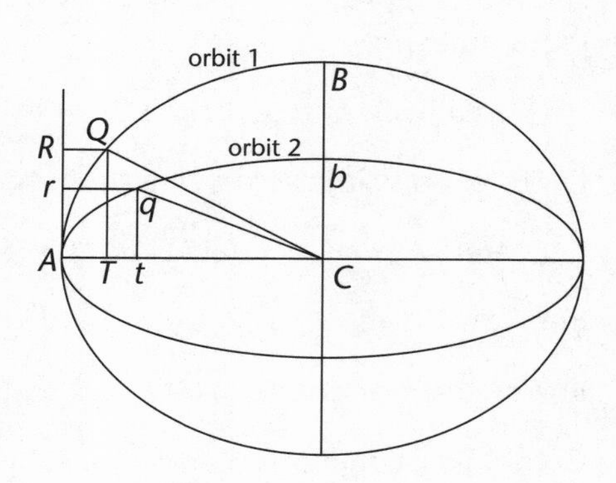

Since $\widehat{AQ}$ and $\widehat{Aq}$ are equal-time arcs, Δt cancels out; this becomes

$$V_A \overset{ult}{\propto} \widehat{AQ} \quad \text{or} \quad \frac{V_{A(1)}}{V_{A(2)}} \overset{ult}{=} \frac{\widehat{AQ}}{\widehat{Aq}}.$$

By Lemma 7 Corollary 1,

$$\widehat{AQ} \overset{ult}{=} QT \quad \text{and} \quad \widehat{Aq} \overset{ult}{=} qt.$$

Therefore,

$$\frac{V_{AQ}}{V_{Aq}} \overset{ult}{=} \frac{QT}{qt}.$$

Here we have a little problem: We need the ratio of ordinates drawn to the same point T of the common axis, but we have two lines QT and qt drawn to two separate points of the axis. Since the arcs AQ and Aq are described in the same time, we cannot immediately assume in the finite case that q lies on QT. (Actually, it *will* turn out to be true in the finite case, though we're not yet justified in assuming it—can you see why?)

Therefore to complete the proof, we need the following little lemma:

Lemmita:

By Proposition 6 Corollary 1, force varies as the subtense divided by a quantity which itself varies with the time in which the arc is traversed. Since $\widehat{AQ}$ and $\widehat{Aq}$ are traversed in the same time, this numerator is constant and the forces are ultimately proportional to the subtenses:

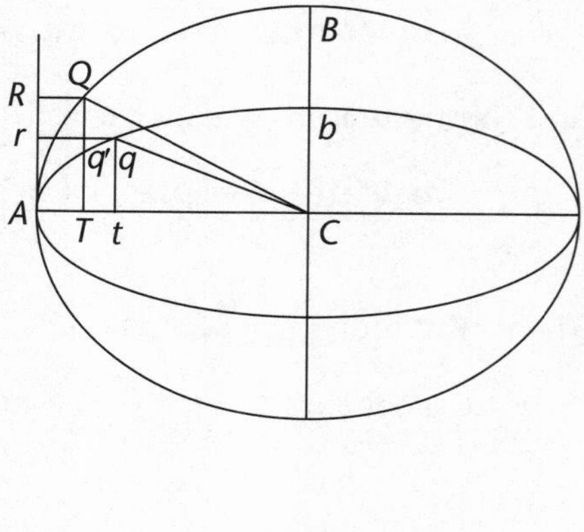

$$\frac{F}{f} \overset{ult}{=} \frac{QR}{qr}.$$

Because both bodies begin at A, the force at point A is the same on each, and therefore

$$QR \overset{ult}{=} qr.$$

Ultimately, therefore, line qt will fall on line QT.

Let q' be the intersection of line QT with the thinner ellipse.

Ultimately qt will coincide with and be equal to $q'T$.

Q.E.D.

Getting back to the proposition, we may now substitute $q'T$ for qt:

$$\frac{V_{AQ}}{V_{Aq}} \stackrel{\text{ult}}{=} \frac{QT}{q'T} \ . \tag{11}$$

Now Equation 10 says

$$\frac{P}{p} = \frac{BC/bC}{V_{A(1)}/V_{A(2)}}.$$

Substituting from Equation 11, we have

$$\frac{P}{p} \stackrel{ult}{=} \frac{BC/bC}{QT/q'T}. \tag{12}$$

(We have to put the "ult" in because we're substituting evanescents for the finite velocities.)

But QT and $q'T$ are the ordinates to the same point T of the common axis AC.

This is the conclusion of Step 3.

Step 4

"... and therefore in the ratio of equality."

Rewriting Equation 3:

$$BC : bC \ :: \ QT : q'T.$$

Here we need to proceed carefully. This proportion is *always* true because q' always lies on QT. Therefore, $BC : bC$ may always be substituted for $QT : q'T$, even when they are evanescents. And since we can substitute the finite ratio, we can drop the "ult" when we substitute this ratio into Equation 12:

$$\frac{P}{p} = \frac{BC/bC}{BC/bC} = \frac{1}{1}.$$

Part 3: Conclusion

We have now proved periodic times equal for similar ellipses (Part 1) and for ellipses with the same major axis (Part 2). We will use these two conclusions to prove that periodic times will be equal for any two ellipses with the same center.

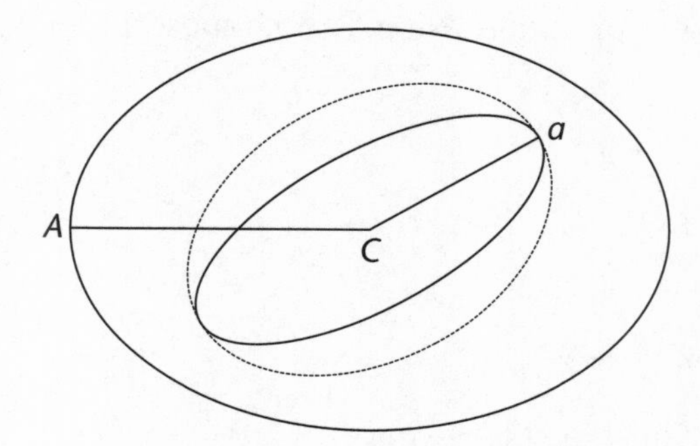

Take any two ellipses around the same center *C*. Let the first have semi-major axis *Ca* and the second semi-major axis *CA*.

Around center *C*, with semi-major axis *Ca*, construct a third ellipse (dashed line) similar to the second ellipse *CA*.

The dashed-line ellipse has the same period as the second ellipse, the one on *CA*, because they are similar, by Part 1.

And the dashed-line ellipse on *Ca* has the same period as the first ellipse *Ca* because they share a major axis, by Part 2.

Therefore the first ellipse *Ca* has the same period as the second *CA*.

Q.E.D.

Scholium

If the center depart to infinity and the ellipse be turned into a parabola, the body will move in this parabola, and the force, now tending to an infinitely distant center, will come out uniform. This is Galileo's theorem. And if (by a change in the inclination of the plane to the cone being cut) the parabolic section of the cone be turned into a hyperbola, the body will move on the perimeter of this with a centripetal force turned into a centrifugal one.

[In the first edition, the scholium ended here. In the second edition, Newton inserted an ungainly and murky sentence further describing the extension of this proposition to other figures. Since it does not add materially to the foregoing, it is not included in the present selection.]

Notes on Scholium after I.10

- Here Newton invites us to use our imagination, to think about modifying the ellipse to get other conics, and to see how Proposition 10 and corollaries would apply. All the conclusions presented here can be proved independently; however, this is not the path Newton chooses. The exercise in imagination he presents instead provides a quicker, and perhaps more vivid, means of understanding the motion of bodies under forces that vary directly with distance from a center.

- Galileo's theorem:

  Under a constant force a body will move in a parabola.

This is the first proposition of "On the Motion of Projectiles" (*Two New Sciences*, Fourth Day).

Ellipse

Attractive force towards *C*, $F \propto r$.

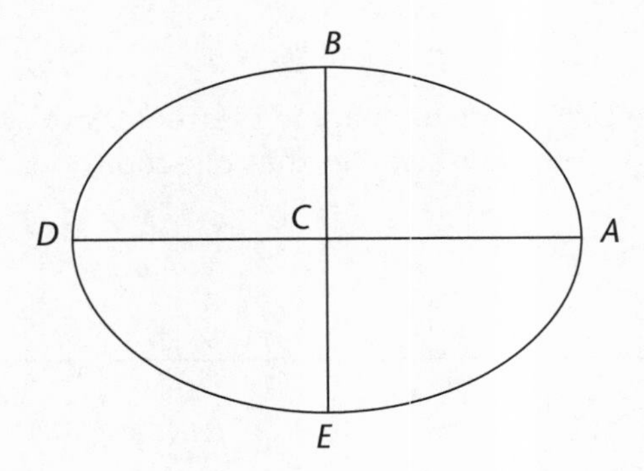

As the body gets farther from *C*, the force is greater and pulls it back in more forcefully. As the body gets closer to *C*, the force is smaller and accelerates the body less. By Proposition 1, the body's velocity will be greatest where its distance from the center of forces is least, at *B* and *E*, and least where its distance is greatest, at *A* and *D*.

Parabola

Now let's stretch out the ellipse along its greater axis. As it stretches, the center moves out, and as the distance to the center from point *D* increases, the force will vary less in proportion to any given change in distance. If we imagine the center at an infinite distance, the force will be constant, and all lines drawn to the center (that is, all diameters) will be parallel. This is the situation in Apollonius I.49, and is also the situation in Proposition 1 of "On the Motion of Projectiles" in Galileo's *Two New Sciences*. The body's velocity component perpendicular to the diameters remains constant, while it decelerates uniformly along the diameters until it gets to *D*, and then accelerates uniformly back towards the center. It is therefore slowest at *D*, and faster as you go out towards *C*.

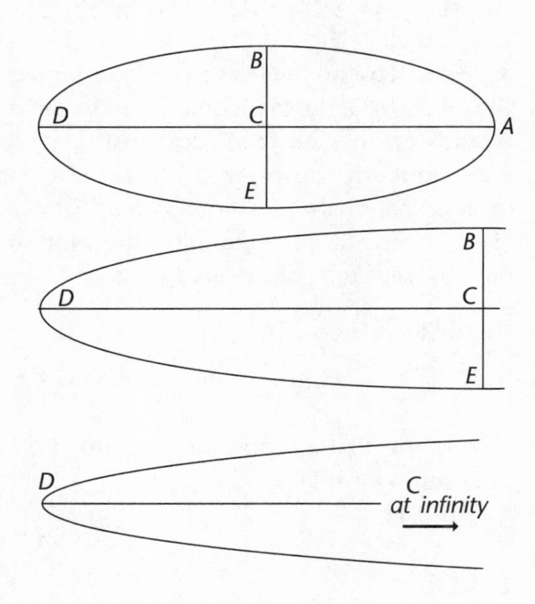

Hyperbola

We now stretch our minds to imagine the center moving *beyond* infinity. Since it can't get any further away in that direction, one might conjecture

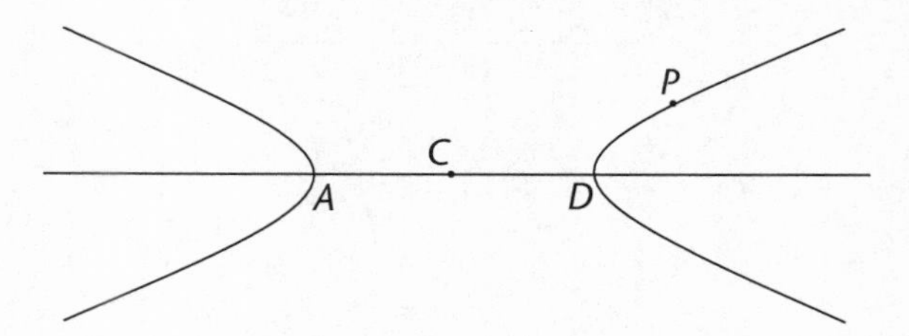

that the center comes back in from the other direction. The force still has the same direction (from left to right in the diagram above), but the lines along which the force acts converge at a point on the other side. Thus the force has been transformed into a repulsive one, and the curve has been stretched into a hyperbola.

A body at *P* begins moving in towards *C* (due to some initial velocity).

As $P \rightarrow$ vertex *D* of the hyperbola, the repulsive force is slowing it down (although with less repulsive force the closer it gets) until at *D* the body has lost all motion towards the center (though keeping some tangential velocity); at this point it starts being pushed away.

As its distance from the center increases, the body at *P* is pushed away more forcefully.

SECTION 3

On the motion of bodies in eccentric conic sections

Section 3 moves into a more focused investigation of the particular orbits and force laws that will turn out to be relevant to our universe. The propositions are still hypothetical, establishing mathematical tools to be used later in conjunction with the phenomena of our world as laid out in Book III. But they are tools that will be specifically useful for the celestial mechanics we will be examining, describing, and explaining in Book III.

Proposition 11

If a body revolve in an ellipse, it is required to find the law of centripetal force tending to the focus [umbilicum] of the ellipse.

Let the focus of the ellipse be S. Let SP be drawn, cutting the diameter of the ellipse DK at E, and the line Qv applied ordinatewise at x; and let the parallelogram $QxPR$ be completed. It is clear that EP is equal to the greater semiaxis AC, since (HI being drawn from the other focus H of the ellipse parallel to EC) ES and EI are equal (because of the equals CS and CH), and therefore EP is half the sum of PS, PI; that is, because of the parallels HI, PR, and the equal angles IPR, HPZ, [half the sum of] PS, PH, which together are equal to the whole axis $2AC$. Let QT be drawn perpendicular to SP, and let L be the principal latus rectum of the ellipse (equal to $\frac{2\,\text{sq.}BC}{AC}$). Then $L \times QR$ will be to $L \times Pv$ as QR to Pv; that is, as PE or AC is to PC; and $L \times Pv$ is to [rectangle] Gv,vP as L to Gv; and [rectangle] Gv,vP [will be] to sq. Qv as sq. PC to sq. CD; and (by Lemma 7 Corollary 2) as points Q and P come together, sq. Qv to sq. Qx is the ratio of equality; and sq. Qx or sq. Qv is to sq. QT as sq. EP is to sq. PF; that is, as sq. CA is to sq. PF, or (by Lemma 12) as sq. CD is to sq. CB. Compounding all these ratios, $L \times QR$ would be to sq. QT as $AC \times L \times \text{sq.}PC \times \text{sq.}CD$, or $2\ \text{sq.}CB \times \text{sq.}PC \times \text{sq.}CD$, is to $PC \times Gv \times \text{sq.}CD \times \text{sq.}CB$, or as $2PC$ is

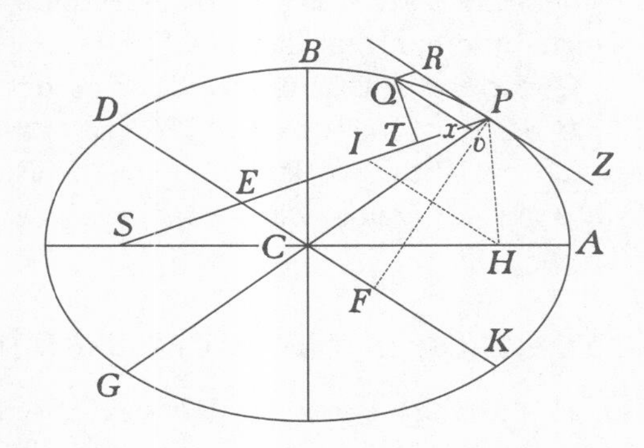

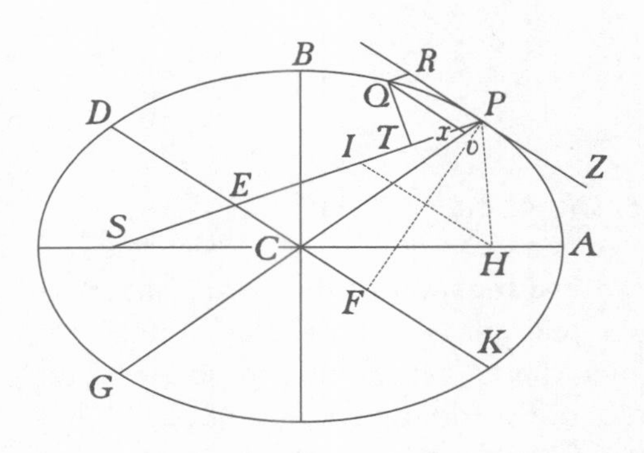

to Gv. But as points Q and P come together, $2PC$ and Gv are equal. Therefore, $L \times QR$ and sq. QT, proportional to them, are also [ultimately] equal. Let these equals be multiplied by $\frac{\text{sq.}SP}{QR}$, and $L \times \text{sq.}SP$ will become equal to $\frac{\text{sq.}SP \times \text{sq.}QT}{QR}$. Therefore (by Proposition 6, Corollary 1 and 5) the centripetal force is inversely as $L \times \text{sq. } SP$, that is, inversely in the duplicate ratio of the distance SP.

Q.E.I.

Notes on I.11

• We begin our more focused investigation of orbits and forces with the ellipse.

In the previous proposition, I.10, we found the force law given the center of forces at the ellipse's center. In that case the force varied directly as the distance of the body from the center of forces. Now we move the center of forces from the center of the ellipse to a focus. The force law, while still varying only as a function of the distance of the body, now varies inversely as the square of that distance.

Question for Discussion: Can you get any intuitive feel for why a simple (and perhaps small) shift of the center would require a change from a law which has the force increasing with distance to one which has the force dropping off rapidly with distance?

Expansion of Newton's Sketch of I.11

"If a body revolve in an ellipse, it is required to find the law of centripetal force tending to the focus [*umbilicum*] of the ellipse."

Given:

1. Body moves in an ellipse;

2. center of forces is at focus of ellipse.

To Find:

Force law for this situation. (It will turn out to be $F \propto 1/SP^2$.)

Proof:

"Let the focus of the ellipse be S. Let SP be drawn, cutting the diameter of the ellipse DK at E, and the line Qv applied ordinatewise at x; and let the parallelogram $QxPR$ be completed."

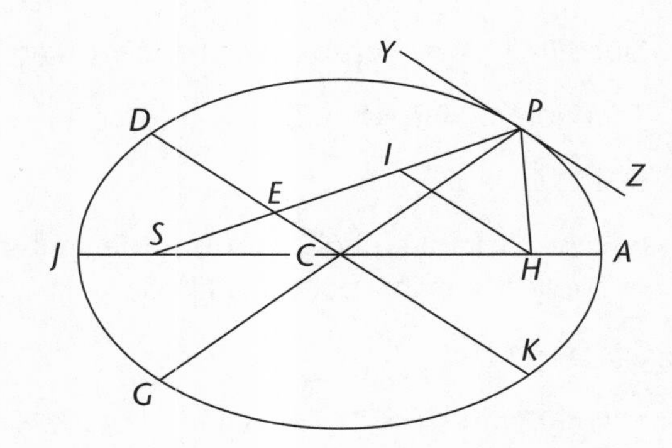

Let C be the center of the ellipse with major axis JA and foci S and H.

Let P be a point on the curve.

Connect P to foci S and H.

Draw diameter PG through C.

Draw line YPZ tangent at P.

Draw conjugate diameter $DK \parallel PZ$. [Ap. I.32 converse]

Where DK crosses PS, label E.

From focus H, drop $HI \parallel DK$ to intersect PS at I.

(We will postpone the constructions of x, v, and parallelogram $QxPR$, so as to have a less cluttered diagram for the first part of the proof.)

"It is clear that EP is equal to the greater semiaxis AC, since (HI being drawn from the other focus H of the ellipse parallel to EC) ES and EI are equal (because of the equals CS and CH), and therefore EP is half the sum of PS, PI; that is, because of the parallels HI, PR, and the equal angles IPR, HPZ, [half the sum of] PS, PH, which together are equal to the whole axis $2AC$."

By construction, $HI \parallel DK$ and $DK \parallel PZ$.

Therefore $HI \parallel PZ$.

Angle YPS = angle PIH. [Eu. I.29] (1)

Angle HPZ = angle PHI. [Eu. I.29] (2)

By Apollonius III.48 applied here:

The same things being so [an ellipse and two "points of application"—foci], the straight lines PS and PH drawn from the point of contact P to the foci S and H make equal angles with the tangent.

Angle YPS = angle HPZ. [Ap. III.48]

But angle YPS = angle PIH by Equation 1;

therefore angle PIH = angle HPZ.

But angle HPZ = angle PHI by Equation 2;

Therefore angle PIH = angle PHI.

Therefore $IP = HP$. [Eu. I.6] (3)

Since $IH \parallel DK$ by construction, in $\triangle SIH$ line EC is parallel to base IH.

Therefore by Euclid VI.2,

$$SE : EI :: SC : CH. \quad (4)$$

But the foci are equidistant from the center by the definition of the "points of application" in Apollonius III.45.

So $SC = CH$.

Therefore, in Equation 4, $SC : CH$ is a ratio of equality; therefore so is $SE : EI$.

$$SE = EI.$$

$$EP = EI + IP.$$

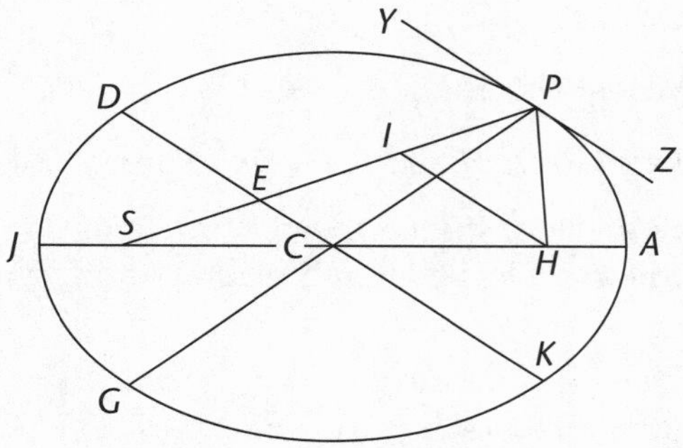

Substituting for EI,

$$EP = SE + IP.$$

Adding EP to both sides,

$$2 \times EP = EP + SE + IP.$$

But $SE + EP = SP$.

Therefore $2 \times EP = SP + IP$.

$$EP = (SP + IP)/2.$$

But by Equation 3, $IP = HP$; substituting:

$$EP = (SP + HP)/2. \quad (5)$$

Apollonius III.52 applied here: If in an ellipse from the points of application S and H straight lines SP and HP are deflected to the line of the section point P, then they will be equal to the axis AJ.

$SP + HP = AJ$. [Ap. III.52]

But $AJ = 2 \times AC$. [Ap. I Def. 6]

So $SP + HP = 2 \times AC$.

Substituting into Equation 5:

$$EP = (2 \times AC)/2.$$

Therefore $EP = AC$. (6)

Setting up Proportions for the Great *Ex Aequali* Being Prepared

Step 1: First Proportion

"Let QT be drawn perpendicular to SP, and let L be the principal latus rectum of the ellipse (equal to $\frac{2\text{sq.}BC}{AC}$). Then $L \times QR$ will be to $L \times Pv$ as QR to Pv; that is, as PE or AC is to PC;..."

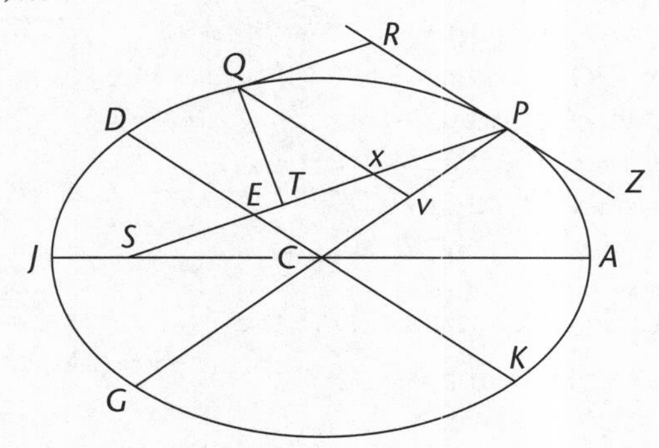

Let $Qv \parallel DK$ be dropped ordinatewise to diameter PG.

[Ap. I Def. 4 and I.15]

$DK \parallel RP$. [Ap. I.32 converse]

Therefore $Qv \parallel RP$.

Construct $QR \parallel SP$.

Let Qv intersect SP in x.

Since $Qv \parallel RP$ and $QR \parallel SP$, figure $QRPx$ is a parallelogram.

$QR = xP$. [Eu. I.34] (7)

Now since $Qv \parallel DC$, line xv is parallel to base EC in $\triangle EPC$, and

$xE : Px :: vC : Pv$. [Eu. VI.2]

$PE : Px :: PC : Pv$. [Eu. V.16, *componendo*]

Inverting and alternating,

$Px : Pv :: PE : PC$.

Substituting from Equation 7: $QR : Pv :: PE : PC$.

By Equation 6, $EP = AC$. Substituting for EP,

$QR : Pv :: AC : PC$.

Multiplying antecedent and consequent of the first ratio by the principal latus rectum L (a constant):

$(L \times QR) : (L \times Pv) :: AC : PC$. (8)

What is L and how do we find it?

In Apollonius I.15 the latus rectum is introduced and defined as the third proportional to a specified pair of conjugate diameters. In our diagram, for example, we may contrive the latus rectum for diameter PG this way:

$$PG : DK :: DK : L_P .$$

Here, subscript $_P$ indicates that L_P belongs to vertex P, an endpoint of diameter PG. But in Proposition 11 Newton is interested in the *principal latus rectum,* which belongs to the *principal vertex A*, endpoint of the *major diameter* $2AC$. We could call this line L_A (since it belongs to vertex A), but Newton calls it L. So, in accordance with Apollonius I.15, contrive

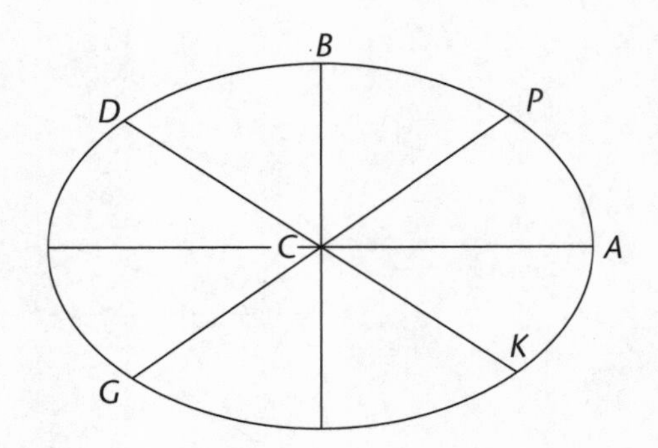

$$2AC : 2BC :: 2BC : L.$$

$$L = \frac{2BC \times 2BC}{2AC}; \quad \text{[Eu. VI.17]}$$

$$L = 2BC^2/AC. \tag{9}$$

Step 2: Second Proportion

"... and $L \times Pv$ is to [rectangle] Gv, vP as L to Gv; ..."

$$L : Gv :: L : Gv. \quad \text{[identity]}$$

Multiply antecedent and consequent of the first ratio by Pv:

$$(L \times Pv) : (Gv \times Pv) :: L : Gv. \tag{10}$$

Step 3: Third Proportion

"... and [rectangle] Gv,vP [will be] to sq. Qv as sq. PC to sq. CD; ..."

By Apollonius I.21 (second part) applied here: If in an ellipse straight lines DC and Qv be dropped ordinatewise to diameter PG, the squares on them DC^2 and Qv^2 will be to each other as the areas contained by the straight lines cut off GC, CP and Gv, vP on diameter PG.

$$DC^2 : Qv^2 :: \text{rect } GC,CP : \text{rect } Gv,vP. \quad \text{[Ap. I.21]}$$

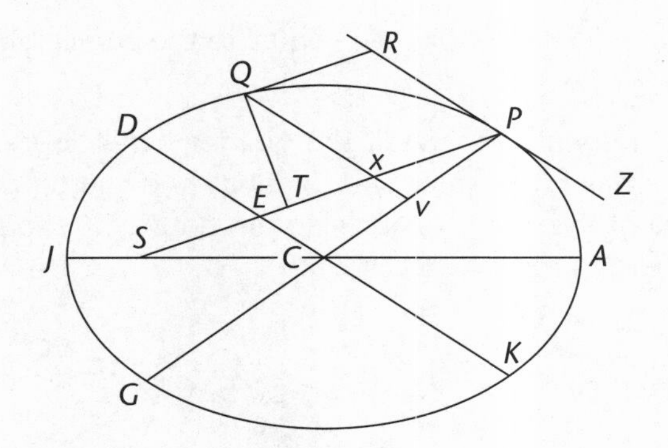

But $GC = CP$ [Ap. I Def. 6], so

$$DC^2 : Qv^2 :: PC^2 : Gv \times vP.$$

Alternating and inverting,

$$Gv \times vP : Qv^2 :: PC^2 : CD^2. \quad (11)$$

Step 4:

"... and (by Lemma 7 Corollary 2) as points Q and P come together, sq. Qv to sq. Qx is the ratio of equality; and sq. Qx or sq. Qv is to sq. QT as sq. EP is to sq. PF; that is, as sq. CA is to sq. PF, or (by Lemma 12) as sq. CD is to sq. CB."

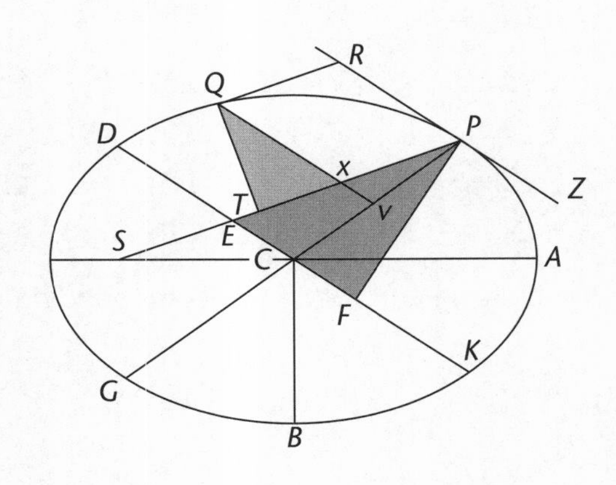

Drop $PF \perp DK$.

Drop $QT \perp SP$.

Angle PFE = angle QTx. [Eu. I Post. 4]

Because $Qv \parallel DK$, cut by transversal PE,

angle PEF = angle QxT. [Eu. I.29]

Therefore $\triangle PEF \sim \triangle QxT$.

$$Qx : QT :: PE : PF. \quad \text{[Eu. VI.4]}$$

But by Equation 6, $EP = CA$.

Substituting, $Qx : QT :: CA : PF. \quad (12)$

By Lemma 12, parallelograms circumscribed about any conjugate diameter of a given ellipse are equal among themselves. Therefore the parallelogram around conjugate diameters *DK, PG* is equal to the parallelogram around the major and minor axes.

Since quarters of equals are equal, the quarter parallelogram on the semi-minor and semi-major axes (rect *CA, CB*) equals the quarter parallelogram on *PC, CD*. The area of the quarter parallelogram on *PC, CK* is found by taking its base times its height, namely, rect *CD, PF*.

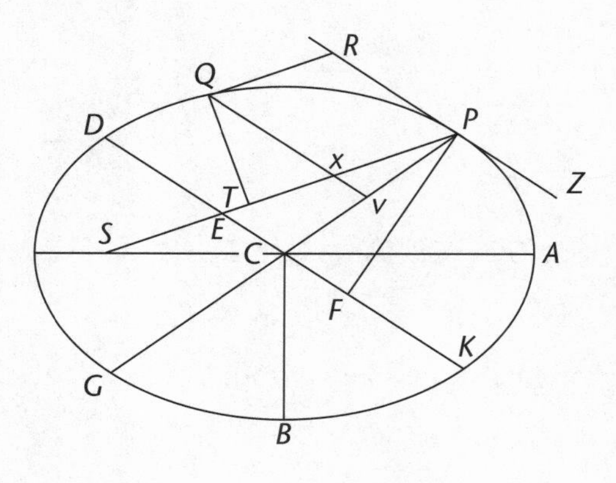

Rect CA, CB = rect CD, PF. [Lem. 12]

$CA : PF :: CD : CB$. [Eu. VI.16]

Substituting from Equation 12,

$Qx : QT :: CD : CB$.

But as $Q \to P$, $Qx \to Qv$.

$Qx \overset{ult}{=} Qv$. [Lem. 7, Cor. 2]

Substituting, $Qv : QT \overset{ult}{::} CD : CB$.

Squaring, $Qv^2 : QT^2 \overset{ult}{::} CD^2 : CB^2$. (13)

Step 5: Now Putting Together the Four Proportions

"Compounding all these ratios, $L \times QR$ would be to sq. QT as $AC \times L \times$ sq. $PC \times$ sq. CD, or 2 sq. $CB \times$ sq. $PC \times$ sq. CD, is to $PC \times Gv \times$ sq. $CD \times$ sq. CB, or as $2PC$ is to Gv."

$L \times QR$	:	$L \times Pv$	::	AC	:	PC.	[by Equation 8]
$L \times Pv$	:	$Gv \times Pv$	::	L	:	Gv.	[by Equation 10]
$Gv \times Pv$	:	Qv^2	::	PC^2	:	CD^2.	[by Equation 11]
Qv^2	:	QT^2	$\overset{ult}{::}$	CD^2	:	CB^2.	[by Equation 13]

Multiplying and canceling (compounding with *ex aequali*):

$$\frac{L \times QR}{QT^2} \overset{ult}{=} \frac{AC \times L \times PC^2}{PC \times Gv \times CB^2};$$

$$\frac{L \times QR}{QT^2} \overset{ult}{=} \frac{AC \times L \times PC}{Gv \times CB^2}. \quad (14)$$

Substituting for L from Equation 9 into the right hand expression of Equation 14:

$$\frac{L \times QR}{QT^2} \overset{ult}{=} \frac{AC \times (2BC^2/AC) \times PC}{Gv \times CB^2}.$$

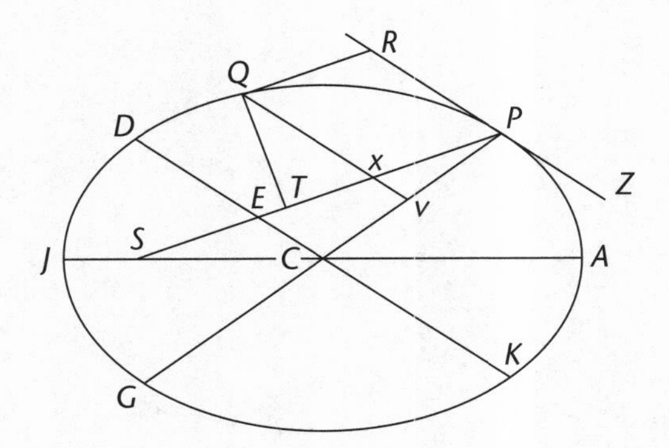

Simplifying,

$$\frac{L \times QR}{QT^2} \stackrel{ult}{=} \frac{2 \times PC}{Gv}.$$

"But as points Q and P come together, $2PC$ and Gv are equal. Therefore, $L \times QR$ and sq. QT, proportional to them, are also [ultimately] equal."

As $Q \to P$, $Gv \to GP$; $2PC \stackrel{ult}{=} Gv$. Substituting for $2PC$,

$$\frac{L \times QR}{QT^2} \stackrel{ult}{=} \frac{Gv}{Gv}.$$

Because $Gv : Gv$ is a ratio of equality, so, ultimately, is $L \times QR : QT^2$.

Therefore $L \times QR \stackrel{ult}{=} QT^2$. (15)

"Let these equals be multiplied by $\frac{\text{sq.}SP}{QR}$, and $L \times \text{sq.}SP$ will become equal to $\frac{\text{sq.}SP \times \text{sq.}QT}{QR}$."

Multiply both sides of Equation 15 by $\frac{SP^2}{QR}$:

$$\frac{L \times QR \times SP^2}{QR} \stackrel{ult}{=} \frac{QT^2 \times SP^2}{QR}.$$

Simplifying,

$$L \times SP^2 \stackrel{ult}{=} \frac{QT^2 \times SP^2}{QR}. \quad (16)$$

"Therefore (by Proposition 6, Corollary 1 and 5) the centripetal force is inversely as $L \times \text{sq.}SP$, that is, inversely in the duplicate ratio of the distance SP.

Q.E.I."

But by Proposition 6 Corollary 1, force $\stackrel{ult}{\propto} 1/\frac{QT^2 \times SP^2}{QR}$.

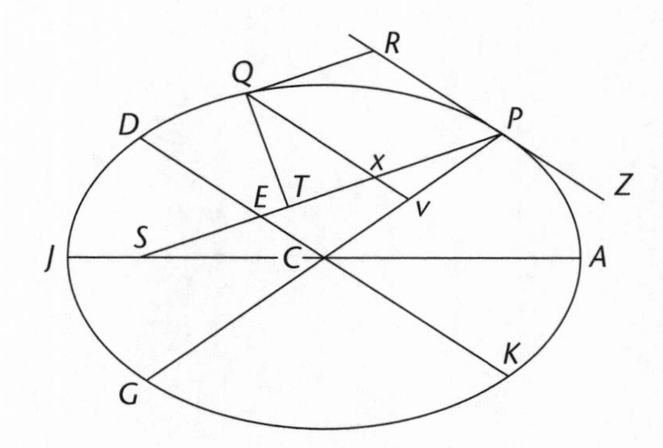

Therefore, substituting from Equation 16,

$$\text{force} \overset{ult}{\propto} \frac{1}{L \times SP^2}.$$

But L in this case is the latus rectum of the major axis ACJ, and so is constant for a given ellipse.

Therefore $F \overset{ult}{\propto} 1/SP^2$.

If a body moves in an ellipse with the center of forces at a focus, the force varies inversely as the square of the distance from the center of forces.

Q.E.I.

Newton's Comment after I.11

It would be possible to carry the proof over to the parabola and the hyperbola with the same brevity we used in Proposition 10. However, because the problem is a worthy one, and because of its use in what follows, it will not be tedious to confirm the other cases with a demonstration.

Proposition 12

If a body move in a hyperbola, it is required to find the law of centripetal force tending to the focus [umbilicum] of the figure.

Let CA, CB be semiaxes of the hyperbola, PG, KD other conjugate diameters, PF perpendicular to the diameter KD, and Qv be applied ordinatewise to the diameter GP. Let SP be drawn, cutting the diameter DK at E, and the line Qv applied ordinatewise at x; and let the parallelogram $QxPR$ be completed. It is clear that EP is equal to the transverse semiaxis AC, since (HI being drawn from the other focus of the hyperbola

H parallel to EC) ES and EI are equal (because of the equals CS and CH), and therefore EP is half the difference of PS, PI; that is, because of the parallels HI, PR, and the equal angles IPR, HPZ, [half the difference of] PS, PH, the difference of

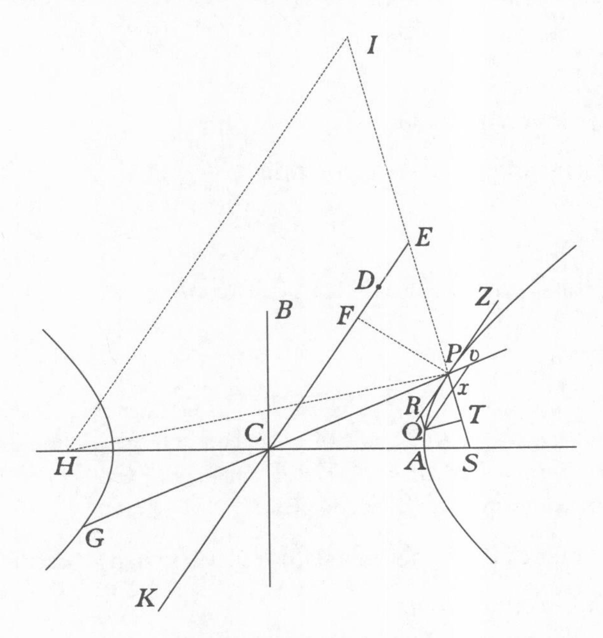

which is equal to the whole axis $2AC$. Let QT be drawn perpendicular to SP. And let L be the principal latus rectum of the hyperbola (equal to $\frac{2\,\text{sq.}\,BC}{AC}$): then $L \times QR$ will be to $L \times Pv$ as QR to Pv; that is, as PE or AC to PC; and $L \times Pv$ will be to [rectangle] Gv,vP as L is to Gv; and (by the nature of conics) [rectangle] Gv,vP will be to sq. Qv as sq. PC to sq. CD; and (by Lemma 7, Corollary 2) as points Q and P come together, sq. Qv to sq. Qx is the ratio of equality; and sq. Qx or sq. Qv is to sq. QT as sq. EP is to sq. PF; that is, as sq. CA is to sq. PF, or (by Lemma 12) as sq. CD is to sq. CB. Compounding all these ratios, $L \times QR$ would be to sq. QT as $AC \times L \times \text{sq.}\,PC \times \text{sq.}\,CD$, or $2\,\text{sq.}\,CB \times \text{sq.}\,PC \times \text{sq.}\,CD$, is to $PC \times Gv \times \text{sq.}\,CD \times \text{sq.}\,CB$, or as $2PC$ is to Gv. But as points Q and P come together, $2PC$ and Gv are equal. Therefore, $L \times QR$ and sq. QT, proportional to them, are also [ultimately] equal. Let these equals be multiplied by $\frac{\text{sq.}\,SP}{QR}$, and $L \times \text{sq.}\,SP$ becomes equal to $\frac{\text{sq.}\,SP \times \text{sq.}\,QT}{QR}$. Therefore (by Proposition 6, Corollaries 1 and 5) the centripetal force is inversely as $L \times \text{sq.}\,SP$, that is, inversely in the duplicate ratio of the distance SP.

Q.E.I.

Note on I.12

This proposition follows the basic strategy of the previous; but certain changes of tactics must be made to accommodate the differences between ellipses and hyperbolas.

Expansion of Newton's Sketch of I.12

"If a body move in a hyperbola, it is required to find the law of centripetal force tending to the focus [*umbilicum*] of the figure."

Given:

1. Body moves in a hyperbola;
2. center of forces is at focus of hyperbola.

To Find:

Force law for this situation. (It will turn out to be $F \propto 1/SP^2$.)

Proof:

"Let *CA*, *CB* be semiaxes of the hyperbola, *PG*, *KD* other conjugate diameters, *PF* perpendicular to the diameter *KD*, and *Qv* be applied ordinatewise to the diameter *GP*. Let *SP* be drawn, cutting the diameter *DK* at *E*, and the line *Qv* applied ordinatewise at *x*; and let the parallelogram *QxPR* be completed."

Let *C* be the center of the hyperbola with transverse axis *AN* and foci *S* and *H*.

Let *P* be a point on the curve.

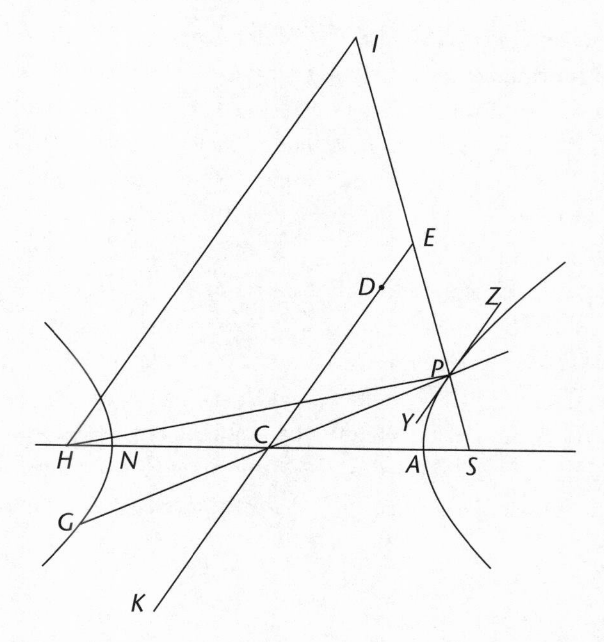

Connect *P* to foci *S* and *H*.

Draw a diameter from *P* through *C* intersecting the other branch in *G*.

Draw line *YPZ* tangent at *P*.

Draw conjugate diameter $DK \parallel PZ$. [Ap. I.32 converse]

Where *DK* crosses *SP* extended, label *E*.

From focus *H*, draw $HI \parallel DK$ to intersect *SP* extended at *I*.

(We will postpone the constructions of *x*, *v*, and parallelogram *QxPR* so as to have a less cluttered diagram for the first part of the proof.)

"It is clear that *EP* is equal to the transverse semiaxis *AC*, since (*HI* being drawn from the other focus of the hyperbola *H* parallel to *EC*) *ES* and *EI* are equal (because of the equals *CS* and *CH*)..."

Since in $\triangle HIS$, line $EC \parallel$ base *IH*,

$SE : EI :: SC : CH$. [Eu. VI.2 and *separando*]

But $SC = CH$. [Ap. III.45]

Because the second ratio is a ratio of equality, so is the first.

$$EI = SE. \qquad (1)$$

"and therefore *EP* is half the difference of *PS*, *PI*..."

$PI = EI + EP$;

$SP = SE - EP$.

I E P S

Subtracting simultaneous equations,

$$PI - PS = (EI + EP) - (SE - EP).$$

Substituting *SE* for *EI* by Equation 1:

$$PI - PS = SE + EP - SE + EP;$$

$$PI - PS = 2EP;$$

$$EP = (PI - PS)/2. \qquad (2)$$

"... that is, because of the parallels *HI*, *PR*, and the equal angles *IPR*, *HPZ*, [half the difference of] *PS*, *PH*..."

By Apollonius III.48 applied here, the angles *YPH* and *YPS*, between the tangent *YPZ* and the lines *HP* and *SP* from each point of application (focus) to the point of tangency, are equal.

Angle YPH = angle YPS. [Ap. III.48]

But angle YPS = angle ZPI. [Eu. I.15]

Therefore angle ZPI = angle HPY. (3)

But by construction, $HI \parallel DK$ and $DK \parallel YPZ$; therefore $HI \parallel YPZ$.

Parallels HI and YPZ are cut by transversal IP, therefore:

angle ZPI = angle PIH. [Eu. I.29] (4)

And since the same parallels are also cut by transversal HP,

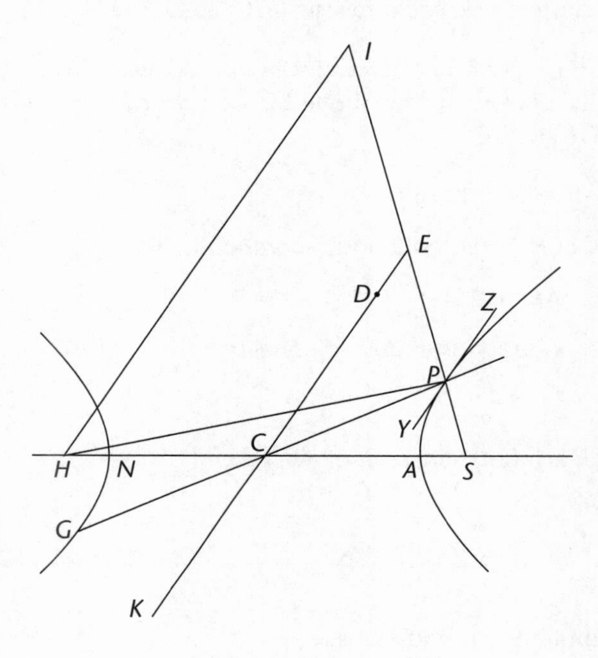

angle HPY = angle PHI. [Eu. I.29] (5)

But since angle HPY = angle ZPI by Equation 3,

and angle ZPI = angle PIH by Equation 4,

therefore angle HPY = angle PIH.

And since angle HPY = angle PHI by Equation 5,

angle PHI = angle PIH.

But these are the base angles of triangle HIP; therefore

$HP = PI$. [Eu. I.6]

Substituting into Equation 2:

$EP = (HP - PS)/2$. (6)

"... the difference of which is equal to the whole axis $2AC$."

By Apollonius III.51 applied here:

Let there be an hyperbola or opposite sections whose axis is NA and center C. Let P be a point on the section. H and S are the points of application (foci), with H being the focus in the opposite section. III.51 proves that $HP = SP + AN$.

Thus $AN = HP - SP$.

But $AN = 2AC$. [Ap. Def. 6]

$$HP - SP = 2AC.$$

Substituting into Equation 6,

$$EP = 2AC/2.$$

$$EP = AC. \quad (7)$$

"Let QT be drawn perpendicular to SP."

We will construct QT when we use it in Step 4.

"And let L be the principal latus rectum of the hyperbola (equal to $\frac{2\,\text{sq.}\,BC}{AC}$)..."

By Apollonius I Definition 11, we get the lengths of the conjugate diameters:

> And let the straight line from the center parallel to an ordinate, being a mean proportional to the sides of 'the figure' [the rectangle contained by the original diameter and the latus rectum], and bisected by the center, be called the second diameter.

$$2AC : 2BC :: 2BC : L.$$

$$\text{Rect}\,(2AC, L) = (2BC)^2. \quad \text{[Eu. VI.17]}$$

$$4BC^2 = (2AC \times L).$$

$$2BC^2 = AC \times L. \quad (8)$$

$$L = 2BC^2/AC.$$

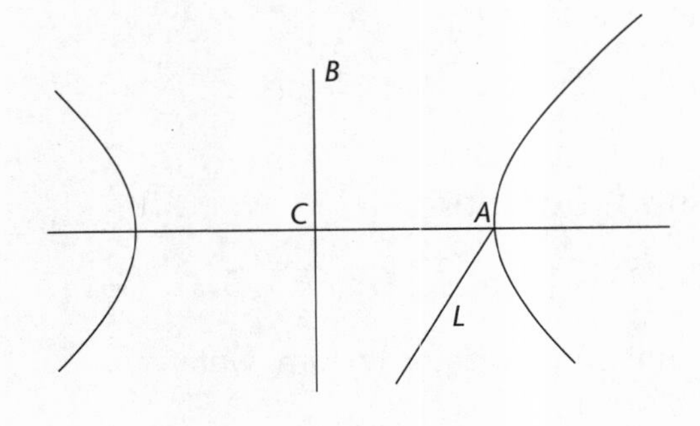

Setting Up the Same Four Proportions As in Proposition 11

Step 1: First Proportion

"... then $L \times QR$ will be to $L \times Pv$ as QR to Pv ; that is, as PE or AC to PC ..."

Take a point Q on the curve where the body will be at a later time.

Drop Qv ordinatewise to semi-diameter CP extended, cutting SP at x.

Draw $QR \parallel PS$.

$Qx \parallel PZ$ by construction; therefore figure $QxPR$ is a parallelogram. And by Euclid I.34,

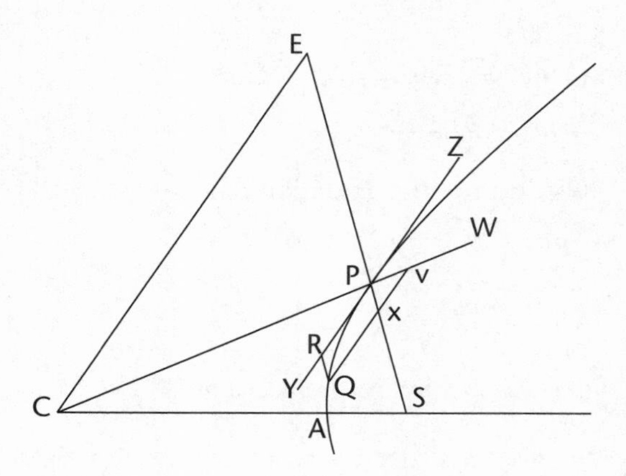

$$QR = xP. \tag{9}$$

Angle EPC = angle vPx. [Eu. I.15]

$Qv \parallel RP$, $PR \parallel CE$, and therefore $CE \parallel xv$.

Parallels CE and xv are cut by transversal CPW, therefore

angle ECP = angle Pvx. [Eu. I.29]

Therefore $\triangle EPC$ is equiangular with $\triangle xPv$.

$Px : Pv :: PE : PC$. [Eu. VI.4]

Substituting from Equation 9,

$QR : Pv :: PE : PC$.

By Equation 7, $PE = AC$; so substituting for PE,

$QR : Pv :: AC : PC$.

Multiply antecedent and consequent of the first ratio by constant L, the latus rectum of the principal vertex of this hyperbola:

$$L \times QR : L \times Pv :: AC : PC. \tag{10}$$

Step 2: Second Proportion

"... and $L \times Pv$ will be to [rectangle] Gv,vP as L is to Gv ..."

Again taking L, the principal latus rectum, write

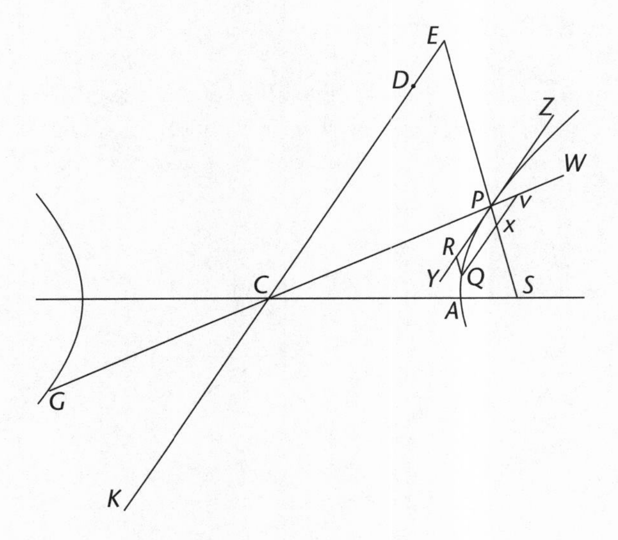

$L : Gv :: L : Gv.$ [identity]

Multiply antecedent and consequent of the first ratio by Pv:

$$L \times Pv : Gv \times Pv :: L : Gv. \quad (11)$$

Step 3: Third Proportion

"... and (by the nature of conics) [rectangle] Gv, vP will be to sq. Qv as sq. PC to sq. CD..."

(Now we must prove that, in the hyperbola, $Gv,vP : Qv^2 :: PC^2 : CD^2$.)

Consider diameter PG and its corresponding latus rectum, L_P. By Apollonius I.21 first part, applied here:

$$Qv^2 : Gv,vP :: L_P : PG. \quad (12)$$

But $L_P : DK :: DK : PG.$ [Ap. I Def. 11]

Or, $L_P : PG :: DK^2 : PG^2.$ [Eu. VI.19 Por.]

$L_P : PG :: (2CD)^2 : (2PC)^2;$

$L_P : PG :: 4CD^2 : 4PC^2;$

$L_P : PG :: CD^2 : PC^2.$

Substituting into Equation 12,

$Qv^2 : Gv,vP :: CD^2 : PC^2.$

Inverting,

$$Gv \times vP : Qv^2 :: PC^2 : CD^2. \quad (13)$$

Step 4: Fourth Proportion

"... and (by Lemma 7, Corollary 2) as points Q and P come together, sq. Qv to sq. Qx is the ratio of equality; and sq. Qx or sq. Qv is to sq. QT as sq. EP is to sq. PF; that is, as sq. CA is to sq. PF, or (by Lemma 12) as sq. CD is to sq. CB."

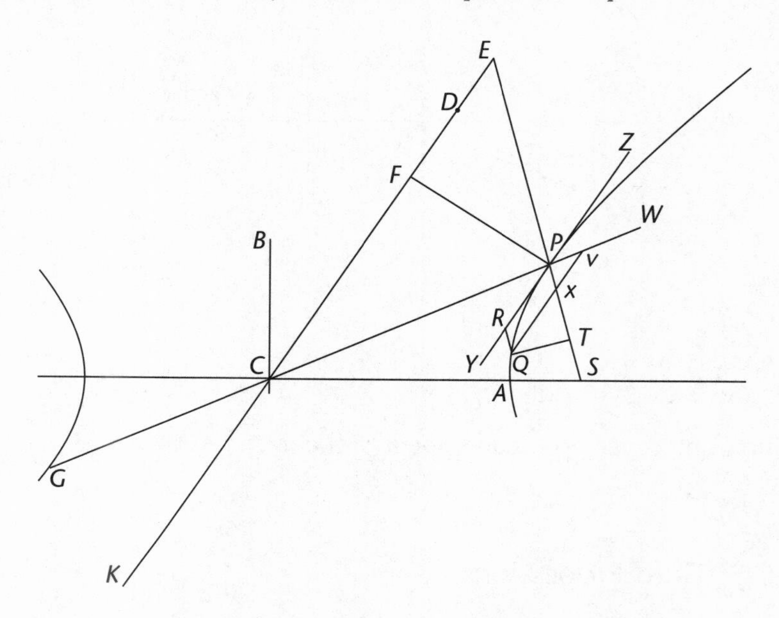

Drop $QT \perp PS$.

Drop $PF \perp$ conjugate diameter DK.

Angle PFE = angle QTx. [Eu. Post. 4]

Since $Qv \parallel CD$ and SE is a transversal,

Angle FEP = angle QxT. [Eu. I.29]

Therefore $\triangle FEP$ is equiangular with $\triangle TxQ$.

$Qx : QT :: PE : PF$. [Eu. VI.4]

But $PE = CA$ by Equation 7; so substituting for PE,

$Qx : QT :: CA : PF$. (14)

By Lemma 12, parallelograms around pairs of conjugate diameters are equal. Therefore also, the quarters of those parallelograms will be equal.

Area CA,CB = area CD,PF.

$CA : PF :: CD : CB$. [Eu. VI.16]

Substituting for $CA : PF$ in Equation 14:

$Qx : QT :: CD : CB$.

But as $Q \rightarrow P$, $Qx \rightarrow Qv$;

$$Qx \overset{ult}{=} Qv.$$

$$Qv : QT \overset{ult}{::} CD : CB.$$

Squaring, $Qv^2 : QT^2 \overset{ult}{::} CD^2 : CB^2$. (15)

Step 5: Putting Together the Four Proportions

"Compounding all these ratios, $L \times QR$ would be to sq. QT as $AC \times L \times$ sq. $PC \times$ sq. CD, or 2 sq. $CB \times$ sq. $PC \times$ sq. CD, is to $PC \times Gv \times$ sq. $CD \times$ sq. CB, or as $2PC$ is to Gv."

$L \times QR$	:	$L \times Pv$	::	AC	:	PC.	[by Equation 10]
$L \times Pv$	:	$Gv \times Pv$	::	L	:	Gv.	[by Equation 11]
$Gv \times Pv$	:	Qv^2	::	PC^2	:	CD^2.	[by Equation 13]
Qv^2	:	QT^2	$\overset{ult}{::}$	CD^2	:	CB^2.	[by Equation 15]

Multiplying and canceling:

$$\frac{L \times QR}{QT^2} \overset{ult}{=} \frac{AC \times L \times PC^2}{PC \times Gv \times CB^2};$$

$$\frac{L \times QR}{QT^2} \overset{ult}{=} \frac{AC \times L \times PC}{Gv \times CB^2}. \quad (16)$$

But by Equation 8, $AC \times L = 2BC^2$

Substituting into Equation 16,

$$\frac{L \times QR}{QT^2} \overset{ult}{=} \frac{2BC^2 \times PC}{Gv \times CB^2}.$$

$$\frac{L \times QR}{QT^2} \overset{ult}{=} \frac{2 \times PC}{Gv}.$$

"But as points Q and P come together, $2PC$ and Gv are equal. Therefore, $L \times QR$ and sq. QT, proportional to them, are also [ultimately] equal."

But as $Q \rightarrow P$, $v \rightarrow P$, and

$$2PC \overset{ult}{=} Gv.$$

$$\frac{L \times QR}{QT^2} \overset{ult}{=} \frac{Gv}{Gv}.$$

Since the second is a ratio of equality, so is the first.

$$L \times QR \overset{ult}{=} QT^2$$

"Let these equals be multiplied by $\frac{\text{sq.}SP}{QR}$, and $L \times \text{sq.}SP$ becomes equal to $\frac{\text{sq.}SP \times \text{sq.}QT}{QR}$. Therefore (by Proposition 6, Corollaries 1 and 5) the centripetal force is inversely as $L \times \text{sq.}SP$, that is, inversely in the duplicate ratio of the distance SP.

Q.E.I."

Multiply both sides by $\frac{SP^2}{QR}$:

$$L \times QR \times \frac{SP^2}{QR} \overset{ult}{=} \frac{SP^2 \times QT^2}{QR};$$

$$L \times SP^2 \overset{ult}{=} \frac{SP^2 \times QT^2}{QR}.$$

But by Proposition 6 Corollary 1, force $\overset{ult}{\propto} 1 / \frac{QT^2 \times SP^2}{QR}$.

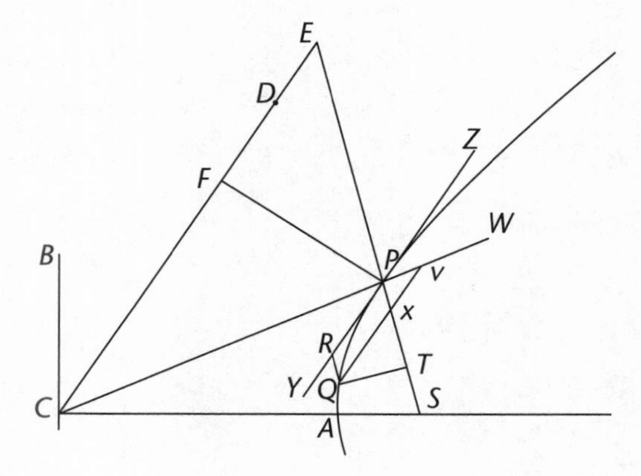

Therefore

$$\text{force} \overset{ult}{\propto} \frac{1}{L \times SP^2}. \tag{17}$$

But L is constant for a given hyperbola, since it is the principal latus rectum.

Therefore

$$F \propto \frac{1}{SP^2}.$$

If a body moves in a hyperbola with the center of forces at a focus, the force varies inversely as the square of the distance from the center of forces.

Q.E.I.

Newton's Comment after I.12

In the same way it is demonstrated that if this centripetal force be changed into a centrifugal force, the body will move in the opposite hyperbola.

This is a similar situation to the hyperbola with center of forces at the center and force law $F \propto SP$, from the scholium after Proposition 10. The body in approaching the center of repulsive force is slowed down, then impelled back the other way.

Lemma 13

The latus rectum of a parabola belonging to any vertex you please is four times the distance of that vertex from the focus of the figure.

This is proved in Conics.

[Newton gives no diagram for this lemma.]

Note on Lemma 13

This lemma states that in a parabola $L_P = 4PS$, where L_P is the latus rectum at the point P (see the diagram given with Lemma 14).

For reasons explained in the note to Lemma 12, we will not pause here for proof of this proposition of conic sections. A full proof can be found in Appendix A if you need or wish to see how it goes.

Lemma 14

The perpendicular drawn from the focus of a parabola to its tangent is the mean proportional between the distances of the focus from the point of tangency and from the principal vertex of the figure.

For let *AP* be a parabola, *S* its focus, *A* the principal vertex, *P* the point of tangency, *PO* applied ordinatewise to the principal diameter, *PM* the tangent meeting the principal diameter at *M*, and *SN* a line drawn from the focus perpendicular to the tangent. Let *AN* be connected,

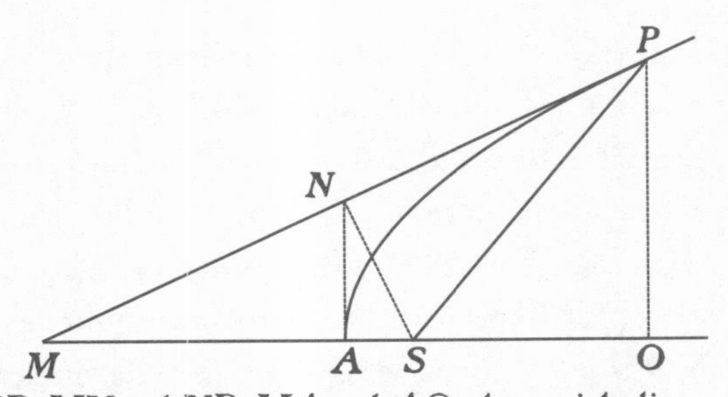

and because of the equals *MS* and *SP*, *MN* and *NP*, *MA* and *AO*, the straight lines *AN* and *OP* will be parallel, and hence the triangle *SAN* will have a right angle at *A*, and will be similar to the equal triangles *SNM*, *SNP*. Therefore, *PS* is to *SN* as *SN* is to *SA*.

Q.E.D.

Lemma 14 Corollary 1

sq. PS is to sq. SN as PS to SA.

[Two other corollaries, not used in Newton's central argument, have been omitted.]

Note on Lemma 14

As with Lemmas 12 and 13, we will not pause here for proof of this proposition of conic sections and its first corollary. Full proofs can be found in Appendix A.

Proposition 13

If a body move in the perimeter of a parabola, it is required to find the law of centripetal force tending to the focus [umbilicus] of this figure.

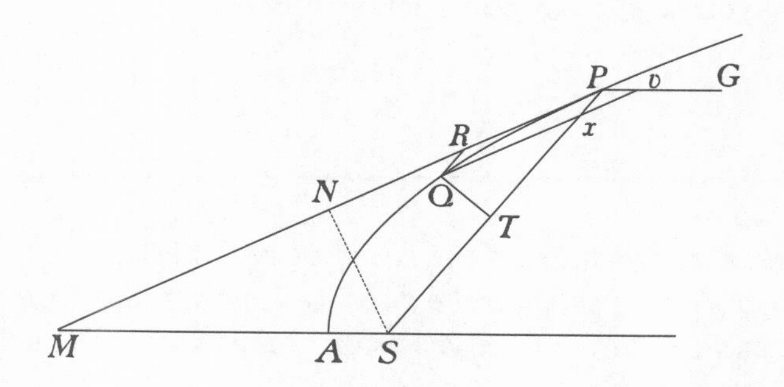

Let the construction of the lemma remain, and let P be a body on the perimeter of the parabola, and from place Q, to which the body moves next, draw QR parallel, and QT perpendicular, to SP, and also draw $Q\upsilon$ parallel to the tangent, meeting the diameter PG at υ and the distance SP at x. Now because of the similar triangles $Px\upsilon$, SPM, and the equal sides SM, SP, of one of them, the sides of the other, Px (or QR) and $P\upsilon$, are equal. But, from conics, the square of the ordinate $Q\upsilon$ is equal to the rectangle contained by the latus rectum and the segment of the diameter $P\upsilon$; that is (by Lemma 13), [the rectangle contained] by $4PS \times P\upsilon$, or $4PS \times QR$; and, points P and Q coming together, the ratio of $Q\upsilon$ to Qx is the ratio of equality (by Lemma 7 Corollary 2). Therefore, sq. Qx in this case is equal to the rectangle $4PS \times QR$. But, because of the similar triangles QxT, SPN, sq. Qx is to sq. QT as sq. PS to sq. SN; that is (by

Lemma 14 Corollary 1) as PS to SA; that is, as $4PS \times QR$ to $4SA \times QR$; and hence (by Book 5 Proposition 9 of the *Elements*) sq.QT and $4SA \times QR$ are equal. Multiply these equals by $\frac{\text{sq.}SP}{QR}$, and $\frac{\text{sq.}SP \times \text{sq.}QT}{QR}$ becomes equal to sq.$SP \times 4SA$; and therefore (by Proposition 6 Corollaries 1 and 5) the centripetal force is inversely as sq.$SP \times 4SA$, that is (since $4SA$ is given) inversely in the duplicate ratio of the distance SP.

Q.E.I.

Expansion of Newton's Sketch of I.13

Given:

1. Body moves in parabola;
2. center of forces at focus.

To Find:

Force law to that center.

"Let the construction of the lemma remain, and let P be a body on the perimeter of the parabola, and from place Q, to which the body moves next, draw QR parallel, and QT perpendicular, to SP, and also draw Qv parallel to the tangent, meeting the diameter PG at v and the distance SP at x."

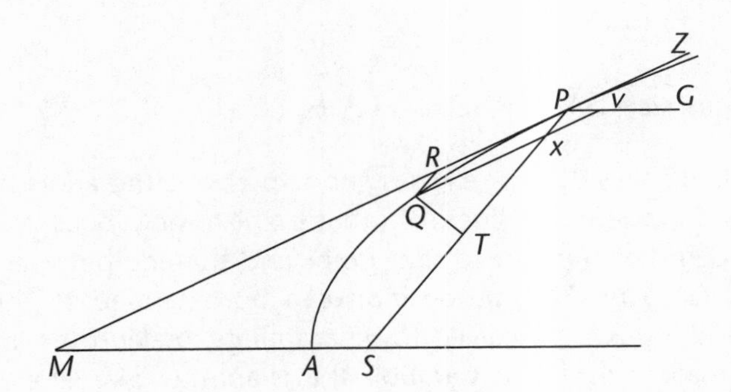

Q and P are on the perimeter of the parabola, with place Q being the next position of the body after P.

Construct $QR \parallel SP$, $QT \perp SP$

Construct a tangent to the parabola at P, intersecting the axis at M. Let this tangent be ZPM.

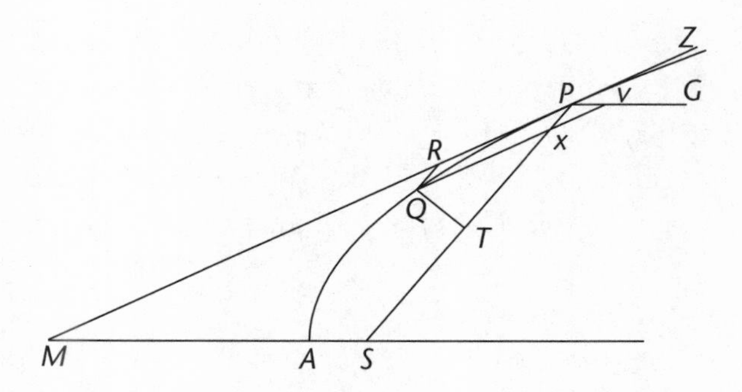

Construct diameter $PG \parallel MS$

Construct a line from Q parallel to tangent MP intersecting diameter PG at v and line SP in x. Qv is an ordinate to diameter PG by Apollonius I.49.

"Now because of the similar triangles Pxv, SPM,..."

$PG \parallel MS$, cut by transversal PS.

Therefore angle GPS = angle PSM. [Eu. I.29]

$MP \parallel Qv$, cut by transversal PS.

Therefore angle MPS = angle Pxv. [Eu. I.29]

Thus $\triangle SPM$ is equiangular with $\triangle Pxv$.

Therefore by Euclid VI.4,

$$SP : SM :: Px : Pv. \quad (1)$$

"...and the equal sides SM, SP, of one of them,..."

Apollonius III.48 says that for ellipses and hyperbolas the angle between the tangent at a point and the line from that point to one focus is equal to the angle between that tangent to that point and the line between that point and the other focus. This proposition can be extended to the parabola, whose second focus is conceived to be at infinity, making the line to it parallel to the major axis. In the parabola this is another diameter through the point of tangency. Applied here, it would say that the angle between the tangent ZPM and the diameter PG through P is equal to the angle between tangent MPZ and PS.

Angle ZPG = angle MPS.

Because $GP \parallel SM$ with line PM as transversal,

angle ZPG = angle PMS. [Eu. I.29]

Therefore angle MPS = angle PMS.

Therefore by Euclid I.6, $MS = SP$. (2)

"...the sides of the other, Px (or QR) and Pv, are equal."

By Equation 1, $SP : SM :: Px : Pv$.

Since $SP = SM$ by Equation 2, the first ratio is a ratio of equality; therefore so is the second.

$$Px = Pv. \quad (3)$$

Since $Qx \parallel RP$ and $QR \parallel Px$, figure $QRPx$ is a parallelogram.

Therefore $QR = Px$. [Eu. I.34]

Substituting from Equation 3, $QR = Pv$. (4)

"But, from conics, the square of the ordinate Qv is equal to the rectangle contained by the latus rectum and the segment of the diameter Pv ;..."

By Apollonius I.11 applied here: In a parabola any straight line Qv which is drawn ordinatewise will equal in square the rectangle contained by the straight line Pv cut off by it on the diameter PG beginning from the section's vertex P and by another straight line L_P which is the latus rectum of that vertex.

That is, $Qv^2 = L_P \times Pv$. (5)

"... that is (by Lemma 13), [the rectangle contained] by $4PS \times Pv$, or $4PS \times QR$ [is equal to the square of the ordinate Qv]..."

By Lemma 13, $L_P = 4PS$.

Substituting into Equation 5, $Qv^2 = 4PS \times Pv$.

Since by Equation 4, $QR = Pv$,

$$Qv^2 = 4PS \times QR. \quad (6)$$

"... and, points P and Q coming together, the ratio of Qv to Qx is the ratio of equality (by Lemma 7 Corollary 2)."

Lemma 7 Corollary 2 applied here: And if through P straight lines Px, Pv, be drawn, cutting the tangent PR and its parallel Qv, the ultimate ratio of all the abscissas Qv, Qx, to one another, will be the ratio of equality.

As $Q \to P$, $Qx \overset{ult}{=} Qv$. [Lem. 7 Cor. 2]

"Therefore, sq. Qx in this case is equal to the rectangle $4PS \times QR$."

Squaring, $Qx^2 \overset{ult}{=} Qv^2$.

Substituting from Equation 6, $Qx^2 \overset{ult}{=} 4PS \times QR$. (7)

"But, because of the similar triangles QxT, SPN, sq. Qx is to sq. QT as sq. PS to sq. SN; that is (by Lemma 14 Corollary 1) as PS to SA; that is, as $4PS \times QR$ to $4SA \times QR$; and hence (by Book 5 Proposition 9 of the *Elements*) sq. QT and $4SA \times QR$ are equal."

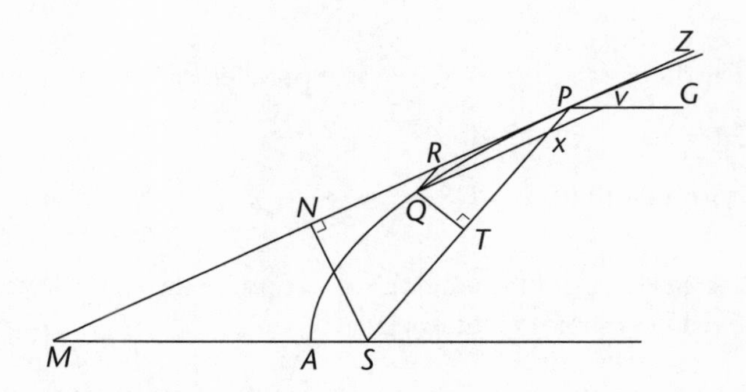

Construct $SN \perp MP$.

Angle QTx = angle SNP. [Eu. Post. 4]

Tangent $NP \parallel$ ordinate Qv, cut by transversal PS. Therefore

angle NPS = angle QxT. [Eu. I.29]

Therefore $\triangle NPS$ is equiangular with $\triangle QTx$.

$Qx : QT :: SP : SN$. [Eu. VI.4]

Squaring each magnitude, $Qx^2 : QT^2 :: PS^2 : SN^2$.

But by Lemma 14 Corollary 1, $PS^2 : SN^2 :: PS : SA$.

Therefore $Qx^2 : QT^2 :: PS : SA$. [Eu. V.11]

Multiply antecedent and consequent of right-hand ratio by $4QR$:

$$\frac{Qx^2}{QT^2} = \frac{4PS \times QR}{4SA \times QR}. \tag{8}$$

Inverting both sides of the equation,

$$\frac{QT^2}{Qx^2} = \frac{4SA \times QR}{4PS \times QR}.$$

But $Qx^2 \overset{ult}{=} 4PS \times QR$. [by Eq. 7]

$$\frac{QT^2}{Qx^2} \overset{ult}{=} \frac{4SA \times QR}{Qx^2}.$$

Therefore by Euclid V.9,

$QT^2 \overset{ult}{=} 4SA \times QR.$

"Multiply these equals by $\frac{\text{sq.}SP}{QR}$, and $\frac{\text{sq.}SP \times \text{sq.}QT}{QR}$ becomes equal to sq. $SP \times 4SA$; and therefore (by Proposition 6 Corollaries 1 and 5) the centripetal force is inversely as sq. $SP \times 4SA$, that is (since $4SA$ is given) inversely in the duplicate ratio of the distance SP."

Multiply both sides by $\frac{SP^2}{QR}$.

$$QT^2 \times \frac{SP^2}{QR} \overset{ult}{=} 4SA \times QR \times \frac{SP^2}{QR}.$$

$$\frac{SP^2 \times QT^2}{QR} \overset{ult}{=} 4SA \times SP^2.$$

But by Proposition 6 Corollary 1,

$$F \overset{ult}{\propto} 1 / \frac{SP^2 \times QT^2}{QR}.$$

Therefore

$$F \propto \frac{1}{4SA \times SP^2}.$$

But $4SA$ is constant, so

$$F \propto \frac{1}{SP^2}.$$

If a body moves in a parabola around a center of forces at the focus, the force varies inversely as the square of the distance from the focus.

Q.E.I.

I.13 Corollary 1

From the last three propositions, the consequence is that if any body you please P go forth from the place P in any straight line you please PR with any velocity whatever, and at the same time be impelled by a centripetal force which is inversely proportional to the square of the distance of places from the center, this body will move in some one of the conic sections having its focus at the center of forces, and conversely. For, if the focus and the point of tangency and the position of the tangent be given, a conic section can be described that will have a given curvature at that point. But the curvature is given from the given centripetal force and velocity of the body; and it is impossible to describe two mutually tangent orbits with the same centripetal force and the same velocity.

Notes on I.13 Corollary 1

• In propositions 11–13, Newton has shown that *if* the orbit is a conic section *then* the force law will be inverse-square. Here, he claims that the converse is true: *if* the force law is inverse-square *then* the orbit will be a conic section.

Questions for Discussion: How does Newton's argument here strike you? Does it seem valid? Does it seem a physical argument, calling on an understanding of how things must work in a real world? Or is it actually mathematical? What argument would you make for this converse?

• This converse is of great importance, since it is used in the culminating step of the first part of Book III (Proposition 13), where the elliptical planetary orbits are established. We must therefore take an interest in whether we have a convincing proof.

Having proved in the last three propositions that orbits following those three different curves resulted in an inverse square force law, we might easily imagine taking as our orbit the Lemniscate of Bernoulli, the Cissoid of Diocles, or some other of the vast menagerie of exotic curves and showing that it, too, would require an inverse square force law. If we were to discover such an instance, we would then have to say that an inverse square law would generate *either* a conic *or* a lemniscate (or whatever). It would be rash of us to say, without further proof, that such a discovery would be impossible.

• In the present corollary, Newton claims that in fact no other orbits could result from an inverse square force law. He argues that

1. a given initial velocity together with a given force will determine the curvature of the orbit, and

2. the curvature, together with the given position of the center of forces, will determine a unique conic section (in Proposition 17 he shows how to tell which section it is, and what shape it has).

But doesn't the second point beg the question? That is, although there is exactly one conic corresponding to the given curvature and other conditions, this does not show that there does not exist any other curve which might also correspond to those conditions.

Newton's argument for this is that "...it is impossible to describe two mutually tangent orbits with the same centripetal force and the same velocity."

This appears to be a *physical* claim. When a body is just beginning to move, it is moving in a tiny bit of curved path with a curvature given from the given centripetal force and velocity of the body. That piece of curve may be indistinguishable from a particular conic at that point. But what happens as it continues to move? Many curves could share that curvature at that point.

But Newton suggests that the body has no means of deciding to go off in some exotic curve that happens to have not only the same curvature but also the same force law and initial velocity as a body in a conic at that point. (Perhaps the body would need some additional guidance to do so, as we saw for example would be required in Proposition 7, a situation unexceptionable mathematically, but disturbing to physical intuition.) This argument has some persuasive power.

But is it any more than intuitively persuasive? Why, ultimately, should there *not* be "two mutually tangent orbits with the same centripetal force and the same velocity"?

It does not seem to me that the discussion Newton gives in connection with I.13 Cor. 1 constitutes a formal mathematical proof. What do you think? Can you extract what satisfies you as a formal proof from his discussion?

• However, not presenting a formal mathematical proof in I.13 Cor. 1 does not mean that Newton couldn't prove it, or that *Principia* doesn't prove it. Newton may have given us the means of constructing such a proof in Section 8 of Book I, a section entitled "The determination of orbits in which bodies will revolve, being acted upon by any sort of centripetal force." Anyone wishing to investigate this question further might explore whether Book I Propositions 39–42 of *Principia* allow one to produce the required mathematical demonstration, perhaps as a corollary to I.41. (A summary of these propositions is provided on pages 293–294.)

• Although Newton doesn't say here as much as we might like about how he understands the cogency of Corollary 1, it may be useful to see what he says elsewhere about it. The following was written in the late 1710's, part of an unfinished and unused draft of a preface to the third edition of *Principia,* in a paragraph he evidently decided not to use and crossed out.

> It has been objected that the Corollary of Proposition XIII of Book 1 in the first edition was not demonstrated by me. For they assert that a body *P*, going off a straight line given in position, at a given speed and from a given place under a centripetal force whose law is given [*viz.* reciprocally proportional to the square of the distance from the force-centre], can describe a great many curves. But they are deluded [*sed hallucinantur*]. If either the

> position of the straight line or the speed of the body be changed, then the curve to be described can also be changed and from a circle become an ellipse, from an ellipse a parabola or hyperbola. But where the position of the straight line, the body's speed and the law of central force stay the same, differing curves cannot be described. And, in consequence, if from a given curve there be determined the centripetal force, there will conversely from the central force given be determined the curve. In the second edition [of *Principia*] I touched on this in a few words merely; but in each [edition] I displayed a construction of this Corollary in Proposition XVII whereby its truth would adequately come to be apparent, and exhibited the problem generally solved in Proposition XLI of Book 1.*

This passage gives us an idea of how Newton himself interpreted this corollary. He apparently saw the necessity as emerging from I.17 and saw a general solution as being offered in I.41. Under an inverse-square force, given any initial condition, some conic section will arise, with the ellipses, parabolas, and hyperbolas covering all the possible parameters. As you work through I.17 yourself, you will be in a position to evaluate whether there might be some initial condition outside these classes. Even if you think I.17 doesn't constitute a proof of I.13 Cor. 1, looking at it through Newton's eyes will enable you to understand the intuition on which you would need to build any genuinely Newtonian proof.

I.13 Corollary 2

If the velocity with which the body goes forth from its place P be that by which the little line PR can be described in some least particle of time, and the centripetal force have the power to move the same body through the space QR in the same time, this body will move in some conic section whose principal latus rectum is that quantity which $\frac{sq.QT}{QR}$ *ultimately has when the little lines PR, QR are diminished* in infinitum. *In these corollaries I consider the circle as an ellipse, and leave out the case where the body descends directly to the center.*

---

* *The Mathematical Papers of Isaac Newton,* Volume VIII, pp. 457–458. Translation from Latin by D. T. Whiteside. I thank Niccolò Guicciardini for suggesting inclusion of this passage here.

Expansion of Newton's Sketch for I.13 Corollary 2

Given:

1. A body starting at point P on a curve moves by its innate (inertial) motion tangentially the distance PR in some evanescent time.

2. In the same evanescent time, the centripetal force would have moved the body along the line RQ.

To Prove:

1. Body will move in a conic section;

2. $L \overset{ult}{=} QT^2/QR$.

Proof:

Part 1

That it will move in a conic section is asserted by Corollary 1.

Part 2

For Ellipse and Hyperbola:

In Propositions 11 and 12, after the four proportions were multiplied, we had:

$$\frac{L \times QR}{QT^2} \overset{ult}{=} \frac{2PC}{Gv}.$$

We also saw that $2PC \overset{ult}{=} Gv$. Substituting,

$$L \times QR \overset{ult}{=} QT^2;$$

$$L \overset{ult}{=} QT^2 / QR.$$

Q.E.D.

For Parabola:

By proposition 13,

$$QT^2 \overset{ult}{=} 4SA \times QR. \tag{1}$$

But by Lemma 13, $L_P = 4PS$.

Or at point A, for the principal latus rectum, $L = 4SA$.

Substituting into Equation 1, $QT^2 \overset{ult}{=} L \times QR$.

$L \overset{ult}{=} QT^2/QR$.

Q.E.D.

Proposition 14

If several bodies revolve about a common center, and the centripetal force be inversely in the duplicate ratio of the distance of places from the center, I say that the principal latera recta of the orbits are in the duplicate ratio of the areas which the bodies describe in the same time by radii drawn to the center.

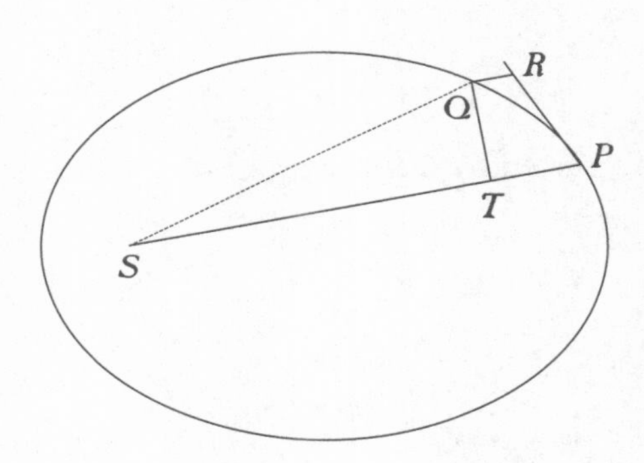

For by Proposition 13 Corollary 2, the latus rectum L is equal to the quantity $\frac{sq.\,QT}{QR}$ which results ultimately, when the points Q and R come together. But the least line QR in a given time is as the generating centripetal force, that is (by hypothesis) inversely as sq.SP. Therefore, $\frac{sq.\,QT}{QR}$ is as sq.$QT \times$ sq.SP, that is, the latus rectum L is in the duplicate ratio of the area $QT \times SP$.

Q.E.D.

Note on I.14

Newton presents what this proposition proves in two ways. In the enunciation, he says that the principal latera recta are in the duplicate ratio of the areas described in a given time. In the sketch he concludes that the principal latera recta are in the duplicate ratio of the area $QT \times SP$.

As brought out in the notes to I.6 Corollary 1 and elsewhere, the areas swept out ($\widehat{\triangle} SPQ$) vary ultimately as $SP \times QT$, twice the area of the rectilinear triangle.

$\widehat{\triangle} SPQ$ is the area described in a given time.

$\widehat{\triangle}SPQ \overset{ult}{=} \triangle SPQ.$ [Lem. 7]

$\triangle SPQ = \frac{1}{2} SP \times QT.$

$\widehat{\triangle}SPQ \overset{ult}{=} \frac{1}{2} SP \times QT.$

$\widehat{\triangle}SPQ \overset{ult}{\propto} SP \times QT.$

It is also the case that $\widehat{\triangle}SPQ$ is proportional to the times, by I.1. Therefore the quantity $SP \times QT$ varies ultimately as the times of description on a particular orbit around a particular center of forces. (See also the proof of I.6 Cor. 1.)

Expansion of Newton's Sketch of I.14

Given:

1. $F \propto 1/SP^2$;

2. a single center of forces;

3. two bodies revolving around this center, each of which, by the first part of the given and Proposition 13 Corollary 1, will be the focus of a conic section.

To Prove:

$L_A \overset{ult}{\propto} QT^2 \times SP^2.$

That is,

$$\frac{L_1}{L_2} \overset{ult}{=} \frac{Q_1 T_1{}^2 \times SP_1{}^2}{Q_2 T_2{}^2 \times SP_2{}^2}.$$

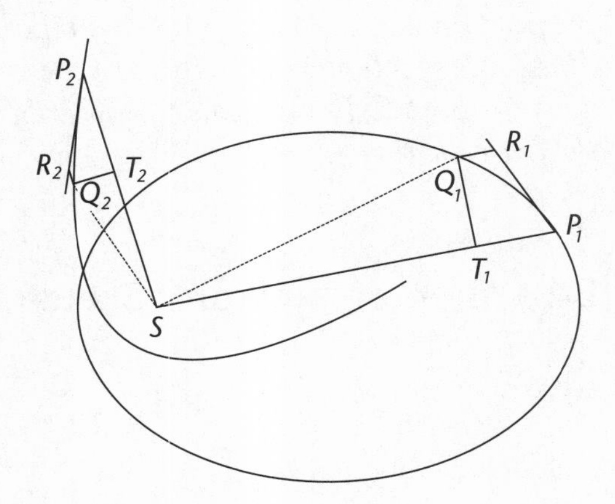

Proof:

"For by Proposition 13 Corollary 2, the latus rectum L is equal to the quantity $\frac{sq.QT}{QR}$ which results ultimately, when the points Q and R come together."

By Proposition 13 Corollary 2, $L \overset{ult}{=} QT^2/QR.$ (1)

"But the least line QR in a given time is as the generating centripetal force, that is (by hypothesis) inversely as sq. SP."

Since QR is the measure of the centripetal force, $F \overset{ult}{\propto} QR$.

But we have given a force law $F \propto 1/SP^2$.

Thus $QR \overset{ult}{\propto} 1/SP^2$.

Or, $1/QR \overset{ult}{\propto} SP^2$. (2)

"Therefore, $\frac{sq.QT}{QR}$ is as sq. $QT \times$ sq. SP, ..."

Multiply both sides of Equation 2 by QT^2:

$$QT^2/QR \overset{ult}{\propto} QT^2 \times SP^2.$$

"that is, the latus rectum L is in the duplicate ratio of the area $QT \times SP$."

Substituting from Equation 1,

$$L \overset{ult}{\propto} QT^2 \times SP^2.$$

Q.E.D.

Corollary to I.14

Hence the whole area of the ellipse, and the rectangle beneath the axes proportional to it, is in the ratio compounded of the subduplicate ratio of the latus rectum and the ratio of the periodic time. For the whole area is as the area $QT \times SP$ which is described in a given time, multiplied by the periodic time.

Expansion of Newton's Sketch of I.14 Corollary

Given:

1. Bodies moving in ellipses;

2. the other conditions of the proposition: inverse square force law, center of forces at the focus.

To Prove:

1. Whole area of ellipse varies as the rectangle under the axes;

2. area of ellipse varies as $\sqrt{L} \times$ periodic time.

Proof:

Part 1

We are to compare two ellipses E and e which share neither axis.

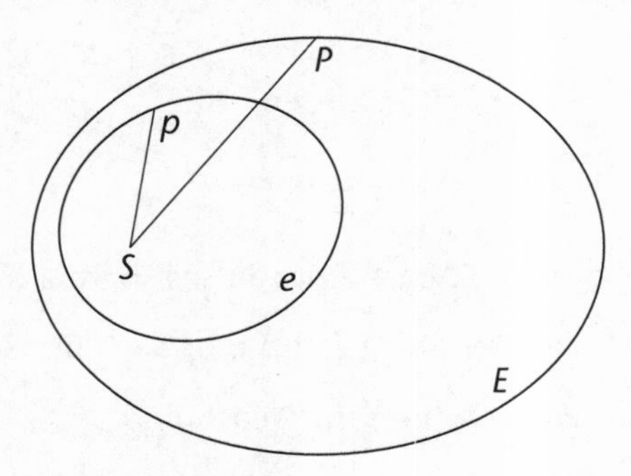

We will do this by constructing an ellipse e' which is congruent with e, and whose axes are respectively collinear with the axis of E (e' therefore has a common center C with E). Anything we prove about the ratio of areas of ellipses E and e' will also be true of E and e, since the areas of e and e' are equal.

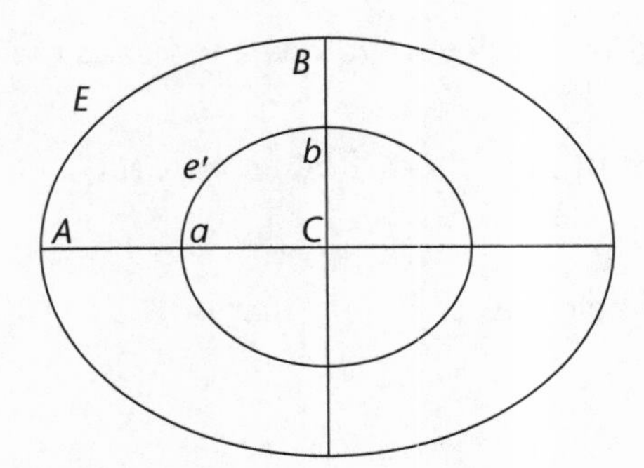

Next, we construct another ellipse ε that shares the major axis of E and the minor axis of e'—the dotted-line ellipse in the diagram.

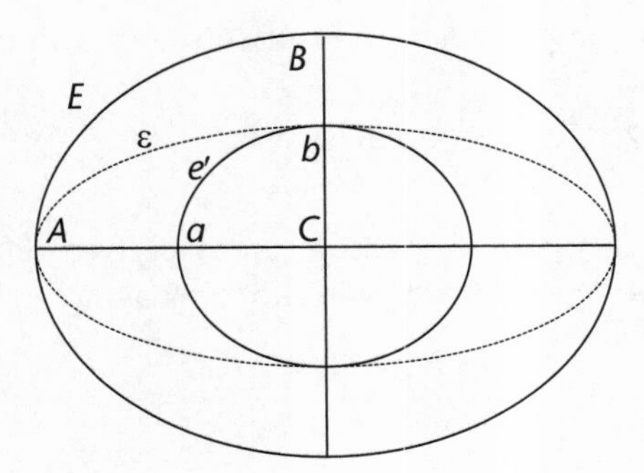

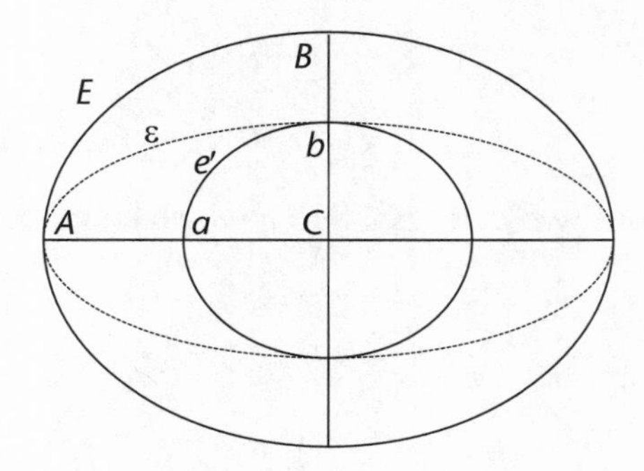

By Proposition 10 Corollary 2 part 2, if ellipses share a major axis,

Whole area E : whole area ε :: semi-minor axis BC : semi-minor axis bC.

If they share a minor axis, by the same reasoning,

Whole area ε : whole area e' :: semi-major axis AC : semi-major axis aC.

Compounding ratios (by the procedure of Euclid VI.23),

Whole area E : whole area e' ::
(semi-minor axis BC : semi-minor axis bC) comp.
(semi-major axis AC : semi-major axis aC).

But by Euclid VI.23, rectangles are to one another as the ratio compounded of the ratios of their sides.

Rect AC, CB : rect aC, Cb :: (AC : aC) comp. (BC : bC).

Therefore by Euclid V.11,

Whole area E : whole area e' or e :: rect AC, CB : rect aC, Cb.

Part 2

By Proposition 1, whole area : $\triangle$area :: periodic time : $\triangle$time.

$\triangle\text{area} \overset{ult}{=} \frac{1}{2}(SP \times QT)$.

Letting P represent the periodic time, substituting, and rearranging:

$$\text{whole area} \overset{ult}{=} \frac{\frac{1}{2}(SP \times QT)}{\text{given time}} \times P.$$

Now we need to compare the whole areas of two different ellipses. The constants $\frac{1}{2}$ and the "given time" will cancel, leaving:

$$\frac{\text{Whole Area } E}{\text{whole area } e} \overset{ult}{=} \frac{SP \times QT \times P}{sp \times qt \times p}.$$

Whole area $\overset{ult}{\propto} SP \times QT \times P$. (1)

By the proposition, $L \overset{ult}{\propto} QT^2 \times SP^2$.

$\sqrt{L} \overset{ult}{\propto} QT \times SP$.

Substituting into Equation 1:

Whole area $\propto \sqrt{L} \times P$.

Q.E.D.

Proposition 15

The same things being supposed, I say that the periodic times in ellipses are in the sesquiplicate ratio of the major axes.

For the minor axis is the mean proportional between the major axis and the latus rectum, and therefore the rectangle contained by the axes is in the ratio compounded of the subduplicate ratio of the latus rectum and the sesquiplicate ratio of the greater axis. But by Proposition 14 Corollary, this rectangle is in the ratio compounded of the subduplicate ratio of the latus rectum and the ratio of the periodic time. Let the subduplicate ratio of the latus rectum be removed on both sides, and what remains will be that the sesquiplicate ratio of the major axis is the same as the ratio of the periodic time.

Q.E.D.

[Newton gives no diagram for this proposition.]

Notes on I.15

- The sesquiplicate ratio is, as in Proposition 4, the $\frac{3}{2}$ power of the magnitudes.

- The relationship known as Kepler's Third Law states that the periodic times of planets in orbits are as the $\frac{3}{2}$ powers of the mean distances, that is, of the average of the aphelial and perihelial distances.

This proposition, of course, is not establishing or even asserting Kepler's law; but it shows a situation which, if its conditions are found to hold, would exhibit that relationship as a consequence. This situation is the one set out in our givens: bodies revolving around a common center of forces which is at a focus of an ellipse, with centripetal forces inversely as the squares of the distances of the bodies from that focus.

Expansion of Newton's Sketch of I.15

Given:

1. Centripetal forces inversely as squares of distances from the center of forces;

2. bodies revolving around a common center; here again, by Proposition 13 Corollary 1, the focus of the conic;

3. path is an ellipse.

To Prove:

$P \propto$ (greater axis)$^{3/2}$.

Proof:

"For the minor axis is the mean proportional between the major axis and the latus rectum, and therefore the rectangle contained by the axes is in the ratio compounded of the subduplicate ratio of the latus rectum and the sesquiplicate ratio of the greater axis."

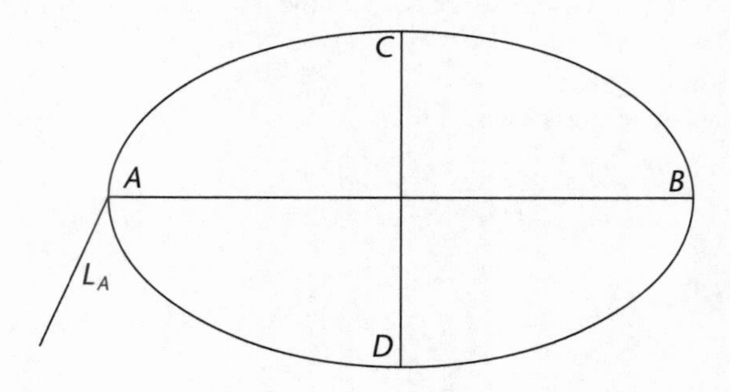

$$AB : CD :: CD : L. \quad \text{[Ap. I.15]}$$

$$CD^2 = AB \times L.$$

$$CD = \sqrt{AB \times L}.$$

Multiply both sides by AB:

$$CD \times AB = AB \times \sqrt{AB} \times \sqrt{L};$$

$$CD \times AB = \sqrt{AB^3} \times \sqrt{L};$$

$$CD \times AB = AB^{3/2} \times L^{1/2}. \quad (1)$$

"But by Proposition 14 Corollary, this rectangle is in the ratio compounded of the subduplicate ratio of the latus rectum and the ratio of the periodic time."

By Proposition 14 Corollary, Part 1 in expanded proof,

$$\text{(whole area of ellipse)} \propto \text{rect } AB, CD.$$

By the same corollary, Part 2 in expanded proof,

$$\text{(whole area of ellipse)} \propto \sqrt{L} \times P.$$

Therefore $\quad AB \times CD \propto \sqrt{L} \times P. \qquad (2)$

"Let the subduplicate ratio of the latus rectum be removed on both sides, and what remains will be that the sesquiplicate ratio of the major axis is the same as the ratio of the periodic time."

By Equation 1, $\quad CD \times AB = AB^{3/2} \times L^{1/2}.$

Combining with Equation 2,

$$AB^{3/2} \times L^{1/2} \propto \sqrt{L} \times P.$$

Dividing both sides by $\sqrt{L}$ or $L^{1/2}$,

$$AB^{3/2} \propto P.$$

That is, the periodic times vary as the $\frac{3}{2}$ powers of the greater axes.

Q.E.D.

I.15 Corollary

Therefore, the periodic times in ellipses are the same as in circles whose diameters are equal to the major axes of the ellipses.

Expansion of Newton's Sketch of I.15 Corollary

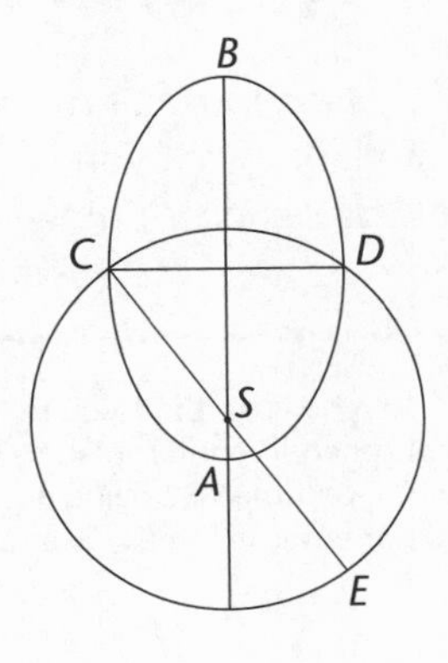

Given:

1. Ellipse having major axis AB and minor axis CD;

2. circle whose diameter CE is equal to the major axis AB of the ellipse;

3. common center of forces S at ellipse's focus and circle's center.

To Prove:

The periodic times around the two orbits will be the same.

Proof:

The focus of the ellipse coincides with the center of the circle. Therefore the vertices of the minor axis of the ellipse must fall on the circle. (This is because, as a special case of Apollonius III.52, $SC = \frac{1}{2}AB = r$.)

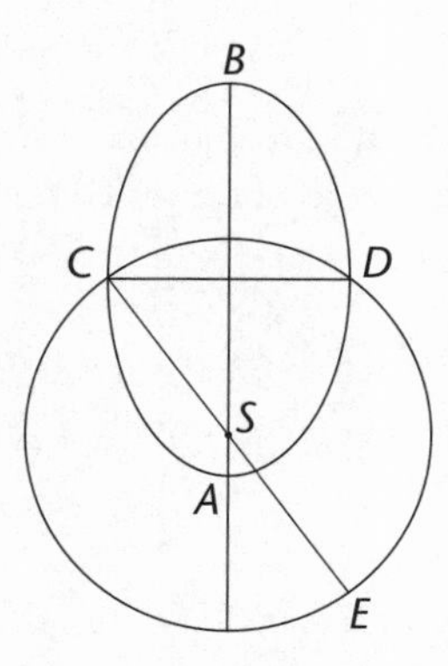

By the proposition, for ellipses and circles (circles are a special kind of ellipse):

$$P \propto (\text{greater axis})^{3/2}.$$

In a circle, the greater axis is the diameter. Therefore,

$$P_{\text{ellipse}} : P_{\text{circle}} :: (\text{greater axis})^{3/2} : (\text{diameter})^{3/2}.$$

Therefore, when the greater axis of an ellipse equals the diameter of a circle, the periodic time for the circle and that for the ellipse are equal.

Q.E.D.

Pause After I.15 Corollary

- It was intriguing, in fact downright strange, to see in Proposition I.10 Corollary 2 that you could squash in the sides of the ellipse all you want and yet the periodic time would stay the same.* The different ellipses could have

---

* Do you say that you're not amazed, because you used up your amazement at this phenomenon contemplating the pendulum? Yes, pendulums are acting according to I.10. But *why*? Don't mistake being accustomed to seeing pendulums act that way for understanding the phenomenon—it really is amazing. We'll look into it more in Book III.

different major axes as well as different minor axes. Analogously with the pendulum, the periodic times will be the same for *all* ellipses with a shared center, if the center of forces is at the center of the ellipses, and the force law one of direct variation.

But notice that here in Proposition 15 Corollary we have something else even more astonishing. What we see is that this apparent indifference to stretching doesn't just hold for ellipses with center of forces at the center. According to this proposition and corollary, it also holds when the center of forces is at a focus and we have an inverse square force law. (In this case, when the center of forces is at a focus, we must hold the major axis fixed and only stretch or squash the minor axis.)

Questions for Discussion: Can you explain why the result of this corollary would be? Why should the minor axis not matter? Why should it not matter any more with an inverse square law around a focus than it did for direct variation with the distance around the center? And why, do you think, does the major axis have to stay constant with inverse square force law around a center of forces at a focus when it didn't matter at all what the major axis was with a direct variation force law and center of forces at the center?

Proposition 16

The same things being supposed, and straight lines being drawn to the bodies, tangent to the orbits in those places, and perpendiculars being dropped from the common focus to these tangents, I say that the velocities of the bodies are in the ratio compounded of the ratio of the perpendiculars inversely, and the subduplicate ratio of the principal latera recta directly.

From the focus S drop SY perpendicular to the tangent PR, and the velocity of the body P will be inversely in the subduplicate ratio of the quantity $\frac{\text{sq.}SY}{L}$. For that velocity is as the least arc PQ described in a given particle of time; that is, by Lemma 7, as the tangent PR; that is, because of the proportionals PR to QT and SP to SY, as $\frac{SP \times QT}{SY}$, or as SY inversely and $SP \times QT$ directly. And $SP \times QT$ is as the area described in a given time, that is (by Proposition 14) in the subduplicate ratio of the latus rectum.

Q.E.D.

[Newton gives no diagram for this proposition.]

Expansion of Newton's Sketch of I.16

Given:

1. Bodies revolving in any conic sections around one common center S (a focus);

2. centripetal forces inversely as the squares of distances from that center;

3. perpendiculars SY to tangents from the common focus.

To Prove:

$V \propto \sqrt{L} / SY.$

Proof:

Let P and Q be points on the orbit with a tangent PZ at P.
Let QR connect point Q to the tangent PZ such that $QR \parallel SP$.
Draw $Qv \parallel RP$. Drop $QT \perp SP$.

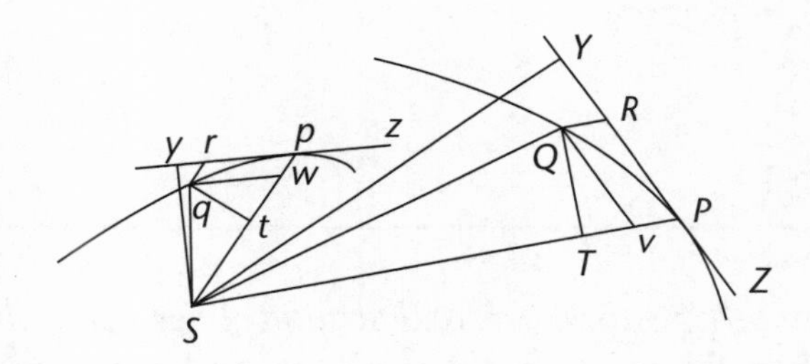

"From the focus S drop SY perpendicular to the tangent PR,..."

Drop SY perpendicular to the tangent at P. It may fall within or to either side of the figure whose elements make "the solid."

"...and the velocity of the body P will be inversely in the subduplicate ratio of the quantity $\frac{\text{sq.}\,SY}{L}$."

This is Newton's statement of what he will prove. The rest of the sketch gives his hints for how to get there.

"For that velocity is as the least arc PQ described in a given particle of time;..."

Average velocity is distance traveled in a given time:

$\text{Vel}_{PQ} = \widehat{PQ} / \triangle t.$

As $Q \to P$, the limit of the average velocity is the instantaneous velocity at P. When $\triangle t$ becomes very small, it will be the "given particle of time."

$$\text{Vel}_P = \lim_{Q \to P} \widehat{PQ} / \triangle t.$$

$$\text{Vel}_P \overset{ult}{=} \widehat{PQ} / \triangle t;$$

Or, $V_P : v_p \overset{ult}{::} (\widehat{PQ} / \triangle t) : (\widehat{pq} / \triangle t)$.

Multiply antecedent and consequent of the second ratio by the given particle times.

$$V \overset{ult}{\propto} \widehat{PQ}.$$

"... that is, by Lemma 7, as the tangent PR; ..."

$$PR \overset{ult}{=} \widehat{PQ}. \quad \text{[Lem. 7]}$$

Substituting, $V \overset{ult}{\propto} PR$. (1)

"... that is, because of the proportionals PR to QT and SP to SY, as $\frac{SP \times QT}{SY}$, or as SY inversely and $SP \times QT$ directly."

$RP \parallel Qv$, crossed by transversal SP, therefore:

angle RPS = angle QvS. [Eu. I.29]

Since $SY \perp PR$ by construction, SYP is a right angle. And, also by construction, $QT \perp SP$, so angle QTv is also a right angle.

Angle SYP = angle QTv. [Eu. Post. 4]

Therefore $\triangle SYP$ is equiangular with $\triangle QTv$.

And by Euclid VI.4,

$$Qv : QT :: SP : SY. \quad (2)$$

But $Qv \parallel RP$ and $QR \parallel vP$, so figure $QRPv$ is a parallelogram and

$Qv = PR$. [Eu. I.33]

Substituting into Equation 2, $PR : QT :: SP : SY$.

$PR = SP \times QT/SY$. [Eu. VI.16]

Therefore substituting into Equation 1,

$$V \overset{ult}{\propto} SP \times QT/SY. \quad (3)$$

Exercise: This proof has followed the diagram in placing Y beyond R. Show how the proof would work if Y fell on the Z side of P.

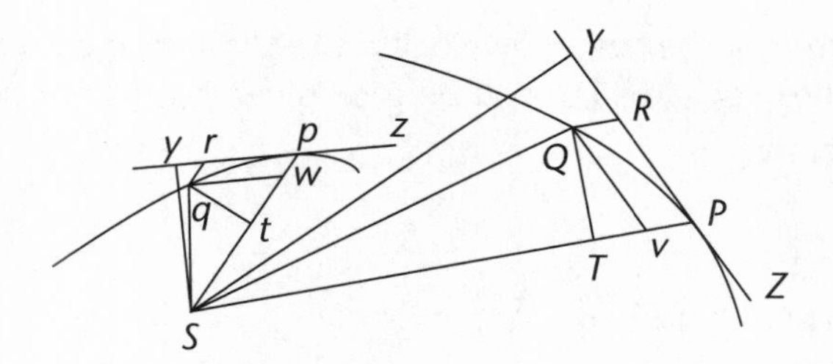

"And $SP \times QT$ is as the area described in a given time, that is (by Proposition 14) in the subduplicate ratio of the latus rectum.

Q.E.D."

A note at the beginning of Proposition 14 reviewed the relationship between the area described in a given time and $SP \times QT$. In short, areas $\widehat{QSP}$ described in a given time vary ultimately as $SP \times QT$.

By Proposition 14,

$$SP \times QT \overset{ult}{\propto} \sqrt{L}. \quad [I.14]$$

Substituting into Equation 3, $V \propto \sqrt{L}\,/\,SY$.

Q.E.D.

I.16 Corollary 1

The principal latera recta are in the ratio compounded of the duplicate ratio of the perpendiculars and the duplicate ratio of the velocities.

This corollary falls out of the proposition by an easy algebraic rearrangement.

I.16 Corollary 2

The velocities of bodies at the greatest and least distances from the common focus, are in the ratio compounded of the ratio of the distances inversely and the subduplicate ratio of the principal latera recta directly. For the perpendiculars are now those distances.

Notes on I.16 Corollary 2

- The "greatest and least distances" are the distances from a focus to the two vertices of the major axis for the ellipse. The proof works identically for either principal vertex, since the latus rectum is the same. For a parabola or hyperbola we would be using the single principal vertex.

- This corollary can be set up with any two conics—ellipse and parabola, parabola and hyperbola, etc., as long as P is at a principal vertex.

Expansion of Newton's Sketch of I.16 Corollary 2

Given:

1. Any conics with center of forces at focus;
2. orbits have a common focus;
3. centripetal forces inversely as the squares of the distances from that center.

To Prove:

Velocities (at greatest and least distances from focus) $\propto \sqrt{L}\,/\,SP$.

Proof:

Take P and P', both at "greatest distance" from the common focus.

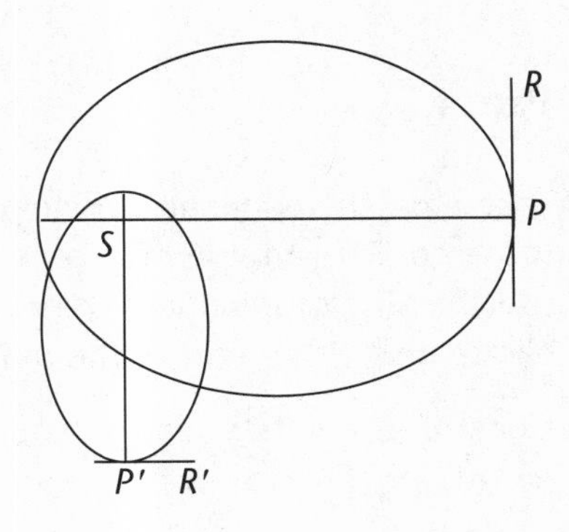

By Apollonius I Definition 7 and I.32, the axis will be perpendicular to the tangent at the principal vertex.

Therefore, when P is at principal vertex A, and we drop $SY \perp$ tangent PR, SY coincides with SP.

By the proposition, $V \propto \sqrt{L}\,/\,SY$

Therefore $V \propto \sqrt{L}\,/\,SP$.

Q.E.D.

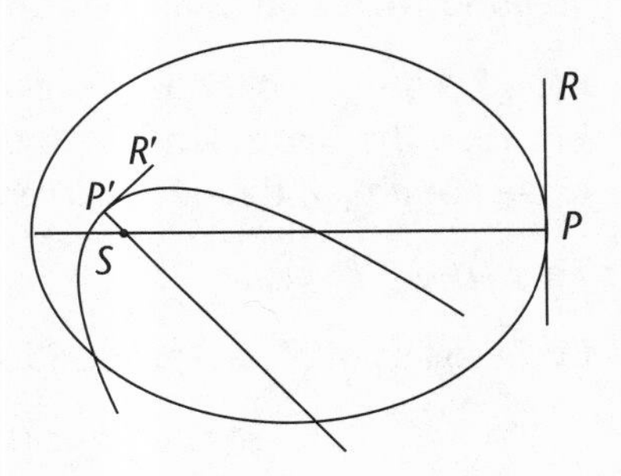

The proof is identical for the parabola and hyperbola, except that, as noted above, the least distance must be taken, since there is no finite greatest distance.

I.16 Corollary 3

And therefore, the velocity in a conic section at the greatest and least distance from the focus is to the velocity in a circle at the same distance from the center in the subduplicate ratio of the principal latus rectum to twice that distance.

Expansion of Newton's Sketch of I.16 Corollary 3

Given:

1. Any conic section;

2. body is at a principal vertex;

3. circle with radius same length as the body's distance from the focus (that is, the circle's center is at S).

To Prove:

$$V_{\text{any conic at principal axis}} : V_{\text{circle}} :: \sqrt{L} : \sqrt{2SP}.$$

Proof:

Take P on the conic at a principal vertex. If the conic is a hyperbola or a parabola, use the single principal vertex. If it is an ellipse, use either end of the major axis.

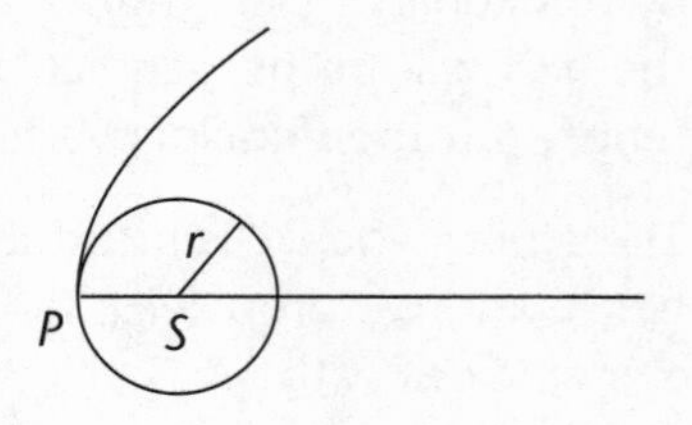

Construct a circle with center S and radius SP.

Let the circle be conceived as an orbit having its center of forces at S.

Treating the circle as an ellipse:

For all ellipses, including the circle, the latus rectum is the third proportional between the two diameters by Apollonius I.15. Since both diameters of the circle are equal, the latus rectum of the circle is equal to its diameter.

Therefore $L_{\text{circle}} = 2r.$ (1)

By Corollary 2 of this proposition, for all ellipses,

$V \propto \sqrt{L} \,/\, SP$ at the principal vertex.

$$V_{\text{conic}} : V_{\text{circle}} :: \sqrt{L_{\text{conic}}}/SP : \sqrt{L_{\text{circle}}}/SP .$$

Multiply antecedent and consequent of second ratio by SP:

$$V_{\text{conic}} : V_{\text{circle}} :: \sqrt{L_{\text{conic}}} : \sqrt{L_{\text{circle}}} .$$

But by Equation 1, $L_{\text{circle}} = 2r$.

$$V_{\text{conic}} : V_{\text{circle}} :: \sqrt{L_{\text{conic}}} : \sqrt{2r} .$$

But this circle's radius by construction = SP.

$$V_{\text{conic at principal vertex}} : V_{\text{circle}} :: \sqrt{L_{\text{conic}}} : \sqrt{2SP} .$$

Q.E.D.

[**Corollaries 4–9** are omitted; they are not used in this guidebook selection.]

Proposition 17

On the supposition that the centripetal force be inversely proportional to the square of the distance of places from the center, and that the absolute quantity of that force be known, it is required to find the line that a body describes in going forth from a given place with a given velocity along a given straight line.

Let the centripetal force tending to the point S be that by which a body p orbits in any given orbit you please pq, and let the velocity of this body at the place p be known. From the place P along the line PR let the body P go forth with a given velocity, and soon thereafter, being acted upon by the centripetal force, let that body turn aside into the conic section PQ. Thus the straight line PR will touch this conic at P. Let some straight line pr likewise touch the orbit pq at p, and if perpendiculars be understood to be dropped from S to those tangents, by Proposition 16 Corollary 1 the principal latus rectum of the conic section will be to the principal latus rectum of the orbit in the ratio compounded of the duplicate ratio of the perpendiculars and the duplicate ratio of the velocities, and therefore is given. Let L be the latus rectum of the conic section. The focus S of the same conic section is given as well. Let the angle RPH be made the supplement of the angle RPS,

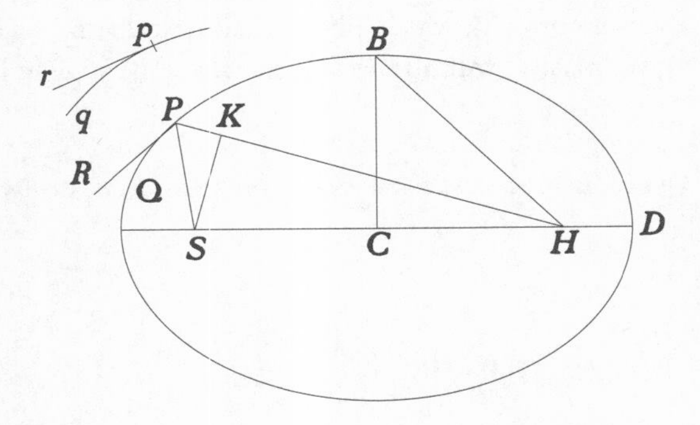

and the line PH, upon which the other focus H is located, will be given in position. SK being dropped perpendicular to PH, let the conjugate semiaxis BC be understood to be set up, and sq. $SP - 2KP.PH$ + sq. PH = sq. SH (by Book II Proposition 13 of the *Elements*) = 4 sq. CH = 4 sq. BH − 4 sq. BC = sq.$(SP+PH)$ − $L \times (SP+PH)$ = sq. $SP + 2SP.PH$ + sq. $PH - L \times (SP+PH)$. Let $2KP.PH$ − sq. SP − sq. $PH + L \times (SP+PH)$ be added to both sides, and the result will be that $L \times (SP+PH) = 2SP.PH + 2KP.PH$, or $SP + PH$ is to PH as $2SP + 2KP$ is to L. Whence PH is given in both length and position. That is, if the velocity of the body at P be such that the latus rectum L would be less than $2SP + 2KP$, PH will fall on the same side of the tangent PR as the line PS, and therefore the figure will be an ellipse, and from the given foci S, H and the principal axis $SP + PH$,* it will be given. If, on the other hand, the velocity of the body be great enough that the latus rectum L would be equal to $2SP + 2KP$, the length of PH will be infinite, and consequently the figure will be a parabola having the axis SH parallel to the line PK, and thence it will be given. And if the body go forth from its place P with a still greater velocity, the length of PH must be taken on the other side of the tangent, and therefore, since the tangent runs between the foci, the figure will be a hyperbola having its principal axis equal to the difference of the lines SP and PH, and thence it will be given. For if the body in these cases revolve in the conic section thus found, it has been demonstrated in Proposition 11, 12, and 13 that the centripetal force will be as the square of the distance of the body from the center of forces S inversely, and therefore the line PQ is shown correctly, which a body shall describe by such a force, in going forth from the given place P with a given velocity along the straight line PR given in position.

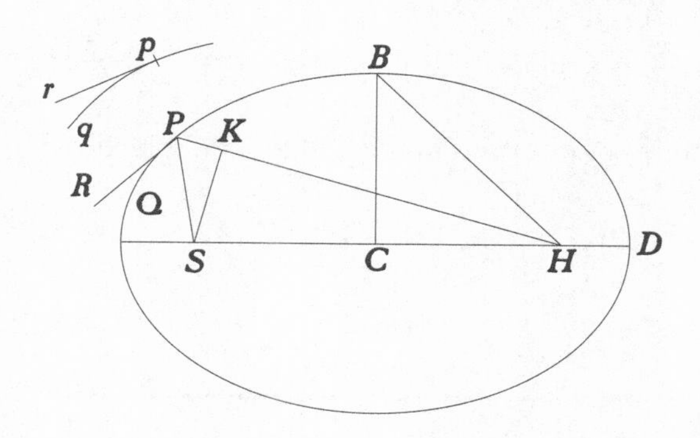

Q.E.F.

[Q.E.F. is an abbreviation for *quod erat faciendum*, "that which was to have been done."]

Notes on I.17

- This proposition is the culmination of the work of this section and one of the high points of *Principia*. Although it isn't being applied to our world (and

* Changed from *AH*.

until we know what force law governs our world, we won't be able to so apply it), it takes a practical turn in showing us how, given a certain force law and certain other bits of information, we can determine an actual orbit.

Example:

Suppose that we establish that the force law operant in our world is inverse square, and that we further know the earth's orbit and velocity at point *P* on the orbit. This gives us a fix on the absolute force towards the sun. Then we can figure out the orbit of another body, such as a new comet, around the sun.

• The method by which the proposition accomplishes its aim is ingenious and demanding; and following it is an adventure that calls upon concentration, disciplined imagination, and geometrical intuition.

• From the givens of this proposition, including the absolute quantity of the force, we compute the actual orbit. This orbit is, in Newton's diagram, indicated by the uppercase letters. The diagram may give a false impression that the full orbit (here an ellipse) is given. This uppercase-letter orbit is not given, and may turn out to be ellipse, parabola, or hyperbola. All we are given is a bit (it might be, but need not be, the whole) of the hypothetical lowercase-letter orbit, along with the tangential velocity component of the body on that orbit. Together these convey information about the absolute quantity of force.

• So far we have worked only with accelerative quantity of centripetal force (Definition 7). Now for the first time we must deal with one of the other categories of force—here absolute quantity of force (Definition 6). The complications of motive quantity of force (Definition 8) still lie in the future.

• How is it that "the absolute quantity of the force" can be "known"?

Newton gives it to us by giving us the small piece of a hypothetical orbit *pq* that would result from the absolute quantity of force given a hypothetical tangential velocity *pr*. (How this gives it to us must be understood. **Exercise:** Can you work it out before looking at the lemmita in the bulleted note below? Hint: Ponder Definitions 6 and 7, and use Proposition 6.)

Newton nowhere in *Principia* gives us any actual value for force; we only deal in proportions. Yet in the enunciation of this proposition he posits the absolute quantity of force as "known." This sounds as if he has somehow told us an actual value; but since that would be contrary to his way of doing things in *Principia*, other interpretations should be sought.

One way that an absolute quantity of force could be "known" would be if we had a proportion in which one magnitude is in ratio to another that has been defined as unity. (The absolute quantity of centripetal force of the earth might be taken as unity, for example, and the absolute quantity of other centripetal forces measured in ratio to that.)

Another way an absolute quantity of force could be "known" is to establish that one absolute force, which we seek, is equal or identical to another, which is given, not as a numerical value but as a phenomenon produced by it—an orbit, for example, or a piece of an orbit. We shall see that Newton has done this and has given us a force which would produce the phenomenon of an orbit with path *pq* and tangential velocity *pr.*

- **A lemmita for finding absolute quantity of force:**

Show that: If orbit *pq* and velocity *pr* are given, the absolute quantity of force is given.

By Proposition 6, for accelerative quantity of force,

$F \overset{ult}{\propto} \text{sagitta}/t^2.$

As shown in the proof for Lemma 11 Corollary 2, $qr \overset{ult}{\propto} \text{sagitta}.$

Therefore, $F \overset{ult}{\propto} qr/t^2.$

But by definition, $v \overset{ult}{=} \overset{\frown}{pq}/t$ or $t \overset{ult}{=} \overset{\frown}{pq}/v.$

Therefore, $F \overset{ult}{\propto} qr/(\frac{\overset{\frown}{pq}}{v})^2;$

$$F \overset{ult}{\propto} \frac{qr}{\overset{\frown}{pq}^2} \times v^2. \qquad (1)$$

Now to go from accelerative quantity of force to the absolute quantity of force:

> *From Definition 6:* The absolute quantity of centripetal force is the measure of the same, greater or less according to the efficacy of the cause propagating it from the center through the encircling regions.

If the absolute quantity of centripetal force should change, the efficacy of the cause by which it is defined, propagating the absolute force from the center through the encircling regions, must change. If the efficacy of the cause changes, the effects, through those encircling regions, must change.

> *From Definition 7:* The accelerative quantity of centripetal force is the measure of the same, proportional to the velocity which it generates in a given time.

The effect is the change in velocity in a given time, that is, the acceleration of the body as a consequence of the absolute centripetal force. This, by Definition 7, is the accelerative quantity of force. The efficacy of the cause is propagated from the center out, and so depends on distance.

The effect is the amount of acceleration. Keeping the distance constant, we can see that the change of the effect would be a change of accelerative quantity of force. Or, a constant effect—a "given" absolute quantity of force—would result in the same "given" accelerative quantity of force.

So if we can show that the accelerative quantity of force is given—determined—at a fixed distance, we know that the absolute quantity of force is also given, that is, fixed.

The difficulty is that Newton hasn't given a way of determining a single, particular force (see previous bulleted note). However, we do have the *measure* provided in Definition 6.

By Equation 1,

$$F \overset{ult}{\propto} \frac{rq}{\widehat{pq}^2} \times v^2 .$$

Looking at this intuitively, we see that rq is the amount of deflection from the tangent pr as the body moves from p to q. The relationship between rq and pr defines the shape of the curve in the neighborhood of p. Thus if we know the velocity at p and the shape of the orbit at p, we have a measure of the accelerative force at p. So we can say the accelerative quantity of force is "given" at distance sp, and consequently, we "know" the absolute quantity of force.

Forces at other places may be found via the inverse square law, or by some property of conics that depends upon the inverse square law. It is this latter course that Newton takes, in using Proposition 16 Corollary 1 to determine the principal latus rectum of the conic.

Expansion of Newton's Sketch of I.17

"Let the centripetal force tending to the point S be that by which a body p orbits in any given orbit you please pq, and let the velocity of this body at the place p be known. From the place P along the line PR let the body P go forth with a given velocity, and soon thereafter, being acted upon by the centripetal force, let that body turn aside into the conic section PQ. Thus the straight line PR will touch this conic at P."

Given:

1. Orbit pq with tangent pr and velocity v at p, center of forces S.

2. Body at P with velocity PR in a given direction and centripetal force turning it into conic PQ from same given focus S at given distance SP (because P is a "given place," it is given in relation to the center of forces).

To Find:

Orbit of body starting at P:

1. Which conic section;
2. what latus rectum;
3. what foci;
4. length of major axis.

Proof:

"Let some straight line pr likewise touch the orbit pq at p, and if perpendiculars be understood to be dropped from S to those tangents, by Proposition 16 Corollary 1 the principal latus rectum of the conic section will be to the principal latus rectum of the orbit in the ratio compounded of the duplicate ratio of the perpendiculars and the duplicate ratio of the velocities, and therefore is given. Let L be the latus rectum of the conic section. The focus S of the same conic section is given as well."

L will be the principal latus rectum of the conic section P is moving in. It must be found first.

By Proposition 16 Corollary 1, $L \propto SY^2 \times v^2$.

$$\frac{L}{l} = \frac{SY^2}{Sy^2} \times \frac{V^2}{v^2}.$$

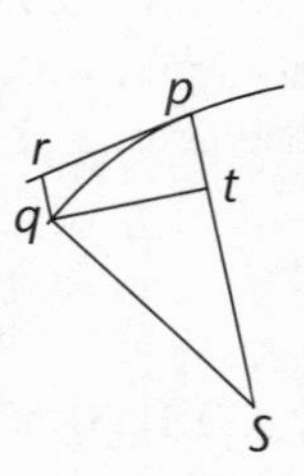

Both velocities are given.

Principal latus rectum l of orbit pq is given by Proposition 13 Corollary 2, which says $l \overset{\text{ult}}{=} qt^2 / qr$.

Since we're given SP and the tangent PR, the perpendicular SY can be found as follows:

In $\triangle SPY$ we know angle YPS because that is the angle the tangent makes with the line to the center, the "given direction" mentioned in the enunciation.

So we know two angles (angle SYP and angle YPS) and a side SP.

Therefore $\triangle SPY$ is a determinate shape, and so side SY is a determinate length.

Similarly we can find perpendicular Sy.

Therefore L is the one unknown in the proportion from Proposition 16 Corollary 1, just above, and so:

L is given. (Conclusion 1)

"Let the angle RPH be made the supplement of the angle RPS, and the line PH, upon which the other focus H is located, will be given in position."

Construct angle RPH supplement to angle RPS.

But angle RPH is supplement to angle GPH. [Eu. I.13]

Therefore angle RPS = angle GPH.

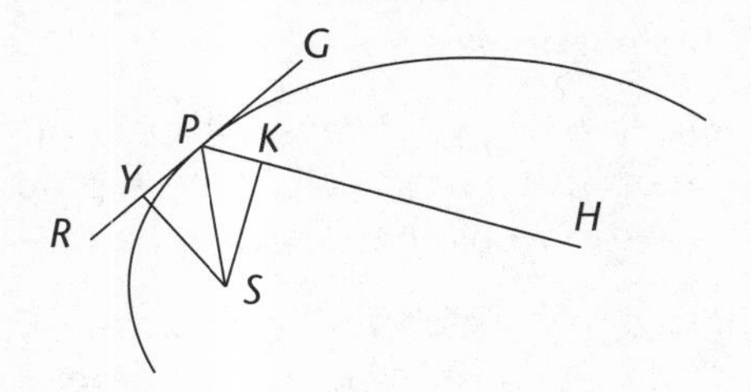

Therefore, by Apollonius III.48, line PH will have the conic's second focus lying on it.

Line PH is given in position with the other focus H lying somewhere on it.

(Conclusion 2)

"SK being dropped perpendicular to PH, let the conjugate semiaxis BC be understood to be set up, and sq. $SP - 2KP, PH$ + sq. PH = sq. SH (by Book II Proposition 13 of the *Elements*) = 4 sq. CH = 4 sq. BH -- 4 sq. BC = sq. $(SP + PH)$ – $L \times (SP + PH)$ = sq. SP + $2SP, PH$ + sq. PH – $L \times (SP + PH)$. Let $2KP, PH$ – sq. SP – sq. PH + $L \times (SP + PH)$ be added to both sides, and the result will be that $L \times (SP + PH) = 2SP, PH + 2KP, PH$, or $SP + PH$ is to PH as $2SP + 2KP$ is to L. Whence PH is given in both length and position."

Paradigm Proof Assuming Ellipse

Now we start a line of thought assuming the conic is an ellipse—later we will see how to modify this for parabola and hyperbola.

Imagine some ellipse with known focus S, with a major axis as yet unknown in position and length, and with a center C and a second focus H as yet unknown in position.

Let's now make a hypothetical diagram by connecting S to the hypothetical H which we know is there somewhere. (Since PH is given in position by Conclusion 2, it is length PH which is unknown, and so length SH remains unknown.)

P may be anywhere along the curve, so the angle at P may be acute, right, or obtuse. Thus $\triangle PSH$ may be acute, right, or obtuse at P.

Drop $SK \perp PH$. In the acute-angle triangle it will intersect PH between P and H. In the right-angle triangle K will fall on P. In the obtuse-angle triangle K will fall beyond P.

At this point we could save ourselves trouble by showing that the Law of Cosines formula will come out the same for acute, right, and obtuse angled triangles for $\triangle PSH$. This will save us from having to make three proofs for three cases of triangles for each conic (ellipse, parabola, and hyperbola), using Euclid II.13 for acute angle, I.47 for right angle, and II.12 for obtuse angle triangles.

But since Newton does not speak of cosines in *Principia*, the use of three-case proofs using only the Euclidean propositions would be truer to a full proof as Newton would have done it had he been pressed.

Newton's sketch for Proposition 17 seems to fall somewhere in between these two possibilities for full proof. He uses the formula from Euclid II.13 for acute-angled triangles and doesn't refer to the other two cases because he knows they will come out the same, once adjustments are made for certain differences between the three types of triangle.

Let's generalize the situation using a Euclidean approach, to show that all three cases can reduce to the acute-angle case which Newton uses.

By Euclid II.13, for an acute-angled triangle,

$$SH^2 + 2(PH,PK) = PH^2 + SP^2.$$

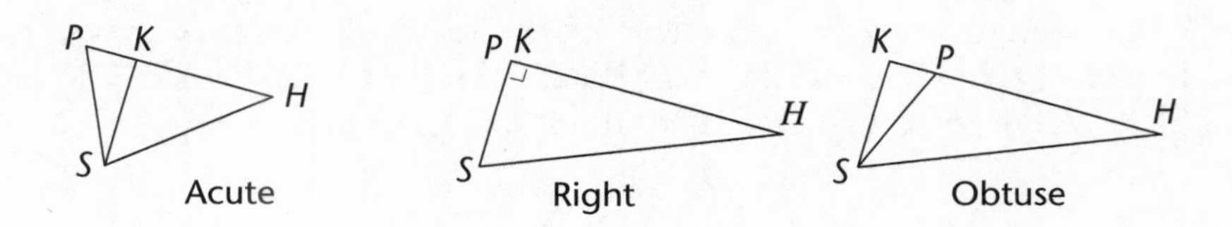

By Euclid I.47, for a right-angled triangle, $SH^2 = PH^2 + SP^2$.

Since SK lies on SP, PK is zero, and rect PH,PK is zero and drops out.

(If we were generalizing the Law of Cosines, we would say in modern terms "cosine of $90^\circ = 0$.")

Thus we can use the formula of Euclid II.13 to cover the right triangle case.

By Euclid II.12, for an obtuse triangle,

$$SH^2 - 2(PH,PK) = PH^2 + SP^2.$$

Since K falls outside the triangle beyond P, line PK has the opposite direction, and thus can be regarded as negative. Therefore rect PH,PK is negative and in subtracting a negative area the formula becomes the same as that for Euclid II.13.

(If we were generalizing the Law of Cosines, we would say, in modern terms, "cosine of an obtuse angle is negative.")

So, now with full generality, we may use:

$$SH^2 = PH^2 + SP^2 - 2(PH,PK). \quad (2)$$

Wherever H may be, we know by Apollonius III.45 that in an ellipse $SC = CH$, so

$$SH = 2CH \quad \text{and}$$

$$SH^2 = 4CH^2. \quad (3)$$

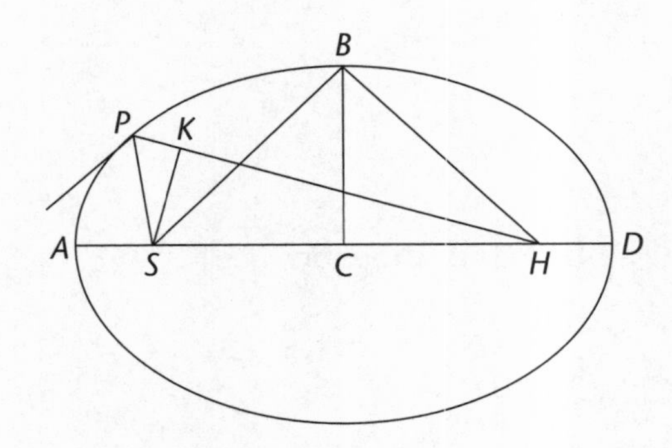

In right triangle BCH, by Euclid I.47, $BH^2 = BC^2 + CH^2$.

$$CH^2 = BH^2 - BC^2;$$

$$4CH^2 = 4(BH^2 - BC^2);$$

$$4CH^2 = (2BH)^2 - (2BC)^2. \quad (4)$$

$BC = BC$. [identity]

$BC \perp DA$. [Ap. I Def. 8]

$HC = CS$. [construction of Ap. III.45]

Therefore $\triangle BHC \cong \triangle BCS$. [Eu. I.4]

So $HB = SB$ and

$$BH + SB = 2BH. \quad (5)$$

By Apollonius III.52, if in an ellipse, from foci H and S, straight lines are deflected to points B and P on the section, then those pairs of lines will each be equal to the axis.

$$BH + SB = 2AC.$$

$$SP + PH = 2AC. \quad (6)$$

Therefore $BH + SB = SP + PH$.

Substituting from Equation 5,

$$2BH = SP + PH.$$

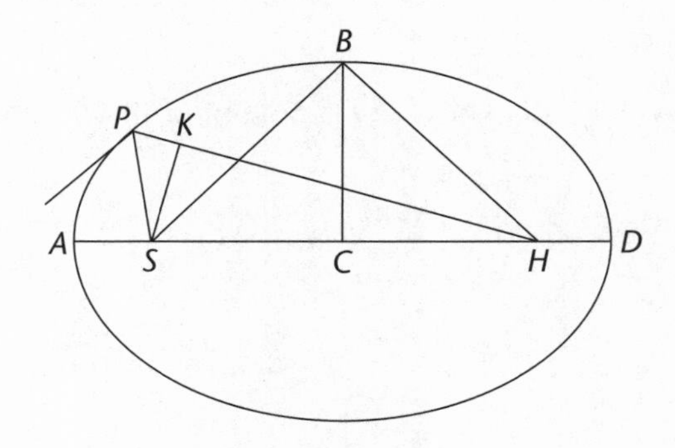

$$(2BH)^2 = (SP+PH)^2. \tag{7}$$

$$L : 2BC :: 2BC : 2AC. \quad \text{[Ap. I.15, converse]}$$

Substituting from Equation 6,

$$L : 2BC :: 2BC : (SP+PH).$$

By Euclid VI.17,

$$(2BC)^2 = L(SP+PH). \tag{8}$$

Combining Equations 2 and 3:

$$4CH^2 = SP^2 - 2(PH \times PK) + PH^2.$$

Substituting for $4CH^2$ from Equation 4:

$$SP^2 - 2(PH \times PK) + PH^2 = (2BH)^2 - (2BC)^2.$$

Substituting for $(2BH)^2$ from Equation 7 and for $(2BC)^2$ from Equation 8:

$$SP^2 - 2(PH \times PK) + PH^2 = (SP+PH)^2 - L(SP+PH). \tag{9}$$

Expanding the binomial in the right-hand expression:

$$(SP+PH)^2 = SP^2 + 2(SP \times PH) + PH^2.$$

Substituting back into Equation 9:

$$SP^2 - 2(PH \times PK) + PH^2 = SP^2 + 2(SP \times PH) + PH^2 - L(SP+PH).$$

Simplifying:

$$L(SP+PH) = 2(SP \times PH) + 2(PH \times PK).$$

Divide both sides by $L \times PH$:

$$\frac{L(SP+PH)}{L \times PH} = \frac{2(SP \times PH) + 2(PH \times PK)}{L \times PH};$$

$$\frac{SP+PH}{PH} = \frac{2SP + 2PK}{L}. \tag{10}$$

Let's look at Equation 10. What do we know about its elements?

Since the positions of S and P are given, SP is given in length.

By Conclusion 1, L is given.

PH is given in position by Conclusion 2, and $SK \perp PH$, therefore length PK can be found as follows:

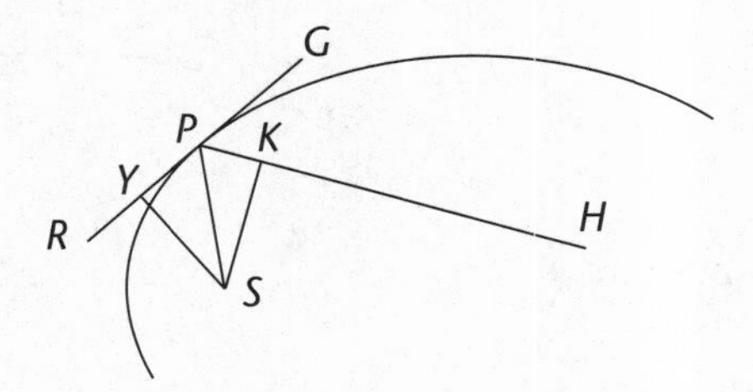

We constructed angle RPH as the supplement of angle RPS; so

angle SPK = angle RPH – angle RPS.

Angle SKP is a right angle.

So in $\triangle PKS$, we know angle SPK and angle SKP, and we know side SP. Thus we know two angles and one side, which is enough to determine the whole triangle in shape. Thus PK is a determinate length. (In modern terms, we can get the actual length of PK by the Law of Sines.)

This leaves length PH as the only unknown in Equation 10. Therefore it can be found and PH is now given in length as well as in position.
(Conclusion 3)

Now consider the other conics as variations on the ellipse.

Let's keep the direction of motion of the body constant at point P but let velocity vary. This direction is identified by the angle of tangent PR. SP is constant, and since the angle SPR is constant, length SY will be constant.

By Proposition 16 Corollary 1,

$$L \propto V^2 \times SY^2.$$

(Remember that all the Ls we have been working with are *principal* latera recta.)

Since SY is being held constant,

$$L \propto V^2.$$

So varying the velocity will vary the principal latus rectum.

Case 1: The Ellipse

"That is, if the velocity of the body at P be such that the latus rectum L would be less than $2SP + 2KP$, PH will fall on the same side of the tangent PR as the line PS, and therefore the figure will be an ellipse, and from the given foci S, H and the principal axis $SP + PH$, it will be given."

By Equation 10,

$$\frac{SP + PH}{PH} = \frac{2\,SP + 2\,PK}{L}.$$

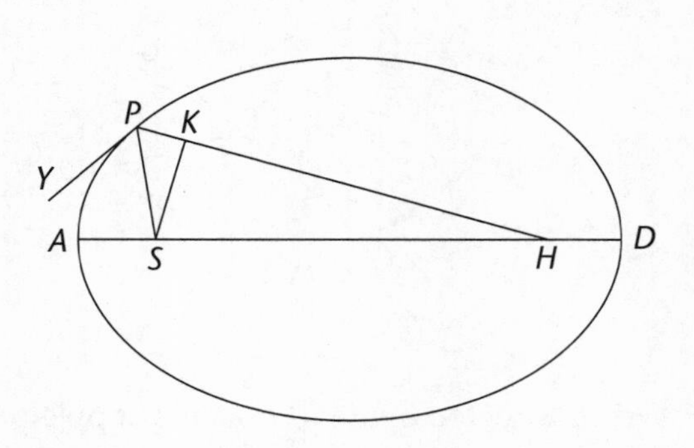

If the velocity of the body at P is such that

$L < 2(SP+PK)$, then

$PH < SP+PH$.
[Eq. 10 and Eu. V.14]

Subtracting PH from both sides of the inequality, we see that SP is positive (greater than zero).

When SP is positive, the figure will be an ellipse and PH will lie on the same side of the tangent PY with SP.

The foci S, H, and the principal axis $AD = SP+PH$ are also given.

Case 2: The Parabola

"If, on the other hand, the velocity of the body be great enough that the latus rectum L would be equal to $2SP + 2KP$, the length of PH will be infinite, and consequently the figure will be a parabola having the axis SH parallel to the line PK, and thence it will be given."

By Equation 10,

$$\frac{SP + PH}{PH} = \frac{2\,SP + 2\,PK}{L}.$$

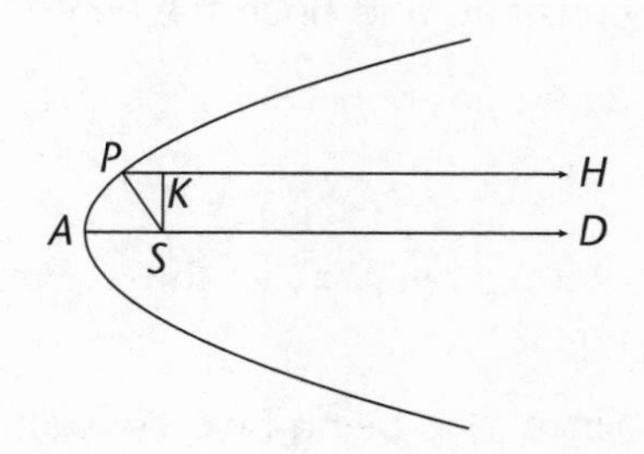

Suppose this equation works for the parabola as well as the ellipse (for which we derived it).

If the velocity is great enough that

$L = 2(SP+PK)$

then $PH = SP+PH$. [Eq. 10 and Eu. V.14]

This equation could only be true if PH is infinite. If PH is infinite, axis AD will be parallel to PH and therefore the figure will be a parabola.

The principal axis AD is also now known (it, too, is infinite).

Because these results are consistent with what we know of the parabola, we have a sort of confirmation that Equation 10 does apply to this conic.

Case 3: The Hyperbola

"And if the body go forth from its place P with a still greater velocity, the length of PH must be taken on the other side of the tangent, and therefore, since the tangent runs between the foci, the figure will be a hyperbola having its principal axis equal to the difference of the lines SP and PH, and thence it will be given."

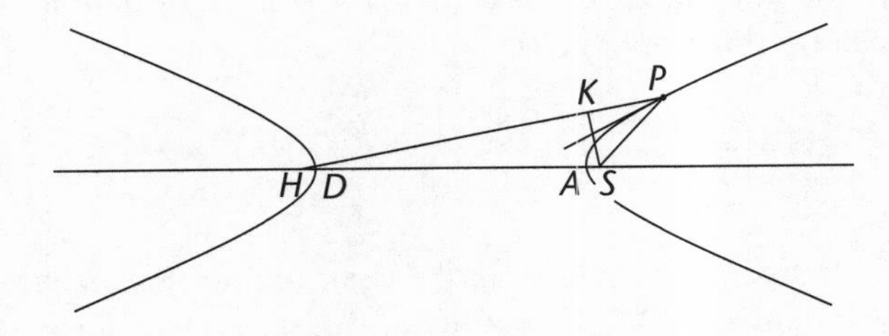

Let's think about this as a variation of the ellipse.

By Equation 10,

$$\frac{SP + PH}{PH} = \frac{2\,SP + 2\,PK}{L}.$$

Suppose this equation works for the hyperbola as well.

If the velocity is even greater than the velocity that would result in a parabola, so that

$L > 2(SP+PK)$, then

$PH > SP+PH$.

Subtracting PH from both sides, we see that SP is less than zero, or "negative," that is, it must be subtracted when it appears in an equation, as it is in Apollonius III.51 for the hyperbola in contrast to III.52 for the ellipse.

In the ellipse, by Apollonius III.52, $SP+PH = AD$.

In the hyperbola, by Apollonius III.51, $PH-SP = AD$.

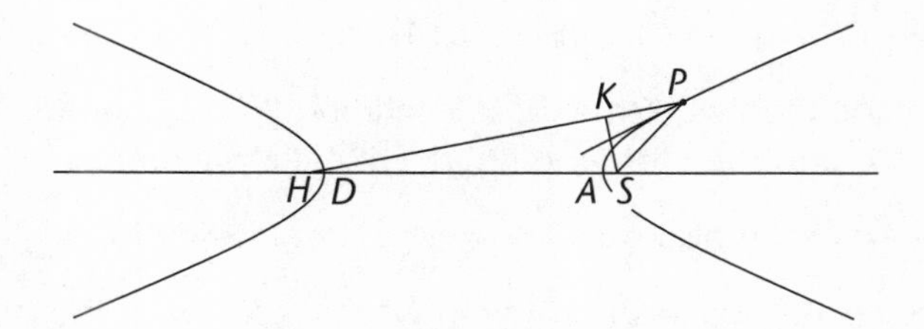

So if SP turns out to be "negative," that is, if it is the *difference* of the lines from the point on the curve to the two foci which is equal to the diameter, then the curve is an inverted ellipse, or hyperbola.

This corresponds to what we know about the hyperbola, giving us a sense of confirmation about the applicability of Equation 10.

The figure will be a hyperbola with PH on the other side of the tangent and the tangent will pass between the foci.

The principal axis will be given ($AD = PH - SP$) by Apollonius III.51, and Equation 10 will be reformulated as:

$$\frac{(PH - PS)}{PH} = \frac{2(PK + PS)}{L}. \tag{11}$$

"For if the body in these cases revolve in the conic section thus found, it has been demonstrated in Propositions 11, 12, and 13 that the centripetal force will be as the square of the distance of the body from the center of forces S inversely, and therefore the line PQ is shown correctly, which a body shall describe by such a force, in going forth from the given place P with a given velocity along the straight line PR given in position. Q.E.F."

Newton seems to be showing here that he's still uncomfortably aware that he hasn't proved the converse to Propositions 11, 12, and 13 (asserted and sketched in Proposition 13 Corollary 1). So we get a little more hand-waving here along this line:

If the body revolve in an ellipse, hyperbola, or parabola, around a center of forces at a focus, then by Propositions 11, 12, and 13, respectively, we will know that it will be moving by an inverse square law. But that's what we've assumed here, so it all fits.

Q.E.F.

Note

Not satisfied? The proofs for parabola and hyperbola depended on the paradigm proof for the ellipse, and in particular on the application of Equation 10 to those figures.

Newton is calling on our intuition about conics here. If you feel that your intuition doesn't justify accepting this, then you will need to construct proofs for independent versions of this proposition for the hyperbola and parabola, using propositions (from Apollonius or another treatise on conic sections) which apply to hyperbolas and parabolas, rather than those we used applying to the ellipse.

Perhaps you are unpersuaded that we can build conclusions for the parabola and hyperbola on a paradigm proof that makes explicit assumptions which apply only to the ellipse. In the cases at the end Newton makes adjustments for characteristics of the other curves, but perhaps, had those characteristics been accounted for earlier, we would have come up with a different Equation 10.

If you suspect this, work through the paradigm proof using relationships and Apollonius propositions appropriate for the parabola and hyperbola and see if you come up with the conclusions of Cases 2 and 3.

For example, here is a way to look at the hyperbola.

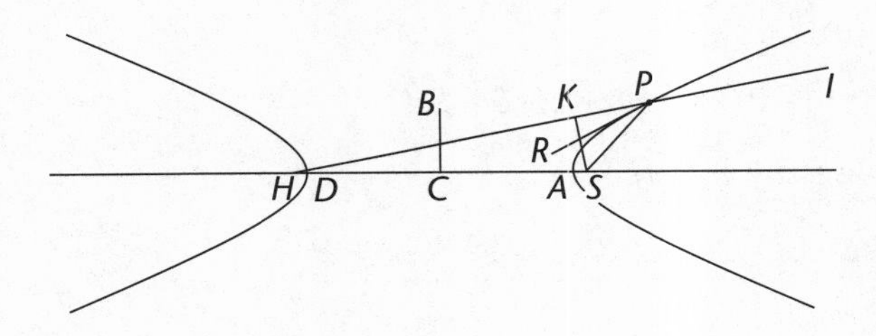

By Apollonius III.45, $\frac{1}{4}AD \times L = HD \times AH$.

But $HD = CH - CD$ and $AH = CH + AC$.

$$\frac{1}{4}AD \times L = (CH - CD)(CH + AC).$$

But $CD = AC$. Substituting AC for CD,

$$\frac{1}{4}AD \times L = (CH - AC)(CH + AC) = CH^2 - AC^2.$$

$$AD \times L = 4(CH^2 - AC^2) = (2CH)^2 - (2AC)^2.$$

But since $2CH = SH$ and $2AC = AD$

$$AD \times L = SH^2 - AD^2. \qquad \text{(N1)}$$

By Apollonius III.51 in hyperbola, $PH - PS = AD$.

Substituting for AD in Equation N1:

$$L(PH - PS) = SH^2 - (PH - PS)^2. \qquad \text{(N2)}$$

Using, with the generality previously established, Euclid II.13 for $\triangle PSH$:

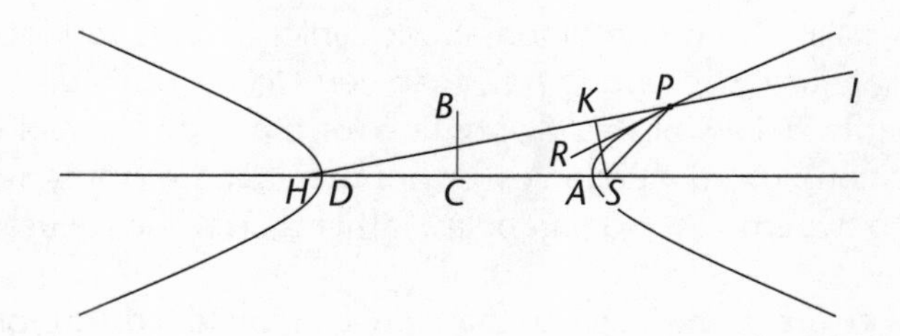

$$SH^2 = SP^2 - 2(PK,PH) + PH^2.$$

Substituting for SH^2 in Equation N2:

$$L(PH - PS) = SP^2 - 2(PK,PH) + PH^2 - (PH - PS)^2.$$

Expanding the binomial in the right-hand expression,

$$L(PH - PS) = SP^2 - 2(PK,PH) + PH^2 - (PH^2 - 2PS \times PH + PS^2);$$

$$L(PH - PS) = SP^2 - 2(PK,PH) + PH^2 - PH^2 + 2(PS \times PH) - PS^2.$$

Combining terms,

$$L(PH - PS) = -2(PK,PH) + 2(PS \times PH).$$

Divide both sides by $L \times PH$

$$\frac{L(PH - PS)}{L \times PH} = \frac{-2(PK, PH) + 2(PS \times PH)}{L \times PH}.$$

$$\frac{(PH - PS)}{PH} = \frac{2(PS - PK)}{L}. \tag{N3}$$

Since $PH - PS < PH, \quad 2(PS - PK) < L.$

Note that we must be very careful with the sign of PK. Depending on the angle SPH, by our generalization which gave us Euclid II.13 for all cases, PK will be treated as positive if K falls between P and H. PK will be treated as negative if P falls between K and H. If K falls on P it will be zero. (If you're not comfortable with treating a distance as negative, you will need a separate proof for each case: acute, obtuse, and right triangles.)

The hyperbola proof must account for the same three cases the ellipse did.

In Equation N3 we have PK subtracted from PS, so if it is being treated as negative its length will be added and vice versa. If P were far out on the hyperbola, SP would get as large as you please, and we would need to subtract PK to counteract that and leave the whole expression less than L. As angle SPH becomes acute, K will fall between P and H and PK will be positive; so in Equation N3 it will indeed be subtracted.

Newton has expressed it in the general form of $L < 2(SP + PK)$.

Exercise: Work this out yourself for the parabola.

I.17 Corollary 1

Hence in every conic section, from the given principal vertex D, latus rectum L, and focus S, the other focus H is given by taking DH to DS as is the latus rectum to the difference between the latus rectum and 4DS. For the proportion SP+PH to PH as 2PS+2KP to L becomes, in the case of this corollary, DS+DH to DH as 4DS to L, and separando, *DS to DH as 4DS−L to L.*

Expansion of Newton's Sketch of I.17 Corollary 1

Given:

1. Principal vertex D,
2. focus S,
3. latus rectum L.

To Find:

Other focus H.

Ellipse

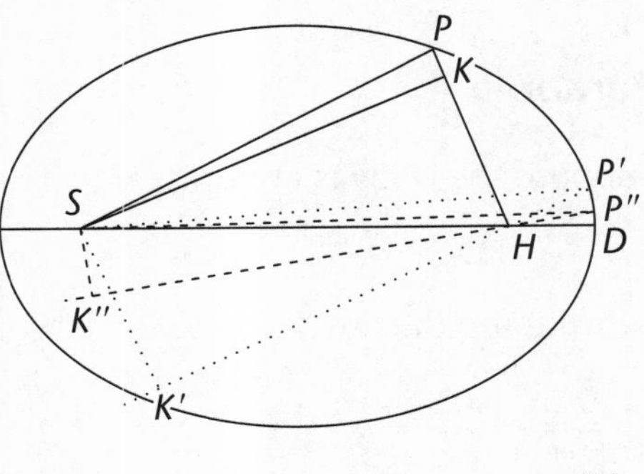

Suppose the body is at a principal vertex D. Then $SP = DS$.

By Equation 10 of the proposition,

$$\frac{SP + PH}{PH} = \frac{2\,SP + 2\,PK}{L}.$$

Substituting,

$$\frac{DS + DH}{DH} = \frac{2\,DS + 2\,PK}{L}. \quad (1)$$

Where would point K, or line PK, fall, when the body is at D?

Remember that SK is constructed as the perpendicular to line PH—therefore line PK always goes through H.

As $P \rightarrow D$, line PH extended pivots around, with SK always falling perpendicular to it.

In the diagram, as P moves to D, K moves around as shown. Triangle SPK is first shown with solid lines, then with dotted lines, then with dashed lines. It can be seen that as P arrives at D, perpendicular SK will have collapsed to S.

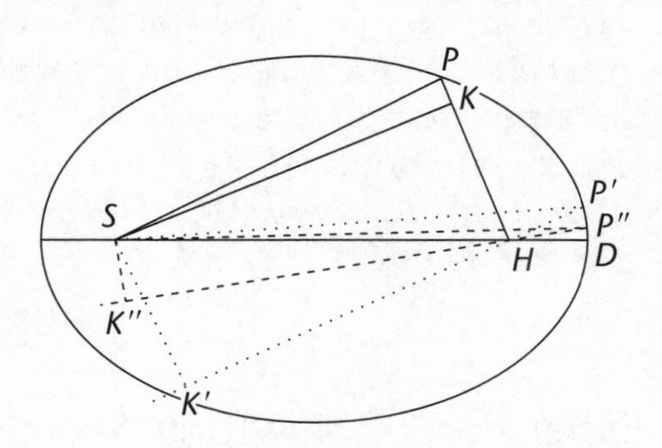

PK ultimately coincides with DS.

$$PK \overset{ult}{=} DS.$$

Therefore $\quad DS + PK \overset{ult}{=} 2DS.$

$$2(DS + PK) \overset{ult}{=} 4DS.$$

Or, when P is *at* D,

$$2(DS + PK) = 4DS.$$

Substituting into Equation 1,

$$\frac{DS + DH}{DH} = \frac{4DS}{L}.$$

By Euclid V.18 (*separando*),

$$\frac{DS}{DH} = \frac{4DS - L}{L}. \tag{2}$$

Since DS is known (since SP is given), and L is known, therefore length DH can be found.

Since D is known, therefore H can be found.

Q.E.D. for ellipse.

Parabola

Suppose the body is at principal vertex D. In a parabola, by Lemma 14,

$$L = 4DS.$$

Substituting into Equation 2,

$$\frac{DS}{DH} = \frac{4DS - 4DS}{L} = \frac{0}{L}.$$

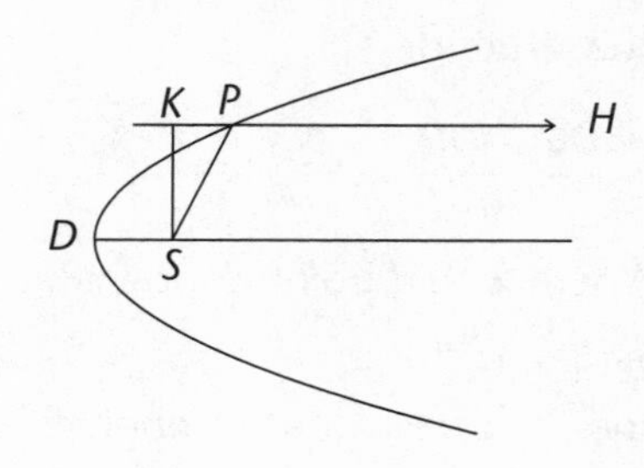

Since DS has a finite value, for this equation to hold DH must be infinite, as we indeed treat it as being in the parabola. That is, there is no other focus H (it is "given" as nonexistent, or as existing at an infinite distance).

Q.E.D. for the parabola.

Hyperbola

By Equation 11 of the proposition,

$$\frac{(PH - PS)}{PH} = \frac{2(PK + PS)}{L}.$$

Suppose the body to be at principal vertex D.

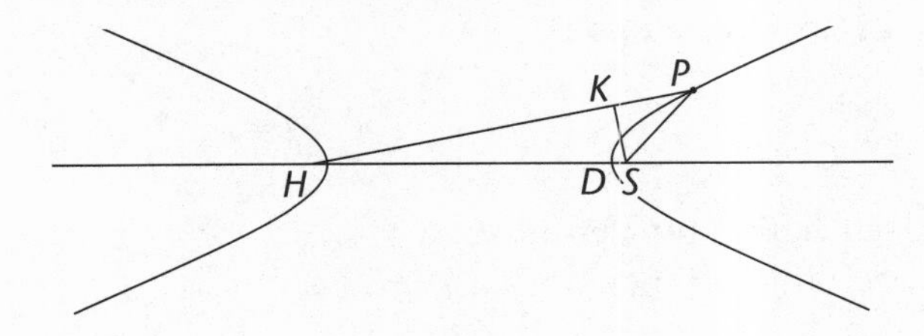

$$\frac{(DH - DS)}{DH} = \frac{2(PK + DS)}{L}.$$

As before, as $P \to D$, $K \to S$. So $PK \to DS$.

> **Question:** Remember that it was in the context of an ellipse that we showed that as $P \to D$, $K \to S$. Is this true for the hyperbola? Sketch what this looks like as $P \to D$.

$$\frac{DH - DS}{DH} = \frac{4DS}{L}.$$

By the reasoning of the end of the proposition (see Case 3 above, in which we found the hyperbola by increasing the velocity), DS, going in the opposite direction from DH, must be thought of as negative. Thus,

$$DH - DS > DH$$

and so $4DS > L$, which is the condition for a hyperbola.

Since DS and L are known, DH can be found, and H can be found.

Q.E.D. for hyperbola.

I.17 Corollary 2

Whence if the velocity of a body at the principal vertex D is given, the orbit is readily found. For one takes its latus rectum to twice the distance DS, in the duplicate ratio of this given velocity to the velocity of a body orbiting in a circle at the distance DS (by Proposition 16 Corollary 3), and then [takes] DH to DS as the latus rectum to the difference between the latus rectum and 4DS.

Expansion of Newton's Sketch of I.17 Corollary 2

Given:

1. Velocity at principal vertex D;
2. everything given in proposition;
3. velocity in a circle with radius DS.

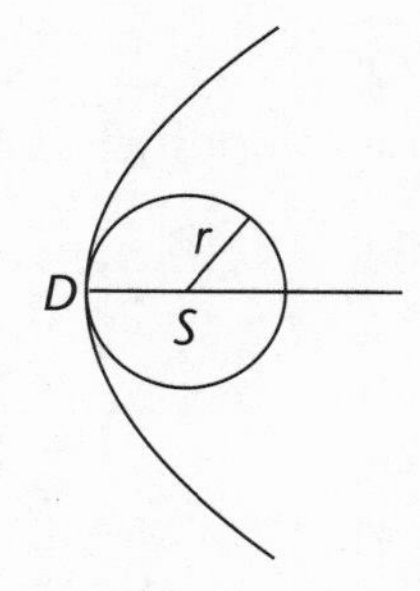

To Find:

The orbit. That is, find the vertex, foci, principal latus rectum, and major axis.

Proof:

By Proposition 16 Corollary 3, at the same distance DS, at the principal vertex D:

$$V_{\text{conic}} : V_{\text{circle}} :: \sqrt{L} : \sqrt{2DS}.$$

$$\left(\frac{V_{\text{conic}}}{V_{\text{circle}}}\right)^2 = \frac{L}{2DS}.$$

Since the rest of the terms are known, we can find L.

Since D and S are already given, by Corollary 1 we can find H.

So we have two foci, L, and the major axis, which is $DH + DS$ for the ellipse, $DH - DS$ for the hyperbola, and infinite for the parabola.

[**Corollaries 3 and 4 and the Scholium** are not used in Newton's central argument.]

Pause After I.17

This ends the selection Newton recommended for us to cover in Book I. He suggested reading "the definitions, laws of motion, and the first three sections"; the third section finishes with I.17. Now he recommends that we carry on to Book III, looking up other propositions from the first two books as they are cited. (This advice appears in the Preface to Book III, page 301 in this guidebook.)

Additional propositions used in Book III.

These propositions are presented in full or summarized at the points they are needed in Book III. Newton suggests that they not be studied until then. They are summarized here in the order they appear in *Principia* for the convenience of readers wishing to refer to them out of the order of use.

SECTION 7

On the rectilinear ascent and descent of bodies.

Proposition 39

Proposition 39 is both the culmination of Section 7 and a preliminary to Section 8. Here, the centripetal force is allowed to vary in any proportion whatever to the distance from the center, and the problem is to find the velocity and time at which a vertically moving body arrives at any height. Newton specifies that the solution to this problem requires knowledge of "the quadratures of curvilinear figures." In practice, this amounts to being able to compute the area contained by a curve (representing the force law), an axis, and two ordinates drawn parallel from the axis to the curve.

This kind of problem (which we would nowadays call "integration") does not usually have a geometrical solution, but was well within the capabilities of Newton's calculus, which he called "the method of fluxions." Thus in stating that the quadratures of curvilinear figures need to be provided, Newton is warning us that we are to some extent stepping outside the bounds of the strict geometrical method.

SECTION 8

On finding the orbits in which bodies revolve when driven by any centripetal forces whatever.

In the series of propositions I.39–I.41 Newton sketches out a procedure by which, with a lot of work, one might determine an orbit from a given force law. A brief summary and discussion of this sequence follows. If you would like to explore these propositions further, especially in relation to I.13 Corollary 1, I recommend several other books that provide more extensive treatments of this sequence.*

Propositions I.40 – I.42

Proposition 40 states that under any kind of centripetal force, if two bodies at the same distance from the center have the same velocities (in any direction whatever), they will have equal velocities at any equal distances whatever from the center. It is a proposition of extraordinary generality, and is equivalent to a statement of the conservation of kinetic and potential energy.

Propositions 41 and 42 point the way toward a general solution of the "inverse problem" raised by I.13 Cor. 1: given any kind of centripetal force, and the initial position and velocity of a body, to find the orbit. These propositions depend on I.39, and thus require calculus techniques (the "method of fluxions") that go beyond what geometry alone can achieve. As stated, the problem is so general that no particular orbits (with one exception, see below) are produced. However, the methods presented can be made to generate a polar equation for an orbit. Although Newton, strangely, does not mention this possibility, it turns out that if the force law were inverse-square, the resulting equation would be the polar equation of one of the conic sections.

I.41 Corollary 3. The exception mentioned above is in I.41 Cor. 3, in which Newton uses lines cut off by a moving tangent to a conic section to construct the spiral that would be generated by an inverse-cube force law. He states that the proof of this follows from "the quadrature of a certain curve;" that is, by the use of his method of fluxions. This point is particularly relevant to I.9, in that it shows that Newton had been working with the equiangular spiral primarily by means of fluxions, but chose not to interrupt the flow of the *Principia* at that early point with an excursion into those techniques.

---

* François De Gandt (*Force and Geometry in Newton's Principia,* pp. 246–264) gives a step-by-step analysis, with very clear diagrams showing the relations of the critical areas. S. Chandrasekhar (*Newton's Principia for the Common Reader,* pp. 161–174) works through the proofs, translating them into modern terms, and provides both a derivation of the orbit under an inverse square law and a speculation as to why Newton did not include such a derivation. Niccolò Guicciardini (*Reading Principia,* pp. 58–60 and pp. 219–223) puts the problem into the historical context of critiques by Bernoulli and others, and cites manuscript evidence of Newton's ability to evaluate the necessary integrals. I. Bernard Cohen provides a thorough review of the steps of I.41 in the introduction to his translation of *Principia,* pp. 334–345; some individual steps are expressed in modern notation.

SECTION 9

On the motion of bodies in moveable orbits; and on the motion of the apsides.

Proposition 45

"It is required to find the motion of the apsides of orbits that are very close to circles."

Proposition I.45 (or rather its first corollary, quoted below) is called upon in Newton's proof of propositon III.2 to determine the force law governing the motion of the planets around the sun. It is quoted and discussed there. We don't take time to follow the full proof of this proposition (which aside from being long and complex in its own right, requires that I.43 and I.44 be proved first). But we will look at it and see what it asserts.

Note that I.45 concerns itself with orbits that are very close to circles. The planets fit this condition.

I.45 investigates the relation of the force law to the motion of the line of apsides. Remember that the line of apsides is the line between the nearest approach of the orbiting body to the center of forces and the most distant point. In Proposition III.2 we're concerned with the line between aphelion and perihelion, the farthest and nearest points of the planet to the sun. The line of apsides corresponds to the major axis of an elliptical path.

Motion of apsides is something we didn't deal with in the first seventeen propositions. We imagined bodies moving in ellipses as following back over their same paths on successive revolutions.

And indeed, it turns out that with an exact inverse square law, that is what will happen. But if the force law should have other terms that are also influencing the orbit, we may still get an ellipse, but one that will rotate in space: the line of apsides will shift gradually (or more quickly) in orientation. This rate of travel of the line of apsides can be predicted if the force law is given. Conversely, the rate of travel can be measured and used to gain insight into the force law.

I.45 Corollary 1:

"Hence, if the centripetal force be as some power of the altitude, that power can be found from the motion of the apsides, and conversely. Specifically, if the whole angular motion by which the body returns to the same apsis be to the angular motion of one revolution, or 360°, as some number m be to another number n, and the altitude be called A, the force will be as that power of the altitude $A^{\frac{nn}{mm}-3}$, whose index is $\frac{nn}{mm}-3$..."

The altitude is the planet's distance, or radius. For example, if the body returns to the same apsis in one revolution (a fixed orbit), n and m will both be 1, and the force law will be the $(\frac{1}{1}-3)$ power of the distance, or inversely as the square of the distance.

In other cases, we have other ratios of $n^2 : m^2$ and can find other force laws.

SECTION 11

On the motion of bodies tending to each other with centripetal forces.

Propositions I.58–I.60

Proposition I.60 is cited in III.4. (I.58 and I.59 are needed to set up I.60.) At the point of the expanded proof of III.4 where I.60 is called on, I explain the application and the relationship among the three propositions. Here is a brief summary.

In I.59, Newton considers what will happen if we suppose a system of two mutually attracting bodies, one of which is fixed with the other orbiting around it, and if we then allow both to move around their common center of gravity under the same force. (Proposition I.58 showed the possibility of such a transformation.) It must be noted that no such relationship has yet been argued for the earth and moon.

Consider the relative distances of the bodies in this hypothetical mutually attracting system. If their distances from each other are to remain the same as before, the periodic time will have to decrease. In I.60 Newton considers what would have to change if the periodic time were to be held constant. The conclusion is that the principal axis must increase according to the following proportion:

$$\frac{\text{New Axis}}{\text{Old Axis}} = \sqrt[3]{\frac{T+M}{T}}.$$

where T is the quantity of matter in the formerly fixed body, and M is that in of the other body.

Proposition 65

Proposition I.65 and some of its corollaries are used in III.6. Both proposition and corollaries are quoted, with explanatory notes, in Appendix B. (Please look there for the full presentation.) Their applications are explained where they are used in III.6.

I.65 brings in the complication of multiple bodies. If "gravity is given towards every planet" as Newton asserts in III.5 Corollary 1, and if this means, as it seems to, that every planet gravitates towards every other planet, then we would expect instability in the universe. The stable ellipses that result from inverse square laws governing two-body systems would be thrown off by the effects of other bodies outside that system. And yet we do find stable elliptical orbits in the planetary system. To justify his claim in III.5 Corollary 1 and its further development, Newton must account for that.

Propositions I.65 and I.66 explore some reasons we might still find stable elliptical orbits. They argue that with many bodies, if they are suitably arranged, it is possible to get the bodies moving, at least approximately, in ellipses with equable description of areas.

Proposition 66

Proposition I.66 is presented and discussed in Appendix B along with I.65. See also the discussion in the last two paragraphs under I.65 just above.

I.66 is used in III.13 where we derive the elliptical orbits of the planets. For example, it is relevant in III.13 that the disturbances from other planets will be less if we consider the sun's actual motion rather than maintaining the assumption that the sun is at rest. Newton gives an illustration of this in the case of Jupiter and Saturn.

Proposition 67

Proposition I.67 is used in III.13. It argues that, when several bodies are mutually attracted by an inverse square law, with two planets orbiting around a larger body, then the outer of the two orbiting planets describes areas more nearly proportional to the times when those areas are figured around the center of gravity of the other two bodies than it does around the larger body as center.

Proposition 69

This proposition is used in the proof of III.7. A full presentation and proof is given between III.6 and III.7.

SECTION 12

On the attractive forces of spherical bodies.

Proposition 70

"...suppose, that to the individual points of a spherical surface there tend equal centripetal forces decreasing in the duplicate ratio of the distances from the points..."

This proposition is referred to in the enunciation of I.71, which leads to I.75, which is used in the transition from III.7 to III.8.

Proposition 71

"The same things being supposed [as in Proposition 70, i.e., that to the individual points of a spherical surface there tend equal centripetal forces decreasing in the duplicate ratio of the distances from the points], I say that a corpuscle placed outside the spherical surface is attracted to the center of the sphere with a force inversely proportional to the square of its distance from the same center."

This proposition is used in the proof of I.74 and thus is built on by I.75 which is used in the transition from III.7 to III.8. A full presentation and proof is given between those two propositions.

I.71 establishes that a spherical shell will attract an outside particle with a force that varies inversely as a square of the distance to the center of the ghost sphere of which this is the shell. The extension to the solid sphere is done by relatively easy steps. We take the conclusion found here about the spherical shell and nest successive shells inside one another to make up our solid sphere.

Proposition 74

"The same things being supposed [as in Proposition 73, namely, that to the individual points of some given sphere there tend equal centripetal forces decreasing in the duplicate ratio of the distances from the points], I say that a corpuscle placed outside the sphere is attracted with a force inversely proportional to the square of its distance from the center of the sphere."

This proposition is used in the proof of I.75 which is used in the transition from III.7 to III.8. A full presentation and proof is given between those two propositions.

Following on the proof of I.71, we now nest enough spherical shells nested together to get a solid sphere, finding that the force on an outside corpuscle will still vary inversely as the square of the distance to the center.

Proposition 75

"If to the individual points of a given sphere there tend equal centripetal forces decreasing in the duplicate ratio of the distances from the points, I say that any other similar sphere you please is attracted by the same [sphere] with a force inversely proportional to the square of the distance of the centers."

This proposition is used in the transition from III.7 to III.8. A full presentation and proof is given between those two propositions.

Again we have as a given that all the particles of a whole sphere are attracting with an inverse square force law. I.74 showed that a *particle* outside this given attracting sphere will be attracted to the sphere's center by inverse square; now we prove that a *whole outside sphere* will be attracted to the given attracting sphere's center by inverse square.

Proposition 76

"If spheres that, in progression outward from the center to the circumference, are dissimilar in any manner as regards density of matter and attractive force, are nonetheless everywhere similar in progression throughout their circumference at every given distance from the center, and if the attractive force of any point decreases in the duplicate ratio of the distance of the attracting body, I say that the whole force by which one sphere of this kind attracts another is inversely proportional to the square of the distance of the centers."

Proposition I.76 is cited by Newton in III.8 as part of the transition from III.7 to III.8. It is presented in its place in the sequence of I.71–I.76 between those two propositions.

Proposition I.76 extends the conclusion of I.75 to spherical bodies that are not homogeneous, provided that the component spherical shells making up each body are themselves homogeneous.

Book II

On the Motion of Bodies ❦ Second Book

SECTION 4

On the motion and resiſtance of bodies suspended by a ſtring.

Proposition 24

This proposition is used in the proof of III.6. A full presentation and proof is given between III.5 and III.6.

Book III

On the System of the World. Third Book.

Newton's Preface to Book III

In the preceding books I have presented the principles of philosophy, not, however, philosophical, but only mathematical; that is, those from which one can argue in philosophical matters. These are the laws and conditions of motions and forces, which pertain to philosophy in the highest degree. Nevertheless, so that these should not appear sterile, I have illustrated them with certain philosophical scholia, treating those things that are most general, and in which philosophy seems most to be poured out, such as the density and resistance of bodies, spaces void of bodies, and the motion of light and of sounds. What remains is that we show from the same principles the constitution of the system of the world. Concerning this theme I had composed the third book using a popular method, so that it might be read by many. But those by whom the established principles might not have been sufficiently well understood will have a very slight perception of the force of the argument, and will not cast aside prejudices to which they have been accustomed for many years. Therefore, so that the matter not be subject to dispute, I have carried over the substance of that book into propositions, in the mathematical manner, so that they may only be read by those who had previously gone through the principles. Nevertheless, because many propositions appear there [i.e., in the first two books] which might take too much time even for those learned in mathematics, I do not wish to propose that everyone read all of them: it would be sufficient for one to read carefully the definitions, laws of motion, and the first three sections of the first book, and then to carry on to this book on the system of the world, looking up at will the remaining propositions of the previous books cited here.

Rules of Philosophizing

Notes on Rules of Philosophizing

- Newton, writing this book in Latin, called these "Regulae Philosophandi." In English, rules of philosophizing, or rules of doing philosophy.

- What are these "rules" intended to be?

For a start, we note that Newton has *not* said that these are rules governing the study of some subject matter. Second, we note that, although the title of this book is *Mathematical Principles of Natural Philosophy,* he has not called this section rules for doing *natural* philosophy but rules for doing philosophy. Perhaps he means to offer this as a set of principles with more general application; perhaps he would apply them to any branch of philosophical thinking.

Rather than intending these "rules" to be something imposed from outside to direct study of a body of knowledge, it seems that Newton sees them as principles that describe the way we actually think if we are thinking philosophically. Applied to natural philosophy, they are standards of sound reasoning about phenomena, causes, and properties of matter. They describe the working of the mind of a careful thinker—in this application, the working of the mind of a natural philosopher, what we would call scientific thinking.

Rule 1

That there ought not to be admitted any more causes of natural things than those which are both true and sufficient to explain their phenomena.

Philosophers state categorically: Nature does nothing in vain, and vain is that which is accomplished with more that can be done with less. For Nature is simple, and does not indulge herself in superfluous causes.

Rule 2

Accordingly, to natural effects of the same kind the same causes should be assigned, as far as possible.

As, for example, respiration in humans and in animals, the descent of stones in Europe and in America, light in a cooking fire and in the sun, the reflection of light on earth and in the planets.

Rule 3

The qualities of bodies that do not suffer intensification and remission, and that pertain to all bodies upon which experiments can be carried out, are to be taken as qualities of bodies universally.

For the qualities of bodies are apprehended only through experience, and are accordingly to be declared general whenever they generally square with experiments; and those which cannot be diminished cannot be removed. It is certain that against the tenor of experiments, dreams are not to be rashly contrived, nor is a retreat to be made from the analogy of nature, since she is wont to be simple and ever consonant with herself. The extension of bodies is apprehended only through the senses, nor is it perceived in all things. But because it belongs to all perceptible bodies, it is affirmed to be universal. We experience many bodies to be hard. Hardness of the whole, moreover, arises from hardness of the parts, and thence we rightly conclude that the undivided particles, not only of these bodies which are perceived, but also of all others, are hard. We conclude that all bodies are impenetrable, not by reason, but by perception. Those that we handle are found to be impenetrable, and thence we conclude that impenetrability is a property of bodies universally. That all bodies are moveable, and that by certain forces (which we call the forces of inertia) they persevere in motion or rest, we gather from these very properties in bodies that we see. Extension, hardness, impenetrability, moveability, and the force of inertia of the whole arise from the extension, hardness, impenetrability, moveability, and forces of inertia of the parts, and thence we conclude that all the least parts of all bodies are extended and hard and impenetrable and moveable and endowed with forces of inertia. And this is the foundation of all of philosophy. Moreover, that parts of bodies that are divided and mutually contiguous can be separated from each other, we come to know from the phenomena, and that undivided parts can by reason be divided up into smaller parts, is certain from mathematics. But whether those distinct and hitherto undivided parts can be divided by the forces of nature and separated from each other, is uncertain. And if it were to be established by but a single experiment that by breaking a hard, solid body, some undivided particle were to suffer division, we would conclude, by the force of this rule, not only that the divided parts are separable, but also that the undivided ones can be divided *in infinitum*.

Finally, if it be established universally by experiments and astronomical observations that all bodies on the surface of the earth are heavy towards the earth, and this according to the quantity of matter in each, and that the moon is heavy towards the earth according to the quantity of its matter, and that our sea in turn is heavy towards the moon, and that all the planets are heavy towards each other, and that the gravity of comets towards the sun is similar, it will have to be said, by this rule, that all bodies gravitate towards each other. For the argument from phenomena for universal gravitation will be even stronger than that for the impenetrability of bodies, concerning which we have absolutely no experiment in the heavenly bodies; nay, not even an observation. Nevertheless, I do not at all assert that gravity is essential to bodies. By "inherent force" [*vis insita*] I understand only the force of inertia. This is inalterable. Gravity is diminished in receding from earth.

Rule 4

In experimental philosophy, propositions gathered from the phenomena by induction are to be taken as true, whether exactly or approximately, contrary hypotheses notwithstanding, until other phenomena appear through which they are either rendered more accurate or liable to exceptions.

This must be done lest an argument from induction be nullified by hypotheses.

Phenomena

Notes on Phenomena

• **Calculations and Planetary Theory.** Despite what might be suggested by their title, these "Phenomena" are not directly observed, but rather are conclusions based on observations.

Therefore they should be worked through with a calculator in order to check, and confirm one's understanding of, how Newton gets from his initial observations to the conclusions in the formal statement of the phenomenon.

They invoke not just observations, but planetary theory in current use by the astronomers of his time. Newton doesn't spell all that out; he assumes we are knowledgeable about contemporary planetary theory. Elements that are required for understanding Newton's steps from observations to Phenomena are included in my notes.

• **Astronomical Units.** In the course of filling out the proofs for these Phenomena and later propositions, it is often necessary to work with distances. As always with Newton's presentation in *Principia*, we have, not absolute distances, but ratios and proportions. But to find distances even in terms of the ratios, we must begin with some value for one of the terms. Since we don't know any of the distances in conventional terrestrial units, we must make astronomical units for ourselves. These can be anything, but the distance from the earth to the sun is frequently used. This unit could be expressed as 1, but for greater precision and convenience we may wish to express the earth-sun distance as some larger number. The other distances, such as Mercury's distance from the sun, are then relative to this unit. With a larger value for the earth-sun distance, the others can more easily be expressed without decimal points.

In these notes we will sometimes just refer to the earth-sun distance as "supposed," we will sometimes define it as 1, sometimes call it 10, and elsewhere, with Newton, let it be 100,000. Distances found by these methods are always in relation to the earth-sun distance as defined by the particular astronomical unit.

• **Approximations.** As we move beyond the mathematical demonstrations of the first two books into working with observations, we begin to use approximations. How do we know when an approximation is valid, and when it might simplify matters at the expense of the truth of celestial mechanics?

Obviously, thoughtfulness and judgment are required here. We can do our best to be alert to these issues, but we aren't in as good a position as the astronomers on whom Newton is relying to make the judgments; we will need to trust that the astronomers have not made any illegitimate approximations in the numbers they have come up with.

There are two sorts of approximations we will be making in our study.

One sort concerns real approximations that are being made by Newton and the astronomers. Most of the questions that will come up for us about the real approximations we will be encountering as we work through Book III can be answered by comparing the approximation to the precision of our measurements.

If, as close as we can measure, an orbit is circular, then we permit ourselves, tentatively, for the purpose of this computation, to treat it so. We may know theoretically that this substitution is an approximation, but to impose by theory added precision that cannot be measured doesn't help the precision of our overall result, and merely adds complexity to the computation.

When we get into the first propositions of Newton's Book III, we will see that, if orbits are close enough to circular that we can make the approximation, we will have the breakthrough to begin applying our mathematical propositions of the first books. Working from that breakthrough, we refine our theory and finally establish the exact orbits.

The other sort of approximation consists in simplifications of more complicated procedures that were actually used by the astronomers. These simplifications will be made in these notes to give the reader the general idea of how a calculation is made. They will get us close to the right numbers, and will give us a sense as to how these things are found, so that we are enabled to engage actively with the calculations of the Phenomenon, rather than passively accepting the numbers Newton cites.

- **"Phenomena."** In the first edition of *Principia* these "Phenomena" were called Hypotheses. Newton changed the word to Phenomena in the Second Edition.

Phenomenon 1

That the planets around Jupiter, by radii drawn to the center of Jupiter, describe areas proportional to the times, and that their periodic times (the fixed stars being at rest) are in the sesquiplicate ratio of the distances from Jupiter's center.

This is established from astronomical observations. The orbits of these planets do not differ perceptibly from circles concentric about Jupiter, and their motions in these circles are found to be uniform. But that the periodic times are in the sesquiplicate ratio of the semidiameters of the orbits, the astronomers are in agreement, and this same thing is manifest from the following table.

Periodic Times of Jupiter's Satellites

$1^d\,18^h\,27'\,34''$ $3^d\,13^h\,13'\,42''$ $7^d\,3^h\,42'\,36''$ $16^d\,16^h\,32'\,9''$

Distance of the Satellites from the Center of Jupiter

From the Observations	1	2	3	4	
of Borelli	$5\frac{2}{3}$	$8\frac{2}{3}$	14	$24\frac{2}{3}$	
of Townley, by micrometer	5.52	8.78	13.47	24.72	Semidiam.
of Cassini, by telescope	5	8	13	23	of Jupiter
of Cassini, by eclipse of satell.	$5\frac{2}{3}$	9	$14\frac{23}{60}$	$25\frac{3}{10}$	
From the Periodic Times	5.667	9.017	14.384	25.299	

Mr. Pound has determined the elongations of Jupiter's satellites and its diameter by means of the best micrometers as follows. The greatest heliocentric elongation of the fourth satellite from the center of Jupiter was taken by a micrometer in a tube fifteen feet long, and came out to be about $8'\,16''$ at Jupiter's mean distance from earth. That of the third satellite was taken by a micrometer in a telescope 123 feet long, and came out to be $4'\,42''$ at the same distance of Jupiter from earth. The greatest elongations of the remaining satellites at the same distance of Jupiter from earth came out to be $2'\,56''\,47'''$ and $1'\,51''\,6'''$, from the periodic times.

The diameter of Jupiter was taken more frequently by a micrometer in a telescope 123 feet long, and, reduced to Jupiter's mean distance from the sun or earth, always came out less than $40''$, never less than $38''$, and most often $39''$. In shorter telescopes this diameter is $40''$ or $41''$. For the light of Jupiter is appreciably spread by unequal refrangibility, and this spreading has a lesser ratio to Jupiter's diameter in longer and more perfect telescopes than in shorter and less perfect ones. The times in which two satellites, the first and third, passed across the body of Jupiter, from the beginning of ingress to the beginning of egress, and from the completion of ingress to the completion of egress, were observed with the help of the same longer telescope. And the diameter of Jupiter at its mean distance from earth came out to be $37\frac{1}{8}''$ by the transit of the first satellite, and $37\frac{3}{8}''$ by the transit of the third. Further, the time in which the shadow of the first satellite passed across the body of Jupiter was observed, and from this the diameter of Jupiter at its mean distance from earth came out to be about $37''$. Let us assume that its diameter is about $37\frac{1}{4}''$ very nearly, and the greatest elongations of the first, second, third, and fourth satellites will be equal to 5.965, 9.494, 15.141, and 26.63 semidiameters of Jupiter, respectively.

[The sesquiplicate ratio is the subduplicate of the triplicate. This is equivalent to taking the square root of the cube—that is, taking the $\frac{3}{2}$ power.]

Notes on Phenomenon 1

• **Note 1: "Heliocentric."** Although Newton refers to "heliocentric elongations," it is not until Phenomenon 3 that he argues from observations that the sun must lie within the planetary orbits. This is not a problem, as none of these earlier conclusions is used in that argument.

• **Note 2: Distance of Jupiter from the Sun.** We want to make our measurements of the elongations of the satellites, which will give us a measure of their distance from the center of Jupiter, at some constant viewing distance, because our distance from the Jupiter-moon pair will affect the measured elongation. The same distance between moon and Jupiter will yield a larger angle if we are closer.

This constant distance that we will use will be the heliocentric distance of Jupiter, that is, its distance measured from the sun. We already know from Ptolemy and Copernicus that Jupiter doesn't have a constant distance from the sun, but we will suppose in what follows that Jupiter is at its mean heliocentric distance. This is a reasonable approximation because its orbital eccentricity is small and its orbit for practical purposes is circular.

But wait! All our observations are measurements made from the earth. How can we say what the distance from the sun is, and why would we even use that as a standard? And what does this have to do with "mean distance from the earth"?

These questions will be addressed in the subsequent bulleted notes.

Note 3 gives a general method of determining the radius of an outer planet. This will be used here (in Note 4) to find the distance of Jupiter from the sun, and it will be used later in the proofs of Phenomena 2 and 4. Notes 5 and 6 address the issues raised in connection with the earth's mean distance.

• **Note 3: Lemmita: To find the radius of the orbit of an outer planet.**

Begin observations when the planet is in opposition with the earth (sun, earth, and Jupiter in line).

Now let Jupiter and earth move around in their respective orbits until Jupiter is 90° from the sun as viewed from the earth ($\theta = 90°$), noting the time taken for this shift.

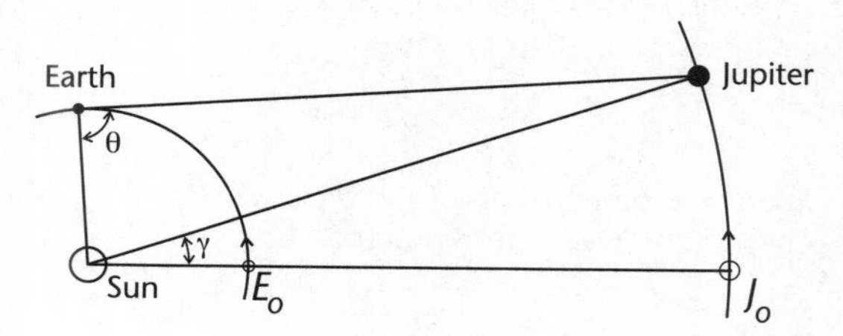

Call this elapsed time Δt.

The change in time from opposition to 90° will be to Jupiter's whole periodic time as the angle γ between Jupiter's original position J_o and Jupiter's position when $\theta = 90°$ is to the full 360°.

$$\gamma : 360° :: \Delta t : P_J.$$

Since the time interval Δt has been noted and Jupiter's full periodic time P_J is known, γ can be found.

Similarly, for the motion of the earth,

$$\delta : 360° :: \Delta t : P_E.$$

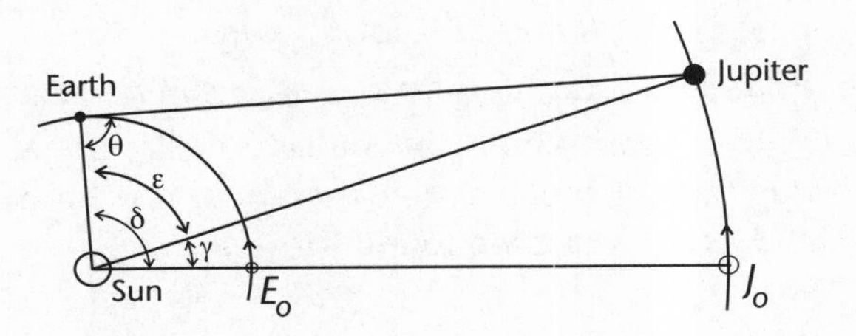

We have the same Δt (because it is the same elapsed time), and we know the periodic time of the earth (one year); therefore we can find δ.

But $\varepsilon = \delta - \gamma$.

Therefore ε can be found.

Let SE be the distance between the earth and the sun and let SJ be the distance between Jupiter and the sun.

Let us suppose the distance between the earth and the sun is 10 units.

$$\text{Cos } \varepsilon = \frac{SE}{SJ} = \frac{10}{SJ}.$$

$$SJ \cos\varepsilon = 10.$$

$$SJ = 10/\cos\varepsilon.$$

Since ε is known, $\cos\varepsilon$ can be found.

Therefore SJ can be found.

Thus we can find the distance of an outer planet from the sun relative to the distance of the earth from the sun. Q.E.D.

- **Note 4: Calculation of Jupiter's Distance from the Sun.** Now we must find what that heliocentric distance is in terms of things we can measure at the earth. What earth-viewable configuration will correspond to our standard viewing distance, and how will we know when we are there?

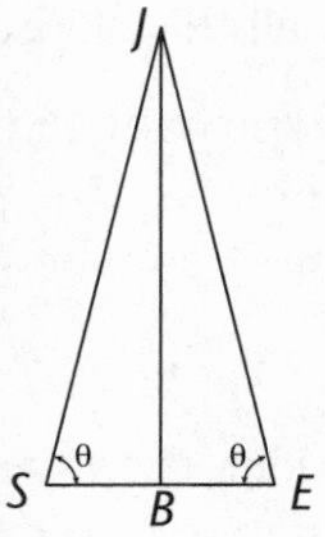

Where distances SJ and EJ are equal, the same angle θ will lie between earth and Jupiter as measured from the sun, and between the sun and Jupiter as measured from the earth. [Eu. I.5]

Furthermore, in that case, a perpendicular dropped from J to SE will bisect SE at B. [Eu. I.10]

$$SB/SJ = \cos\theta$$

If we suppose some distance for SE (see the note on astronomical units in Notes on Phenomena), SJ can be found by Kepler's planetary theories (see lemmita in Note 3 to Phenomenon 1). SB is half our supposed value for SE. Thus we can suppose both SB and SJ to be known.

Therefore $\cos\theta$ can be found; and from that, θ can be found.

So we will take the measurement of the satellites' elongations and the measurement of Jupiter's diameter when the angle on the earth between Jupiter and the sun has the value we found for θ.

• **Note 5: Earth's Mean Distance from Jupiter.** We said that by observation it is known that Jupiter's orbit is nearly circular and only slightly eccentric. Suppose both it and the earth's orbit were circles concentric around the sun.

Then we can see that the earth's mean distance from Jupiter—the distance halfway between its closest distance, when it is just between the sun and Jupiter, and its farthest distance, when it is exactly on the other side of the sun from Jupiter—is the heliocentric distance of Jupiter.

If both orbits were perfectly circular and concentric, then our mean distance would be exactly the heliocentric distance. In fact, they are a bit off, but we can approximate Jupiter's heliocentric distance by using the earth's mean distance.

As we see in Newton's sketch here, "heliocentric distance" and "earth's mean distance" are used interchangeably. The two aren't identical, and we can assume that Pound and Townley would have used one measure consistently.

• **Note 6: Observations from Other Than Earth's Mean Distance.** How would we correct an observation made elsewhere than at our standard heliocentric distance or its close approximation, the earth's mean distance? Or must we always wait till we get around to our mean distance to take our measurements?

Suppose we measure our greatest elongation of Jupiter's moon from Jupiter at point E on our orbit, where we are not at Jupiter's heliocentric distance SJ. That is, $EJ \neq SJ$.

Suppose angle JEM to be our measurement of greatest elongation. Continue line ME to E' such that $E'J$ is the distance of Jupiter from the sun SJ.

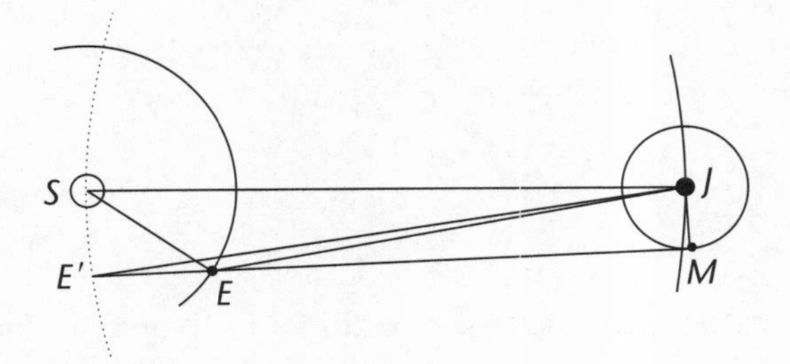

The line EM will be tangent to the orbit because any line cutting the orbit will be closer to Jupiter and thus not the greatest elongation. If the orbits of the moons are circular (this will be proved in the body of Phenomenon 1), we may use Euclid III.18 to show that angle EMJ will be a right angle.

$$\text{Sin}\, JEM = \frac{JM}{JE} \quad \text{or} \quad JM = JE \sin JEM.$$

$$\text{Sin}\, JE'M = \frac{JM}{JE'} \quad \text{or} \quad JM = JE' \sin JE'M.$$

Therefore $JE \sin JEM = JE' \sin JE'M$.

$$\text{Sin}\, JE'M = \frac{JE}{JE'} \sin JEM.$$

Since $JE' = SJ$,

$$\text{Sin}\, JE'M = \frac{JE}{SJ} \sin JEM. \qquad (1)$$

Since Kepler, astronomers have plotted actual orbits for the earth and planets, and so we can know for a given date where the earth is on its orbit and where Jupiter is on its orbit. So we can know the relative distances EJ and SJ for the date of our observation of the greatest elongation.

Therefore we know the ratio $EJ : SJ$ in Equation 1.

Angle JEM is our measured angle and therefore we know its sine in Equation 1.

Therefore we know the sine of the angle $JE'M$, the remaining unknown in Equation 1.

Therefore we can find the angle $JE'M$.

Since $E'M$ is the tangent, angle $JE'M$ will be the greatest elongation viewed from E' as well. But $E'J = SJ$, the heliocentric distance of Jupiter.

Thus we have a maximum elongation at the heliocentric distance, calculated from an observation of maximum elongation at a different distance.

Q.E.D.

Expansion of Newton's Sketch of Phenomenon 1

"That the planets around Jupiter, by radii drawn to the center of Jupiter, describe areas proportional to the times, and that their periodic times (the fixed stars being at rest) are in the sesquiplicate ratio of the distances from Jupiter's center."

To Prove:

1. Jupiter's moons describe areas proportional to their times around Jupiter as the geometric center.

2. The periodic times of the moons are as the $\frac{3}{2}$ powers of their distances from Jupiter.

Proof:

Part 1: Jupiter's moons describe areas proportional to their times around Jupiter.

"This is established from astronomical observations. The orbits of these planets do not differ perceptibly from circles concentric about Jupiter, and their motions in these circles are found to be uniform."

Newton notes that:

(a) Observations show orbits do not differ perceptibly from circles concentric to Jupiter.

(b) Observations show motions in those circles uniform.

By (a) and (b) we would have uniform circular motion, in which bodies describe areas proportional to the times. But how would we observe these things?

A: Circular Concentric Motion

How do we observe that the orbits of Jupiter's moons do not differ perceptibly from circles concentric to Jupiter?

After all, we don't see circles at all from the earth. All we see of Jupiter's moons is an oscillation back and forth from one side of Jupiter to the other. Furthermore, those oscillations look irregular, as we watch them from our various vantage points around our orbit.

All observations must be corrected to those points when (at various orbital configurations) the earth is at the same distance from Jupiter. Note that at a given moment we get one observation; that elongation may not be the maximum. To find the greatest elongation requires a series of observations over time, and during that time the earth has moved. To correct for this, use Note 6.

But when the necessary corrections are made, we find that the maximum elongations of the moons' orbits are always the same. That is, the moon always moves just as far to one side of Jupiter before it starts going back (its "maximum elongation") as it moves to the other side before starting back.

Further, if we correct for our different distances from Jupiter (which Note 6 shows us how to do) we can take measurements from all points around our orbit. The fact that we find the same maximum elongations to either side from all these angles strongly suggests that we are dealing with circular concentric orbits.

If the orbits were not circular, or not concentric, we would expect to see different maximum elongations at some viewing angles or at some times. That is, we would expect to see the moon move farther to one side of Jupiter than to the other side. But it always seems to move exactly the same distance to one side as to the other.

B: Uniform Motion

One way to find whether the motion is uniform would be to see how long the moon takes to get from Jupiter out to maximum elongation on one side, and then on the other, at various vantage points as the earth moves around its orbit. Should all the times be found to be the same, this would strongly suggest that the motion is uniform. If the moons were slowing down or speeding up at some parts of the orbit, we would expect to see that at least from some vantage points.

Of course, as usual, timing this is not simple: measuring the time to maximum elongation means knowing just when the moon has finished going out and is starting to come back—and it appears to be moving very slowly in its oscillation at that point. Newton is working with more accuracy than we would probably get just by watching and guessing when one turnaround happened and then doing the same on the other side.

What astronomers actually did was use eclipses of the moons by Jupiter's shadow. They predicted times of the eclipses on the assumption of regular circular motion. Any deviation from regular circular motion would throw off the predictions at some point as Jupiter moves around its orbit about the sun. As it turned out, the observations were consistently in accord with the predictions.

In making these predictions they had to make allowance for the fact that both the earth and Jupiter were moving, and that light takes time to travel from the moons of Jupiter. The way the timing is actually done is more complicated than we want to get into here: let's just appreciate the painstaking work of these astronomers and acknowledge the tremendous amount of theory that goes into "measurements."

Q.E.D. for Part 1.

Part 2: The periodic times of the moons are as the $\frac{3}{2}$ powers of their distances from Jupiter.

"and that their periodic times (the fixed stars being at rest) are in the sesquiplicate ratio of the distances from Jupiter's center."

Strategy

We will take the periodic time of each of the four moons, determined by observation. For convenience, we will convert all the times to hours.

Then we will compute each satellite's distance from the center of Jupiter measured in semidiameters of Jupiter. Newton first gets elongations for the orbits of the moons. Then he gets an angular magnitude for the diameter of Jupiter, and expresses the radii of the satellite orbits in terms of jovian semidiameters.

These measurements of course must be at the same distance. Our common distance here is the distance of Jupiter and its moons from the sun, which is sometimes referred to by the distance that approximates it, the mean distance of the earth from Jupiter. We must show how we get the heliocentric distance, why it is approximated by the earth's mean distance, and how to correct for that distance if we don't happen to be at it when we take our measurements. Notes 3 through 6 give us the tools for showing those things.

We then may tabulate the relative distances of the moons from Jupiter and compare those numbers to the periodic times, to see whether the relationship Newton claims is present.

Because we are looking for a proportionality, we will check this by seeing whether the ratio between the periodic time and the $\frac{3}{2}$ power of the distance from Jupiter remains consistent for the different moons, that is, whether the one divided by the other yields a consistent number, a constant of proportionality.

"But that the periodic times are in the sesquiplicate ratio of the semidiameters of the orbits, the astronomers are in agreement, and this same thing is manifest from the following table.

Periodic Times of Jupiter's Satellites

$1^d\ 18^h\ 27'\ 34''$ $3^d\ 13^h\ 13'\ 42''$ $7^d\ 3^h\ 42'\ 36''$ $16^d\ 16^h\ 32'\ 9''$ "

To make it easier to calculate with them, let's convert these periodic times to hours.

Periodic Times of the Moons by Observation, in Hours

Moon	*Period from Newton*	*Period in hours*
1	$1^d\,18^h\,27'\,34''$	42.4594
2	$3^d\,13^h\,13'\,42''$	85.2283
3	$7^d\,3^h\,42'\,36''$	171.7100
4	$16^d\,16^h\,32'\,9''$	400.5358

Now we proceed to the computation of the distance of Jupiter's satellites in terms of semidiameters of Jupiter.

"Distance of the Satellites from the Center of Jupiter

From the Observations	1	2	3	4	
of Borelli	$5\frac{2}{3}$	$8\frac{2}{3}$	14	$24\frac{2}{3}$	
of Townley, by micrometer	5.52	8.78	13.47	24.72	Semidiam.
of Cassini, by telescope	5	8	13	23	of Jupiter
of Cassini, by eclipse of satell.	$5\frac{2}{3}$	9	$14\frac{23}{60}$	$25\frac{3}{10}$	
From the Periodic Times	5.667	9.017	14.384	25.299	

Mr. Pound has determined the elongations of Jupiter's satellites and its diameter by means of the best micrometers as follows. The greatest heliocentric elongation of the fourth satellite from the center of Jupiter was taken by a micrometer in a tube fifteen feet long, and came out to be about $8'\,16''$ at Jupiter's mean distance from earth."

Measured from the earth when the earth was at Jupiter's heliocentric distance (or corrected for that distance), Mr. Pound found that the angle between Jupiter and its fourth moon came out to about $8'\,16''$.

"That of the third satellite was taken by a micrometer in a telescope 123 feet long, and came out to be $4'\,42''$ at the same distance of Jupiter from earth. The greatest elongations of the remaining satellites at the same distance of Jupiter from earth came out to be $2'\,56''\,47'''$ and $1'\,51''\,6'''$, from the periodic times."

The elongation of the third satellite was similarly measured. Mr. Pound may have done the other two in the same manner; Borelli's, Townley's, and Cassini's distances in the table are said to be based on observations.

But the values Newton gives here are presented not from observation but working backwards from the periodic times. Since this depends on what we're trying to establish, we'll pass over these and use the values actually observed for the fourth and third satellite to check the values that appear in Newton's table.

But before we can make the next step in our calculation, we must find a value for the diameter of Jupiter, so we can put the elongations we found for the moons in relation to it.

"The diameter of Jupiter was taken more frequently by a micrometer in a telescope 123 feet long, and, reduced to Jupiter's mean distance from the sun or earth, always came out less than 40″, never less than 38″, and most often 39″. In shorter telescopes this diameter is 40″ or 41″. For the light of Jupiter is appreciably spread by unequal refrangibility, and this spreading has a lesser ratio to Jupiter's diameter in longer and more perfect telescopes than in shorter and less perfect ones. The times in which two satellites, the first and third, passed across the body of Jupiter, from the beginning of ingress to the beginning of egress, and from the completion of ingress to the completion of egress, were observed with the help of the same longer telescope. And the diameter of Jupiter at its mean distance from earth came out to be $37\frac{1}{8}''$ by the transit of the first satellite, and $37\frac{3}{8}''$ by the transit of the third. Further, the time in which the shadow of the first satellite passed across the body of Jupiter was observed, and from this the diameter of Jupiter at its mean distance from earth came out to be about 37″. Let us assume that its diameter is about $37\frac{1}{4}''$ very nearly..."

Jupiter's diameter is measured directly by micrometer (always at the mean distance or corrected for that distance) and also by timing the passage of satellites in front of Jupiter at the same distance.

The satellite's transit is measured from the time it begins to pass in front of the disk of Jupiter ("beginning of ingress") to the time it begins to emerge ("beginning of egress").

At a constant distance from the earth, and for small angles, which we have here, the time of transit will be to the total time of the moon's orbit, as the distance it traveled across Jupiter (Jupiter's diameter) is to the distance of the whole orbit. The radius of the whole orbit and the whole periodic time is known by observation, and so the distance traveled in transit can be found. This yields a value for the diameter of Jupiter.

Taking the more accurate observations from the transit method with the longest telescopes, Newton settles on $37\frac{1}{4}''$ for the diameter of Jupiter.

Now we're in a position to put this elongation measurement together with the elongation measurements for the radii of the satellite orbits.

Because we want the radii in jovian semidiameters, we first halve the figure for the full diameter, getting 18.625″.

By Mr. Pound's observations,

fourth satellite: $8'16''/18.625'' = 26.63$ semidiameters

third satellite: $4'42''/18.625'' = 15.14$ semidiameters

"...and the greatest elongations of the first, second, third, and fourth satellites will be equal to 5.965, 9.494, 15.141, and 26.63 semidiameters of Jupiter, respectively."

The numbers Newton cites here for the third and fourth satellites are exactly the ones we got.

These numbers are similar to (although a bit higher than) the numbers based on the observations of the other astronomers Newton puts in his table:

"Distance of the Satellites from the Center of Jupiter

From the Observations	1	2	3	4	
of Borelli	$5\frac{2}{3}$	$8\frac{2}{3}$	14	$24\frac{2}{3}$	
of Townley, by micrometer	5.52	8.78	13.47	24.72	Semidiam.
of Cassini, by telescope	5	8	13	23	of Jupiter
of Cassini, by eclipse of satell.	$5\frac{2}{3}$	9	$14\frac{23}{60}$	$25\frac{3}{10}$	
From the Periodic Times	5.667	9.017	14.384	25.299 "	

The numbers on the last line, "From the Periodic Times," in contrast to the ones above, "From the Observations," are based on assuming the relationship stated in the Phenomenon and working backwards. Let's see how Newton has done this.

We began by converting the periodic times of each moon's orbit to hours; we'll use those numbers here. Let's assume that the second assertion of the phenomenon, that the periodic times are in the sesquiplicate ratio of the distances from Jupiter's center, holds.

This means that, for any pair of moons,

$$P_1 : P_2 :: D_1^{3/2} : D_2^{3/2}.$$

Or, since these are number ratios,

$$\frac{P_1}{P_2} = \frac{D_1^{3/2}}{D_2^{3/2}}$$; from which,

$$\frac{P_1}{D_1^{3/2}} = \frac{P_2}{D_2^{3/2}}.$$

Remember that we're following Newton here, seeing how he got his last line in the table, "From the Periodic Times." His way of checking the assertion of this Phenomenon, as presented here, is to calculate distances using the periodic times and the conclusion of the assertion, and to check those distances against the ones found through observation. His distances in the last row are checked against the ones in the previous rows, "From the Observations."

So now instead of taking the $\frac{3}{2}$ power of the distance, let's take the $\frac{2}{3}$ power of the periodic times.

We check whether:

$$\frac{P_1^{2/3}}{D_1} \stackrel{?}{=} \frac{P_2^{2/3}}{D_2} \stackrel{?}{=} \text{constant.}$$

We must start with one particular observed distance, since we must have a starting ratio with which to put the others into same ratio.

Let's then take the distance for the first moon. This seems to be the one Newton uses, because his value "from the periodic times" for the first moon seems to be simply picking up the observation value that appears twice in the table, $5\frac{2}{3}$. The other numbers in the last row seem to be results of computations.

Then we take the $\frac{2}{3}$ power of the periodic time for the first moon, divide it by the distance and get the constant of proportionality. We then multiply that times the $\frac{2}{3}$ power of each of the other periodic times to get values for the remaining distances. We then compare them to the values for that moon found by observation.

Periodic times of the moons by observation were:

1	$1^d\,18^h\,27'\,34''$	42.4594 hours
2	$3^d\,13^h\,13'\,42''$	85.2283 hours
3	$7^d\,3^h\,42'\,36''$	171.7100 hours
4	$16^d\,16^h\,32'\,9''$	400.5358 hours

Taking the $\frac{2}{3}$ power of each:

1 12.1707

2 19.3667

3 30.8933

4 54.3368

Dividing the first by the distance from observation of 5.667 semidiameters to get the constant of proportionality:

$$12.1707 \div 5.667 = 2.1476$$

The distance should equal the $\frac{2}{3}$ power of the periodic time divided by the constant of proportionality.

$$D_2 \stackrel{?}{=} P_2^{2/3} \div 2.1476 = 19.3667 \div 2.1476 = 9.0176$$

$$D_3 \stackrel{?}{=} P_3^{2/3} \div 2.1476 = 30.8933 \div 2.1476 = 14.3847$$

$$D_4 \stackrel{?}{=} P_4^{2/3} \div 2.1476 = 54.3368 \div 2.1476 = 25.3007$$

These numbers are very close to the ones listed in the table:

From the Observations	1	2	3	4	
of Borelli	$5\frac{2}{3}$	$8\frac{2}{3}$	14	$24\frac{2}{3}$	
of Townley, by micrometer	5.52	8.78	13.47	24.72	Semidiam.
of Cassini, by telescope	5	8	13	23	of Jupiter
of Cassini, by eclipse of satell.	$5\frac{2}{3}$	9	$14\frac{23}{60}$	$25\frac{3}{10}$	
From the Periodic Times	5.667	9.017	14.384	25.299	

This confirms the relationship asserted.

Q.E.D. for Part 2.

Note:

If you are suspicious that Newton's procedure might be circular, and want to check the statement of the Phenomenon in a more straightforward way, just take the $\frac{3}{2}$ power of each distance in the table and see if the ratio of any two distances is the same as the ratio of the corresponding periodic times.

For example, taking the Townley numbers (which may be the most accurate since he was using the micrometer):

Moon	Distance	$\frac{3}{2}$ Power of Distances	Periodic Times
1.	5.52	12.97	42.4594 hours
2.	8.78	26.02	85.2283 hours
3.	13.47	49.44	171.7100 hours
4.	24.72	122.91	400.5358 hours

Check whether:

12.97 ÷ 26.02 = 42.4594 hours ÷ 85.2283 hours

$0.4985 \stackrel{?}{=} 0.4982$

Yes, the ratio holds very well for the first two. Go ahead on your own to check the rest. Some of the others don't match as exactly as this one, but they are as close to Newton's in "From the Periodic Times" in his table as Newton's are to Townley's measurements.

Phenomenon 2

That the planets around Saturn, by radii drawn to Saturn, describe areas proportional to the times, and that their periodic times, the fixed stars being at rest, are in the sesquiplicate ratio of the distances from Saturn's center.

Cassini expressly stated their distances from the center of Saturn and their periodic times, from his observations, as follows.

Periodic Times of the Saturnian Satellites

$1^d 21^h 18' 27''$ $2^d 17^h 41' 22''$ $4^d 12^h 25' 12''$ $15^d 22^h 41' 14''$ $79^d 7^h 48' 0''$

Distances of the Satellites from the Center of Saturn in Semidiameters of the Ring

From the Observations	$1\frac{19}{20}$	$2\frac{1}{2}$	$3\frac{1}{2}$	8	24
From the Periodic Times	1.93	2.47	3.45	8	23.35

[In the third edition Newton added a paragraph commenting on the numbers in the table, based on some observations made with a longer telescope, and suggesting more accurate values. Since this has no bearing on the assertion of Phenomenon 2, or the use he makes of it in this book, it has been omitted here.]

Expansion of Newton's Sketch of Phenomenon 2

To Prove:

1. Saturn's moons describe areas proportional to their times.

2. The periodic times of the moons are as the $\frac{3}{2}$ powers of their distances from Saturn.

Proof:

Part 1: Saturn's moons describe areas proportional to their times.

Because Newton says nothing by way of evidence for the first assertion of Phenomenon 2, we must assume that he means the proof to be identical to that for Jupiter's moons; and that the same evidence is to be brought forward to support the proof. This requires that we prove that the moons of Saturn move in circular orbits.

As it turns out, the orbits of Saturn's satellites are indeed highly circular. The five mentioned by Newton differ only slightly, if at all, from circular orbits. Titan, which diverges the most, has an eccentricity of only 0.029 by modern measurements. Titan is Newton's "fourth satellite from Saturn's center." Tethys, Newton's "first satellite," has no observable eccentricity even by modern measurement.

From this we can conclude that observations at Newton's time would have found maximum elongations to either side of Saturn consistent when viewed from all configurations (when corrected for the varying distance of the earth).

Observations were presumably made (and necessary calculations done) for the time the satellites took for equal portions of each moon's orbit and these times were found to be constant.

Part 2: The periodic times of the moons are as the $\frac{3}{2}$ powers of their distances from Saturn.

Strategy

The approach is the same as that used for Phenomenon 1.

We will take the periodic time of each of the five moons, determined by observation. For convenience, we will convert all the times to hours.

Then we will compute each satellite's distance from the center of Saturn measured in semidiameters of Saturn's ring.

These measurements of course must be at the same distance or corrected for the same distance. Notes 3 through 6 of Phenomenon 1 give us the tools for showing those things.

We then may tabulate the relative distances of the moons from Saturn and compare those numbers to the periodic times, to see whether the relationship Newton claims is present.

Newton's method as presented here is to take one distance, apparently Titan's, assume the relationship predicted by the Phenomenon, calculate the constant of proportionality, and then use this constant of proportionality to compute the other distances from the periodic times based on the asserted relationship. (The asserted relationship is that the periodic times vary as the $\frac{3}{2}$ power of the distances.)

Newton computes the distances from the periodic times according to this formula and then compares those computed distances to the distances calculated from observations.

Calculations

First we convert the periodic times to hours.

Periodic Times of the Moons by Observation, in Hours

Moon	D.h.m.s	Decimal	
1	$1^d 21^h 18' 27''$	45.3075	hours
2	$2^d 17^h 41' 22''$	65.6894	hours
3	$4^d 12^h 25' 12''$	108.42	hours
4	$15^d 22^h 41' 14''$	382.6872	hours
5	$79^d 7^h 48' 00''$	1903.8	hours

Now we compute the satellites' distance.

As we did with Phenomenon 1 for Jupiter's moons, we must first get values for the elongations of the satellites from Saturn's center, then get an angular value for the semidiameter of Saturn's ring, and place the former in units of the latter. Here Newton only gives us the results of the last step.

"Distances of the Satellites from the Center of Saturn
in Semidiameters of the Ring

Moon	1	2	3	4	5
From the Observations	$1\frac{19}{20}$	$2\frac{1}{2}$	$3\frac{1}{2}$	8	24
From the Periodic Times	1.93	2.47	3.45	8	23.35 "

Now let's take the $\frac{3}{2}$ powers of the distances in the table that are listed as being from observations:

Moon	Distances	$\frac{3}{2}$ Power of Distances
1	$1\frac{19}{20}$	2.72
2	$2\frac{1}{2}$	3.95
3	$3\frac{1}{2}$	6.55
4	8	22.6
5	24	118

The most straightforward check of the assertion of Phenomenon 2 can be done now, and you might want to do it. That is to take the ratio of $\frac{3}{2}$ powers of two distances and the ratio of the corresponding periodic times and see if they are in same ratio, repeating this through the table.

Newton's way, however, works backwards. He takes the periodic time of 382.6872 for Titan (the fourth satellite in his table) and divides it by the $\frac{3}{2}$ power of Titan's observed distance of 8 semidiameters of Saturn's ring. This yields a constant of proportionality that he uses to calculate distances for the other four moons. These are the numbers in the second line of his table.

We may then compare these to the observed distances for these other four moons. Looking at his numbers, we see a close correlation.

Now let's check whether we get his numbers if we follow this procedure.

$$382.6872 \div 8^{3/2} = 16.91$$

Our constant of proportionality k is found to be 16.91.

Now our formula will be:

$(\text{periodic time}/k)^{2/3} = \text{distance}$

1st Moon = 1.93

2nd Moon = 2.47

3rd Moon = 3.45

5th Moon = 23.32

Except for the hundredths place in the distance for the fifth moon, we get his numbers exactly.

Phenomenon 3

The five primary planets, Mercury, Venus, Mars, Jupiter, and Saturn, enclose the sun in their orbits.

That Mercury and Venus revolve around the sun is demonstrated from their lunar phases. When they shine with a full face, they are located beyond the sun, when halved they are even with the sun, and when sickle-shaped they are this side of the sun, sometimes passing across its disk in the manner of sunspots. Further, from Mars's full face near conjunction with the sun, and its gibbous face in the quadratures, it is certain that it encompasses the sun. The same is also demonstrated of Jupiter and Saturn from their ever full phases, for it is manifest from the shadows of satellites cast upon them that these shine with light borrowed from the sun.

Notes on Phenomenon 3

- The assertion of this Phenomenon does not include a claim that the earth moves around the sun. The sun, with its five encompassing planets, could revolve around the earth. This was the theory of the Danish astronomer Tycho Brahe (1546–1601).

- In order to think through the argument of this Phenomenon, we must understand how a planet would look given different angles between us observing from earth, the sun providing the light, and the planet.

Lemmita: How the Planet Looks

The following assumes that the light we see from the planets is reflected light of the sun. If we believed that the light might be coming from the planet, all this would change. In that case, however, it would be hard to make sense of the phase observations for Mercury, Venus and Mars. And Newton ends this Phenomenon with a demonstration that the light we see from Jupiter and Saturn is borrowed from the sun.

Given: Sun S, earth E, and planet P.

To Prove:

1. If angle EPS is right, the planet will show as half phase.
2. If angle EPS is acute, the planet will show as gibbous.
3. As angle $EPS \to 0$, phase $\to$ full.
4. If angle EPS is obtuse, the planet will show as crescent.

Proof:

We will consider possibilities for any configuration of sun, earth, and planet.

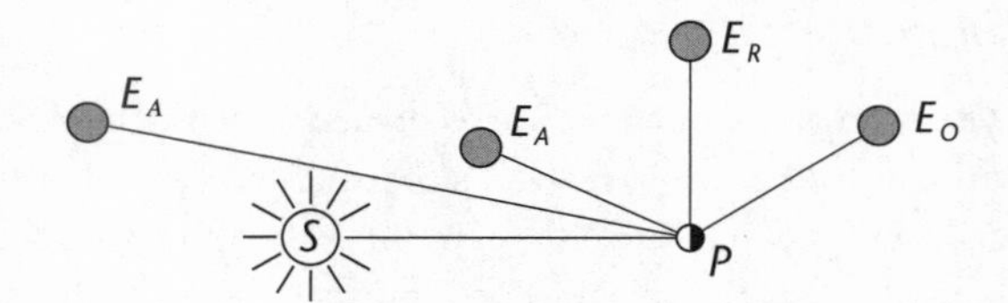

We may have the earth at points E_A where the angle EPS is acute, either with the earth farther from the sun than the planet, or nearer.

Or, we may have the earth at a point such as E_R, at whatever distance, such that angle EPS is a right angle.

Finally, we may have the earth at a point such as E_O, at whatever distance, such that angle EPS is obtuse.

Case 1: First Consider Right Angle EPS

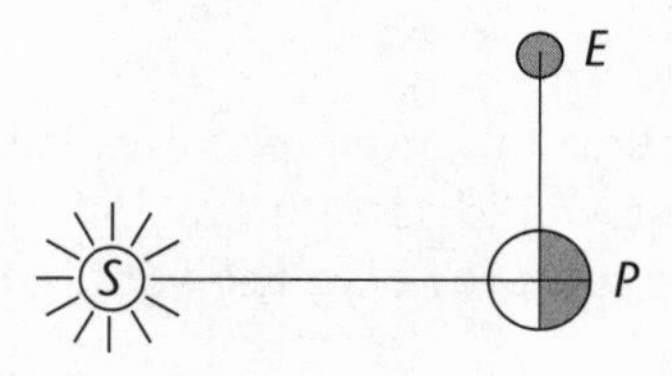

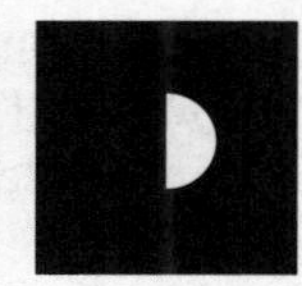

What do we see? (We must imagine this in three dimensions, and shift from the diagram's view from above the three bodies to a view from the surface of the earth.)

We would see half phase.

Case 2: Now Consider Acute Angle *EPS*

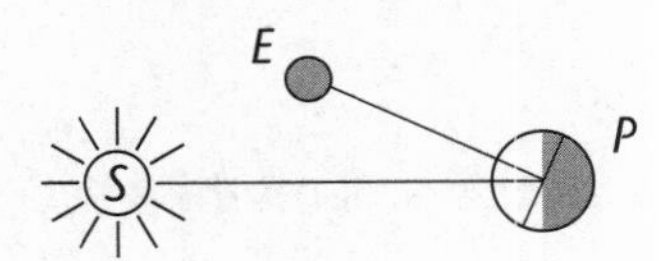

Adding the third dimension and shifting 90° to look at this from our earth's eye view, what do we see?

We would see gibbous.

Case 3: What Happens when Acute Angle is Nearly 0°?

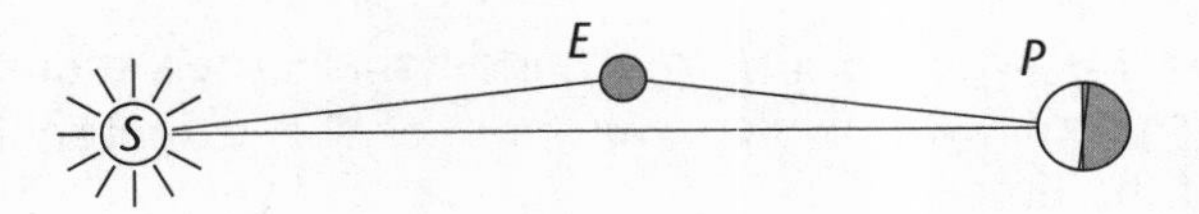

What would we see from the earth?

We would see the planet as nearly full.

Case 4: Now Consider an Obtuse Angle *EPS*

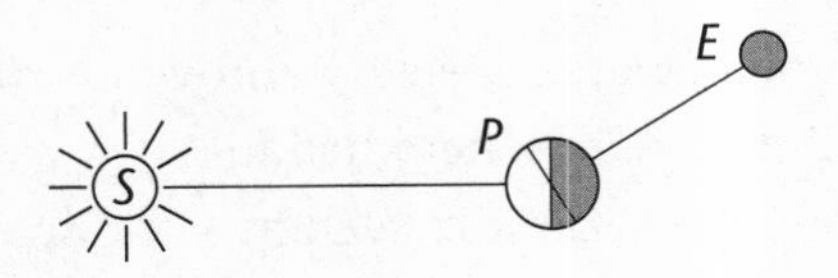

Shifting 90° into the paper to imagine this viewed from the earth, what would we see?

We would see a crescent.

Q.E.D.

Expansion of Newton's Sketch of Phenomenon 3

To Prove:

1. That Mercury and Venus revolve around the sun.
2. That Mars encompasses the sun.
3. That the same is demonstrated for Jupiter and Saturn.

Proof:

Part 1: Mercury and Venus

"That Mercury and Venus revolve around the sun is demonstrated from their lunar phases. When they shine with a full face, they are located beyond the sun, when halved they are even with the sun, and when sickle-shaped they are this side of the sun, sometimes passing across its disk in the manner of sunspots."

We must take as given the following additional observational conclusion Newton doesn't cite, but which was part of the common astronomical knowledge of the time:

General Observation:

The maximum elongation of Venus and Mercury is a small acute angle (48° for Venus, even less for Mercury). The elongation is the angle between the sun and the planet as we view it from the earth, angle *SEP* in these diagrams.

Thus we know for these two planets that angle *SEP* is always acute.

We also may consider the following particular observations, which Newton does cite:

Particular Observations:

Observation 1: We sometimes see Mercury and Venus near to full.
Observation 2: Sometimes we see them half full.
Observation 3: At other times they are seen as sickle-shaped (crescent).
Observation 4: Sometimes the planets are seen like spots traversing the sun's disk.

Step 1:

"That Mercury and Venus revolve around the sun is demonstrated from their lunar phases. When they shine with a full face, they are located beyond the sun..."

Now suppose by Observation 1 we find the planet near to full.

What must the configuration be then?

It could be close to full in this configuration:

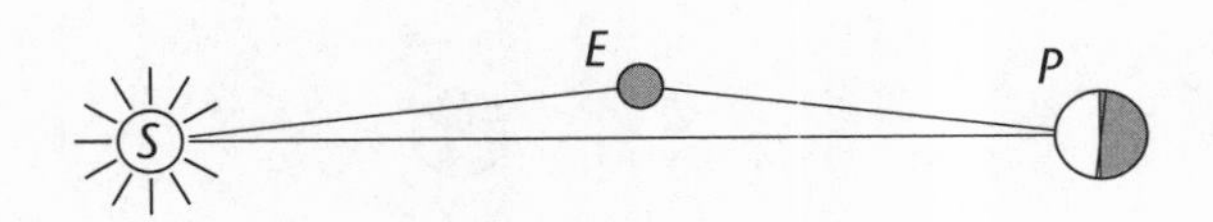

By the lemmita entitled How the Planet Looks, in the notes above, if phase is approaching full, angle *EPS* is approaching 0, as it does in this configuration.

But in this configuration angle *SEP* would not be small acute, contrary to the General Observation above.

Therefore this isn't the configuration.

It would also be near to full in this configuration:

Here angle $EPS \to 0$ as required by the lemmita. But here angle SEP also $\to 0$.

Therefore, when we see Mercury or Venus near to full, the planet is beyond the sun.

Q.E.D. for Step 1 of Part 1.

Step 2

"...when halved they are even with the sun, ..."

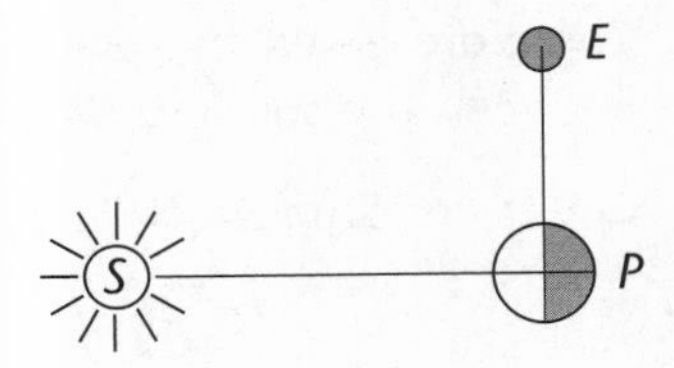

By Observation 2, sometimes they are half full.

What configuration would give us this?

By the lemmita, angle *EPS* = a right angle.

The diagram below illustrates the two possible configurations, which are mirror images of one another. In both cases angle *SEP* is acute, consistent with the General Observation above.

Therefore, from the point of view of the earth, when the planet is half full, it is "even with the sun," to one side or the other, as shown:

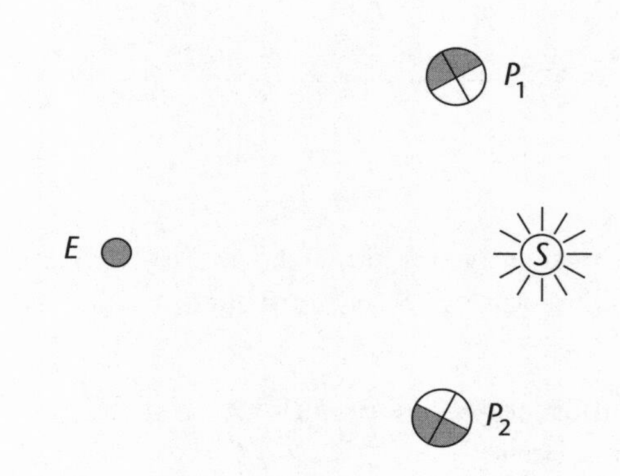

Q.E.D. for Step 2 of Part 1.

Step 3a

"...and when sickle-shaped they are this side of the sun,..."

By the lemmita, the planet would appear sickle-shaped or crescent when angle *EPS* is obtuse.

This can only happen in this configuration:

Here angle *SEP* is still acute, so it fits the General Observation.

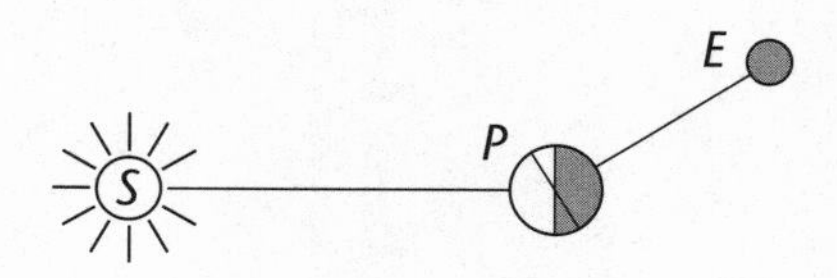

Therefore when the planet is seen as crescent, it is on our side of the sun, between the sun and earth.

Q.E.D. for Step 3a of Part 1.

Step 3b

"...sometimes passing across its disk in the manner of sunspots."

By Observation 4, the planets sometimes are seen like spots traversing the sun's disk.

In this case again they must be passing between the sun and earth.

Q.E.D. for Step 3b of Part 1.

Conclusion of Part 1

Thus we have seen that Venus and Mercury are sometimes between us and the sun (Steps 3a and 3b), sometimes at the sides of the sun (Step 2) and sometimes beyond the sun (Step 1).

From this we conclude that Mercury and Venus revolve around the sun.

Q.E.D.

Note:

The alert reader has already observed that the above arguments do not *prove* that the planets revolve around the sun. The set of configurations could be equally explained by the sun and Mercury standing still while the earth moves around both, or by some other odd arrangement, such as Mercury moving in a small circle at some distance from the sun and the earth moving around the sun and the orbit of Mercury.

However, Newton is not trying to disprove every conceivable hypothesis including ones no astronomer would think of putting forward. Astronomers had proposed that either the sun or the earth was the center of the planetary system. In the *Dialogue on the Two Chief World Systems,* Galileo argued from the phenomena that "the center of the celestial rotation for the five planets, Saturn, Jupiter, Mars, Venus, and Mercury, is the sun."* What Newton is doing, here for the inner planets and in the following section for the outer planets, is formalizing what Galileo had argued earlier.

* Stillman Drake translation, University of California Press. The quotation is on page 322. The discussion in the dialogue is in The Third Day on pages 321–326.}

Part 2: Mars Encompasses the Sun

"Further, from Mars's full face near conjunction with the sun, and its gibbous face in the quadratures, it is certain that it encompasses the sun."

Step 1

"... full face near conjunction with the sun, ..."

We observe that Mars is near full when it is near conjunction, that is, when it is on the same side of the earth as the sun. What configurations would put it on the same side?

Here is one configuration in which Mars and the sun are on the same side of the earth:

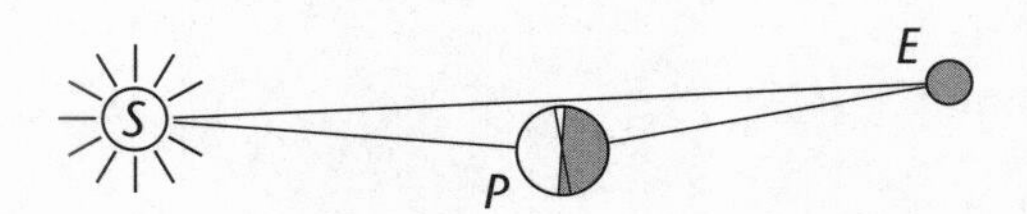

This is conjunction, but it wouldn't be full here, but "new." Angle *EPS* is obtuse, and by the lemmita, the planet would be crescent.

In the configuration shown below, Mars and the sun are also on the same side.

Angle *EPS* is acute approaching 0°; therefore Mars's phase is approaching full.

Therefore it must be this configuration; Mars is *beyond* the sun.

Q.E.D. for Step 1 of Part 2.

Step 2

"... and its gibbous face in the quadratures, ..."

We observe also that when Mars is at quadrature, that is, when angle *SEP* = 90°, it is gibbous. By the lemmita, angle *EPS* would be acute.

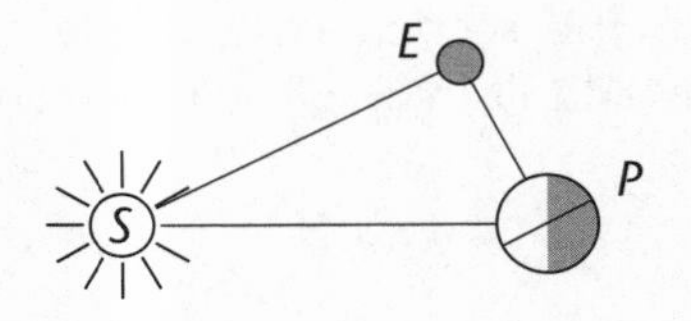

From its quadrature configuration we know that distance *SP* opposite the right angle is the greatest side of the triangle and Mars is farther from the sun than the earth is from the sun.

Because Mars comes to quadrature, it is (at least at that point) farther from the sun than the earth is.

Q.E.D. for Step 2 of Part 2.

Conclusion of Part 2

"…it is certain that it encompasses the sun."

Thus we see that Mars is sometimes to the side of us (Step 2) and sometimes on the other side of the sun from us.

Therefore Mars's orbit encompasses the sun.

Q.E.D. for Part 2.

(See the note at the end of Part 1.)

Part 3: Jupiter and Saturn

"The same is also demonstrated of Jupiter and Saturn from their ever full phases, …"

Our observation here is that they are full in all situations.

Consider conjunction, quadrature, and opposition.

Step 1: Conjunction

At conjunction Jupiter sets with the sun. There are two cases:

In the first case, angle *EPS* is small acute. By the lemmita, Jupiter could look full.

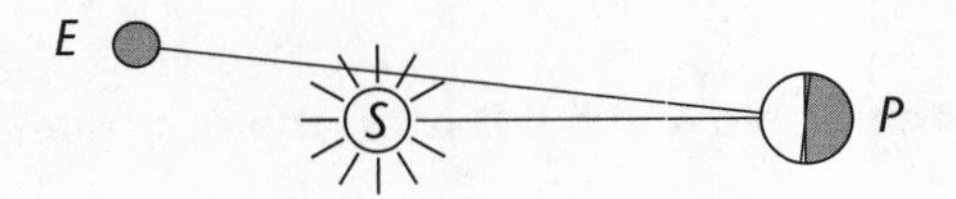

In the second case, angle EPS is obtuse. By the lemmita above, Jupiter wouldn't look full but crescent.

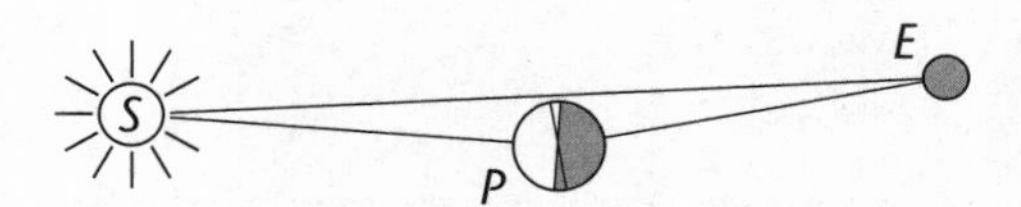

Thus if it looks full at conjunction, it must be in the first configuration.

Therefore Jupiter must be beyond the sun at conjunction.

Q.E.D. for Step 1 of Part 3.

Step 2: Quadrature

At quadrature angle $SEP = 90°$. The observation is that in this configuration Jupiter looks full.

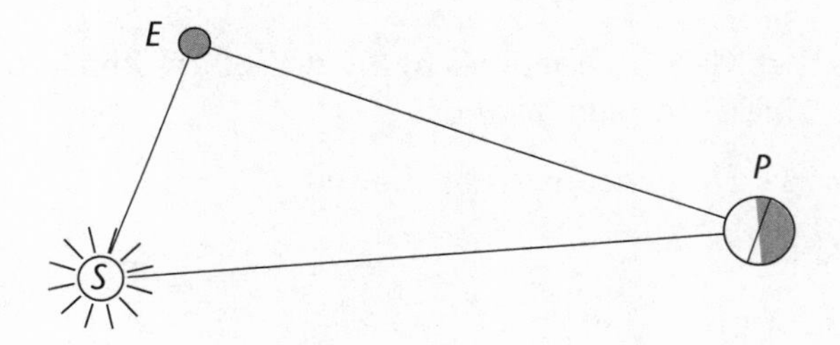

In Part 2, when Mars was at at quadrature, we said that it would appear gibbous. Jupiter, however, is so far away that it appears full even when angle $SEP = 90°$. This is because EP and SP are distances so great in relation to SE that angle EPS is small acute and so by the lemmita Jupiter would appear near to full. It is not visibly gibbous.

Q.E.D. for Step 2 of Part 3.

Step 3: Opposition

At opposition Jupiter rises as the sun sets.

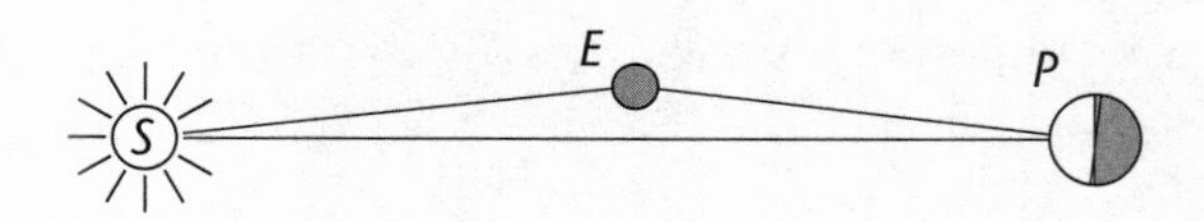

Angle $EPS \rightarrow 0$ and the planet is full if it is on the opposite side of the earth from the sun.

(If Jupiter fell between earth and sun, angle *EPS* would be obtuse and we would not see the planet full.)

Q.E.D. for Step 3 of Part 3.

Conclusion of Part 3

Therefore, since they are sometimes beyond the sun and sometimes on our side of the sun, Jupiter and Saturn encompass the sun.

Q.E.D. for Part 3.

(See note at the end of Part 1.)

"... for it is manifest from the shadows of satellites cast upon them that these shine with light borrowed from the sun."

We see the moons of Jupiter and Saturn crossing the disks of the planets leaving shadows such that we can see both the moon and the shadow. This might be best seen at quadrature; the moon crosses into Jupiter before the shadow or vice versa.

This proves that Jupiter and Saturn are not their own light source.

Phenomenon 4

That the periodic times of the five primary planets, and that of the sun around the earth or of the earth around the sun (the fixed stars being at rest), are in the sesquiplicate ratio of the mean distances from the sun.

This ratio discovered by Kepler is acknowledged by all. The periodic times are necessarily the same, as well as the dimensions of the orbits, whether the sun revolve around the earth or the earth around the sun. And concerning the measure of the periodic times, there is unanimity among all astronomers. But Kepler and Bullialdus have determined the magnitudes of the orbs from the observations most diligently of all, and the mean distances which correspond to the periodic times, do not differ perceptibly from the distances which they have found, and for the most part fall in between them, as may be seen in the following table.

The Periodic Times of the Planets and the Earth about the Sun
With Respect to the Fixed Stars, in Days and Decimal Parts of a Day

Saturn	Jupiter	Mars	Earth	Venus	Mercury
10,759.275	4,332.514	686.9785	365.2565	224.6176	87.9692

Mean Distances of the Planets and the Earth from the Sun

	According to Kepler	According to Bullialdus	According to the Periodic Times
Saturn	951,000	954,198	954,006
Jupiter	519,650	522,520	520,096
Mars	152,350	152,350	152,369
Earth	100,000	100,000	100,000
Venus	72,400	72,398	72,333
Mercury	38,806	38,585	38,710

Concerning the distances of Mercury and Venus from the sun there is no room for dispute, since these are determined by their elongations from the sun. Further, concerning the distances of the superior planets from the sun, all dispute is removed by eclipses of Jupiter's satellites. For by those eclipses is determined the position of the shadow that Jupiter casts, and thereby is obtained Jupiter's heliocentric longitude. But from the heliocentric and geocentric longitudes compared with each other, the distance of Jupiter is determined.

Notes on Phenomenon 4

• **Note 1:** The argument for Phenomenon 4 is similar to that for Phenomena 1 and 2, although there are significant complications (see Note 2). Explanation of the general strategy and necessary planetary theory already given in notes and expanded proof for Phenomenon 1 are not repeated; you may want to refer back to them as you fill out the proof of Phenomenon 4. In addition, we need a means of calculating the distances of the inner planets; Note 3 gives this.

• **Note 2:** Showing that a planetary orbit is circular, or establishing just what that orbit is, turns out to be a much larger undertaking than to establish the orbits of the moons of Jupiter or Saturn.

The method used for determining the orbital radii of Jupiter's moons cannot be used for determining the orbital radii of the outer planets, because we are not outside those orbits, as we are outside the orbits of the moons of Jupiter and Saturn.

We *are* outside the orbits of the inner planets, and can begin as we did with Jupiter's moons. But the first problem we encounter in getting elongations for the inner planets is that, unlike our situation with Jupiter and its moons, we can't see both the planet and the sun at the same time.

This must be addressed by having a solar theory that tells us where the sun is at a given time. We must have one that shows the location of the sun on the current date, projected against the fixed stars as seen from the earth. To this we compare our observed position for the planet as projected against the fixed stars. The difference in celestial longitude gives us our angle.

Suppose that we have done that, and that we have corrected for the distance of the earth and the eccentricity of the earth's orbit. Even after a correction is made for the earth's eccentricity, we find with Venus and Mercury that we have a residual eccentricity for the planet. The greatest elongation is not constant.

From this we can conclude that the orbits are not concentric circular. But we still don't have an easy way of determining whether the residual eccentricity is due to its being an ellipse, an eccentric circle, or some other sort of eccentric oval. This is largely why Kepler's treatise establishing the orbit of Mars runs to 665 pages (in the English translation).

However, if, as here, we don't need to establish the exact orbit, but only want to know the *mean distance*, we can use the following simplification.

Remember that when we take observations, we measure an elongation that is the angle between a planet and its moon, or the sun and a planet. When we take the next observation, that elongation will have changed, as the moon or planet moves in its orbit. If we continue to watch the satellite, we can catch the angle of greatest elongation, the angle between the satellite and the body it is orbiting when that angle is as large as we see it get.

But if we observe another revolution around the orbital path, the greatest elongation for a planet not moving in a circular concentric orbit may be different, because we are viewing it from a different angle. If we keep observing carefully and diligently over time, we will find the *greatest* maximum elongation and the *least* maximum elongation, at least as we are able to observe from our various vantage points around our orbit and in the various configurations of the planet moving in its orbit.

It may take many earth years to get into a viewing position to measure the greatest "greatest elongation." In the case of an inner planet moving in an ellipse, this would occur when the planet was at a principal vertex when the earth was in a position such that the line *EP* was tangent to the orbit.

The greatest and least "greatest elongations" thus observed will be the true elongations of the satellite only if earth's orbit is circular concentric, or if we have corrected for our eccentricity. Suppose we have done that, and now want to use these calculated elongations to get a value for the mean distance. The lemmita of Note 3 will tell us how to do this.

Correction for the eccentricity of the earth's orbit involves using the Copernican or Keplerian theory of the earth's motion (or the Ptolemaic solar theory) to figure what the elongation would have been if it had been observed from a fixed distance (such as the mean sun-earth distance). Review of this is beyond the scope of this note. In the interests of getting on with Newton, we will let it suffice that the methodology to correct for earth's eccentricity has been in the hands of astronomers at least since Ptolemy's time.

• **Note 3:** The lemmita in Note 3 to Phenomenon 1 showed how to find the mean distance from the sun for an outer planet. But how would we go about finding the mean distance from the sun of an inner planet? The following lemmita shows how we can do that. (Within the demonstration is a so-called sub-lemmita which, in addition to supplying steps of the proof of the main lemmita, shows things that we will need elsewhere in the course of our work; it will be referred back to from later points.)

Lemmita: To Find the Mean Distance of an Inner Planet from the Sun.

Let θ be the greatest angular distance of the planet from the sun, measured from the earth ("maximum elongation" as Newton terms it in this Phenomenon). *EP* will be tangent to the orbit since any line cutting the curve will fall closer to the sun and thus be less than maximum elongation.

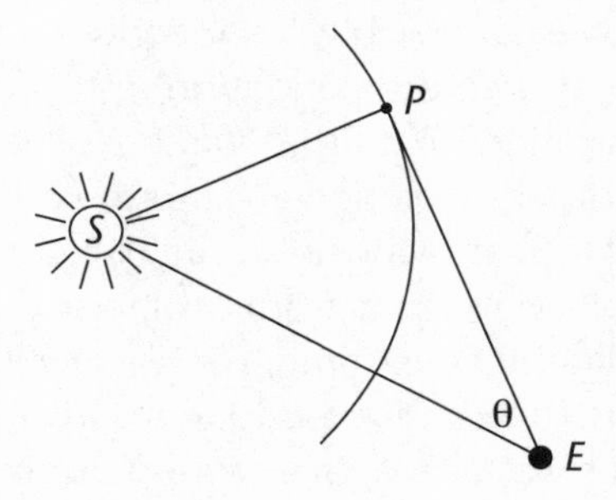

If planetary orbits were circular, with sun at the center, the distance *SP* would be constant for each planet and all maximum elongations for a given planet would be equal to one another. But in an elliptic or eccentric orbit the distance *SP* varies between a least and a greatest value. We wish to determine the mean distance.

Our first step is to relate the mean distance to the geometry of an elliptical orbit. The following sub-lemmita will do this.

Sub-lemmita: Mean Distance

In this sub-lemmita we establish three things for elliptical orbits. First, that the mean distance is equal to the semi-major axis of the ellipse. Second, that the mean distance is an average of the planet's maximum and minimum distances from the sun. Third, that the mean distance is also equal to the line from the focus to an end of the minor axis.

To Prove:

1. *Mean distance* $= AC$.
2. *Mean distance* $= \frac{1}{2}(SA + SD)$.
3. *Mean distance* $= SB$.

Proof:

Let *DA* be major axis; *BF* minor axis; *C* center; *H* and *S* foci.

1. The mean distance of a planet *P* from focus *S* is the average of all the distances *SP* around the orbit.

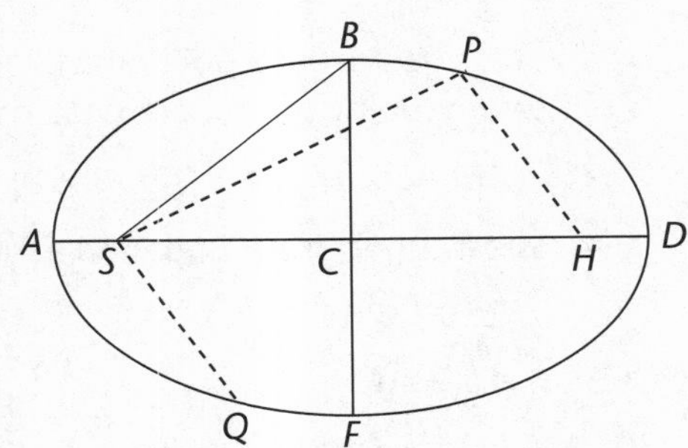

Apollonius III.52 applied here states: If in an ellipse, from foci *H* and *S*, straight lines are deflected to point *P* on the section, their sum will be equal to the axis; that is,

$SP + HP = AD$. [Ap. III.52]

Consider any position *P* on the upper half of the ellipse. From point *S* draw a line *SQ* equal and parallel to *HP* but in the opposite direction (toward the lower half of the ellipse). Since the ellipse is symmetrical about both the major and minor axes, *Q* will lie on the ellipse. And since *SQ* was made equal to *HP*,

$SP + SQ = AD$.

Then let the planet move around its orbit. To every position *P* that it occupies on the upper half of the ellipse, there corresponds a position *Q* that it will at some other time occupy on the lower half of the ellipse. Thus all the distances which the planet can have from *S* can be divided into pairs. Each of these pairs sums to *AD* and, therefore, each pair averages to $AD/2$ or *AC*.

Then the average of *all* the distances SP, that is, the mean distance, must likewise be equal to the semi-major axis, *AC*; that is,

Mean distance $= AC$. (Q.E.D. for Conclusion 1.)

2. Now from the diagram, $SA + SD = AD$. Therefore

$\frac{1}{2}(SA + SD) = \frac{1}{2}AD = AC =$ *Mean distance*

(Q.E.D. for Conclusion 2.)

3. Finally, since

$BC \perp DA$ [Ap. I. Def. 7]

$HC = CS$
[construction of Ap. III.45]

$BC = BC$

Therefore $\triangle BHC \cong \triangle BCS$
[Eu. I.4]

So $HB = SB$;

therefore $HB + SB = 2SB$ (1)

But, again from Apollonius III.52, the sum of distances from the foci to any point on the ellipse is equal to the major axis *AD*, so

$HB + SB = AD$ [Ap. III.52]

Substituting from Equation 1, $2SB = AD$

$2SB = 2AC$

$SB = AC$

But *AC* was shown to be equal to the mean distance; therefore SB is likewise equal to the mean distance (Q.E.D. for Conclusion 3).

[End of sub-lemmita]

The paths of the inner planets, Mercury and Venus, approximate ellipses as closely as our measurements are precise. Therefore let *A* and *D* be the principal vertices of the inner planet's orbit. And suppose the planet is at *D*

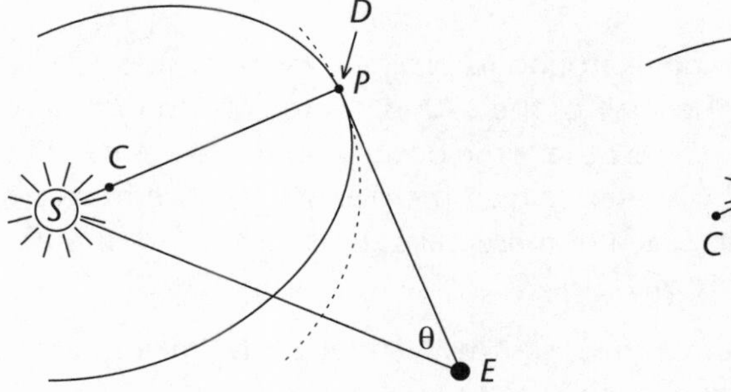

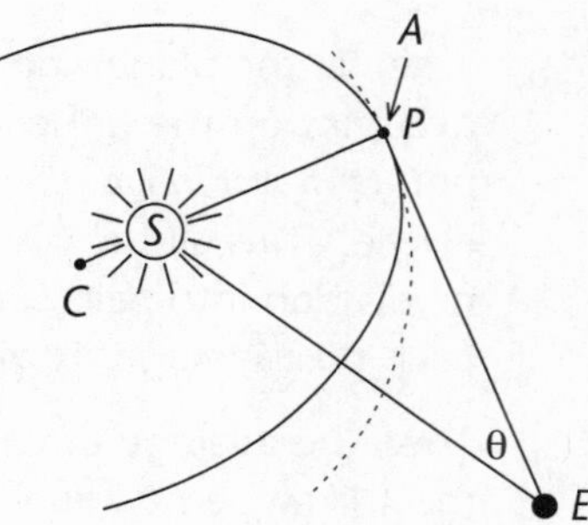

(furthest from the sun) at the same time as our line of sight *EP* is tangent to the orbit. Since *EP* is tangent we will observe a maximum elongation; and it will be the *greatest* maximum elongation since the angle *SEP* subtends the greatest distance, *SD*.

Likewise, if the planet is at *A* (closest to the sun) at the same time as our line of sight EP is tangent to the orbit, we will observe the *least* maximum elongation since angle SEP will then subtend the least distance, SA.

At both A and D, by Apollonius I.32 and I. Def. 7,

Angle $SPE = 90°$.

Now let's call angle PES, the greatest or least maximum elongation, angle θ.

Let's assign distance SE to be 10 units, as before.

$$\frac{SP}{SE} = \frac{SP}{10} = \sin\theta.$$

$$SP = 10\sin\theta.$$

Therefore the radius of the planet's orbit is ten times the sine of the angle of greatest or least maximum elongation, in these units.

Then by the sub-lemmita, if we calculate SP for both greatest and least maximum elongations and take the average, we shall obtain the mean distance.

Q.E.I.

[End of Lemmita to Find the Mean Distance of an Inner Planet from the Sun.]

Expansion of Newton's Sketch of Phenomenon 4

Given:

Periodic times and distances for Saturn, Jupiter, Mars, earth, Venus, Mercury.

To Prove:

The periodic times of all these planets are as the $\frac{3}{2}$ powers of their mean distances from the sun.

Proof:

"Concerning the distances of Mercury and Venus from the sun there is no room for dispute, since these are determined by their elongations from the sun."

The distances of Mercury and Venus can be determined using the lemmita in Note 3 of this Phenomenon.

The distances of Mars, Jupiter, and Saturn can be determined using the lemmita in Note 3 of Phenomenon 1.

"Further, concerning the distances of the superior planets from the sun, all dispute is removed by eclipses of Jupiter's satellites. For by those eclipses is determined the position of the shadow that Jupiter casts, and thereby is obtained Jupiter's heliocentric longitude. But from the heliocentric and geocentric longitudes compared with each other, the distance of Jupiter is determined."

Newton here gives a more exact method:

Calculation of Jupiter's Heliocentric Longitudes and Distances Using Eclipses

This method makes no assumption about the orbit of Jupiter. It may be circular, elliptical, or clover-leaf. It *does* assume the concentric circularity of the orbit of Jupiter's moon, which was established in Phenomenon 1.

Calculating the Heliocentric Longitude of Jupiter

Pick some moon of Jupiter. Observe its maximum elongation (angle JEM_2) and where it enters Jupiter's shadow (M_1).

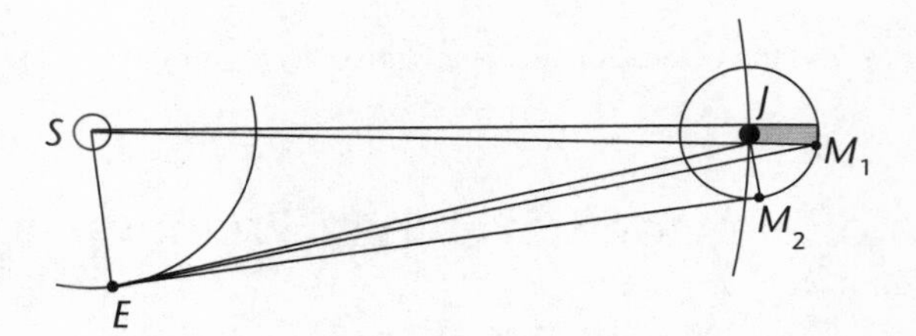

Observation 1: Maximum elongation of moon from Jupiter, angle JEM_2.

Observation 2: Elongation of same moon from Jupiter when it enters Jupiter's shadow, angle JEM_1.

Observation 3: Position of line EJ: that is, observed position of Jupiter against the fixed stars at a particular date.

Take distance SE, the distance of the earth from the sun, as the unit.

By the law of sines, in $\triangle EJM_1$,

$$\frac{JM_1}{\sin JEM_1} = \frac{JE}{\sin EM_1 J}. \tag{1}$$

Because we are entirely outside the moon's orbit, this situation corresponds to observations of an inner planet around the sun. At greatest elongation $EM_2 \perp JM_2$.

By definition of sine,

$$\frac{JM_2}{JE} = \sin JEM_2.$$

$JM_2 = JE \sin JEM_2.$

But $JM_2 = JM_1$ because the moon's orbit is circular.

$JM_1 = JE \sin JEM_2.$

Substituting into Equation 1,

$$\frac{JE \sin JEM_2}{\sin JEM_1} = \frac{JE}{\sin EM_1J}.$$

Divide both sides by JE

$$\frac{\sin JEM_2}{\sin JEM_1} = \frac{1}{\sin EM_1J}.$$

Inverting,

$$\sin EM_1J = \frac{\sin JEM_1}{\sin JEM_2}.$$

But we know angle JEM_1 by Observation 2 and angle JEM_2 by Observation 1.

Therefore we know $\sin JEM_1$ and $\sin JEM_2$.

Therefore we know $\sin EM_1J$.

And so we know angle EM_1J. (Conclusion 1)

By Euclid I.32, angle SJE = angle M_1EJ + angle EM_1J.

We know angle M_1EJ by Observation 2 and angle EM_1J by Conclusion 1.

Therefore we know angle SJE. (Conclusion 2)

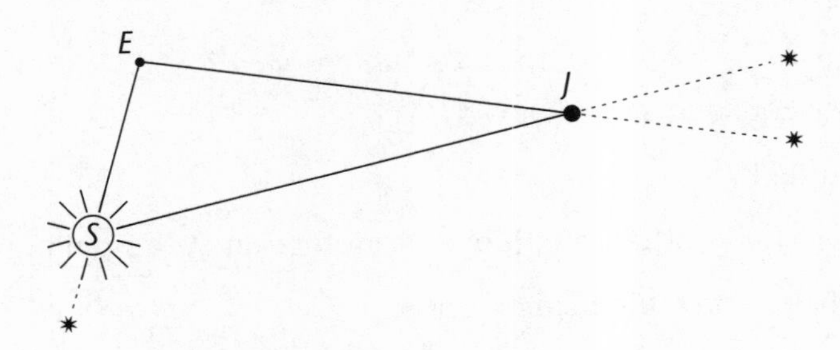

Now Observation 3 gives us the direction of line EJ; that is, we know what fixed star Jupiter is lined up with at that point. By adding or subtracting the angle SJE just found by Conclusion 2, we can calculate the direction of SJ—that is, what star Jupiter would be lined up with when seen from the *sun*.

This is the heliocentric longitude of Jupiter.

Calculating the Heliocentric Distance of Jupiter

Next, we use the heliocentric longitude to find Jupiter's distance at this longitude.

In $\triangle JES$ by the law of sines,

$$\frac{SJ}{\sin JES} = \frac{SE}{\sin SJE} .$$

By Euclid V.16 (*alternando*),

$$\frac{SJ}{SE} = \frac{\sin JES}{\sin SJE} . \tag{2}$$

Using Observation 3, the position of Jupiter against the fixed stars, we can find the elongation of Jupiter from the sun, angle JES when the moon is at position M_1.

(Note that we need to derive this angle from Observation 3 rather than observing it directly, since we can't see both the sun and Jupiter at the same time. We observe what fixed star Jupiter is lined up with at our time of observation, and we know from our solar theory what fixed star the sun would seem to be lined up with from the point of view of the earth at this date in the year. This gives us the elongation between Jupiter and the sun.)

Thus we know angle JES.

Therefore we know $\sin JES$.

Because we know angle SJE by Conclusion 2, therefore we know $\sin SJE$.

So we know

$$\frac{\sin JES}{\sin SJE} .$$

Therefore by Equation 2, we know SJ/SE.

But SE is our unit.

So we know SJ, the distance of Jupiter from the sun at this point.

This is the heliocentric distance of Jupiter.

(Conclusion of calculation to find Jupiter's heliocentric longitudes and distances using eclipses.)

Calculation of Longitudes and Distances for Saturn and Mars

The same method that was used for calculating Jupiter's heliocentric longitudes and distances can be used for Saturn.

We can't use the same method for Mars because we can observe no moons for Mars. For Mars we must use the lemmita of Note 3 of Phenomenon 1. That lemmita finds the values for γ and ε in the third diagram using a concentric circle. They can be gotten more exactly by using an eccentric circle.

The actual distances that appear in Newton's table, however, are calculated by applying Kepler's complete planetary theory, using ellipses. Kepler's method first figures the actual elliptical orbit for the planets and then based on that cites the distance *SP* when the body *P* is at the vertex of the minor axis. As we know from the sub-lemmita in Note 3 above, this will be the mean distance.

This is premature for Newton here, so he shows us how to compute them without assuming ellipses. Even using the methods suggested by Newton, we would find a close approximation of the periodic times to the $\frac{3}{2}$ powers of the distances. But Newton doesn't give us the actual observations to check that.

His not bothering to be consistent on this is significant. Newton is not concerned with what figure we approximate the orbit here: it may be a circle, an ellipse, an eccentric circle, or perhaps an eccentric oval. Whatever it is, if there is an eccentricity, the procedure exists for correcting for it to compute the mean distance.

And in fact he's only looking for distance, or mean distance, here, not making any assertion about the exact orbit. That will be established later, in Proposition III.13.

Newton gives the periodic times as follows:

"The Periodic Times of the Planets and the Earth about the Sun
With Respect to the Fixed Stars, in Days and Decimal Parts of a Day

Saturn	Jupiter	Mars	Earth	Venus	Mercury
10,759.275	4,332.514	686.9785	365.2565	224.6176	87.9692"

And he gives the mean distances as:

"Mean Distances of the Planets and the Earth from the Sun

	According to Kepler	According to Bullialdus	According to the Periodic Times
Saturn	951,000	954,198	954,006
Jupiter	519,650	522,520	520,096
Mars	152,350	152,350	152,369
Earth	100,000	100,000	100,000
Venus	72,400	72,398	72,333
Mercury	38,806	38,585	38,710"

As in Phenomena 1 and 2, Newton checks this by assuming it true for the earth and finding the constant of proportionality from that ratio, and then using that constant to calculate distances for the planets. These distances "according to the periodic times" appear in the last line of his table and can be compared to the distances found by Kepler's planetary theory, which uses observations.

So we take the distance of 100,000 for the earth, which defines our astronomical unit. Then we take the $\frac{2}{3}$ power of the periodic time for the earth and divide it into that distance to get the constant of proportionality.

Our periodic time, from the table, is 365.2565. The $\frac{2}{3}$ power of that is 51.0974. Dividing into our distance we get 1,957.0467.

Taking the $\frac{2}{3}$ power of the periodic times of the planets (second column) and multiplying by the constant of proportionality 1,957.0467 to get the distance (third column):

Planet	*$\frac{2}{3}$ Powers of Times*	*Distances*
Saturn	487.3661	953798.22
Jupiter	265.7623	520109.23
Mars	77.8566	152369.00
Venus	36.9513	72315.42
Mercury	19.7797	38709.80

These numbers are close to the distances based on observation.

Q.E.D.

Alternate Proof

If you prefer a more straight-forward check, take the $\frac{3}{2}$ powers of the distances and see if they vary as the corresponding periodic times.

Let's average the distances found by Bullialdus and Kepler and take the $\frac{3}{2}$ power of each.

Planet	*Average Distance*	*$\frac{3}{2}$ Powers*
Saturn	952,599	929,747,848.2
Jupiter	521,085	376,151,551.6
Mars	152,350	59,465,310.08
Earth	100,000	31,622,776.6
Venus	72,399	19,480,444.01
Mercury	38,695.5	7,611,858.284

Note that these distances are in terms of the earth-sun distance as 100,000.

Is the ratio of periodic time to the $\frac{3}{2}$ power of the distance constant?

Planet	*Ratio*	*Quotient*
Saturn	10,759.275 : 929,747,848.2	1.1572×10^{-5}
Jupiter	4,332.514 : 376,151,551.6	1.1518×10^{-5}
Mars	686.9785 : 59,465,310.08	1.1553×10^{-5}
Earth	365.2565 : 31,622,776.6	1.1550×10^{-5}
Venus	224.6176 : 19,480,444.01	1.1530×10^{-5}
Mercury	87.9692 : 7,611,858.284	1.1557×10^{-5}

(Numbers followed by "$\times 10^{-5}$" must be multiplied by 10^{-5} (= .00001). So, for example, the number for Saturn's quotient came out to be .000011572, but it was multiplied by 100,000 for convenience in comparing the six values.)

We see that the values are very close, and so may conclude that the periodic times vary as the $\frac{3}{2}$ powers of the distances.

Q.E.D.

Phenomenon 5

That the primary planets, by radii drawn to the earth, describe areas by no means proportional to the times, but by radii drawn to the sun, do traverse areas proportional to the times.

For with respect to the earth, they now progress, now are stationary, now even retrogress; but with respect to the sun, they always progress, and do so with a motion very nearly uniform, but a little faster at perihelia and slower at aphelia, so as to make the description of areas equable. The proposition is very well known to astronomers, and in Jupiter particularly it is demonstrated by eclipses of the satellites, by which eclipses we have said that the heliocentric longitudes of this planet and its distances from the sun are determined.

Expansion of Newton's Sketch of Phenomenon 5

"That the primary planets, by radii drawn to the earth, describe areas by no means proportional to the times, but by radii drawn to the sun, do traverse areas proportional to the times."

To Prove:

That Mercury, Venus, Mars, Jupiter, and Saturn traverse areas proportional to the times around the sun as center, but not around the earth as center.

"For with respect to the earth, they now progress, now are stationary, now even retrogress; but with respect to the sun they always progress, and do so with a motion very nearly uniform, but a little faster at perihelia and slower at aphelia, so as to make the description of areas equable."

The only serious contenders for the center are the earth and the sun. Newton is not undertaking to disprove all hypotheses that could be invented by rashly contrived dreams, against which he cautions us in the commentary to the third Rule of Philosophizing. Let us first consider the first serious contender, the earth.

Observation: From the earth the planets seem sometimes to move forward, sometimes to remain stationary, sometimes to retrogress.

If in one time period Δt the planet moves some angular distance as viewed from the earth, it must have swept out some finite area around the earth as center. If in the next time period Δt it moves no angular distance, but either stands still or moves in or out along the radius, it is sweeping out no area around the earth as center. Therefore it cannot be sweeping out equal areas around the earth as center.

Phenomenon 3, showing that the five planets encompass the sun, eliminated Ptolemy's view. That leaves the views of Copernicans (all planets plus earth in circular or nearly circular orbits around the sun), Tychonians (planets move in circular or nearly circular orbits around the sun which moves around the earth), and Keplerians (planets plus earth move in ellipses around the sun).

For all these theories, either (a) the planets' orbits are concentric circular, or (b) if eccentric, they go faster at perihelion, slower at aphelion (according to some device or other). This is what we would need to have happening if we were to maintain areas proportional to the times around the sun as a center.

Thus the proportionality of the areas to their times is maintained at least approximately.

"The proposition is very well known to astronomers, and in Jupiter particularly it is demonstrated by eclipses of the satellites, by which eclipses we have said that the heliocentric longitudes of this planet and its distances from the sun are determined."

As shown in Calculation of Jupiter's Heliocentric Longitudes and Distances using Eclipses, in Phenomenon 4, we can get both heliocentric longitudes of Jupiter and distances of Jupiter at various points around the orbit. Please refer back to those calculations; they will not be repeated here.

The same procedure can be used for Saturn. We can find the same things for Mars, although perhaps not with quite the same accuracy.

If we know distances at respective heliocentric longitudes all around the orbit, along with the times, we can say how fast the planet is moving at each distance. By this means we find that it moves fastest at perihelion.

Q.E.D.

Phenomenon 6

That the moon by a radius drawn to the center of the earth describes an area proportional to the time.

This is clear from a comparison of the moon's apparent motion with its apparent diameter. However, the lunar motion is perturbed somewhat by the sun's force, but in these phenomena I neglect minute, imperceptible errors.

Expansion of Newton's Sketch of Phenomenon 6

The adjacent diagram shows the earth-moon system, drawn to scale. The moon *M* is shown at its mean distance (slightly greater than 60 earth radii). *A* is its apogee, *P* its perigee, and *C* the center of its orbit. The orbit (whose eccentricity is .0549) differs from circularity by less than the thickness of the lines in this diagram.

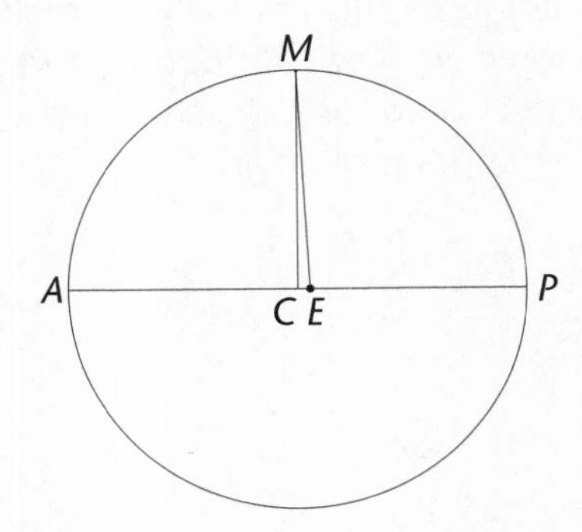

"This is clear from a comparison of the moon's apparent motion with its apparent diameter."

The exact shape of the moon's orbit is very difficult to determine from observations taken from the earth, since we are so near its center, and the various lunar theories current in Newton's day were not in agreement about this. They did, however, agree on the position and motion of apogee and perigee, and could predict where the moon would be against the fixed stars for any date.

The moon's apparent diameter can be measured with a telescope and micrometer. Its speed is measured as arc traversed per unit time and can be read from the published tables. The apparent diameter is compared to the speed that it is moving at that time.

The moon's apparent diameter varies from about $29\frac{1}{2}'$ to about $33\frac{1}{2}'$ as it moves from its farthest point, apogee *A*, to its nearest point, perigee *P*. That is, we can see that it is closer when it is at point *P* by the fact that it looks larger.

It is also found to move more swiftly at *P* than at *A*, that is, it covers more degrees of arc in the same time. Thus it is moving faster through that part of the orbit in which it is closer to us.

If there were equable description of areas, and we were at the center of forces but not at the center of the orbit, then we would *need* the moon to go faster when it was closer to us, as we found it does, in order to keep up areas proportional to the times.

Newton doesn't give us the numbers to check this, but the numbers we need to confirm it at least approximately may be assumed to be agreed upon by astronomers of his time.

"However, the lunar motion is perturbed somewhat by the sun's force, but in these phenomena I neglect minute, imperceptible errors."

There are many lunar anomalies that create extra wobbles. One of these wobbles is related to the moon's position in relation to the sun, but the magnitude of this anomaly is small in relation to the dimensions of the orbit depicted here. Although Newton investigates this anomaly elsewhere, the effect is not of sufficient magnitude to affect our observation of the equable description of areas.

Q.E.D.

Book III Propositions

Proposition 1

That the forces by which the planets around Jupiter are perpetually drawn back from rectilinear motions and held back in their orbits, look to the center of Jupiter, and are inversely as the squares of the distances from the same center.

The former part of the proposition is clear from the first Phenomenon and the second or third Proposition of the first Book, and the latter part is clear from the first Phenomenon and the sixth Corollary of the fourth Proposition of the same Book.

The same is understood of the planets that accompany Saturn, from the second Phenomenon.

Expansion of Newton's Sketch of III.1

"That the forces by which the planets around Jupiter are perpetually drawn back from rectilinear motions and held back in their orbits, look to the center of Jupiter, and are inversely as the squares of the distances from the same center. ... The same is understood of the planets that accompany Saturn..."

To Prove:

1. That the forces on the moons tend towards Jupiter's center;

2. that the forces on the moons are inversely as the squares of the distances from Jupiter.

3. Both parts may be shown for Saturn as well.

"The former part of the proposition is clear from the first Phenomenon and the second or third Proposition of the first Book,..."

Proof:

Part 1

By Phenomenon 1, the moons sweep out areas proportional to the times about Jupiter as a center.

If Jupiter were at rest or moving with uniform rectilinear motion we could use Proposition I.2, and would be able to conclude that the center of Jupiter is the center of forces towards which the moons are being impelled by the centripetal force.

Proposition I.2 says:

> Every body that moves in some curved line described in a plane, and, by a radius drawn to a point that either is immobile or proceeds uniformly in a straight line, describes areas about that point proportional to the times, is urged by a centripetal force tending to the same point.

But Jupiter is not at rest, nor is it even moving uniformly in a straight line. Jupiter is, by Phenomenon 3, encompassing the sun in its own curved orbit. (We don't at this point know whether the sun might be at any kind of center for Jupiter's orbit; it's enough to know that to encompass the sun Jupiter will not be moving in a straight line.)

So we must go to Proposition I.3 for our situation here. I.3 says:

> Every body that, by a radius drawn to the center of another body moved in any way whatever, describes areas about that center proportional to the time, is urged by a force compounded of a centripetal force tending toward that other body, and of all the accelerative force by which that other body is urged.

We are told here that the moons, which describe areas around the center of Jupiter proportional to the times, are urged by forces compounded of those which impel them to Jupiter and all the forces by which Jupiter is urged.

If we subtract out the common accelerative force which is impelling a moon-Jupiter system, we are left with the force which is impelling that moon toward the center of Jupiter.

Part 2

"... and the latter part is clear from the first Phenomenon and the sixth Corollary of the fourth Proposition of the same Book."

The "latter part" is Part 2 of our to-prove: that the forces on the moons are inversely as the squares of the distances from Jupiter.

Part 1 of Phenomenon 1 shows that Jupiter's moons travel in equable circular concentric orbits. (We ignore the motion common to the Jupiter-moons system.) This means that we can invoke Proposition I.4, which deals with motion in such circles.

Part 2 of Phenomenon 1 shows that their periodic times are as the $\frac{3}{2}$ powers of their radii. This means that we can use I.4 Corollary 6, which says that, if the periodic times vary as the $\frac{3}{2}$ powers of the radii, the centripetal forces will be inversely as the squares of the radii.

We now know something about the centripetal force causing the moons of Jupiter to depart from their tangential rectilinear inertial motion and curve into orbits around Jupiter: we know that the force acts inversely as the square of the distances of the moons from the center of Jupiter.

That is,

$$f \propto \frac{1}{SP^2}.$$

Part 3

"The same is understood of the planets that accompany Saturn, from the second Phenomenon."

We may follow the same reasoning to the same conclusion for the moons of Saturn, substituting the conclusions of Phenomenon 2 for those of Phenomenon 1.

Therefore, for Saturn's moons as well as Jupiter's,

$$f \propto \frac{1}{SP^2}.$$

Q.E.D.

Proposition 2

That the forces by which the primary planets are perpetually drawn back from rectilinear motions, and are held back in their orbits, look to the sun, and are inversely as the squares of the distances from its center.

The former part of the proposition is clear from the fifth Phenomenon and the second Proposition of the first Book, and the latter part is clear from the fourth Phenomenon and the fourth Proposition of the same Book. However, this part of the Proposition is demonstrated with greatest accuracy by the aphelia being at rest. For (by Corollary 1 Proposition 45 Book I) the least departure from the duplicate ratio would be bound to effect a noticeable motion of the apsides in single revolutions, and an enormous one in many revolutions.

Note on III.2

Apsides: The line of apsides is the line between perihelion, the point on the path closest to the sun, and aphelion, the point on the path farthest from the sun. In an elliptical path, with the sun at the focus, this line will be the major axis.

Expansion of Newton's Sketch of III.2

"That the forces by which the primary planets are perpetually drawn back from rectilinear motions, and are held back in their orbits, look to the sun, and are inversely as the squares of the distances from its center."

To Prove:

1. That the sun is the center of the centripetal force which turns the planets Mercury, Venus, Mars, Jupiter, and Saturn off from their tangential inertial motion and retains them in orbits.

2. That these centripetal forces act inversely as the squares of the distances of these planets from the center of the sun.

"The former part of the proposition is clear from the fifth Phenomenon and the second Proposition of the first Book,..."

By Phenomenon 5, the primary planets sweep out areas proportional to the times around the sun as center. Newton uses I.2 to establish that the sun is at the center of forces.

By this use of I.2, we may conclude that Newton sees the the sun as being at rest or moving uniformly in a straight line. If this is taken to be the case, the sun could not be revolving around the earth, and the theory of Tycho Brahe will have been set aside. (Recall that the Tychonian theory was that the sun with its orbiting planets itself orbits the earth. See the first note to Phenomenon 3.) If we wanted to continue to hold open the possibility of an orbiting sun, we would have to use Proposition I.3 here, as Newton did in III.1 but doesn't do here.

"... and the latter part is clear from the fourth Phenomenon and the fourth Proposition of the same Book."

We now have to prove that the force law is inverse square.

Phenomenon 4 established that the periodic times of the planets are as the $\frac{3}{2}$ powers of their distances from the sun.

The orbits of the planets approximate circles, so if we will be satisfied with an approximation, we can invoke Proposition I.4 Corollary 6 as we did for Jupiter and Saturn, and have our result. This is Newton's first suggestion.

"However, this part of the Proposition is demonstrated with greatest accuracy by the aphelia being at rest. For (by Corollary 1 Proposition 45 Book I) the least departure from the duplicate ratio would be bound to effect a noticeable motion of the apsides in single revolutions, and an enormous one in many revolutions."

We can do better than the previous suggestion, however, by consulting I.45, which Newton proves later in Book I. We won't take time to follow the full proof of this proposition (which aside from being long and complex in its own right, requires that I.43 and I.44 be proved first). But we will look at it and see what it asserts.

> *Proposition I.45:* It is required to find the motion of the apsides of orbits that are very close to circles.

Note that I.45 concerns itself with orbits that are very close to circles. The planets fit this condition.

I.45 deals with the relation of the force law to the motion of the line of apsides. Remember that the line of apsides is the line between the nearest approach of the orbiting body to the center of forces and the most distant point. In Proposition III.2 we're concerned with the line between aphelion and perihelion, the farthest and nearest points of the planet to the sun.

Motion of apsides is something we didn't deal with in the first seventeen propositions. We imagined bodies moving in ellipses as following back over their same paths on successive revolutions.

And indeed, it turns out that with an exact inverse square law, that is what will happen. But if the force law should have other terms which are also influencing the orbit, we may still get an ellipse, but one which will rotate in space: the line of apsides will shift gradually (or more quickly) in orientation. This rate of travel of the line of apsides can be predicted if the force law is given. Alternatively, that rate of travel can be measured and used to gain insight into the force law.

> *I.45 Corollary 1:* Hence, if the centripetal force be as some power of the altitude, that power can be found from the motion of the apsides, and conversely. Specifically, if the whole angular motion by which the body returns to the same apsis be to the angular motion of one revolution, or 360°, as some number m be to another number n, and the altitude be called A, the force will be as that power of the altitude $A^{\frac{nn}{mm}-3}$, whose index is $\frac{nn}{mm}-3$. ...

The altitude is the planet's distance, or radius. For example, if the body returns to the same apsis in one revolution (our fixed orbit), n and m will both be 1, and the force law will be the $(\frac{1}{1}-3)$ power of the distance, or inversely as the square of the distance.

In other cases, we have other ratios of $n^2 : m^2$ and can find other force laws.

Here in Proposition III.2, Newton gives us an unnumbered phenomenon, telling us that the aphelia of the planets do not move. (Newton knows that this isn't strictly true; however, the motion is very slow, and for these purposes it may be neglected.)

From this, using I.45 Corollary 1, we can conclude that the force law must be inverse square.

Q.E.D.

Proposition 3

That the force by which the moon is held back in its orbit looks to the earth, and is inversely as the square of the distance of places from its center.

The former part of the assertion is clear from the sixth Phenomenon and the second or third Proposition of the first Book, and the latter part by the very slow motion of the lunar apogee. For that motion, which in single revolutions is but three degrees and three minutes forward, can be disregarded. For it is clear (by Corollary 1 Proposition 45 Book I) that if the distance of the moon from the center of the earth be called D,* the force by which such a motion would arise would be inversely as $D^{2\frac{4}{243}}$, that is, inversely as that power of D whose index is $2\frac{4}{243}$, or in other words, in a ratio of the distance a little greater than the duplicate inversely, but which approaches closer by $59\frac{3}{4}$ parts to the duplicate than to the triplicate. It arises, moreover, by the action of the sun (as will be said below), and for that reason is to be neglected here. It therefore remains that that force which looks to the earth is inversely as D^2.** A fact which will also be more fully established by comparing this force with the force of gravity, as it is in the following Proposition.

Notes on III.3

There are lots of little twists and wobbles in the moon's orbit, but it is basically an ellipse whose line of apsides has a slow forward motion. The ellipse is practically indistinguishable from a circle, but it is noticeably eccentric: see the scale drawing and accompanying explanation under Phenomenon 6.

- **Terminology:**

The aphelia are the points of greatest distance from the sun in the path of a body orbiting the sun. They are "at rest" if the body returns to the same point at each aphelion.

The line of apsides of the moon is the line between perigee (nearest approach to the earth) and apogee (greatest distance from the earth) in the moon's orbit. The line of apsides corresponds to the major axis of an elliptical path.

Because the moon's line of apsides is moving, there are two different measures of the moon's average motion. Motion in longitude is measured

---

* The translation follows the first edition here. Later editions needlessly complicate the definition of *D*.

** For this sentence, the translation follows the first edition. In later editions, Newton replaced this with a recomputation based upon other data regarding the sun's disturbing force on the moon. Since the computation is difficult to follow and does not add materially to the argument, it is omitted.

with respect to the fixed stars starting from the spring equinox point, while motion in anomaly is measured with respect to the apsides, starting from apogee.

- **Strategy for Proof:**

Unlike our situations in III.1, for Jupiter and Saturn and their moons, and III.2, for the sun and its planets, here in III.3, with the earth and its moon, we have only one revolving body. As a consequence we can't compare the periodic times of different bodies revolving around the central body, and so can show no $\frac{3}{2}$ power relationship between different orbiting satellites. This means we can't use I.4 Corollary 6, as we did in III.1, even if we approximate the moon's orbit by a circle.

But because we have an orbit very close to a circle, we can use I.45 Corollary 1 as we did in III.2.

Our use here, however, will not be so easy. There we had the happy observation for the planets that the aphelia were at rest. Here Newton gives us another unnumbered phenomenon which notes that the line of apsides of the moon moves by 3°3′ in each revolution.

The motion of the apsides was known quite accurately to the ancient Babylonians, who are cited by Ptolemy in the *Almagest* Book IV Chapter 3. Ptolemy's numbers for the daily mean motion in longitude and in anomaly, converted to decimals, are 13.17638221515° and 13.06498376882°, respectively (from *Almagest* IV.3). Dividing the latter into 360° gives the number of days in a full cycle of anomaly. Multiplying this by the mean longitudinal motion gives the total longitudinal motion in one cycle of anomaly, which comes out to about 363°4′. Later astronomers refined this to 3°3′ beyond a full cycle, as Newton says.

- **Sun's Influence:**

Newton is jumping the gun a bit here to speak of the disturbing influence of the sun. We have no reason yet to think that the sun would have any effect on the moon. But this is another case of looking forward to where he is going. He will later show that all matter is impelled towards all other matter, and it will follow that the sun and moon should be be included in this.

Here he is saying something like this: "The motion of the apsides will turn out to be caused by the sun. Later we will explain it. For now, it is sufficient to know that if the motion is related to anything outside the earth-moon system, it must be subtracted if we are to isolate the centripetal force of the earth-moon relationship."

Expansion of Newton's Sketch of III.3

"That the force by which the moon is held back in its orbit looks to the earth, and is inversely as the square of the distance of places from its center."

To Prove:

1. That the earth is at the center of the centripetal force urging the moon to turn out of its tangential inertial motion into orbit.

2. That the centripetal force acts inversely as the square of the distance of the moon from the center of the earth.

Proof:

Part 1

"The former part of the assertion is clear from the sixth Phenomenon and the second or third Proposition of the first Book,..."

By Phenomenon 6, the moon describes areas proportional to the times of description.

Therefore we may apply I.3, as discussed in the expanded proof for III.1. If we remove from consideration any force common to the earth-moon system, we may conclude that the center of the earth is the center of centripetal force for the moon.

(Consult the Expanded Proof for I.3 for the bridge from I.3 back to I.2 which allows us to use I.3 as an extension of I.2.)

Part 2

"... and the latter part by the very slow motion of the lunar apogee. For that motion, which in single revolutions is but three degrees and three minutes forward, can be disregarded."

By I.45, if there is any motion of the apsides, we know that the force law will not be exactly inverse square. And we are told that there is motion of the moon's apsides. But is the force operating on the moon approximately inverse square?

Newton says that the motion of the apsides of the moon is small enough to be neglected, and supports this by calculating the force law to which this movement would correspond.

"For it is clear (by Corollary 1 Proposition 45 Book I) that if the distance of the moon from the center of the earth be called D, the force by which such a motion would arise would be inversely as $D^{2\frac{4}{243}}$, that is, inversely as that power of D whose index is $2\frac{4}{243}$, ..."

Recall the procedure for calculating the force law from the movement of the apsides in I.45 Corollary 1:

> Hence, if the centripetal force be as some power of the altitude, that power can be found from the motion of the apsides, and conversely. Specifically, if the whole angular motion by which the body returns to the same apsis be to the angular motion of

one revolution, or 360 degrees, as some number m be to another number n, and the altitude be called A, the force will be as that power of the altitude $A^{\frac{nn}{mm}-3}$, whose index is $\frac{nn}{mm}-3$. ...

Let us apply this to the movement of apsides of the moon's orbit.

$$m = 363°3'$$

$$n = 360°$$

$$\frac{n^2}{m^2} = 0.9832$$

$$\frac{n^2}{m^2} - 3 = -2.0167$$

$$f \propto r^{-2.0167}$$

$$f \propto 1/r^{2.0167}$$

But this amount by which we diverge from inverse square, 167/10,000, is equivalent to the fraction of a power Newton gives for the divergence, $\frac{4}{243}$.

$$\frac{167}{10{,}000} = \frac{4}{243}$$

Therefore

$$f \propto 1/r^{2\frac{4}{243}}$$

We have thus found the same result which Newton's calculation yields.

"...or in other words, in a ratio of the distance a little greater than the duplicate inversely, but which approaches closer by $59\frac{3}{4}$ parts to the duplicate than to the triplicate."

We can see that $1/r^{2\frac{4}{243}}$ is much closer to inverse square than to inverse cube.

How much closer is it? Newton answers by calculating the ratio of the distance away from the inverse triplicate to the distance away from inverse duplicate.

How close is it to inverse square? Inverse square would be $2\frac{0}{243}$; so it's off by 4 out of 243, or $\frac{4}{243}$.

How close is it to inverse cube? Inverse cube would be $2\frac{243}{243}$; so it's off by 239 out of 243, or $\frac{239}{243}$.

$$\frac{239}{243} : \frac{4}{243} :: 59\frac{3}{4} : 1.$$

"It arises, moreover, by the action of the sun (as will be declared later), and for that reason is to be neglected here. It therefore remains that that force which looks to the earth is inversely as D^2."

Newton says that this deviation, for which he added a calculation in a later edition, is due to the sun. Newton will take up the explanation of how it is due to the sun later, after he has shown that all matter is impelled towards all other matter. In III.22, which is not included in this guidebook, he shows qualitatively how universal gravitation accounts for all the lunar anomalies.

However, if it is due to the sun, it must be subtracted out to get the force of the moon towards the earth. (See the note before the expanded proof.) Thus, though out of place here, Newton can't resist mentioning it.

When we do subtract out this movement of the apsides, we have an inverse square force law by I.45 Corollary 1.

"A fact which will also be more fully established by comparing this force with the force of gravity, as it is in the following Proposition."

The next proposition will prove this result in another and quite dramatic way. It is a crucial proposition and perhaps the most thrilling demonstration in *Principia*, as well as an important milestone in the history of science.

[A corollary to Proposition 3, which was added in the second edition, adds little to the proposition, and so is omitted.]

Proposition 4

That the moon gravitates towards the earth, and is always drawn back from rectilinear motion, and held back in its orbit, by the force of gravity.

The moon's mean distance from the earth in the syzygies, in terrestrial semidiameters, is 59 according to Ptolemy and most astronomers; 60 according to Wendelin and Huygens, $60\frac{1}{3}$ according to Copernicus, $60\frac{2}{5}$ according to Streete, and $56\frac{1}{2}$ according to Tycho. But Tycho and those who follow his tables of refraction in setting a greater refraction—by as much as four or five minutes—for the sun and moon than for the fixed stars (in complete opposition to the nature of light), had increased the parallax of the moon by the same number of minutes; that is, by as much as the twelfth or fifteenth part of the whole parallax. Let this error be corrected, and the distance will come out to be $60\frac{1}{2}$ terrestrial semidiameters, more or less, about what was assigned by the others. Let us assume that the mean distance is sixty semidiameters at the syzygies, and that the lunar period with respect to the fixed stars amounts to 27 days, 7 hours, and 43 minutes, as is stated by the astronomers; and that the circumference of the earth is 123,249,600 Paris feet, as is established by the measuring Frenchmen. If the moon be supposed to be deprived of all motion and dropped, so as to descend towards the earth, under the influence of all that force by which (by Proposition 3 Corollary) it is held back in its orbit, it will in falling traverse $15\frac{1}{12}$ Paris feet in the space of one minute. This conclusion comes from a computation based either upon Proposition 36 of the first Book or (what amounts to the same thing) the ninth Corollary of the fourth Proposition of the same Book. For the versed sine of that arc which the moon in its mean motion describes in the time of one minute at a distance of sixty terrestrial semidiameters, is about $15\frac{1}{12}$ Paris feet, or more accurately, 15 feet 1 inch and $1\frac{4}{9}$ lines. Whence, since in approaching the earth that force increases in the inverse of the duplicate ratio of the distance, and is thus greater at the surface of the earth by 60×60 parts than at the moon, a body, in falling by that force in our regions, ought to describe a space of $60 \times 60 \times 15\frac{1}{12}$ Paris feet in the space of one minute, and in the space of one second, $15\frac{1}{12}$ feet, or more accurately, 15 feet 1 inch and $1\frac{4}{9}$ lines. And heavy bodies on earth do in fact descend with the same force. For the length of a pendulum oscillating in seconds, at the latitude of Paris, is three Paris feet $8\frac{1}{2}$ lines, as Huygens has observed. And the height which a heavy body traverses in falling in the time of one second, is to half the length of this pendulum, in the duplicate ratio of the circumference of the circle to its diameter (as Huygens has also pointed out). It is therefore 15 Paris feet 1 inch $1\frac{7}{9}$ lines. And because the force which holds the moon back in its orbit, if it should descend to the surface of the earth, comes out equal to our force of gravity, therefore (by Rules 1 and 2) it is that very force which we are accustomed to call gravity. For if gravity were different from it, bodies, in seeking the earth with the two forces conjoined, would descend twice as fast, and in falling in the space of one second would describe $30\frac{1}{6}$ Paris feet, in complete opposition to experience.

This computation is based upon the hypothesis that the earth is at rest. For if the earth and the moon should move around the sun, and should also at the same time move around their common center of gravity, the law of gravity remaining the same, the distance of the centers of the moon and the earth from each other will be about $60\frac{1}{2}$ terrestrial semidiameters, as will be clear to anyone undertaking the computation. And the computation can be undertaken by Proposition 60 of Book I.

Notes on III.4

- **Note 1**

As noted in the expanded proof to the last proposition, III.4 is a crucial proposition in the development of universal gravitation and our modern cosmology, an important milestone in the history of science, and perhaps the most thrilling demonstration in *Principia*.

Remember that, before *Principia*, gravity has meant "terrestrial heaviness" and will continue to mean no more than that until we can show that it does. The enunciation asserts that the moon is drawn off from its tangential motion by force of gravity.

It must be appreciated that this is not the same thing we have been saying all along when we spoke of a centripetal force. It is in fact an assertion that the same power that makes rocks fall on earth is responsible for keeping the moon in orbit.

This notion was startling, and really quite new. And it is still somewhat startling today if one really thinks about it, that is, sets aside what one "knows" because one has been told or read certain things. Look up at that orb hanging in the sky! A rock? It's too round, too luminous! It can be heart-stoppingly beautiful. It has a poetical, even mystical power that has never ceased to have an effect on poets, lovers, and werewolves, despite Newton's work.

Traditionally, the matter of celestial bodies had been regarded as different in kind from terrestrial matter, and this created fundamentally different expectations of the moon and a terrestrial rock. Then, in addition, most theories of celestial mechanics had accounts of why and how the bodies moved in orbits, accounts that relied on very different propensities and mechanisms.

Descartes and Kepler each saw the matter of the moon and the earth as similar and so were in that sense closer to seeing things in the revolutionary way Newton did. But in both cases their ideas of gravity were significantly different from Newton's. Kepler understood terrestrial heaviness as a tendency for like matter to clump together, and he saw the moon (but nothing beyond) as matter like the earth's. However, for Kepler, the moon was retained in its orbit not by that tendency of like matter to clump together, but because that's where God told it to be: it was given its proper distance from the earth according to archetypal principles, and it was given a velocity such that it plays its part in a harmonious whole with other motions in the universe. He would have been far from subscribing to Newton's view.

Descartes saw both the orbit of the moon and the descent of rocks as consequences of the workings of vortices: differential pressures of subtle fluids. This hypothesis encompassed an enormous range of phenomena. It was nevertheless unable to provide quantitative predictions, and mathematicians such as Huygens and Leibniz who tried to get the details of planetary motion out of it failed. (For Newton's remarks on this problem, see the General Scholium at the end of Book III, page 485.)

Galileo also believed the moon was a rock, but this was because it *looked* like a rock: for example, light reflected off it the way light reflected off a rock.

Newton, in contrast, gave us, in III.4 here, both the startling claim that the moon is held in orbit by gravity and proof of that claim. It is the first step of the richly-articulated product of rational mechanics that is his theory of universal gravitation as laid out in this book—a theory he uses to predict the regular motions of celestial bodies and all their known anomalies. This is the understanding of gravity that has become a basic component of our common-sense understanding of our physical universe.

- **Note 2**

This astonishing, wonderful demonstration deserves careful and orderly proof. Each part should be separately and completely argued so that when they fit together the full force of the evidence and thesis is appreciated.

The first part determines the accelerative force on the moon in orbit, and calculates how far the moon would fall in one minute of time if it were deprived of all tangential motion.

The second part determines the accelerative force that would be urging the moon at the earth's surface, assuming an inverse square force law, and calculates how far the moon would fall in one second at the surface.

The third part determines the accelerative force on a *rock* at the earth's surface, which we can measure most accurately using a pendulum. Here we call upon another unnumbered phenomenon. Using a proposition of Christian Huygens (see Note 7), Newton is able to calculate the distance a rock would fall in one second at the surface.

Finally we note that these latter two distances are almost identical, and by Rule 2, to the same effect we assign the same cause.

- **Note 3**

A passive read-through of this proposition will rob you of much of the excitement. It is important to work through all the calculations on your own calculator. Pause to ponder the coincidence—or try to explain it!—that gives the same number for the moon's fall in a minute from orbit and in a second at the surface.

And ask yourself whether you really would have believed (at least if you had not been taught so all your life) that the moon would fall as fast as a rock. Or is the amazing thing that it falls as slowly as a rock? What would even give someone the idea of dropping the moon and a rock together, or to suspect that the moon and a rock might obey the same laws of nature?

- **Note 4: Syzygies**

"Syzygies" means "places of being yoked together," namely, conjunction and opposition, when the sun, earth, and moon are in line.

In opposition, the moon is on the opposite side of the earth from the sun, and is seen as full. In conjunction, the moon is on the same side of the earth as the sun, and is seen as "new."

We need some particular place to take our measurements. Syzygy is a place popular with astronomers because that's where eclipses occur. (Among the useful opportunities offered by eclipses is the chance to observe the moon at conjunction.) This means that distances at syzygy had been measured much more often and much more carefully than at other places on the orbit. Lunar theories reflected this: some (most notably Ptolemy's) gave wildly erroneous distances at the quadratures, though they were nearly correct at the syzygies. So if Newton had not specified that distances were to be taken at syzygies, he would not have found anything approaching consensus on the mean distance.

As we know from the previous proposition, the moon's line of apsides (containing apogee and perigee) moves slowly around the zodiac. The place of the syzygies (i.e., the place of full and new moons) also moves around the zodiac, but more quickly: since it follows the sun, its period is one year. This difference in speeds means that the distance varies from syzygy to syzygy. Thus the distances at syzygies represent fairly accurately the range of distances to be found anywhere on the orbit (the sun distorts the orbit somewhat, but this effect is very slight). Different astronomers had different ways of finding the mean distance at syzygy. Rather than go into what these ways are, it will be enough to note the impressive consensus among different theories.

• Note 5: Estimating Distance Using Parallax and Refraction

Parallax

Parallax is the apparent angular displacement of the observed body that results from change in the position of the observer. For example, if we view the moon when it's straight overhead, then we're right in line with center of the earth. In this configuration, shown in Figure 1, the moon is seen lined up with a certain fixed star.

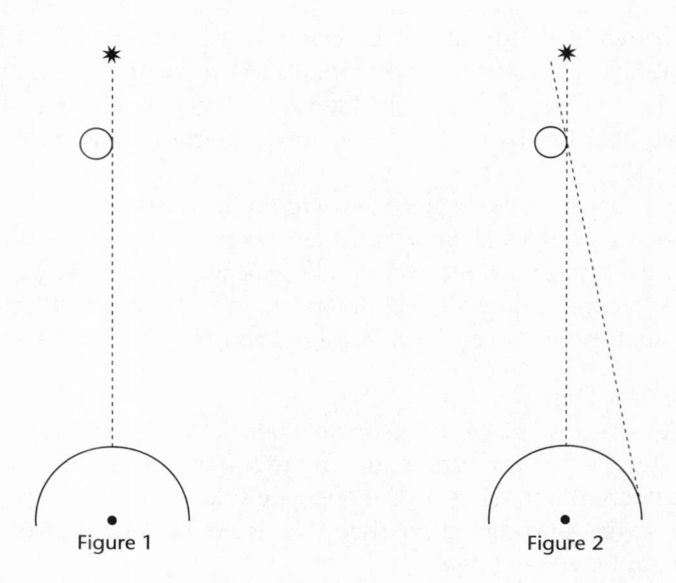

Figure 1 Figure 2

But if someone a quarter of the way around the globe were viewing the moon at the same moment, seeing it rising or setting, that other observer is off the line between the center of earth and the moon, and so will not see the moon lined up with the same fixed star. In a configuration like the one in Figure 2, it will appear lower in the sky than it would appear if viewed from the center of the earth. (The same observer can make both observations, using lunar tables to correct for the movement of the moon against the fixed stars over the hours between measurements.)

The angle between the apparent position of the moon when directly overhead to its apparent position at rising or setting, is the angle of parallax.

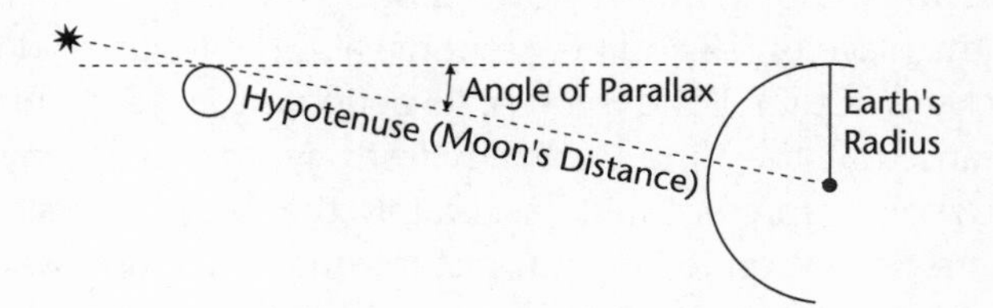

The sine of that angle is the radius of the earth divided by distance of the moon from the center of the earth at setting, by the definition of sine. Because we know the angle, we know the sine of the angle. If we have a value for the radius of the earth (it could be unity, "one earth radius"), we can calculate the distance of the moon in the same units.

However, there is a possible complication. We might have to take refraction into account, depending on how we measured the moon's position.

Refraction

Refraction is the apparent displacement of a celestial body owing to the bending of light by the earth's atmosphere. For example, as light from the moon enters the earth's atmosphere, a denser medium, it is bent downwards towards the vertical. Thus in a viewing situation as illustrated in Figure 3, the moon will appear to be higher in the sky than it really is.

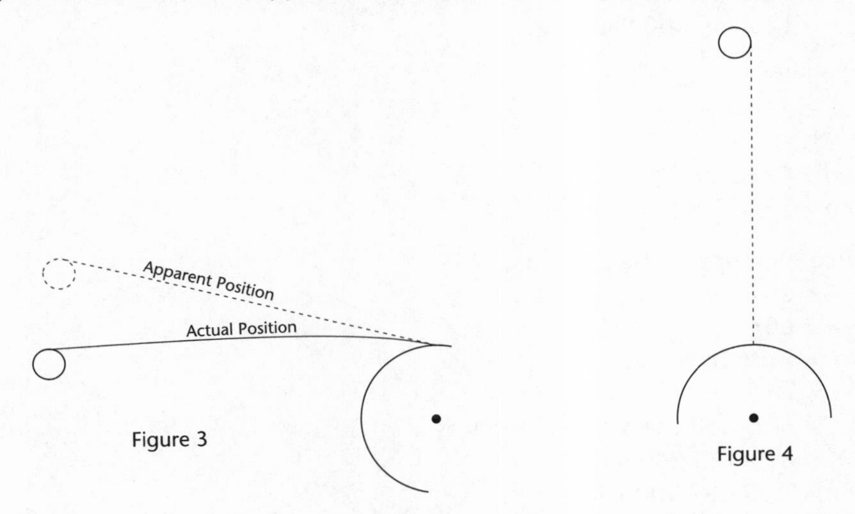

Figure 3

Figure 4

If we are viewing the moon when it is directly overhead as in Figure 4, the light from the moon is coming in along the normal and so is not refracted. The moon is indeed where we are seeing it.

But why should we care about refraction? If we measure the moon's position relative to nearby stars, wouldn't they all be refracted by the same amount?

This would be (nearly) true if the moon's position were always obtained relative to nearby stars. But there are other ways to take the moon's position, and such measurements often included determining the moon's altitude above the horizon, or its angular distance from the sun. These measurements would be affected by refraction and so the refraction would have to be taken into account.

But there is an added complication, which Newton addresses in his proof. Tycho Brahe had believed that the moon's light was subject to greater refraction than the stars' light, and that had led him to give the moon a correspondingly greater parallax than was generally accepted. (From our explanations above, it can be seen that parallax and refraction have opposite effects.) Since a greater parallax means a closer distance, Newton corrected Brahe's error, thus bringing Brahe's distance into agreement with those given by other astronomers.

- **Note 6: Some Useful Numbers**

Mean distance of the moon from the earth = radius of the moon's orbit = 60 earth radii.

$R = 60r$

Period of the moon's revolution against the fixed stars:

$P = 27^d\, 7^h\, 43' = 39{,}343'$

Circumference of the earth:

$c = 123{,}249{,}600$ Paris feet

Diameter of the earth:

$c = 2\pi r = \pi d$

$d = c/\pi$

$d = 123{,}249{,}600/\pi$

$d = 39{,}231{,}500$ Paris feet

Circumference of the moon's orbit:

$C = 2\pi R$

$R = 60r$

$C = 2\pi(60r) = 60(2\pi r)$

$c = 2\pi r$

$C = 60c = 60$ earth circumferences

$C = 60 \times 123{,}249{,}600$

$C = 7{,}394{,}976{,}000$ Paris feet

Diameter of the moon's orbit:

$D = 60d = 60 \times 39{,}231{,}500$ Paris feet

$D = 2{,}353{,}890{,}000$ Paris feet

Conversions

1 Paris foot = 0.3248 meters = 1.066 English feet

A "line" is a twelfth of an inch.

12 Paris lines = 1 Paris inch

- **Note 7: Timing the Fall of a Rock**

Why did Newton invoke inconvenient pendulum measurements and a pendulum equation to get the rate of fall of a rock at the earth's surface? Why didn't he just go out on the balcony and drop a rock and time it?

Newton and contemporary investigators lacked modern laboratory aids such as photo-gate timers or even stopwatches. In fact they had no clocks sufficiently accurate to time the fall of a rock with the precision Newton needs here for determining the acceleration of gravity.

But by using a pendulum it was possible to get the necessary accuracy. The investigator would time the movement of the sun over many degrees in many minutes of time while the pendulum was going through many oscillations. Then any error in the measurement of sun time would be spread out, with a very small error in terms of each second.

By adjusting the string length of the pendulum, the investigator could get it oscillating with a period of exactly two seconds. Then, using theorems derived for pendulums, conclusions about the force on the pendulum bob could be drawn with sufficient precision.

Newton cites the work of Christian Huygens, who performed many experiments with pendulums and formulated propositions on gravitational fall in his book *The Pendulum Clock*, published in 1673.

Huygens was the first to find the true value for the constant of gravitational acceleration, a number almost exactly the modern accepted value for his latitude. To do this he had to invent a clock accurate enough to do the measurements and he had to formulate an original set of mathematical theorems on both circular and gravitational motion for the calculations.

Expansion of Newton's Sketch of III.4

"That the moon gravitates towards the earth, and is always drawn back from rectilinear motion, and held back in its orbit, by the force of gravity."

To Prove:

The force of terrestrial heaviness and the centripetal force have the same effect at the same distance and may therefore be concluded to be identical.

Proof:

"The moon's mean distance from the earth in the syzygies, in terrestrial semidiameters, is 59 according to Ptolemy and most astronomers; 60 according to Wendelin and Huygens, $60\frac{1}{3}$ according to Copernicus, $60\frac{2}{5}$ according to Streete, and $56\frac{1}{2}$ according to Tycho."

See Note 4 for explanation of syzygies, and why the moon's mean distance was often calculated based on the distances measured there. Note that most astronomers, starting with Ptolemy, place it just around 60; Tycho Brahe, who gets a number significantly lower, is the exception.

"But Tycho and those who follow his tables of refraction in setting a greater refraction—by as much as four or five minutes—for the sun and moon than for the fixed stars (in complete opposition to the nature of light), had increased the parallax of the moon by the same number of minutes; that is, by as much as the twelfth or fifteenth part of the whole parallax. Let this error be corrected, and the distance will come out to be $60\frac{1}{2}$ terrestrial semidiameters, more or less, about what was assigned by the others."

See Note 5 above for the explanation of parallax and refraction, and for the method of measuring distance using parallax.

Here Newton is saying that Tycho's measurement of the moon's distance using parallax was inaccurate because his figure for the refraction was too great by four or five minutes. By adding back that too-large number, the parallax measurement came out too large by the same amount. This came out to a twelfth or fifteenth part of the whole parallax, a significant amount, throwing off his calculation (and the calculations of all who used his tables of refraction) for the distance of the moon.

Using the correct number for the refraction, the distance comes out to about $60\frac{1}{2}$ terrestrial semidiameters, more or less.

"Let us assume that the mean distance is sixty semidiameters at the syzygies,..."

The final result of these considerations is the conclusion that the mean distance of the moon is 60 earth radii.

"...and that the lunar period with respect to the fixed stars amounts to 27 days, 7 hours, and 43 minutes, as is stated by the astronomers; and that the circumference of the earth is 123,249,600 Paris feet, as is established by the measuring Frenchmen."

Newton here gives us more numbers we will need to do the calculations of this proposition. There are a few we will need in addition, which can be derived from the ones Newton gives. Newton gives us the mean distance of the moon from the earth in earth radii, the period of the moon's revolution, and the circumference of the earth. We will also need the diameter of the earth, the circumference of the moon's orbit, and the diameter of the earth's orbit.

Note 6 gathers these in one list.

Part 1: First, We Calculate the Accelerative Centripetal Force on the Moon in Orbit.

"If the moon be supposed to be deprived of all motion and dropped, so as to descend towards the earth, under the influence of all that force by which (by Proposition 3 Corollary) it is held back in its orbit, it will in falling traverse $15\frac{1}{12}$ Paris feet in the space of one minute."

This is the statement of Part 1 of our three-step demonstration, the part that gives the accelerative force of the moon in orbit.

We will use a measure of this force, namely how far the moon will fall starting from the distance of its orbit in a given time, here one minute.

"This conclusion comes from a computation based either upon Proposition 36 of the first Book or (what amounts to the same thing) the ninth Corollary of the fourth Proposition of the same Book."

I.36 does get to exactly the conclusion of I.4 Corollary 9 by a more complicated route, deriving it as an exploration of consequences from things he develops later about conics. We will use I.4 Corollary 9, which we have proved.

"For the versed sine of that arc which the moon in its mean motion describes in the time of one minute at a distance of sixty terrestrial semidiameters, is about $15\frac{1}{12}$ Paris feet, or more accurately, 15 feet 1 inch and $1\frac{4}{9}$ lines."

Step 1: How far will the moon travel in orbit in one minute?

$$\frac{\text{distance of full orbit}}{\text{time of full orbit}} = \frac{\text{distance in one minute}}{\text{one minute}}.$$

$$\frac{7{,}394{,}976{,}000 \text{ Paris ft}}{39{,}343 \text{ minutes}} = \frac{x}{1 \text{ minute}}.$$

$x = 187{,}961$ Paris feet.

Distance traveled in orbit in one minute = 187,961 Paris feet.

Step 2: If deprived of tangential motion, how far would it fall in the same time under the influence of the same centripetal force?

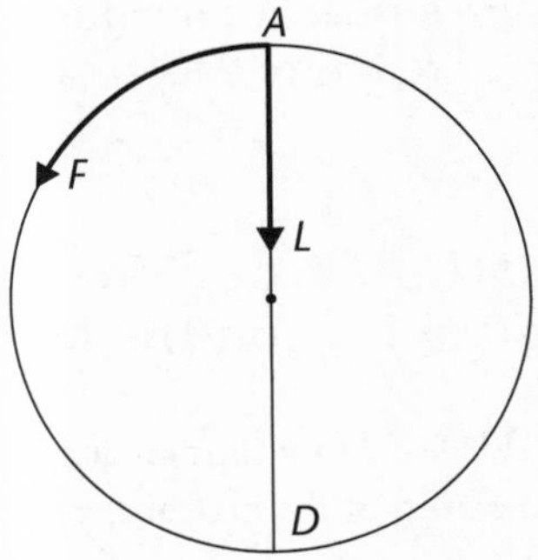

By I.4 Corollary 9, if the moon were released from a point in orbit, and fell towards the earth under the same force that held it in orbit, it would travel a distance such that:

$$\frac{AL}{\widehat{AF}} = \frac{\widehat{AF}}{AD}$$

that is,

$$\frac{\text{distance would fall } AL}{\text{distance traveled in orbit } \widehat{AF}} = \frac{\text{distance in orbit } \widehat{AF}}{\text{diameter of orbit } AD}.$$

For one minute:

$$\frac{AL}{187{,}961 \text{ Paris feet}} = \frac{187{,}961 \text{ Paris feet}}{2{,}353{,}890{,}000 \text{ Paris feet}}.$$

$AL = 15.0089161$ Paris feet, or 15 Paris feet, 0 inches, $1\frac{3}{11}$ lines.

This is the distance the moon would fall in one minute under the force that holds the moon in orbit.

This measures the accelerative force on the moon in orbit.

Conclusion of Part 1.

But wait! We have a problem here. The accelerative force on the moon in orbit may be more than just the accelerative force exerted towards the earth: what if the moon is also being accelerated towards the sun? Our measurement of the time it takes to move around its orbit would reflect both accelerations, and we really want just the acceleration towards the earth.

In fact, Newton is one step ahead of us in this, and the reason his number, 15 Paris feet 1 inch $1\frac{4}{9}$ lines, differs from ours, 15 Paris feet 0 inches $1\frac{3}{11}$ lines, is that he has taken into account the effect of the sun's force on the moon. He's been all through the detective story we're following, and knows how it comes out; and he can calculate, using Book I Proposition 45 Corollary 2, how much the sun perturbs the motion of the moon. It turns out that the action of the sun in drawing the moon away from the earth is to the centripetal force of the moon as 1 to $178\frac{29}{40}$.

He can also work his calculations backwards and subtract out the influence of the sun. So he multiplies the result (15.0089161 Paris feet per minute) by $178\frac{29}{40} / 177\frac{29}{40}$ to get 15.093366, or 15 Paris feet, 1 inch, 1.44 or $1\frac{4}{9}$ lines.

Without the sun drawing the moon away from the earth (at conjunction) and the earth away from the moon (at opposition), the moon would fall even farther in a minute than we observe it to do, by a little more than eight-hundredths of a foot, or more than an inch.

Part 2: Now We Consider What the Accelerative Force Would Be on the Moon at the Earth's Surface.

"Whence, since in approaching the earth that force increases in the inverse of the duplicate ratio of the distance, and is thus greater at the surface of the earth by 60×60 parts than at the moon, a body, in falling by that force in our regions, ought to describe a space of $60 \times 60 \times 15\frac{1}{12}$ Paris feet in the space of one minute, and in the space of one second, $15\frac{1}{12}$ feet, or more accurately, 15 feet 1 inch and $1\frac{4}{9}$ lines."

We're going to do a thought experiment here.

Let's assume that the centripetal force holding the moon in orbit obeys an inverse square law, as demonstrated in Proposition III.3.

If the relation $f \propto 1/r^2$ can be imagined to apply to different moon orbits at different radii, we could find different centripetal accelerative forces operating at those different distances.

So if we supposed the moon had an orbit with a radius twice as great as it currently has, we would expect the accelerative quantity of centripetal force acting on it there to be one quarter of what it is in its orbit of 60 earth radii.

But Newton challenges our imagination to bring it the other way. Suppose we brought the moon down to an orbit of just one earth radius. That is, imagine it to be in orbit just above the earth's surface. Then we will follow the same procedure we used in Part 1 and imagine it deprived of its tangential inertial motion, and let it just drop under the influence of that accelerative centripetal force. Just as we figured out how far it would drop in a given time from its orbit of 60 earth radii in Part 1, we can figure out how far it would drop in a given time from its new orbital distance of one earth radius.

$$\frac{\text{force at orbit}}{\text{force at earth's surface}} = \frac{(\text{radius at surface})^2}{(\text{radius at orbit})^2}.$$

$$\frac{f_o}{f_s} = \frac{r^2}{(60r)^2} = \frac{1}{60^2}.$$

Therefore the force should be 60^2 times greater at the earth's surface.

Now by Lemma 10 Corollary 4,

> And therefore the forces are directly as the spaces described in the very beginning of the motion, and inversely as the squares of the times.

This proportionality was derived for the ultimate ratios. It will only apply over short times and distances—times and distances for which the force will not vary. Now, our derivation of the distance of fall in Part 1 assumed a constant force in applying I.4 Corollary 9. So from that perspective, we need not be concerned about the force varying. But because we're doing a thought experiment, it is worthwhile to see whether the assumption of a constant force would be justified here.

For this we appeal to experience: the acceleration of gravity at the earth's surface does not vary measurably over, say, 15 feet. At a distance of 60 earth radii the proportional distance would be 15×60 or 900 feet. Then if we are working with an inverse square law, a 900-foot change of altitude at 60 earth radii would vary the gravitational force only $1/60^2$ as much as does a 15-foot change of height at the earth's surface. Thus the conditions for use of the proportionality hold even better at the height of the moon than they do at the surface of the earth.

$$F \overset{ult}{\propto} d/t^2. \quad \text{[Lem. 10 Cor. 4]}$$

But we also know that $f \propto 1/r^2.$ [III.3]

Therefore

$$d/t^2 \overset{ult}{\propto} 1/r^2.$$

$$d \overset{ult}{\propto} t^2/r^2. \tag{1}$$

Let d_m be distance fallen at moon's orbit height

Let d_e be distance fallen at earth's surface height

Let t_m be time of fall at moon's orbit height

Let t_e be time of fall at earth's surface

Rewriting Equation 1 as a proportion,

$$\frac{d_m}{d_e} \overset{ult}{=} \frac{(t_m{}^2/r_m{}^2)}{(t_e{}^2/r_e{}^2)}. \tag{2}$$

Because Newton uses minutes at the moon's orbit height but converts to seconds at the earth's surface here, we must reduce the time by a factor of 60.

$$t_m = 60 \times t_e.$$

And we must reduce the radius by 60 to go from the moon's orbit radius to the earth's radius.

$$r_m = 60 \times r_e.$$

Substituting into Equation 2:

$$\frac{d_m}{d_e} \overset{ult}{=} \frac{(60t_e)^2/(60r_e)^2}{t_e{}^2/r_e{}^2}.$$

$$\frac{d_m}{d_e} \overset{ult}{=} \frac{(60)^2(t_e)^2/(60)^2(r_e)^2}{t_e{}^2/r_e{}^2}.$$

$$\frac{d_m}{d_e} \overset{ult}{=} \frac{t_e{}^2/r_e{}^2}{t_e{}^2/r_e{}^2}.$$

Since the second ratio of the proportion is a ratio of equality, so is the first, and

$$d_m = d_e.$$

In one *second*, the moon should fall 15 Paris feet, 1 inch, $1\frac{4}{9}$ lines at the surface of the earth, just as it falls 15 Paris feet, 1 inch, $1\frac{4}{9}$ lines in one minute at the distance of the moon's orbit.

This is what we would expect, at any rate, if the same accelerative centripetal force that is now holding the moon in its orbit would still be operating on it if it were brought to a point (a very low imaginary orbit) close to the surface of the earth.

Part 3: What is the Accelerative Force on a *Rock* at the Earth's Surface?

"And heavy bodies on earth do in fact descend with the same force."

We will show this using a pendulum bob at the surface of the earth. See Note 7 on the use of the pendulum for accurate timings for falling bodies.

"For the length of a pendulum oscillating in seconds, at the latitude of Paris, is three Paris feet $8\frac{1}{2}$ lines, as Huygens has observed."

We are given an observation, an unnumbered phenomenon, that at least at one point of the surface of the earth, near sea level, the string length associated with a pendulum whose period is two seconds is 3 Paris feet, $8\frac{1}{2}$ lines.

Converting to decimal, $L = 3.059$ Paris feet.

"And the height which a heavy body traverses in falling in the time of one second, is to half the length of this pendulum, in the duplicate ratio of the circumference of the circle to its diameter (as Huygens has also pointed out). It is therefore 15 Paris feet 1 inch $1\frac{7}{9}$ lines."

$$\frac{d}{L/2} = \frac{(2\pi r)^2}{(2r)^2} = \pi^2 .$$

$$d = \frac{L}{2}\pi^2 .$$

Since L was 3.059 Paris feet,

$$d = 15.0956 \text{ Paris feet.}$$

Thus in one second a rock at the surface of the earth would fall 15.096 Paris feet, or 15 feet, 1 inch, $1\frac{7}{9}$ lines.

(In English feet, that comes out to 16.09 feet in one second, in case you want to compare it to the modern value.)

"And because the force which holds the moon back in its orbit, if the moon were to descend to the surface of the earth, comes out equal to our force of gravity, therefore (by Rules 1 and 2) it is that very force which we are accustomed to call gravity."

The moon at the surface of the earth, under the centripetal force that pulls it into orbit, would fall (by Part 2) 15 Paris feet, 1 inch, $1\frac{4}{9}$ lines in one second.

A rock at the surface of the earth, under the force of terrestrial heaviness, would fall (by Part 3) 15 Paris feet, 1 inch, $1\frac{7}{9}$ lines in one second.

By Rule 2, to the same effect, assign the same cause.

Therefore since by Rule 2 we would think that the moon would fall under the same accelerative force as a rock, the moon falls by gravity (terrestrial heaviness).

Notice that Newton didn't invoke Rule 2 at the start to say "something is attracting the moon towards the earth and something is attracting rocks to the earth, so to the same effect assign the same cause." He made sure the effect was not just generally the same, but the same as close as Huygens could make measurements. Newton's use of Rule 2 is not careless or crude, but very well-considered and supported by a line of argument that inclines one to think, "Yes, to deny the same cause here runs counter to the way we reason."

Then, of course, we might invoke Rule 4 and say: "And we will hold to this conclusion from induction despite any hypothesis that may be devised to circumvent it." We might need to do this because certainly someone could argue that lunar matter really is different, and if brought to the surface would fall $15\frac{1}{12}$ feet by centripetal force, but not at all by terrestrial heaviness, since it isn't terrestrial. And a rock brought to the distance of the moon might even fall $15\frac{1}{12}$ feet in a minute by terrestrial heaviness, but not be affected by the centripetal force that governs the celestial bodies.

"For if gravity were different from it, bodies, in seeking the earth with the two forces conjoined, would descend twice as fast, and in falling in the space of one second would traverse $30\frac{1}{6}$ Paris feet, in complete opposition to experience."

Once we have figured out, in Part 2, that the moon, if its orbit were brought down near the surface of the earth, would fall by centripetal force about $15\frac{1}{12}$ feet in a second, and have determined by Part 3 that a rock would fall the same distance in the same time by its terrestrial heaviness, if we weren't willing to say that the two forces are the same force, we would have to suppose that the moon would fall twice as fast, under the influence of both the centripetal force and its terrestrial heaviness.

In fact, we don't have experience of this, since we have not brought the moon to that low orbit and dropped it except in our thought experiment.

Newton may be suggesting here a thought experiment going the other way, in which we take bodies known to have terrestrial heaviness and imagine taking them up to the moon's orbit. Then presumably they would be moved by centripetal force just as the moon is. Then if we brought them back down we would have the same rocks our experience tells us fall only at a speed corresponding to one of the accelerations.

Again, of course, we do not have a physical experiment and Newton probably goes too far in saying "in complete opposition to experience."

Perhaps the scholium following this proposition is the thought experiment that he intends will justify this claim.

"This computation is based upon the hypothesis that the earth is at rest. For if the earth and the moon should move around the sun, and should also at the same time move around their common center of gravity, the law of gravity remaining the same, the distance of the centers of the moon and the earth from each other will be about $60\frac{1}{2}$ terrestrial semidiameters, as will be clear to anyone undertaking the computation. And the computation can be undertaken by Proposition 60 of Book I."

> **Note:** This consideration is a bit out of place here. Although Newton knows where he is headed, and that the gravitation of moon to earth will turn out to be a mutual gravitation of each body to the other, that has not yet been demonstrated. And Proposition I.60, which he suggests we use, is a hypothetical proposition which assumes mutual attraction. Since we don't yet know that we can meet that condition, we cannot legitimately use I.60.
>
> The place for this refinement would, it seems, be after Newton has proved III.7, in which he shows that the attractions are mutual in our world. Then he could ask whether we need to rethink, or even discard, the conclusion reached here in III.4. He could then go through this line of thought and show, using I.60, that the amount the earth is moving towards our common center of gravity with the moon is less than we can measure, just as it is for the earth and the rock or pendulum bob.

In I.59, Newton considers what will happen if we suppose a system of two mutually attracting bodies, one of which is fixed with the other orbiting around it, and allow both to move around their common center of gravity under the same force. (Proposition I.58 showed the possibility of such a transformation.) It must be noted that no such relationship has yet been argued for the earth and moon.

Consider the relative distances of the bodies in this hypothetical mutually attracting system. If their distances from each other are to remain the same as before, the periodic time will have to decrease. In I.60 he considers what would have to change if the periodic time were to be held constant. The conclusion is that the principal axis must increase according to the following proportion:

$$\frac{\text{New Axis}}{\text{Old Axis}} = \sqrt[3]{\frac{T+M}{T}}$$

where T is the quantity of matter in the formerly fixed body, and M is that in of the other body. It is this proposition, rather than I.59, that applies to the present case, because the moon's period is very precisely established, while its distance is less so.

Newton's evident aim is to show that the correction in the distance required by I.60 would not be inconsistent with the range of distances measured by astronomers. For the quantities of matter, we must anticipate some of the conclusions to which Newton is leading us, by way of Universal Gravitation. (We must nonetheless keep firmly in mind that Universal Gravitation is *not* established yet!)

In III.37, by comparing lunar and solar effects upon the tides, he concludes that the moon's force is to the sun's force of as 4.4815 to 1. In Corollary 3 to that proposition, he uses this ratio together with the apparent diameters of the bodies to calculate the ratios of densities of sun to moon and moon to earth. Finding the latter to be about 11 to 9, he notes, "Therefore the body of the moon is more dense and more earthly than the earth itself." In the next corollary (Corollary 4) he combines this with the ratio of the earth's and moon's diameters (365 to 100, according to the astronomers) to find that the ratio of quantities of matter is 1 to 39.788.

We can use this number provisionally in the above proportion to compute the adjustment in the distance:

$$\frac{\text{New Axis}}{60} = \sqrt[3]{\frac{1 + 39.788}{39.788}}$$

As Newton says, this comes out to about $60\frac{1}{2}$.

Therefore, even if we do need to adjust the distance to allow for the earth's motion, the adjustment will be small enough to ignore.

Q.E.D.

Scholium

The demonstration of the proposition can be more amply displayed thus. If many moons were to revolve around the earth, exactly as in the system of Saturn or Jupiter, their periodic times (by an argument of induction) would observe the law of the planets discovered by Kepler, and therefore, their centripetal forces would be inversely as the squares of the distances from the center of the earth, by Proposition 1 of this Book. And if the lowest of these were small, and were nearly to touch the peaks of the highest mountains, its centripetal force, which keeps it in its orbit, would (by the foregoing computation) be very nearly equal to the gravities of bodies on the peaks of those mountains. It would thus come to pass that if the same small moon were deprived of all the motion by which it proceeds in its orbit, it would descend to earth because of the loss of the centrifugal force by which it had remained in its orbit, and would do this with the same velocity with which heavy bodies fall on the peaks of those mountains, because of the equality of the forces with which they descend. And if that force by which that small lowest moon descends were different from gravity, and that small moon were also heavy towards the earth in the manner of the bodies on the peaks of the mountains, the same small moon under the two conjoined forces would descend twice as fast. Therefore, since both forces—the latter ones, of the heavy bodies, and the former ones, of the moons—look to the center of the earth, and are similar and equal to each other, these same [forces] (by Rules 1 and 2) will have the same cause. And consequently, that force by which the moon is kept back in its orbit, will be that very force which we usually call "gravity," and this must above all be true lest the small moon at the peak of the mountain either lack gravity, or fall twice as fast as heavy bodies usually fall.

Note on Scholium After III.4

It seems Newton still wasn't content that he had said enough that the force of Rule 2 would compel us to the conclusion that the centripetal force was gravity. So far from crudely relying on Rule 2, he wants the inductive step to be inescapable.

In addition to appreciating his efforts to satisfy us that we have a rigorous demonstration, we may be grateful for his scruples because they have given us this really delightful thought experiment of the little moons orbiting just at the mountain peaks: a charming finish to an exhilarating proposition.

Pause After III.4

- **Universality of Inverse Square Law**

In our thought experiment in this proposition, we bring the moon down to the distance from the center of one earth's radius, and predict what its rate of fall would be based on an assumption that an inverse square law holds at that distance.

Remember how we got here. In I.4 Corollary 6, for circular orbits, we established that if we found a sesquiplicate ratio of periodic times to radii, the force holding those bodies in orbit would be inverse square. Then in III.1 and III.2 we applied that to sesquiplicate ratios found for the moons of Jupiter and Saturn and to the primary planets orbiting the sun to say that the forces on those bodies were according to inverse square. For Earth's moon, which we deal with in III.4 here, we used I.45 Corollary 1 to find the force law that would fit with a circular or near-circular orbit and a given rate of movement of apsides. In I.45, as in I.4, we were given an orbit and we drew certain conclusions from it. The important thing to notice is that no assumption has been made, nor any argument advanced, that another body in another orbit at, perhaps, another distance, would have an inverse square relationship with the previous bodies.

As discussed in the Notes on I.4 Corollaries 3–7, having found certain bodies which exhibit the sesquiplicate relationship, we only know something about forces on *those* bodies at *those* distances. We don't know what other bodies we might place at various distances would show for their periodic times.

We are accustomed to thinking of the sesquiplicate ratio as a natural principle that would apply to any body put into a circular orbit at any distance. But as we have noted in following Newton's argument, he has not been making such an assumption.

However, to go on from here, Newton needs a generalization that gives us an inverse square law *everywhere* in the universe, without having to prove that a $\frac{3}{2}$ power relationship exists, or a certain motion of apsides occurs, for every point.

The direct way of showing that the ratio is a consequence of a continuously varying inverse square force law, and not (as Kepler thought) just the contingent consequence of several factors, would be to put something into orbit and see how its periodic time relates to its height. This way was not open to Newton, so he had to pile up the evidence, showing that a view such as Kepler's would require us to abandon the evidence of III.4 and to suppose an improbably large number of coincidences in the establishment of orbits of all kinds.

That the inverse square relationship is continuous would be a simple and elegant and plausible natural law. But sometimes it does turn out in nature that the conditions governing a result are complex and particular to the context. Thus, it would be precipitous to *assume* that if the moon were moved to the surface of the earth, the force on it could be calculated by taking the inverse of the square of the distance.

However, as a thought experiment, we could suppose that the law was continuous and see what results. This is what Newton does here. When he discovers that this assumption leads to a rate of fall of the moon which is identical to the rate of fall of a rock, he has a kind of confirmation not just of the sameness of the moon and the rock and the sameness of the centripetal force and the force of gravity, but also of the continuity of the inverse square relation.

So, once again, we are struck by the extraordinary significance of III.4. Now we discover another dimension to its power by taking its reasoning conversely. Suppose we assume that the moon is held in its orbit by gravity, that is, by the power that makes terrestrial bodies heavy. We then consider whether the celestial bodies are not just being accelerated inversely as the square of their distances from their central bodies out there at those particular, and celestial, distances, but whether it might be "inverse square all the way down."

This is a necessary step for Newton's work in Proposition III.6 and beyond. It will allow us to say there that accelerative force is constant at equal distances and therefore that weights vary as quantities of matter at equal distances, which then leads us to the coincidence of inertial and gravitational mass.

- **Universality of Gravitation**

We might (if very precipitous) imagine that by proving that the moon is kept in orbit by gravity we practically have universal gravitation. After all, we have broken the barrier between terrestrial and celestial matter, and we can see it all unfolding before us.

In fact, including the moon with terrestrial matter is just the first step. Newton himself had speculated that it was gravity that kept the moon in orbit well before *Principia*'s formal mathematical proof. By contrast, Newton was suspicious of the full universality of the inverse square attraction, even *after* he had derived ellipses from the inverse square law. And there were good reasons for resisting this.

First, he could show that our world exhibits stable elliptical orbits and equable description of areas. But universal gravitation would lead to complicated relationships among many bodies which would destroy the simple inverse square law forces, and seemingly destroy the stability of the universe. In Proposition I.66 Newton explores this, and many of the corollaries deal with why the universe might appear stable, and the orbits of planets very close to elliptical, despite assuming the "universality" of the influence of bodies on one another. (See Appendix B.)

Second, there is a significant stumbling block to the proof that all particles attract all other particles: it requires that a sphere of such particles acts as if the whole force were concentrated at the center. This he proved in Propositions I.71 through I.75, and we will look at that development.

So let's be appropriately thrilled by III.4, and look forward to watching the rest of the development fall out, but know that Newton still has some tricky sledding ahead.

Pause Before III.5

By the end of III.4 we can say that the moon is falling under the same accelerative force as the bob. And since we say that the pendulum bob (or a rock at the surface of the earth) is understood to fall by gravity, we should be able to conclude that the moon falls by gravity.

Proposition III.4 establishes, dramatically and persuasively, that the moon, a celestial body, is operated on by the laws of terrestrial heaviness that we call gravity on earth. The laws for the "sub-lunar world" apply to the moon also; in fact, the moon's terrestrial heaviness is the very force that is turning it out of its straight line inertial path and into its orbit.

Now Newton has an important step to make. He wants to assert that it is also the same gravity that makes the moons of Jupiter and Saturn tend towards those planets, and that makes the circumsolar planets tend towards the sun, as made our moon and terrestrial rocks tend towards the earth.

But wait! Gravity traditionally was understood as a tendency towards the center of the earth. The moons of Jupiter and Saturn, as he showed in III.1, tend not towards the center of the earth, but towards the center of those planets; and the circumsolar planets similarly tend towards the center of the sun.

Furthermore, the celestial bodies had traditionally been seen as different in essence from terrestrial bodies. Even Kepler, who suggested that the moon was held in orbit by terrestrial gravity,* made no such supposition about the planets in their orbits around the sun. Kepler had other ways of explaining those motions.

Newton takes on this important step in III.5. He mentions Rule 2 again: to the same effect we ought, in the absence of other indicators, to assign the same cause. The "same effect" is that the revolutions of these other moons about their respective centers of forces and the planets about the sun are appearances of the same sort as the revolution of our moon about the earth.

Newton spells out the "sameness" of the effects by citing them specifically, showing that the appearances are more than approximately similar. First, by III.1 and III.2, the centers towards which the orbiting bodies tend are the planets and the sun, just as the center of forces towards which the moon tends is the earth. Second, by the same propositions, the forces, "in receding from Jupiter, Saturn, and the sun, decrease in the same ratio and by [the same] law by which the force of gravity decreases in recession from earth."

The first part of the first corollary to III.5 draws the full conclusion: "Gravity is therefore given towards all planets without exception. For no one doubts that Venus, Mercury, and the rest are bodies of the same kind as Jupiter and Saturn." The second corollary follows up with the conclusion that: "Gravity, which looks to each planet" (the sun here taking its traditional place in this category) "is inversely as the square of the distances of places" (and of satellites in those places) "from its center." The third corollary adds: "All planets are mutually heavy towards each other by Corollaries 1 and 2. ..."

The Pause After Proposition III.5 considers where this leaves us.

Proposition 5

That the planets around Jupiter gravitate towards Jupiter, the planets around Saturn towards Saturn, and the planets around the sun towards

---

* Johannes Kepler, *New Astronomy*, Cambridge University Press, 1992, pp. 55–56.

the sun, and by the force of their gravity are always drawn back from rectilinear motions and confined to curvilinear orbits.

For the revolutions of the planets around Jupiter about Jupiter, of the planets around Saturn about Saturn, and of Mercury and Venus and the rest of the planets around the sun about the sun, are phenomena of the same kind as the revolution of the moon about the earth, and for that reason (by Rule 2) depend upon causes of the same kind. This is primarily because it is demonstrated that the forces upon which those revolutions depend, look to the centers of Jupiter, Saturn, and the sun, and in receding from Jupiter, Saturn, and the sun, decrease in the same ratio and by the [same] law by which the force of gravity decreases in recession from earth.

Note on III.5

Planets: Traditionally the "seven planets" were Mercury, Venus, Mars, Jupiter, Saturn, the sun, and the earth's moon.

When Newton says "planets," he may include the sun or earth's moon; judge by the context. Newton also at times refers to the moons of Jupiter and Saturn as "planets." Sometimes he makes the distinction by calling the moons "secondary planets" as opposed to "primary planets."

Note that traditionally, and in Newton, the earth is not called a planet, although there are places where Newton states general conclusions about "planets" which we have proved apply to the earth.

Expansion of Newton's Sketch of III.5

"That the planets around Jupiter gravitate towards Jupiter, the planets around Saturn towards Saturn, and the planets around the sun towards the sun, and by the force of their gravity are always drawn back from rectilinear motions and confined to curvilinear orbits."

To Prove:

That the satellites of Jupiter, Saturn, and the sun are all held in orbit by gravity.

Proof:

"For the revolutions of the planets around Jupiter about Jupiter, of the planets around Saturn about Saturn, and of Mercury and Venus and the rest of the planets around the sun about the sun, are phenomena of the same kind as the revolution of the moon about the earth, and for that reason (by Rule 2) depend upon causes of the same kind."

Since they behave with respect to their centers of forces as the moon behaves with respect to the earth, by Rule 2 we would assign the same effect to the same cause.

But note that Newton doesn't rely on a hand-wave, or use that "rule" as authority: he proceeds to list the exact behaviors that are the same, so that the evidence will lead *us* to reason according to that rule.

"This is primarily because it is demonstrated that the forces upon which those revolutions depend, look to the centers of Jupiter, Saturn, and the sun, and in receding from Jupiter, Saturn, and the sun, decrease in the same ratio and by the [same] law by which the force of gravity decreases in recession from earth."

Specifically, their behavior is identical to that of the moon around the earth in two ways:

1. Proposition III.1 demonstrated that the centers of forces upon which Jupiter's and Saturn's moons depend are Jupiter and Saturn respectively, and Proposition III.2 that the center of forces on which the circumsolar planets depend is the sun, just as the center of forces on which the moon depends is the earth (by III.3).

2. Again by Propositions III.1 and III.2, the respective forces are inversely as the squares of the distances of the places of those planets from their centers, just as the force on the moon varies as the moon's distance from the earth (by III.3).

By Proposition III.4, the force on the moon was shown to be the same as the force of terrestrial heaviness on a stone, and therefore centripetal force, the force that is inversely as the distance from the earth, is gravity.

Since the centripetal force has been shown to be gravity for our moon, it must also be gravity for the other satellites that exhibit the same behavior.

Q.E.D.

Note on III.5 Corollaries

The corollaries of this proposition challenge our alertness to distinguish the two sorts of commentary Newton gives us, because his excitement is spilling over at this point.

The key to success in this is the practice of being scrupulous about proving every proposition by our own mental efforts, and not just swallowing Newton's claims and sketches as if we were passive and gullible consumers. As active adventurers laying for ourselves the foundations of universal gravitation, we are in a position to recognize the difference between drawing out implications already present in what we have just done (true corollaries) and views ahead to what is yet to be proved (conclusions that cannot be used until they are proved). We *can* distinguish the two, but only by keeping careful track of what we have actually established by formal proof.

III.5 Corollary 1

Gravity is therefore given towards all planets without exception. For no one doubts that Venus, Mercury, and the rest are bodies of the same kind as Jupiter and Saturn. And since by the Third Law of Motion all attraction is mutual, Jupiter will gravitate towards all of its satellites, Saturn towards all of its [satellites], the earth towards the moon, and the sun towards all the primary planets.

Expansion of Newton's Sketch of III.5 Corollary 1

The first part of this is a true corollary, the second seems to be more of a scholium, looking ahead to what Newton will prove in the upcoming propositions.

Part 1

"Gravity is therefore given towards all planets without exception. For no one doubts that Venus, Mercury, and the rest are bodies of the same kind as Jupiter and Saturn."

Having made the hardest steps, to see that the moon acts like a rock being drawn to the earth as its gravitational center, that Jupiter's and Saturn's moons are drawn to Jupiter and Saturn as gravitational centers, and that the planets are likewise drawn to the sun by gravity, it may feel like an easy step to say that Venus, Mercury, and Mars are "bodies of the same kind."

But let's see the full implications of this, lest we be guilty of the too-hasty generalizations of Rules 2 and 3 that Newton works so carefully to avoid.

We began with the most familiar experience of gravity: rocks falling to earth. Earth is clearly a gravitational center for the rocks. Where this gravitational power is to be found is a matter for some debate. For Aristotle, the tendency to the gravitational center was in the rocks. Modern readers of Newton might imagine the power of gravity as existing in the body at the center of forces. But Newton's many disclaimers about not hypothesizing any cause remind us that we are not to assume that, and we know (if we have been careful) that this is not something we have proved.

But whatever the power, whatever the mechanism, gravity certainly exists. Rocks do fall. Now we ask whether that same power, whatever the mechanism, operates more broadly in the cosmos. Proposition III.4 has argued from the astonishing agreement of effects that the power that holds our moon in orbit is that same power of gravity.

But we see that the moons of Jupiter and Saturn behave just like our moon—same force law around the planet as a center of forces. Newton is attempting to persuade us that the same cause is operating here too.

Perhaps those planets possess a power of gravity for their moons (and therefore rocks at their surface) just like the earth's. Or perhaps the mechanism doesn't lie in the earth or planet, but in the tendency of the rocks and moons to seek the planet. Or perhaps there's an intermediate cause. Maybe, for example, something is happening in the medium filling space around them: it might be that the bodies are pushed in because the ether is denser farther from the central body. We cannot, and, fortunately, need not, decide now what the cause or mechanism of centripetal force or gravity is.

Rule 1 said "there ought not to be admitted any more causes of natural things than those which are both true and sufficient to explain their phenomena." That is, the conviction we share as philosophers that "Nature...does not indulge herself in superfluous causes" leads us to be content with the one cause, whatever it may be.

And if this force is operating between the earth, Jupiter, and Saturn, and bodies in their vicinities, why would one not suppose the same for Mercury, Venus, and Mars? Admittedly, we can observe no moons or falling rocks in their vicinity, but we may not have seen a rock fall on some remote desert island either, yet still believe that it would tend to fall there too, as Newton says in Rule 2.

If it were a quality or power or mechanism of the *matter* of the planet to attract, or of the rocks and moons to be attracted, then by Rule 3 we would incline to that belief that "gravity is given towards" Mercury, Venus, and Mars.

Even if gravity is not a power or mechanism of matter, our sense of the simplicity of Nature, as formalized in Rule 1, would disincline us to believe that rocks would fall by a different cause or principle on Mercury than on Jupiter.

Part 2

"And since by the Third Law of Motion all attraction is mutual, Jupiter will gravitate towards all of its satellites, Saturn towards all of its [satellites], the earth towards the moon, and the sun towards all the primary planets."

This is something that would appear to follow from the Rules of Philosophizing. If Jupiter orbits the sun by gravity, and Jupiter's moons orbit Jupiter by gravity, wouldn't a rock orbit a jovian moon by gravity? And if a rock is impelled towards the moon, why wouldn't Jupiter also be impelled towards the moon? (Kepler pointed out that a large magnet will be pulled towards a small one.)

Newton, however, does not argue it from Rules of Philosophizing, but from the third Law of Motion. He invokes this law as justification for the mutuality of gravitational attractions again in Proposition III.7. Some questions to be considered about this are raised in the notes in the Pause After III.7.

III.5 Corollary 2

Gravity, which looks to each planet, is inversely as the square of the distances of places from its center.

Expansion of Newton's Sketch of III.5 Corollary 2

The "inverse square" part of this corollary is demonstrated for Jupiter and Saturn in III.1, for the primary planets in III.2, and for the moon in III.3. The "gravity" part is shown in III.4 and III.5.

III.5 Corollary 3

All planets are mutually heavy towards each other by Corollaries 1 and 2. And hence, Jupiter and Saturn, attracting each other near conjunction, perturb each others' motions perceptibly; the sun perturbs the lunar motions; the sun and moon perturb our sea; as will be explained in what follows.

Expansion of Newton's Sketch of III.5 Corollary 3

The first sentence generalizes the claim of Corollary 1. It also turns the focus of attention around since Corollary 1 has gravity being given towards the attracting body and Corollary 3 has it acting on the attracted body. This perspective of Corollary 3 is further supported by the argument of the first part of III.6.

The second sentence seems to be more a scholium or preview of things to be demonstrated later than a true corollary. (He's going to show later in Book III that the perturbations can all be explained as gravity.) Newton tips us off about this preview function by the final words "as will be explained in what follows."

Scholium

Hitherto we have called that force by which the celestial bodies are confined to their orbits "centripetal." It is now established that the same [force] is gravity, and for that reason we shall hereafter call it "gravity." For the cause of that centripetal force by which the moon is confined to [its] orb is obligated to extend to all planets, by Rules 1, 2, and 4.

Note on Scholium After III.5

This scholium summarizes what we have arrived at in the last two propositions and announces that from now on we will use the term gravity for centripetal force.

Question for Discussion: The wording of the last sentence is intriguing. The Rules of Philosophizing have seemed up to now to be statements of the workings of what we might call scientific thinking. (See notes on the Rules of Philosophizing.) The responsible natural philosopher is constrained by this discipline of reasonable thought.

The odd wording of the Scholium that the cause "is obligated" (the Latin is "*debet*") under Rules 1, 2, and 4 to extend to all planets sounds more as if it is Nature, or the Creator, who is being constrained. *Debet* is a legal word; the cause is obligated under law to extend to all planets. This may fit with the idea of laws of nature. Or does it?

Pause After III.5

Principia's readers are now in a new world. Gravity has cut loose from terrestrial heaviness and is—what? The new terrain is still unfolding, but gravity is clearly now something quite different from a tendency of bodies to seek one natural place of rest as Aristotle (and those so heavily influenced by him through the middle ages and into the Renaissance) had thought of it. Newton has not only left his readers without a cause, he has taken away their traditional understanding of what gravity is. We must proceed very carefully here. And if, as present day readers, we imagine we know the cause or at least have a common-sense intuition of how it all works, we must proceed even more carefully.

So let us look at exactly what we have proved and what we have not yet proved.

What we *have* established is:

(a) "That the circumjovial planets gravitate towards Jupiter, the circumsaturnian planets towards Saturn, and the circumsolar towards the sun, and by the force of their gravity are always drawn back from rectilinear motions and confined to curvilinear orbs" (by III.5).

(b) That by extension, what is true of these bodies should also be true of all the planets, so that there is a power of gravity towards all the planets, including the sun and moons, and the force of gravity towards those bodies is inversely as the square of the distance from them (by III.5 Cor. 1 first part).

(c) By further extension this power of gravity should operate not only on satellites of the bodies and on rocks near their surfaces, but on all bodies at whatever distance, including other planets. So Jupiter will gravitate towards Saturn, and towards its own moons (III.5 Cor. 1 second part, III.5 Cor. 3).

The scholium says that we will hereafter call centripital force "gravity." By extension, we are equally entitled to substitute the word "gravity" in any proposition we draw from Book I where in the original form we saw "centripetal force".

In addition we might note that we are now dealing with a body at the center of forces, and we now speak of gravity tending towards that body. This came by way of seeing that the center of forces appeared to be at the center of the body to which other(s) tended.

What have we *not* proved? To start with, we have not proved "universal gravitation." Universal gravitation requires that each particle of every body attract each particle of every other body, all inversely as the square of the distance between the particles.

And although we're now speaking of a body at the center of forces, nothing ties the formal effect of "gravitation" to masses. It could still be angelic intelligences moving the planets around centers based on a plan of their own (or of God) having nothing to do with the masses of either the orbiting body or the body being "looked to" or tended towards.

Flashback to II.24 for Use in III.6

In III.6, Newton will at last bring the mass of attracted bodies into our purview. Newton is going to prove that the gravitational attractions of bodies—their weights—are precisely proportional to their inherent inertial forces—their masses—at any given distance from the gravitational center. As far as we can see, there is no reason why this must be so, why the universe had to be so arranged that a body that is twice as hard to push around is also attracted towards the center of the earth, or of the sun, exactly twice as much. Couldn't the proportion have been different? Would there have to be any fixed relationship at all? There is no such relationship for other forces, such as electricity: the attraction of charged bodies towards each other has nothing to do with their masses. Why does it happen with gravity?

For more on this subject, which is central to Einstein's Theory of General Relativity, see the Pause After III.6. It is mentioned here because before he can prove III.6, Newton needs to have a way to explore the relationship between weight and mass. This has to be done experimentally, because of the remarkable fact that we can't tell whether weight and mass are proportional without actually looking at our universe. .

II.24 gives a very precise and accurate way of finding out whether weight and mass on earth are always proportional. By setting up pendulums with equal string lengths and equal weights, and comparing the times of their swings, we can discover the exact proportion, if any, between weight and mass. It turns the pendulum into a device for determining the mass by the measurement of weight, length, and time. Although the mathematical argument is rather long, the simplicity and beauty of the device makes going through the proof well worth the trouble. If there isn't time to do the whole proof, it can be made considerably simpler by taking the conclusion of Part 2 of the expanded proof as given, and proving the rest.

Book II Proposition 24

The quantities of matter in pendulous bodies whose centers of oscillation are equally distant from the center of suspension are in the ratio compounded of the ratio of the weights and the duplicate ratio of the times of oscillation in a vacuum.

For the velocity which a given force can generate in a given matter in a given time is as the force and the time directly, and the matter inversely. To the extent that the force is greater or the time is greater or the matter is less, a greater velocity will be generated. This is manifest through the Second Law of Motion. But now if the pendulums are of the same length, the motive forces in places equidistant from the perpendicular are as the weights, and therefore if two bodies in oscillating describe equal arcs, and those arcs be divided into equal parts, since the times in which the bodies describe the individual corresponding parts of the arcs are as the times of the whole oscillations, the velocities in the corresponding parts of the oscillations will be to each other as the motive forces and the whole times of the oscillations directly and the quantities of matter inversely, and consequently the quantities of matter are as the forces and the times of oscillations directly and the velocities inversely. But the velocities are inversely as the times, and consequently the times directly and the velocities inversely are as the squares of the times, and therefore the quantities of matter are as the motive forces and the squares of the times, that is, as the weights and the squares of the times.

Q.E.D.

Notes on II.24

This proposition, though demanding some concentration to see the algebraic development through, is one of the most important in *Principia* and illustrates the impressive breadth and originality of Newton's insight and the power of the method he is bringing to bear in *Principia*.

- **Hypothetical:** The first thing to note is that this proposition, like all those in Books I and II, is a hypothetical mathematical demonstration. It makes no assumptions about a particular physical universe. Whether, and in what way, it applies to our universe must be established by reference to physical observations; this will be done in III.6. Like the other propositions of the first two books of *Principia*, this is a tool in the toolbox, but it illustrates the plan particularly well. Newton doesn't assume what relation (if any) string length will have to periodic times. He makes no assumption about how different materials in the pendulum bobs might affect the periodic times. He makes no assumption about how things might be different in different locations. In this proposition and its corollaries, he gives us the general mathematical relationships into which to fit the results of our tests of the nature of our world, so that we may then draw conclusions from what we find.

When we later perform those tests and apply these mathematical conditionals to the results, we are going to find such amazing things that it will be grounding to have seen just how they came to fall out.

The section called Pause After II.24 suggests some things to ponder or discuss after working through this proposition, in preparation for going on to the astonishing results of III.6.

• **Motive Force:** Note that we're dealing with another measure of centripetal force here, motive force. Through our study of Book I we dealt entirely with accelerative quantity of force, except for a cameo appearance in I.17 of absolute quantity of force.

Let's explore this new measure of force a little, and see its relation to the accelerative force we have been using. Jumbling of these two measures of force is common at this stage, and it can be confusing and frustrating for the student.

As we go on from here we will be complicating our simple propositions about accelerative force by introducing the mass of the attracted body. Then we will get even more complicated by adding consideration of the mass of the attracting body. You can imagine that not being sure what definition of force we are using will be a nuisance you don't need; let's take a moment now to review.

> *Definition 8:* The motive quantity of centripetal force is the measure of the same proportional to the motion which it generates in a given time.
>
> *Law 2:* The change of motion is proportional to the motive force impressed; and is made in the direction of the right line in which that force is impressed.

Therefore motive quantity of force varies as the motion generated in a given time.

> *Definition 2:* The quantity of motion is the measure of the same arising from the velocity and the quantity of matter conjointly.

So quantity of motion $\propto v \times m$.

Substituting into our definition for motive force,

$f_m \propto \Delta mv$ in equal times.

$$f_m \propto \frac{\Delta mv}{\Delta t}.$$

That is,

$$\frac{F_m}{f_m} = \frac{\Delta(MV)/T}{\Delta(mv)/t},$$

where one pendulum is represented by capitals and the other by lower case letters.

Within each quantity of motion is a body whose mass remains constant. Therefore we may bring the constant mass out:

$$\frac{F_m}{f_m} = \frac{M\Delta V/T}{m\Delta v/t}.$$

$$f_m \propto m\frac{\Delta v}{\Delta t}. \qquad \text{(N1)}$$

But by Definition 7, "The accelerative quantity of centripetal force is the measure of the same, proportional to the velocity which it generates in a given time."

$$f_a \propto \frac{\Delta v}{\Delta t}.$$

Therefore $f_m \propto m \times f_a$. (N2)

That is,

$$\frac{F_m}{f_m} = \frac{M}{m} \times \frac{F_a}{f_a}. \qquad \text{(N2')}$$

Or, solving Equation (N2) for f_a,

$$f_a \propto \frac{f_m}{m}, \qquad \text{(N3)}$$

that is,

$$\frac{F_a}{f_a} = \frac{F_m/M}{fm/m} = \frac{F_m m}{f_m M}. \qquad \text{(N3')}$$

We will refer back to Equations N2 and N3 and the steps of their derivation from propositions to come.

The motive force and the accelerative force in this proportion are both forces *on* the body of mass m. No assumption is made of any body at the center of forces.

- **Weight:** Weight is an example of motive force, as given by Definition 8.
  What we understand as weight is quantity of matter times the acceleration of gravity, or mass times the velocity generated in a given time. It is the measure of a force, a motive force, the aggregate tending downward toward

the center of the earth, or towards some other center, of a certain quantity of matter.

Expansion of Newton's Sketch of II.24

Given:

Pendulums with equal string length, swinging in a vacuum.

To Prove:

The masses of the pendulum bobs vary as their weights multiplied by the squares of the periodic times.

$m \propto w \times P^2$.

Part 1: Forces and motions in the pendulum

"For the velocity which a given force can generate in a given matter in a given time is as the force and the time directly, and the matter inversely. To the extent that the force is greater or the time is greater or the matter is less, a greater velocity will be generated. This is manifest through the Second Law of Motion."

Solving Equation N1, in the note entitled Motive Force, for Δv,

$$\Delta v \propto \frac{f_m \Delta t}{m} .$$

"But now if the pendulums are of the same length, the motive forces in places equidistant from the perpendicular are as the weights, ..."

The motive forces f_m are the forces tending to pull the pendulum bobs in to the central position.

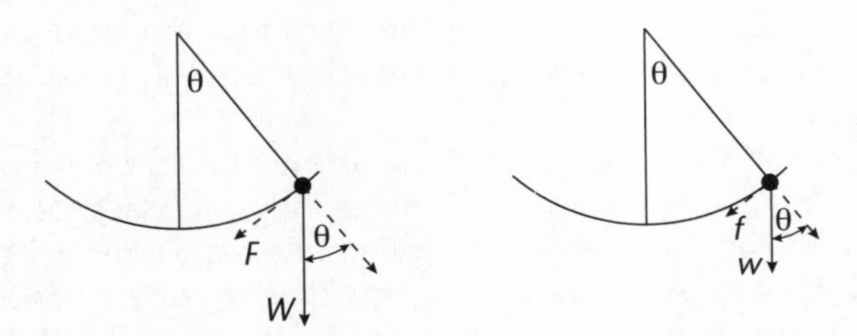

In the above diagram, they are represented by F and f. By Corollary 2 of the Laws of Motion, the weights W and w of the two "pendulous bodies" are each resolved into a pair of component forces, represented by the broken lines. The components that act along the string serve only to produce tension in the string, and have no effect upon the motion of the bodies. The

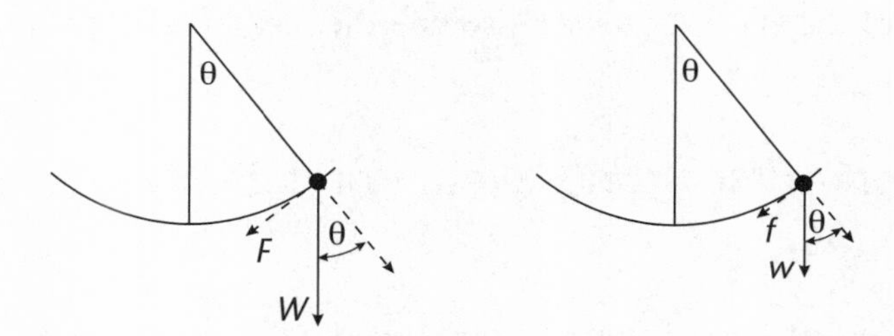

other components, F and f, act along the direction of travel, and are thus the active components in this situation. By trigonometry,

$$F = W \sin\theta \quad \text{and} \quad f = w \sin\theta.$$

That is, dividing simultaneous equations and cancelling $\sin\theta$,

$$F/f = W/w \quad \text{or} \quad f \propto w.$$

Part 2: Proof that the times to traverse equal small parts of arcs on two pendulums are proportional to the whole times.

Although the truth of Newton's claim here might seem intuitively clear, the proof is complicated and full of traps. In order to find that the periodic times are in a given ratio, we divide the swing of each pendulum into a large number of equal parts, with the intention of showing that the times on corresponding pairs of parts are in a fixed ratio. But the key to doing this is the constant of proportionality that we are calling k, which is in effect the ratio of weights of two pendulum bobs of equal mass, and this constant turns out to relate the times on the small parts to the changes in velocity on those parts. So to find the times, we must find the changes in velocity over each part. But that turns out not to be easily accessible; instead, we can find the ratio of the *final* velocities at the end of any number of parts, and can show that that ratio is constant, and can use that constancy to show that the *changes* in velocity must also be in a constant ratio. This then gives us the ratio of the times over the small parts, which in turn gives the ratio of the periodic times.

The general idea, however, can be put into this simple form which, though not a fruitful basis for a proof, makes the claim plausible. If we have two equal pendulums, and divide the swings into a great many equal parts, each one being approximately straight, the time over each corresponding pair of these parts is going to be in a fixed ratio dependent only on the weights, because everything else (the length and incline of the parts) is the same for each pair of parts.

"... and therefore if two bodies in oscillating describe equal arcs, and those arcs be divided into equal parts, since the times in which the bodies describe the individual corresponding parts of the arcs are as the times of the whole oscillations, ..."

We are given two bodies oscillating in equal arcs. Let the equal arcs be divided into equal parts. We will prove that times for the full oscillation will be as the times for the little parts.

Step 1

Although in actuality the force experienced by a pendulum bob depends on the displacement, let the parts into which the equal arcs were divided be taken so small that the force can be treated as constant throughout the extent of each little part.

By Law 2, motive force $\propto \dfrac{\Delta mv}{\Delta t}$ (see second bulleted note above);

or, since the mass is constant, over any single little part we have

$$f \propto m\frac{\Delta v}{\Delta t}$$

or, comparing two little parts of the arcs of two different pendulums,

$$\frac{F}{f} = \frac{M\Delta V/\Delta T}{m\,\Delta v/\Delta t},$$

where ΔT and Δt represent the times required to traverse the little parts, respectively, and where ΔV and Δv represent the changes in speed which the bodies experience in traversing those parts.

By Part 1, the forces at any equal angles of displacement are as the weights:

$$f \propto w \quad \text{or} \quad \frac{F}{f} = \frac{W}{w}.$$

Substituting,

$$\frac{W}{w} = \frac{M\Delta V/\Delta T}{m\,\Delta v/\Delta t}.$$

Multiply both sides by m/M:

$$\frac{\Delta V/\Delta T}{\Delta v/\Delta t} = \frac{Fa}{f_a} = \frac{W\,m}{w\,M}.$$

(Note that this is an instance of Equation N3′, with the weights appearing as the motive forces).

But both the mass and the weight are constant; therefore:

$$\frac{W\,m}{w\,M} = k \quad \text{or} \quad \frac{Fa}{f_a} = k. \tag{1}$$

Thus at any equal angles of displacement,

$$\frac{\Delta V/\Delta T}{\Delta v/\Delta t} = k\,,$$

where k is the constant of Equation 1. Note that k is the same for *any* pair of equal displacements.

Multiply both sides by $(1/k)(\Delta T/\Delta t)$:

$$\frac{\Delta T}{\Delta t} = \frac{1}{k}\frac{\Delta V}{\Delta v}\,. \tag{2}$$

By this equation, if we can show that either the Δt's or the Δv's have a fixed ratio for all intervals, we can also conclude that the other pair has a fixed ratio.

Step 2: Ratio of Changes in Velocity

Since each of the little equal parts of the oscillation has been taken sufficiently small that the force may be considered as constant throughout its length, the velocity on such a part will be changing nearly uniformly. Therefore, we may use an extended version of Galileo's first proposition on accelerated motion. This proposition states:

> The time in which a certain space is traversed by a moveable in uniformly accelerated movement from rest is equal to the time in which the same space would be traversed by the same moveable carried in uniform motion whose degree of speed is one-half the maximum and final degree of speed of the previous, uniformly accelerated motion.

To apply this to motions that begin with some initial velocity, we must take the initial velocity into account in finding the mean speed. Thus, instead of half the final velocity, we take half the sum of the initial and final velocities. The distance travelled can be expressed as the product of this half-sum (the average velocity) and the time:

$$s = \frac{v_f + v_i}{2}\Delta t\,.$$

Multiply this by the accelerative force f_a:

$$f_a s = f_a \Delta t \frac{v_f + v_i}{2}\,.$$

By Definition 7, the accelerative force is measured by the change in velocity over a given time:

$$f_a = \frac{\Delta v}{\Delta t}\,.$$

Substitute this on the right side of the previous equation:

$$f_a s = \frac{\Delta v}{\Delta t} \Delta t \, \frac{v_f + v_i}{2} \, .$$

Since the Δt's cancel, and $\Delta v = v_f - v_i$,

$$f_a \, s \; = \; \tfrac{1}{2} (v_f - v_i)(v_f + v_i);$$

$$2f_a \, s \; = \; v_f{}^2 - v_i{}^2 .$$

Solving for $v_f{}^2$,

$$v_f^2 \; = \; v_i^2 + 2f_a \, s. \qquad (3)$$

For the first small interval, the initial velocity is 0, so Equation (3) becomes

$$v_f^2 = 0 + 2f_{a_1} s \; = \; 2s \, f_{a_1},$$

where f_{a_1} is the accelerative force for the first interval.

For the second small interval, the initial velocity is the same as the final velocity for the first interval:

$$v_f^2 \; = \; 2s \, f_{a_1} + 2s \, f_{a_2} \; = \; 2s \, (f_{a_1} + f_{a_2}) .$$

For the third small interval,

$$v_f^2 \; = \; (2s \, f_{a_1} + 2s \, f_{a_2}) + 2s \, f_{a_3} \; = \; 2s \, (f_{a_1} + f_{a_2} + f_{a_3}) .$$

Therefore, for the nth interval,

$$v_f^2 \; = \; 2s \, (f_{a_1} + f_{a_2} + \cdots + f_{a_n}) .$$

Now let's compare the final velocities in the two pendulums after equal numbers of equal intervals have been traversed. Using capitals for one pendulum, lower case for the other, as before,

$$\frac{V_f{}^2}{v_f{}^2} = \frac{2s(f_{A_1} + f_{A_2} + \cdots + f_{A_n})}{2s(f_{a_1} + f_{a_2} + \cdots + f_{a_n})} = \frac{(f_{A_1} + f_{A_2} + \cdots + f_{A_n})}{(f_{a_1} + f_{a_2} + \cdots + f_{a_n})} .$$

Even though the accelerative force is continually changing as the pendulums swing, nevertheless, by Equation 1, the accelerative forces for any pair of corresponding intervals on the two pendulums are in the constant ratio k. Therefore, by Euclid V.12 since all the parts are in the same ratio, the wholes will be in the same ratio:

$$\frac{V_f^2}{v_f^2} = k, \qquad \text{or} \qquad \frac{V_f}{v_f} \; = \; \sqrt{k} \, . \qquad (4)$$

Remember that after deriving Equation 2 we observed that if we could find that the ratio $\Delta V/\Delta v$ was constant, we could conclude that the ratio $\Delta T/\Delta t$ was constant too. Finding whether the ratio of ΔV's is constant is the work of Step 2, in which we are engaged. We will need to find the ratio of $\triangle T$'s to be fixed to conclude that the little arcs are in the same ratio as the whole oscillations, the work of Part 2, of which this is a step. Part 3 will need to know that the $\triangle V$'s and $\triangle T$'s are in a certain ratio, and we need the ratio of $\triangle V$'s to set that up. So what about the ratio of the $\triangle V$'s?

Equation 4 tells us that after any equal number of small path intervals have been traversed, the final velocities are in the same ratio. But each final velocity is the initial velocity of the next interval. Thus the ratio of the initial velocities is the same as shown in Equation 4 as well.

$$\frac{V_f}{v_f} = \frac{V_i}{v_i} = \sqrt{k}\,. \tag{5}$$

Alternating,

$$\frac{V_f}{V_i} = \frac{v_f}{v_i}\,.$$

But by Euclid V.17,

$$\frac{V_f - V_i}{V_i} = \frac{v_f - v_i}{v_i}\,.$$

The pairs V_f and V_i and v_f and v_i are the initial and final velocities over two corresponding segments of the arcs of the two pendulums. Since these are evanscent arcs, the force over them may be taken as constant, and so the difference between the initial and final velocities is equal to the $\triangle V$ or $\triangle v$ of Step 1 above, and the time taken to traverse the evanescent arc is $\triangle T$ or $\triangle t$.

That is,

$V_f - V_i = \triangle V$ and $v_f - v_i = \triangle v$.

Therefore,

$$\frac{\triangle V}{V_i} = \frac{\triangle v}{v_i}\,.$$

Alternating again,

$$\frac{\triangle V}{\triangle v} = \frac{V_i}{v_i}\,.$$

But by Equation 5, the ratio of final velocities is the same as the ratio of initial velocities. Therefore

$$\frac{\triangle V}{\triangle v} = \frac{V_i}{v_i} = \frac{V_f}{v_f}\,.$$

By Equation 4,

$$\frac{V_f}{v_f} = \sqrt{k}\,.$$

Therefore

$$\frac{\Delta V}{\Delta v} = \sqrt{k}\,. \tag{6}$$

Step 3

Equation 2 stated:

$$\frac{\Delta T}{\Delta t} = \frac{1}{k}\frac{\Delta V}{\Delta v}\,.$$

Substituting from Equation 6 for the ratio of ΔV's,

$$\frac{\Delta T}{\Delta t} = \frac{1}{k}\sqrt{k} = \frac{1}{\sqrt{k}}\,.$$

Now we add up all the ΔT's and Δt's to make up the periodic times P and p, respectively. Since the small parts of the times, taken two by two, are in the same ratio, the wholes P and p will be in the same ratio, by Euclid V.12:

$$\frac{P}{p} = \frac{\Delta T}{\Delta t} = \frac{1}{\sqrt{k}} = \sqrt{\frac{f_a}{F_a}}\,. \tag{7}$$

P and p are the sums of all the intervals as the pendulum bobs swing from the starting points across to the opposite high points and back to the starting points again. They are what Newton calls the "times of the whole oscillations" and what we would call the periodic times of the pendulums.

Part 3: Putting it all together

"... the velocities in the corresponding parts of the oscillations will be to each other as the motive forces and the whole times of the oscillations directly and the quantities of matter inversely, ..."

By Equation N1 as rewritten in Part 1,

$$\Delta v \propto \frac{f_m \times \Delta t}{m}\,.$$

By Equation 7, we may substitute the ratio of periodic times for the ratio of $\triangle t$ times:

$$\triangle v \propto \frac{f_m \times p}{m} .$$

"...and consequently the quantities of matter are as the forces and the times of oscillations directly and the velocities inversely."

Solve for m:

$$m \propto \frac{f_m \times p}{\triangle v} . \qquad (8)$$

"But the velocities are inversely as the times,..."

By Equation 6,

$$\frac{\triangle V}{\triangle v} = \sqrt{k} .$$

And by Equation 7,

$$\frac{P}{p} = \frac{1}{\sqrt{k}} .$$

Combining these two,

$$\frac{\triangle V}{\triangle v} = \frac{p}{P} \quad \text{or} \quad \triangle v \propto \frac{1}{p} . \qquad (9)$$

"... and consequently the times directly and the velocities inversely are as the squares of the times,..."

Divide both sides of Equation 9 by p:

$$\frac{\triangle v}{p} \propto \frac{1}{p^2} ;$$

Inverting,

$$p^2 \propto \frac{p}{\triangle v} . \qquad (10)$$

"and therefore the quantities of matter are as the motive forces and the squares of the times, that is, as the weights and the squares of the times.

Q.E.D."

By Equation 8,

$$m \propto f_m \times \frac{p}{\Delta v}.$$

Substituting for $p/\Delta v$ from Equation 10:

$$m \propto f_m p^2.$$

But $f_m \propto$ weight.

Therefore,

$$m \propto \text{weight} \times \text{square of periodic times}.$$

Q.E.D.

II.24 Corollary 1

Therefore if the times are equal, the quantities of matter in individual bodies will be as the weights.

Note on II.24 Corollary 1

Newton doesn't waste any time getting to the possibility that may have had the student hopping in eagerness to get out of the hypothetical mode.

If it should be found to be the case that periodic times for pendulums of equal string length are equal, we can state the consequence: the weights will be as the quantities of matter.

By the proposition, $m \propto w \times p^2$.

If p is constant,

$$m \propto w.$$

We will explore this further in III.6.

II.24 Corollary 6

But also in a nonresisting medium, the quantity of matter of the pendulum is as the comparative weight and the square of the time directly and the length of the pendulum inversely. For the comparative weight is the motive force of a body in any heavy medium you please, as I have explained above. Thus it acts the same in such a nonresisting medium as does the absolute weight in a vacuum.

Note on II.24 Corollary 6

We may bypass the part about pendulum length, which comes from previous corollaries, because we're only going to be interested in equal string length pendulums when we use this corollary later.

Comparative weight is introduced in II.20 Corollary 6:

> Therefore, the gravity of bodies placed in fluids is twofold: one is true and absolute, the other apparent, common, and comparative. Absolute gravity is the whole force by which the body tends downward; relative and common gravity is the excess of gravity by which the body tends downward more than the surrounding fluid.

Newton will refer to this in III.6 when he gives results of pendulum experiments. He will be concerned to account for the fact that these pendulums are swinging in air, not in a vacuum. This could possibly affect the results.

There are two possible problems with air: resistance and buoyancy. We can get a measure of resistance by watching the bobs swing, and we can see that it is very low for air: the bob keeps swinging for a long time without slowing down.

Buoyancy is another matter, for if a body is weighed in a buoyant fluid it will not register its "true and absolute" weight (as it would if weighed in a vacuum). Rather it will display its "apparent, common, and comparative" weight which, as Newton explains, is equal to its true weight, less the weight of the displaced fluid. Thus an experiment performed in air can only directly reveal the relationship between comparative weight to mass, not that between absolute weight and mass, which is required.

As we shall see in Proposition III.6, however, Newton's experiment compares pendulums made "entirely similar as to ... shape." His reference to "shape" here includes size also, so the effects of buoyancy are the same on both pendulums. Therefore, when the pendulums are adjusted to equality in apparent or comparative weight, they will be equal in absolute weight also; and the comparison between absolute weight and mass can proceed.

In addition, pendulums that are identical in both shape and size are equal in their susceptibility to air resistance. Thus, even though that resistance is evidently small, Newton does not ignore it but instead equalizes even that small effect as much as possible.

II.24 Corollary 7

And thus there is an evident procedure both for comparing bodies among themselves, as to the quantity of matter in each, and for comparing the weights of the same body in different places, to discover the variation of gravity. Moreover, by the most accurate experiments I have found that the quantity of matter in individual bodies is always proportional to their weight.

Expansion of Newton's Sketch of II.24 Corollary 7

"And thus there is an evident procedure both for comparing bodies among themselves, as to the quantity of matter in each..."

The conclusion of the proof of the proposition was that:

$$\frac{\text{quantity of matter in body A}}{\text{quantity of matter in body B}} = \frac{\text{weight of A} \times {P_A}^2}{\text{weight of B} \times {P_B}^2}$$

where P is the "time of the whole oscillation," or the periodic time.

"...and for comparing the weights of the same body in different places, to discover the variation of gravity."

Since we know that while accelerative force may be constant at a given distance, it may vary with distance, and so we can expect weights to vary with distance.

By the proposition, $m \propto w \times P^2$.

Taking the same body at two different places for the proportion:

$$\frac{\text{quantity of matter in A at place 1}}{\text{quantity of matter in A at place 2}} = \frac{\text{weight A at place 1}}{\text{weight A at place 2}} \text{ comp } \frac{{P_A}^2 \text{ at place 1}}{{P_A}^2 \text{ at place 2}}.$$

But the quantity of matter in the same body is constant in the two places.

Therefore

$$\frac{\text{weight A at place 1}}{\text{weight A at place 2}} \propto \frac{{P_A}^2 \text{ at place 2}}{{P_A}^2 \text{ at place 1}}.$$

That is, the ratio of weights is inversely as the ratio of the squares of the periodic times.

"Moreover, by the most accurate experiments I have found that the quantity of matter in individual bodies is always proportional to their weight."

Newton couldn't resist spilling the beans of III.6 here; the application of this corollary was just too exciting. And to that proposition we will turn after this Pause.

Pause After II.24

What does this mean? II.24 opens a possibility of an infinite number of universes of different proportionalities between mass and weight.

For equal string lengths, $m \propto w \times P^2$.

We can easily imagine a world in which we could have two pendulums side by side, same string length, with one swinging fast and the other slowly. In fact, if we haven't played with pendulums, we might assume that our own world was like this.

Consider the different ways we might imagine changing the period. One of the first that springs to mind would be to change the weight. The weight certainly seems as if it ought to make a difference: a heavier bob should move faster. Or should it move slower?

But even if periodic time didn't vary with mass, it might vary with material. Some materials might be more attracted than others. Some materials take more electrical charge for a given mass than others, and so have more electrical attraction. Could it happen that some materials might have more weight for a given mass? Maybe the type of matter, say wood as opposed to gold, would make a difference.

What else might you imagine could make a difference?

This proposition gives us a means of finding out which world we're living in. (It's like the way I.4 set us up with a whole menu of different possible relationships of periodic times to distances, associating each with a force law. Then in the Phenomena and early Book III propositions, we are able to find out what force law we were dealing with by checking the ratio of distances to periodic times.)

We continue this exploration in III.6.

Proposition 6

That all bodies gravitate to each of the planets, and that their weights towards whichever particular planet you please, at equal distances from the center of the planet, are proportional to the quantity of matter in each.

That the descent of all heavy bodies to earth (if the unequal retardation arising from the air's very slight resistance be subtracted) occurs in equal times, others have observed for some time now, and the equality of times can be noted very accurately indeed in pendulums. I have made a trial with gold, silver, lead, glass, sand, common salt, wood, water, [and] wheat. I prepared two equal and round wooden boxes. I filled one with wood, and suspended the same weight of gold at the center of oscillation of the other (as exactly as

I could). When the boxes were hung from equal threads of eleven feet, they constituted pendulums entirely similar as to weight, shape, and air resistance, and, placed side by side, they went to and fro together for a very long time with equal oscillations. Accordingly, the amount of matter in the gold (by Book II Proposition 24 Corollaries 1 and 6) was to the amount of matter in the wood as the action of the motive force on all the gold was to the action of the same on all the wood, that is, as weight to weight. And thus it was in the others. In bodies of the same weight, a difference in matter of no greater than the thousandth part of the whole matter could have been plainly perceived in these experiments.

But now there is no doubt that the nature of gravity in the planets is the same as on earth. For let these terrestrial bodies be imagined to be raised right up to the orbit of the moon and to be dropped along with the moon, deprived of all motion, so as to fall to earth at the same time. Through what has already been shown, it is certain that in equal times they will describe equal distances with the moon, and therefore that they are to the quantity of matter in the moon as their weights are to its weight. Further, since the satellites revolve in times that are in the sesquiplicate ratio of the distances from the center of Jupiter, their accelerative gravities towards Jupiter will be inversely as the squares of the distances from the center of Jupiter, and consequently, at equal distances from Jupiter their accelerative gravities will come out equal. Further, in falling from equal altitudes in equal times, they would describe equal spaces, exactly as happens with heavy bodies in this our earth. And by the same argument, the planets around the sun, dropped from equal distances from the sun, would in their descent to the sun describe equal spaces in equal times.

But the forces by which unequal bodies are equally accelerated are as the bodies; that is, the weights are as the quantities of matter in the planets. Further, that the weights of Jupiter and its satellites towards the sun are proportional to the quantities of their matter is evident from the motion of the satellites, as regular as can be, by Book I Proposition 65 Corollary 3. For if some of them were to be pulled towards the sun more in proportion to their quantity of matter than the others, the motions of the satellites (by Book 1 Proposition 65 Corollary 2) would be perturbed by the inequality of the attraction. If, at equal distances from the sun, some satellite were heavier towards the sun in proportion to its quantity of matter than Jupiter in proportion to its quantity of matter, in any given ratio (say, d to e), the distance between the center of the sun and the center of the satellite's orbit would always be greater than the distance between the center of the sun and the center of Jupiter in the subduplicate ratio, approximately, as I have found by entering upon a certain computation. And if the satellite were less heavy towards the sun in that ratio d to e, the distance of the center of the satellite's orbit from the sun would be less than the distance of Jupiter's center from the sun in that subduplicate ratio. And thus if at equal distances from the sun, the accelerative gravity of any satellite you please towards the sun be greater or less than the accelerative gravity of Jupiter towards the sun, by only the thousandth part of the whole gravity, the distance of the center of the satellite's orbit from the sun would be greater or less than the distance of Jupiter from the sun by the $\frac{1}{2000}$ part of the whole distance, that is, by the fifth part of the outermost satellite's distance from the center of Jupiter—an eccentricity of the orbit that would be very easily perceptible. But the orbits of the satellites are concentric with Jupiter, and consequently the accelerative gravities of Jupiter and of the satellites towards the sun are equal among themselves. And by the same argument, the weights of Saturn and its companions towards the sun, at equal distances from the sun, are as the quantities of matter in them, and the weights of the moon and the earth towards the sun are either nothing, or accurately proportional to their masses. But they are something, by Proposition 5 Corollaries 1 and 3.

And furthermore, the weights of individual parts of any planet to any other whatever are to each other as the matter in the individual parts. For if certain parts were to gravitate more, others less, than according to the quantity of matter, the whole planet, according to the kind of parts with which it mostly abounds, would gravitate more or less than according to the quantity of matter of the whole. Nor does it matter whether those parts be external or internal. For if, for example, the terrestrial bodies which are with us were imagined to be raised to the orbit of the moon, and compared with the body of the moon, if their weights were to the weights of the external parts of the moon as the quantities of matter in the same bodies, but were to the weights of the internal parts in a greater or less ratio, the same [weights] would be to the weight of the whole moon in a greater or less ratio, contrary to what was shown above.

Notes on III.6

- **Motive Force and Weight**

We are now dealing with motive quantity of centripetal force (Definition 8). Weight is a kind of motive force. A note at the start of II.24 entitled Motive Force explored how this new force relates to the accelerative force we have heretofore used. Notes and expanded proof here assume familiarity with the derivations and implications spelled out there.

- **Inertial and Gravitational Mass**

This proposition (using Proposition II.24 and an unnumbered phenomenon) gives us the discovery of the first part of the coincidence of inertial and gravitational mass.

The coincidence and its strangeness are explored and discussed in the Pause After III.6, which comes between the expanded proof for the proposition and the statements of the corollaries.

To get this coincidence we need to establish the general applicability of the inverse square law, which we began in III.4. Then, comparing different bodies at equal distances, we find that at equal distances, weights vary as quantitites of matter.

- **Continuity of Inverse Square Law**

We must now have the application of the inverse square gravitational force law to bodies at all distances explicitly and universally. See the note in the Pause After III.4 for discussion of how that proposition gives confirmation to the idea that it applies everywhere, not just to the cases in which we have found it to hold for certain bodies at particular distances.

Newton seems also to have in mind the third Rule of Philosophizing:

> The qualities of bodies that do not suffer intensification and remission, and that pertain to all bodies upon which experiments can be carried out, are to be taken as qualities of bodies universally.

If bodies everywhere we find them act under an inverse square law, and we find none that seem not to do so, we are justified in taking this to be true of bodies universally until contrary evidence calls it into question.

Expansion of Newton's Sketch of III.6

"That all bodies gravitate to each of the planets, and that their weights towards whichever particular planet you please, at equal distances from the center of the planet, are proportional to the quantity of matter in each."

To Prove:

1. All bodies gravitate to each of the planets.

2. The weights of bodies towards any particular planet, at equal distance from the center of that planet, are proportional to the quantity of matter in each.

3. The weights of individual parts of the planets are as their respective quantities of matter.

Part 1: "That all bodies gravitate to each of the planets."

By III.5 Corollary 1 part 1, "Gravity is therefore given towards all planets without exception."

Questions for Discussion: Consider the transition from (a) there being a power of gravity towards every planet to (b) all bodies gravitating towards each planet. Are we just bringing out something implicit in III.5 Corollary 1? Do we need to invoke a Rule of Philosophizing?

Note the seeming implication here, a potentially troubling one, that bodies are not only attracted to the planet in whose sphere of influence they fall, but to all planets. Newton has suggested this already in the corollaries to III.5, but it might be otherwise. Each planet could have had its own sphere of influence over matter like itself; its influence would extend over its own matter spread out around it, some in the form of rocks, some in the form of moons. The sun, being larger and greater, could be understood as having a larger sphere of influence.

This way of seeing each major body as having spheres of influence and kindred matter would have been more traditional and less revolutionary, since it would echo ideas of Kepler. It would also have avoided the problems that come from the influence of all bodies on all other bodies. Unfortunately, it would not have explained everything it turns out are there to be seen in the heavens.

Part 2: "The weights of bodies towards any particular planet, at equal distance from the center of that planet, are proportional to the quantity of matter in each."

This means two things.

2a. The weights of two stones or pendulum bobs towards the earth, at equal distances, are proportional to the relative quantities of matter in the stones or bobs.

2b. The weights of two planets towards the sun, at equal distances, are proportional to the relative quantities of matter each planet contains.

Part 2a

Let's look at the implication of 2a first.

"That the descent of all heavy bodies to earth (if the unequal retardation arising from the air's very slight resistance be subtracted) occurs in equal times, others have observed for some time now, and the equality of times can be noted very accurately indeed in pendulums. I have made a trial with gold, silver, lead, glass, sand, common salt, wood, water, [and] wheat. I prepared two equal and round wooden boxes. I filled one with wood, and suspended the same weight of gold at the center of oscillation of the other (as exactly as I could)."

When Newton set up pendulums with equal string lengths and equal weights, he observed that the periodic times of the pendulums were equal, and remained equal whatever material he used, as long as he arranged that the pendulums were entirely similar as the weight, shape, and air resistance.

"When the boxes were hung from equal threads of eleven feet, they constituted pendulums entirely similar as to weight, shape, and air resistance, and, placed side by side, they went to and fro together for a very long time with equal oscillations."

Newton tells us that the effect of air resistance seems to be negligible, because "they went to and fro for a very long time with equal oscillations," that is, oscillations equal in extent or amplitude. (This does not mean the oscillations were equal for the two pendulums. It means that, for each pendulum, one oscillation was equal in maximum displacement to the subsequent ones.) If the air resistance were great, we would expect the maximum displacement for each swing to decrease quickly.

"Accordingly, the amount of matter in the gold (by Book II Proposition 24 Corollaries 1 and 6) was to the amount of matter in the wood as the action of the motive force on all the gold was to the action of the same on all the wood, that is, as weight to weight. And thus it was in the others. In bodies of the same weight, a difference in matter of no greater than the thousandth part of the whole matter could have been plainly perceived in these experiments."

Newton cites II.24 Corollary 6 to cover the objection that his results might be flawed by the buoyancy of air. My note on that corollary describes the measures Newton took to deal with the effects of buoyancy.

Having accounted for possible discrepancies due to the effect of swinging the pendulums in air, he concludes that the equal periodic times he observes may be trusted. This observation, an unnumbered phenomenon, tells us that our world fits the hypothetical conditions of II.24 Corollary 1 and allows us to use that corollary.

II.24 Corollary 1 states that for equal string lengths and equal periodic times,

$$\frac{\text{quantity of matter in A}}{\text{quantity of matter in B}} = \frac{\text{motive force on A}}{\text{motive force on B}} = \frac{\text{weight of A}}{\text{weight of B}}.$$

Newton says his experiments have confirmed this to within 0.1%.

Part 2b

We will show that the weights of two planets towards the sun is proportional to the relative quantities of matter each planet contains.

Step 1: Comparing Gravity on Planets at Equal Distances

"But now there is no doubt that the nature of gravity in the planets is the same as on earth. For let these terrestrial bodies be imagined to be raised right up to the orbit of the moon and to be dropped along with the moon, deprived of all motion, so as to fall to earth at the same time. Through what has already been shown, it is certain that in equal times they will describe equal distances with the moon,..."

If we establish that the force law is inverse square everywhere, we can argue that a rock raised to the moon's orbit will have the same accelerative forces on it as does the moon itself, and therefore that both the moon and the rock will drop equal distances in equal times.

But Newton begins this section by saying "Now there is no doubt that the nature of gravity in the planets is the same as on earth": it looks as though this is as much argument as we are going to get here about the continuity and universality of the inverse square law. So what other evidence, if any, do we have for the universal applicability of that law?

The reasoning in Proposition III.4 is a strong first step. We there assumed continuity of the inverse square law and calculated, on that basis, how fast the moon would fall if it were brought down near the earth's surface and then released there: our result was identical to the measured rate of fall of stones at the earth's surface.

Then, from Newton's reference to the nature of gravity being the same on the planets as on earth, we may surmise that he is thinking of the third Rule of Philosophizing. That rule says that we expect that each planet is not created and placed uniquely, and that if the force law is continuous in the sublunar realms then it will be continuous around Jupiter and everywhere else. (See the Pause After III.4 for a thorough consideration of this question, and third bulleted note above for a summary.)

From now on, then, we will consider the continuity and universality of the inverse square law as established.

"...and therefore that they are to the quantity of matter in the moon as their weights are to its weight."

Comparing motive forces using Equation N2 from the notes in II.24:

$$\frac{F_{m_1}}{f_{m_2}} = \frac{m_1}{m_2} \times \frac{F_{a_1}}{f_{a_2}}.$$

But at equal distances, accelerative forces are equal, so:

$$\frac{F_{m_1}}{f_{m_2}} = \frac{m_1}{m_2}.$$

Now we begin a series of steps that explore this idea by looking at Jupiter and its satellites as planets (a "primary planet" and some "secondary planets") that he will argue gravitate towards the sun according to their quantities of matter.

Step 2:

"Further, since the satellites revolve in times that are in the sesquiplicate ratio of the distances from the center of Jupiter, their accelerative gravities towards Jupiter will be inversely as the squares of the distances from the center of Jupiter, and consequently, at equal distances from Jupiter their accelerative gravities will come out equal."

Jupiter's satellites have been shown to have accelerative gravities towards Jupiter inversely as the squares of the distances by III.1. Invoking the continuity and universality of the inverse square law, we will say that at equal distances the accelerative forces on them will be equal.

"Further, in falling from equal altitudes in equal times, they would describe equal spaces, exactly as happens with heavy bodies in this our earth."

By Rule 3, we assume that they will act like our moon and rocks with respect to the earth, as considered in Step 1, namely that at equal distances, weights (motive forces) will vary as the quantities of matter.

"And by the same argument, the planets around the sun, dropped from equal distances from the sun, would in their descent to the sun describe equal spaces in equal times."

We saw that the circumsolar planets have accelerative gravities towards the sun inversely as the squares of their distances by III.2. And by the Rules of Philosophizing, we expect them to act like our moon and rocks as well.

Weight is proportional to motive force, as explained in Newton's commentary to Definition 8. (See also the notes on motive force and weight before II.24.)

So at equal distances, by Step 1, weights (motive forces) of circumsolar planets will vary as the quantities of matter.

"But the forces by which unequal bodies are equally accelerated are as the bodies; that is, the weights are as the quantities of matter in the planets."

Now Newton says this another way. Motive forces that equally accelerate unequal bodies vary as those bodies. By Equation N2 in the Motive Force note to II.24,

$$F_m \propto m \times F_a .$$

That is,

$$\frac{F_{m_1}}{f_{m_2}} = \frac{m_1}{m_2} \times \frac{F_{a_1}}{f_{a_2}} .$$

Thus if the accelerations are equal and the quantities of matter are unequal, the motive forces must be unequal in the same ratio. That is, for the planets,

$$\frac{m_1}{m_2} = \frac{w_1}{w_2}$$

How we have come to this point and where we go next: Part 2a found weight varying as quantity of matter at the surface of the earth, as shown in pendulum experiments. Part 2b deals generally with earth and the planets: Step 1 took a converse approach to III.4 to argue that the moon would be accelerated according to an inverse square law at *any* distance and then went on to show that at any particular distance motive force would vary as quantity of matter.

Step 2 expands Step 1 to apply to the planets as well as to the earth. The proof in Step 2 relies on the Rules of Philosophizing. But again, Newton doesn't invoke his Rules of Philosophizing to argue from authority, nor does he wheel them in as a *deus ex machina*. He suggests that scientific thinking points in this direction, but then proceeds in the remaining steps to give some more direct proofs for the conclusions of Step 2.

Step 3

"Further, that the weights of Jupiter and its satellites towards the sun are proportional to the quantities of their matter is evident from the motion of the satellites, as regular as can be, by Book I Proposition 65 Corollary 3."

I.65 brings in the complication of multiple bodies. If "gravity is given towards every planet" as Newton asserts in III.5 Corollary 1, and if this means, as it seems to, that every planet gravitates towards every other planet, then we would expect instability in the universe. The stable ellipses that result from inverse square laws governing two-body systems would be thrown off by the effects of other bodies outside that system. And yet we do find stable elliptical orbits in the planetary system. To justify his claim in III.5 Corollary 1 and its further development, Newton must account for that. Propositions I.65 and I.66 explore some reasons we might still find stable elliptical orbits. They argue that it is possible with many bodies, if they are suitably arranged, to get the bodies moving, at least approximately, in ellipses with equable description of areas. (See Appendix B for a full statement and further discussion of I.65.)

I.65 deals with three bodies and, with its corollaries 3 and 2 mentioned by Newton in this proposition, essentially restates I.3, which we proved and can use instead if we prefer. I.65 Corollary 3 says that if we have ellipses or circles with equable description of areas, at least approximately, then either the central and orbiting bodies are not urged, or they are very little urged, by accelerative forces outside; or they are urged equally, by approximately parallel lines.

Here our outside force is urging Jupiter and its moons towards the sun. Since we see equable description of areas and circular motion, either the influence of the sun is very small or it falls equally on Jupiter and moons, in approximately parallel lines. Since we know it to be great enough to hold Jupiter in orbit (Proposition III.1), it cannot be that its effect is too small to distort the orbit. Therefore we will conclude that its influence is equal on both bodies, and the moons have the same acceleration towards the sun that Jupiter has.

Reasoning from I.3:

I.3 says that if a body describes equal areas in equal times around another body, howsoever moved, it is urged by a force compounded of the centripetal force tending to that other body, compounded with all the accelerative forces by which that other body is impelled.

We have seen in III.1 that Jupiter's satellites describe equal areas in equal times around Jupiter as a center. So we may conclude by I.3 that they have the same acceleration towards the sun as Jupiter has towards the sun.

$F_m \propto F_a \times m$. [see Equation N2 in the Motive Force note at start of II.24]

Since the distances are essentially equal, acceleration is the same, and $F_m \propto m$ for both Jupiter and its moons.

Therefore weights vary as quantities of matter.

Step 4

"For if some of them were to be pulled towards the sun more in proportion to their quantity of matter than the others, the motions of the satellites (by Book 1 Proposition 65 Corollary 2) would be perturbed by the inequality of the attraction."

I.65 Corollary 2 says the motions will get "disturbed" or "confused"—this means we won't have ellipses and equable description of areas—if bodies are not accelerated by the same amount at the same distance. (See Appendix B for full text of I.65 and its corollaries, with notes.)

Therefore, since we don't see this with Jupiter and its moons, we can assume equal accelerative forces on Jupiter and its moons towards the sun.

Again we will be led to conclude that weights vary as quantities of matter.

Step 5

But Newton isn't content to rest on I.65 Corollary 2 either; he gives us an argument to prove this by *reductio ad absurdum*.

Suppose that some were more strongly attracted, that some satellite, in proportion to its quantity of matter, gravitates towards the sun with a motive force greater or less than Jupiter does in proportion to its quantity of matter.

Let this ratio of accelerative forces be $d{:}e$, that is,

$$\frac{f_{m\,\text{moon}}}{m_{\text{moon}}} = \text{acceleration } d \text{ of satellite towards sun.}$$

$$\frac{f_{m\,\text{Jupiter}}}{m_{\text{Jupiter}}} = \text{acceleration } e \text{ of Jupiter towards sun.}$$

$$\frac{\text{accel}_{(\text{moon})}}{\text{accel}_{(\text{Jupiter})}} = \frac{d}{e}.$$

In this proof we are establishing that the accelerative forces urging Jupiter and its moons towards the sun are the same, by showing that any assumption that they are not the same leads to a conclusion contrary to what we observe.

Given:

"If, at equal distances from the sun,..."

$r_J = r_m.$

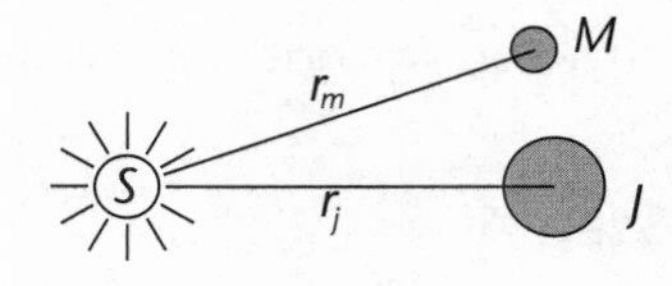

To Prove:

"[T]he accelerative gravities of Jupiter and of the satellites towards the sun are equal among themselves."

That is, $a_J = a_m.$

Proof:

For, if possible, suppose that $a_J \neq a_m.$

That is, where

$$\frac{\text{accel}_{(\text{moon})}}{\text{accel}_{(\text{Jupiter})}} = \frac{d}{e} \neq \frac{1}{1}.$$

First case: If possible, let $d > e$.

"If, at equal distances from the sun, some satellite were heavier towards the sun in proportion to its quantity of matter than Jupiter in proportion to its quantity of matter, in any given ratio (say, d to e), the distance between the center of the sun and the center of the satellite's orbit would always be greater than the distance between the center of the sun and the center of Jupiter in the subduplicate ratio [of $d : e$], approximately, as I have found by entering upon a certain computation."

We know that the moon has a stable orbit around Jupiter. What would the distance from the sun to the center of the satellite's orbit have to be if the accelerative force on the satellite were, as we are supposing here, *greater* than the accelerative force on Jupiter?

If the force on the moon towards the sun were greater, as supposed, something would have to balance this out to maintain a stable orbit. An eccentric orbit for the moon, such that the center of the orbit were out beyond Jupiter, would do it. When the moon lies between the sun and Jupiter, Jupiter will be closer to it by enough to counteract the greater accelerative force towards the sun, drawing it back away from the sun and keeping it in a stable orbit.

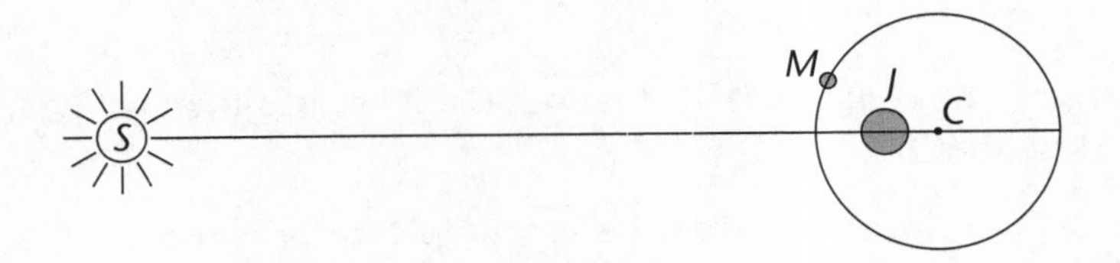

At the opposite end of its orbit, the greater distance from Jupiter, where the acceleration towards Jupiter would be joining the acceleration towards the sun, would counterbalance the greater force towards the sun.

But we don't see such an eccentric orbit, we see an orbit centered on Jupiter, by Phenomenon 1.

Therefore $d \not> e$.

Second case: If possible, let $d < e$.

"And if the satellite were less heavy towards the sun in that ratio d to e, the distance of the center of the satellite's orbit from the sun would be less than the distance of Jupiter's center from the sun in that subduplicate ratio."

Similarly, for there to be the stable orbit around Jupiter that we see, if the force of the sun on the moon were less than on Jupiter, the distance to the center of the moon's orbit would have to be less.

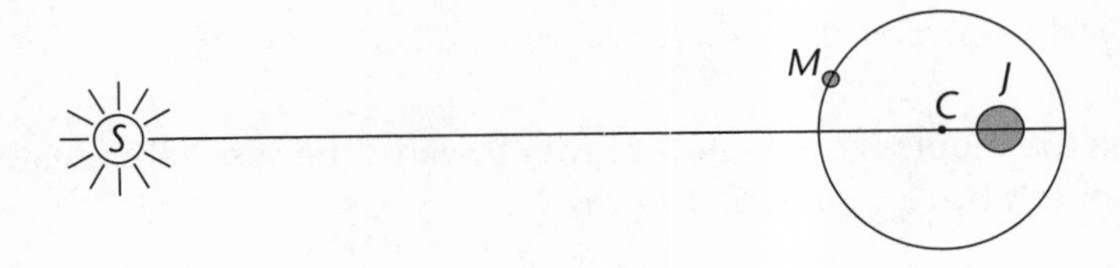

So when the moon is between the sun and Jupiter, the moon will be farther from Jupiter than when it is beyond Jupiter.

Thus it will be urged back towards Jupiter by a lesser amount, that reduction making up for the supposed relative deficiency of accelerative force on the moon towards the sun compared to that on Jupiter towards the sun.

On the opposite side of Jupiter, it is closer to Jupiter, so it will be urged more towards Jupiter, and that extra impulse will add to the impulse towards the sun to make up for the supposed deficiency.

However, we don't see such an eccentric orbit, we see a circular concentric orbit around Jupiter by Phenomenon 1.

Therefore $d \not< e$.

Conclusion:

Since we have now shown that d cannot be either less than or greater than e, it must be that $d = e$. Q.E.D

Newton says in the sketch that the center will be off "in the subduplicate ratio, approximately, as I have found by entering upon a certain computation." He says nothing in the preceding sketch to suggest what this mysterious computation might be. Since it involves the gravitational interaction of three bodies, it may well have been derived from the model presented in I.66 and its corollaries, especially Corollary 6. In any case, it was certainly a complicated computation, which may be why Newton chose not to present it here. Computations using modern algebraic methods come out with very nearly the same numbers that Newton found for the predicted eccentricity of the moon's orbit.

Step 6

Newton now gives a numerical example based on his "certain computation" to show that if the ratio were off by even a small amount, the distortion of the orbit would be noticeable.

"And thus if at equal distances from the sun, the accelerative gravity of any satellite you please towards the sun be greater or less than the accelerative gravity of Jupiter towards the sun, by only the thousandth part of the whole gravity, the distance of the center of the satellite's orbit from the sun would be greater or less than the distance of Jupiter from the sun by the $\frac{1}{2000}$ part of the whole distance,..."

Suppose the moon's accelerative gravity towards the sun were greater or less than Jupiter's by $\frac{1}{1000}$ of Jupiter's gravity:

$$\frac{d}{e} = \frac{f_a \text{ moon}}{f_a \text{ Jupiter}} = \frac{1001}{1000} \text{ or } \frac{999}{1000}$$

at a given distance from the sun.

Then the distance of the center of the satellite's orbit from the sun can be shown to be greater or less than the distance of Jupiter from the sun by $\frac{1}{2000}$ of the whole distance.

First, suppose the accelerative gravity of Jupiter's moon towards the sun is greater by 1/1000.

By Step 5, first case,

> "...the distance between the center of the sun and the center of the satellite's orbit would always be greater than the distance between the center of the sun and the center of Jupiter in the subduplicate ratio [of $d:e$]..."

Let r_c be the distance of the center of the moon's orbit from the sun, and r_j be the distance of the center of Jupiter's orbit from the sun. Then, by the "certain computation,"

$$\frac{r_c}{r_j} = \sqrt{\frac{d}{e}} = \sqrt{\frac{1001}{1000}} \approx \frac{2001}{2000}.$$

If on the other hand we suppose the accelerative gravity of Jupiter's moon towards the sun is less by 1/1000, our ratio will be

$$\frac{r_c}{r_j} = \sqrt{\frac{d}{e}} = \sqrt{\frac{999}{1000}} \approx \frac{1999}{2000}$$

Thus in either case the difference in distance will be about $\frac{1}{2000}$ as Newton says.

"...that is, by the fifth part of the outermost satellite's distance from the center of Jupiter—an eccentricity of the orbit that would be very easily perceptible. But the orbits of the satellites are concentric with Jupiter, and consequently the accelerative gravities of Jupiter and of the satellites towards the sun are equal among themselves."

But $\frac{1}{2000}$ of the radius of Jupiter's orbit can be shown to be $\frac{1}{5}$ of the outermost moon's orbit, as follows.

By Phenomenon 1, the greatest heliocentric elongation of the fourth satellite from Jupiter's center, with micrometer, at earth's heliocentric distance from Jupiter, was 8′ 16″.

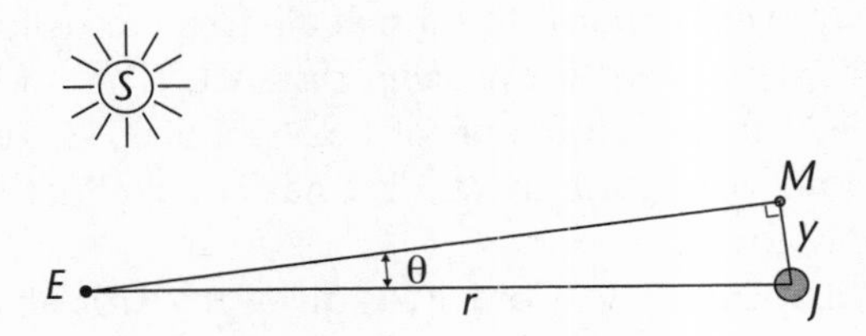

What is the ratio of Jupiter's orbital radius to the fourth moon's orbital radius?

$$\frac{r}{y} = \csc\theta = \csc(8'16'') = 415.9 \approx 416.$$

But r is Jupiter's orbital radius and y is the moon's orbital radius.

Therefore Jupiter's distance from the sun is 416 times the moon's distance from Jupiter. Or, if we call Jupiter's distance 416, the moon's distance from Jupiter will be 1.

Suppose $d>e$, with *C*, the center of the moon's orbit, beyond Jupiter as shown in the following figure. If the ratio of distances is off by $\frac{1}{2000}$, the distance of *C* beyond *J* would be $\frac{1}{2000}$ of 416, or 0.2.

But the moon's orbit was 1 of those units, so the eccentricity of the moon's orbit is $\frac{1}{5}$.

This eccentricity would be easily perceptible; since we don't see it, it must be that the accelerative forces do not differ by even as much as $\frac{1}{1000}$.

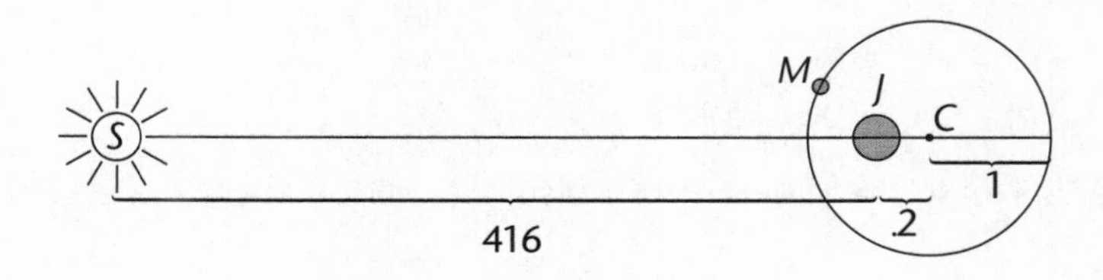

Therefore because the accelerative forces are equal at equal distances, the weights of Jupiter and its satellites are as their several quantities of matter.

"And by the same argument, the weights of Saturn and its companions towards the sun, at equal distances from the sun, are as the quantities of matter in them, and the weights of the moon and the earth towards the sun are either nothing, or accurately proportional to their masses. But they are something, by Proposition 5 Corollaries 1 and 3."

Saturn and its companions (using Phenomenon 2) and the earth and our moon (using Phenomenon 6) could be shown by the argument of Step 6 to be, if impelled by gravity towards the sun at all, to be impelled with a weight to mass ratio that is constant for a given distance. The "if impelled at all" probably goes back to the language of I.65 Corollary 3, where either the outside body has no significant effect, or it has equal effect on both bodies. (It could also apply to I.3.)

And since by III.5 Corollary 1, gravity is given towards all planets without exception, gravity will be given towards the sun, and Saturn and its companions, and earth and our moon, will all gravitate towards the sun.

Therefore the weights of Saturn and its moons towards the sun will be as their quantities of matter, and the weights of earth and our moon towards the sun will be as their respective quantities of matter.

Part 3: "The weights of parts of these planets are as the matter of those parts."

"And furthermore, the weights of individual parts of any planet to any other whatever are to each other as the matter in the individual parts. For if certain parts were to gravitate more, others less, than according to the quantity of matter, the whole planet, according to the kind of parts with which it mostly abounds, would gravitate more or less than according to the quantity of matter of the whole."

Now, although we have established that the planets as a whole, at the same distances, gravitate with weights varying as the quantities of matter, perhaps different parts have different weight-to-mass ratios. Perhaps cows weigh more for their quantities of matter than trees, and rocks and oceans have their own weight-to-mass ratios. Perhaps fire, air, water, and earth have different weight-to-mass ratios. Perhaps the plastic peanuts used as packing material have their own ratio, maybe a negative one, causing them to exhibit levity.

Planets with more cows than trees, or ones that were mostly rock or mostly ocean, would gravitate according to a weight-to-mass ratio reflecting the substance of which they are primarily constituted.

Newton finds it implausible, and expects us to find it so, that all these mixes would happen to average out so that it everywhere looked as if the ratio were constant.

"Nor does it matter whether those parts be external or internal. For if, for example, the terrestrial bodies which are with us were imagined to be raised to the orbit of the moon, and compared with the body of the moon, if their weights were to the weights of the external parts of the moon as the quantities of matter in the same bodies, but were to the weights of the internal parts in a greater or less ratio, the same [weights] would be to the weight of the whole moon in a greater or less ratio, contrary to what was shown above."

We know that the rocks at the earth's surface fall at the same rate as the moon, and therefore have the same $f_m : m$ ratio. But what if the earth had a core that was different?

Then the whole would be different and earth and moon would accelerate differently towards the sun. But if that were the case, by I.3 or by I.65 Corollary 3, we would not see equable description of areas, which we know by Phenomenon 6 that we in fact have.

We may argue similarly against a supposition that the core of the moon might be different. We have not, of course, proved the impossibility of it all just happening to balance out. Newton only wants us to acknowledge it as unlikely; he's not undertaking to disprove every "rashly contrived dream" (see his commentary to Rule 3).

Pause After III.6

• After watching this careful application of the gravitation of Jupiter's moons both to Jupiter and to the sun, and seeing the way I.3 or I.65 is applied there, we begin to get a feel for bodies gravitating towards "every" planet. And if the moon can be gravitating towards both Jupiter and the sun, we have a better picture of Jupiter gravitating towards both the sun and Saturn. Thus our application of Rule 2 for the mutual attractions of Jupiter and Saturn in III.5 Corollary 3 is better supported, and we have made another step in the universalization of gravitation.

• Let's look now at the implications of the discovery of the proportionality of mass and weight at the same distance from a center of forces. Newton recognized that there is no reason that these should be proportional, even though they had been conflated by previous natural philosophers and by many students since. We have here the first part of a coincidence of gravitational and inertial mass. (The second part will come with III.7.)

By Definition 3:

> The inherent force of matter is the power of resisting, by which each and every body, to the extent that it can, perseveres in its state either of resting or of moving uniformly in a straight line. This is always proportional to its body…

But by Definition 1,

> The quantity of matter is the measure of the same arising from its density and magnitude conjointly. … Further, in what follows, by the names "body" or "mass" I everywhere mean this quantity.

Definition 3 says inertia is proportional to body, which by Definition 1 means quantity of matter; so inertia is proportional to quantity of matter.

We have shown in III.6 that the amount by which rocks are impelled or gravitate towards the center of the earth is proportional to their quantities of matter and therefore (by Definition 3) to their inertias, that is, to their resistance to change of motion.

Now the force of gravity working on heavy objects is a force attempting to change their motion. This means that what we would call inertial mass is proportional to weight at equal distances from a center of forces.

The earth has more to pull on in bodies with more quantity of matter, but that tendency towards greater acceleration is exactly foiled and counterbalanced by the greater resistance to change of motion in bodies with greater quantity of matter.

The coincidence is that:

1. bodies are urged by gravitational force in exact proportion to the quantities of their matter; and

2. bodies resist having their motion changed by this force in exact proportion to those quantities of matter.

There seems no evident necessity for this to be the case, that is, for all bodies at the same height to accelerate equally no matter what their mass.

This may even seem counter-intuitive. Doesn't it come as a source of wonder to people when they first learn that lead and wood balls of the same size, or large and small iron balls, fall at the same rate from a tower? Many will protest against the possibility. It was so contrary to reason that Aristotle denied it.

And suppose one queried a group of people who had not seen the experiment or heard the claim that the quantity of matter has no bearing on the acceleration. If asked what they would expect if a small iron ball and a cannonball were dropped simultaneously from a tower, what would be their prediction? Most would probably expect that the ball with greater quantity of matter would fall faster. But an energetic minority would probably argue that the more massive object would fall more slowly since its resistance to being set in motion is greater: after all, we can easily push a child's wagon, but to set a car in motion by pushing takes longer.

That these two considerations counter-balance is not, perhaps, surprising, when both are brought to mind together. But that matter's resistance to being set in motion should exactly equal the gravitational attraction between it and the earth is mysterious. Newton proved that they are, but why they are remains a puzzle and wonder. The nagging of this wonder led Einstein into his General Theory of Relativity.

III.6 Corollary 1

Hence the weights of bodies do not depend upon their shapes and textures. For if the weights could vary with shapes, they would be greater or less, according to the variety of shapes, in an equal matter, entirely contrary to experience.

III.6 Corollary 2

All bodies that are around the earth are heavy towards the earth, and the weights of all that stand at the same distance from the center of the earth are as the quantities of matter in them. This is a quality of all [bodies] upon which experiments can be carried out, and accordingly must be affirmed universally, by Rule 3. If the aether or any other body whatever either were entirely lacking in gravity, or were to gravitate less in proportion to its quantity of matter, then, because (in the opinion of Aristotle, Descartes, and others) it does not differ from other bodies except in the shape of matter, the same [body], by a gradual alteration of shape, could be transformed into a body of the same condition as those that gravitate most in proportion to the quantity of matter. And in turn, the heaviest bodies, by gradually assuming the shape of the former [body], would be able gradually to shed their gravity. And thus the weights would depend upon the shapes of bodies, and would be capable of changing with the shapes, contrary to what is proved in the preceding corollary.

Notes on III.6 Corollaries 1 and 2

Newton is addressing the ideas of Descartes and his followers in these corollaries, and intends the conclusions of III.6 to refute both Descartes' conception of matter and his way of doing natural philosophy.

Descartes claims matter is just extension, since that's all one can clearly and distinctly perceive, and the only thing matter always has (see Descartes, *Principles of Philosophy* II.4).

Weight for Descartes comes from the arrangement, sizes, and shapes of particles, how many pores the particle has, and so on. These arrangements determine where it will go in the whirlpool. "Heavy" things have a shape that makes them go to the center of the terrestrial vortex. (See *Principles*, II.23 and III.48–85, etc.)

The question is whether you can find out anything about the essential nature of things by building mathematical arguments on observations, without first determining what the causal mechanisms must be.

Descartes says no, that it's only by reasoning about things that are clear and distinct that you can find out anything about the essential nature of things; and only extension is clear and distinct. Descartes says that to do otherwise is to return to "occult qualities"—Aristotelian qualities like the tendency to seek the center of the earth. Descartes would say that Newton's mathematical construction introduces concepts like force and weight that aren't indubitably clear. Then, Descartes would say, Newton manipulates these unclear and arbitrarily-defined concepts, in order to come to his conclusions.

Newton says yes, you can find out important truths about the essential nature of things by working with numbers as he does here in Book III. And he shows how to do it, giving us a dazzling display of its power when part of the right methodology.

III.6 Corollary 3

All spaces are not equally full. For if all spaces were equally full, the specific gravity of the fluid that fills the region of the air would concede nothing to the specific gravity of quicksilver or gold or any other extremely dense body, because the matter would be of the greatest density; and consequently, neither gold nor any other body whatever could descend in air. For unless bodies in fluids are specifically heavier, they do not descend in the least. And if the quantity of matter in a given space could be diminished somewhat through rarefaction, why might it not be diminished in infinitum?

III.6 Corollary 4

If all the solid particles of all bodies be of the same density, and not be capable of being rarefied without pores, the vacuum is given. I declare to be of the same density, those things whose forces of inertia are as [their] magnitudes.

Notes on III.6 Corollaries 3 and 4

Newton is still addressing Descartes and his followers, who believed that space was full of matter and that there was no void or vacuum (*Principles*, II.16).

III.6 Corollary 5

The force of gravity is different in kind from the magnetic force. For magnetic attraction is not as the matter attracted. Some bodies are pulled more, others less, most are not pulled. And the magnetic force in one and the same body can be intensified and remitted, and is not infrequently far greater in proportion to the quantity of matter than the force of gravity, and in withdrawal from the magnet it decreases, not in the duplicate, but nearly in the triplicate ratio of the distance, as far as I have been able to ascertain by some rather crude observations.

Note on III.6 Corollary 5

It is important that gravity differs in significant ways from magnetism; we must not assume generally that what we may observe or calculate to be true for one sort of attraction will be true for another.

In particular, a given body may have more or less magnetic power at different times, and bodies with the same quantity of matter may have different degrees of magnetism. The material the body is made of makes a difference; some materials seem unaffected by magnetic force.

Another difference, which Newton doesn't mention here, but might have, in anticipation of what is coming up next, is that in magnetism a body may have the susceptibility to be attracted but not the power to attract other bodies to itself. An unmagnetized metal paperclip will be attracted to a magnet, but will not attract another unmagnetized paperclip.

Flashback to I.69 for Use in III.7

Having established that bodies are attracted in exact proportion to their mass, Newton turns his attention to the attracting bodies, that is, to the bodies that are at the centers of forces towards which the attracted bodies are pulled. To prove that the forces towards these bodies are likewise proportional to their masses, he needs another hypothetical proposition from Book I, namely, I.69. This proposition, unlike II.24, does not require any observational input but deduces, from the Third Law alone, that if bodies attract one another according to an inverse-square law (or any function of distance), their absolute attractive powers must be as the bodies themselves. We will prove I.69, and then return to apply it in III.7.

Book I Proposition 69

In a system of many bodies A, B, C, D, etc., if some body A pulls all the others B, C, D, etc., with accelerative forces that are inversely as the squares of the distances from the pulling body, and another body B also pulls the rest A, C, D, etc., with forces that are inversely as the squares of the distances from the pulling body, then the absolute forces of the pulling bodies A, B will be to each other as are the bodies A, B themselves, whose forces they are.

For the accelerative attractions of all the bodies *B*, *C*, *D*, towards *A*, at equal distances, are equal to each other by hypothesis, and likewise, the accelerative attractions of all the bodies towards *B*, at equal distances, are equal to each other. Now the absolute attractive force of the body *A* is to the absolute attractive force of the body *B* as the accelerative attraction of all the bodies towards *A* is to the accelerative attraction of all the bodies towards *B* at equal distances, and thus is the accelerative attraction of the body *B* towards *A* to the accelerative attraction of the body *A* towards *B* [in this ratio]. But the accelerative attraction of the body *B* towards *A* is to the accelerative attraction of the body *A* towards *B* as the mass of the body *A* to the mass of the body *B*, for the reason that the motive forces, which (by Definitions 2, 7, and 8) are as the accelerative forces and the attracted bodies conjointly, here are equal to each other (by the third Law of Motion). Therefore, the absolute attractive force of the body *A* is to the absolute attractive force of the body *B* as the mass of the body *A* is to the mass of the body *B*.

Q.E.D.

Notes on I.69

• In III.7 we will start considering the effect of the mass of the *attracting* body. To prepare for this, we need to lay some groundwork, which Newton does in I.69. So we will dip back for some mathematical preparation which, if we can meet its conditions in the real world, we can apply to prove III.7.

This will be the fullest level of complication for two bodies. At first we happily limited ourselves to accelerative forces independent of either the mass of the attracting body or the mass of the attracted body. Then we ease in the mass of the attracted body in III.6, and finally now we are getting ready for the most complicated case, considering the effects of the masses of two mutually attracting bodies.

Having taken things step by step, we are ready for the most complex case, but we must take great care if we are not to lose track of whether a force is being viewed as coming from a body or operating on a body. Shortcuts in notation are dangerous here. Another danger is carelessness born of a complacency that it's all really very obvious and simple, just this formula we learned in grade school. Look out! We don't yet know what that formula will be, nor do we know whether or not it will be simple. And keeping track of what is acting on what will demand all your concentration as you work through the proof here.

• We are given in the enunciation that each body "pulls" all the other bodies in the system. Newton evidently intends these hypothetical pulls we are given in this enunciation to be mutual third law actions since he invokes the Third Law in the proof (where he is calling them attractions). And he is fully entitled to suppose anything he pleases. Remember that we are dipping back into Book I, where everything is hypothetical. If this proposition is to be applied to gravitational attractions in the real world, we would need to justify the assumption that the *gravitational* "pulls" to which I.69 is being applied are Third Law mutual actions. We'll think about this further when we get to that point.

• In this proposition Newton may appear to have abandoned all pretence of keeping the options open on mechanisms for the cause of gravity, by speaking exclusively of bodies "pulling" and "attracting" other bodies. However, this is not the case: he immediately follows this proposition with a scholium again reminding us that when he uses the word attraction, he means *any* tendency of the bodies to approach one another, whether that tendency is a mutual seeking or the action of some sort of medium (corporeal or incorporeal) pushing the bodies floating in it towards each other.

Expansion of Newton's Sketch of I.69

"In a system of many bodies A, B, C, D, etc., if some body A pulls all the others B, C, D, etc., with accelerative forces that are inversely as the squares of the distances from the pulling body, and another body B also pulls the rest A, C, D, etc., with forces that are inversely as the squares of the distances from the pulling body, then the absolute forces of the pulling bodies A, B will be to each other as are the bodies A, B themselves, whose forces they are."

Given:

1. A system of many bodies;

2. each pulls all the others with accelerative forces that are inversely as the squares of the distances from the pulling body.

To Prove:

The absolute forces of the pulling bodies will be to each other as their respective quantities of matter.

Proof:

"For the accelerative attractions of all the bodies B, C, D, towards A, at equal distances, are equal to each other by hypothesis, and likewise, the accelerative attractions of all the bodies towards B, at equal distances, are equal to each other."

We establish in III.4 and III.6 that this applies to our world; but in this proposition, I.69, we are still in the hypothetical Book I and these attractions are, as Newton says, given by hypothesis.

"Now the absolute attractive force of the body A is to the absolute attractive force of the body B as the accelerative attraction of all the bodies towards A is to the accelerative attraction of all the bodies towards B at equal distances, and thus is the accelerative attraction of the body B towards A to the accelerative attraction of the body A towards B [in this ratio]."

$$\frac{\text{absolute attractive force of A}}{\text{absolute attractive force of B}} = \frac{\text{efficacy of cause } f_a \text{ towards A}}{\text{efficacy of cause } f_a \text{ towards B}}$$

$$= \frac{f_a \text{ on any other body at same distance towards } A}{f_a \text{ on any other body at same distance towards } B}$$

$$= \frac{\text{accelerative attraction of body } B \text{ towards } A}{\text{accelerative attraction of body } A \text{ towards } B} \text{ at the same distance.} \quad (1)$$

"But the accelerative attraction of the body B towards A is to the accelerative attraction of the body A towards B as the mass of the body A to the mass of the body B, for the reason that the motive forces, which (by Definitions 2, 7, and 8) are as the accelerative forces and the attracted bodies conjointly, here are equal to each other (by the Third Law of Motion)."

$$\underset{B,\ C,\ \text{etc., towards } A}{\text{motive force}} \propto \underset{\text{in } B,\ C,\ \text{etc.}}{\Delta \text{ quantity of motion.}} \qquad \underset{\text{(for a given time)}}{[\text{Definition 8}]}$$

$$\text{quantity of motion} \propto \text{velocity} \times \text{quantity of matter} \qquad \underset{\text{(for a given time)}}{[\text{Definition 2}]}$$

Therefore,

$$\underset{B,\ C,\ \text{etc., towards } A}{\text{motive force}} \propto \Delta\ (\underset{\text{of } B,\ C,\ \text{etc.}}{\text{velocity}} \times \underset{\text{in } B,\ C,\ \text{etc.}}{\text{quantity of matter}}).$$

Also,

$$\text{accelerative force} \propto \underset{\text{(for a given time)}}{\Delta \text{ velocity}}$$

Therefore,

$$\underset{\text{towards } A}{\text{motive force}} \propto \underset{\text{of } B \to A}{\text{accelerative force}} \times \underset{\text{in } B}{\text{quantity of matter}}$$

The enunciation gives us that A attracts B and B attracts A. Newton is evidently (the evidence appearing in his proof) giving us hypothetical actions that are Third Law mutual attractions.

Law 3 says that in mutual actions, motive forces are equal and opposite.

Motive force of A towards B = motive force of B towards A.

By Euclid VI.16,

$$\frac{\text{accelerative force on } B \text{ towards } A}{\text{accelerative force on } A \text{ towards } B} = \frac{\text{quantity of matter in } A}{\text{quantity of matter in } B} \qquad (2)$$

"Therefore, the absolute attractive force of the body A is to the absolute attractive force of the body B as the mass of the body A is to the mass of the body B."

By Equation 1, at the same distance,

$$\frac{\text{absolute attractive force of } A}{\text{absolute attractive force of } B} = \frac{\text{accelerative attraction of body } B \text{ towards } A}{\text{accelerative attraction of body } A \text{ towards } B}.$$

Substituting from Equation 2,

$$\frac{\text{absolute attractive force of } A}{\text{absolute attractive force of } B} = \frac{\text{quantity of matter in } A}{\text{quantity of matter in } B}.$$

Q.E.D.

I.69 Corollary 1

Hence if the individual bodies A, B, C, D, etc., in a system, considered separately, pull all the rest with accelerative forces that are inversely as the squares of the distances from the pulling body, the absolute forces of all those bodies to each other are as the bodies themselves.

Note on I.69 Corollary 1

This corollary says that I.69 will work for any number of bodies attracted by an inverse square law. It extends the demonstration from the two bodies *A* and *B* to all the bodies in the system.

[Corollaries 2 and 3 have been omitted. Corollary 2 considers force laws other than inverse square. Corollary 3 mentions elliptical orbits, which have not yet been established as of Book III Proposition 7, where we will be using this proposition.]

Scholium following Proposition I.69

By these propositions we are led by the hand to the proportionality between centripetal forces and the central bodies towards which those forces are apt to be directed. For it is in conformity with reason that the forces that are directed towards bodies depend upon the nature and magnitude of the same bodies, as it is in magnetic bodies. And whenever instances of this sort occur, the attractions of the bodies are to be estimated by assigning the appropriate forces to their individual particles and gathering together the sums of the forces. The word "attraction" I here use generally for any attempt whatever of bodies to approach one another, whether that attempt arise from the action of the bodies (whether of mutually seeking one another or of setting each other in motion by emitted spirits), or whether it arises from the action of the aether, or of air, or of any medium whatsoever (corporeal or incorporeal) in any way pushing bodies floating in it towards each other. I use the word "impulse" [*impulsus*] in the same general sense, as in this treatise I am considering, not species of forces and physical qualities, but mathematical quantities and proportions, as I have explained in the Definitions. In mathematics, one must consider those quantities and ratios of forces which follow from any given conditions whatever, and thereafter, when it comes down to physics, these ratios are to be compared with the phenomena, that it may become known which particular characteristics of forces belong to the individual kinds of attracting bodies. And then at last it will be permissible to discuss the species, causes, and physical reasons [*rationes*] more safely. Let us now see by what forces spherical bodies, consisting of the attracting particles just described, ought to act upon one another, and what sorts of motions may follow thence.

Proposition 7

That gravity is given towards bodies universally, and that it is proportional to the quantity of matter in each.

That all the planets are mutually heavy towards each other, we have now already proved, as well as that gravity towards any one of them, considered separately, is inversely as the square of the distance of places from the center of the planet. And the consequence of this (by Book I Proposition 69 and its corollaries) is that gravity towards all is proportional to the matter in the same bodies.

Further, since all the parts of any planet you please A are heavy towards any planet you please B, and the gravity of any part is to the gravity of the whole as the matter of the part is to the matter of the whole, and to every action there is an equal reaction (by the Third Law of Motion), [therefore] the planet B will in turn gravitate towards all the parts of the planet A, and its gravity towards any particular part will be to its gravity towards the whole as the matter of the part to the matter of the whole.

Q.E.D.

Notes on III.7

- **Third Law Mutuality:**

In the course of the proof of this proposition, gravitational attractions are treated as mutual actions in the sense of the Third Law of Motion; that is, it is assumed that when each of two bodies is heavy towards the other, the heaviness of each is the other's Third Law reaction.

This demonstration invokes Law 3 directly for Steps 7 and 8 and indirectly in Step 3 by using I.69. It depends upon our being able to apply this law to gravitational attractions.

Law 3 seems most obviously to apply to mechanical, direct-contact, pushes and pulls: the examples in Newton's commentary to the law are someone pushing a stone with a finger and a horse pulling a stone tied to a rope. In the second edition, Newton added a sentence at the end of the commentary: "This law applies to attractions as well, as will be proven in the next scholium." The scholium that follows the Laws of Motion includes some thought experiments that suggest persuasively—but perhaps not conclusively—that gravitational attractions in our world seem to be equal and opposite. (See the discussion in the Note on Law 3 and Scholium to the Laws, page 45.)

Questions for Discussion: Newton speaks of the bodies considered in this proposition being mutually heavy. What does mutual mean? When do two simultaneous and adjacent actions become mutual? Is the "mutual" of common speech different from the "mutual" of Law 3? If my friend and I spot each other and joyously move towards each other, the attraction and the resulting motion will probably be called mutual in common speech, but are they Third Law "reactions" or are they two independent actions arising out of the promptings of feelings of friendship taking place independently in each of our souls? Or, for a purely physical example, consider two cue sticks wielded by two different people directing two billiard balls towards each other. Each ball is impelled towards the other and in outward form the two actions might look mutual (indeed, if one didn't see the cue sticks or the people, one might think one was seeing a gravitational attraction at work). But would they be Third Law mutual reactions?

Although every action may entail an equal and opposite reaction, it does not follow that any two actions one might fasten upon are each other's mutual reactions. What do you think about the applicability of the Third Law to gravitational attractions? What arguments could you make for its plausibility? Could it be proved logically? Have we proved it indirectly by finding that the universe obeys Newton's laws of celestial mechanics that in turn depend on this assumption? How is attraction the same as and different from the pushes and pulls that the Third Law seems to be primarily concerned with? How is gravitational attraction different from other kinds of attractions, such as inclinations of the soul?

• **Universal Gravitation?** This proposition appears to establish that every particle of matter attracts every other particle by the force of gravity with an inverse square distance law. Having worked through this proof, we will surely, finally, be able to say that we have demonstrated universal gravitation as we know it and take it for granted today. But Newton was not satisfied, and the snag that made him resist and even doubt the possibility of particle-to-particle universal gravitation must be dealt with before we will really be safely at the goal of this great adventure. The Pause After III.7 will take up that snag, and the following propositions will deal with it.

Expansion of Newton's Sketch of III.7

To Prove:

"That gravity is given towards bodies universally, and that it is proportional to the quantity of matter in each."

Proof:

We must set up the conditions for invoking I.69; this is done in Steps 1 and 2.

Step 1

"That all the planets are mutually heavy towards each other, we have now already proved,..."

The first condition for I.69 is that each body attracts all other bodies. Proposition III.6 established that all planets gravitate towards each other: that is, that each body is attracted to all other bodies. If each body is attracted to all, the bodies to which it is attracted must be "attracting," whatever that might mean. Thus all bodies must be "attracting."

Step 2

"... as well as that gravity towards any one of them, considered separately, is inversely as the square of the distance of places from the center of the planet."

The second condition for invoking I.69 is that these attractions be all taking place according to an inverse square force law.

Props III.1, 2, 5, and 6 established that accelerative force of gravity towards Jupiter, Saturn, earth, and the sun varies inversely as the square of the distances from the centers of the "attracting" planets.

Thus we have the required force law for Jupiter in relation to its moons, Saturn in relation to its moons, the sun in relation to the primary planets, and the earth in relation to its moon. By the third Rule of Philosophizing,

> The qualities of bodies ... upon which experiments can be carried out, are to be taken as qualities of bodies universally.

And in the commentary following Rule 3,

> Finally, if it be established universally by experiments and astronomical observations that all bodies on the surface of the earth are heavy towards the earth, and this according to the quantity of matter in each, and that the moon is heavy towards the earth according to the quantity of its matter, and that our sea in turn is heavy towards the moon, and that all the planets are heavy towards each other, and that the gravity of comets towards the sun is similar, it will have to be said, by this rule, that all bodies gravitate towards each other.

This reasoning applies equally to the universality of heaviness in bodies and to the universality of the inverse square force law. Thus we have met the two explicit conditions of I.69.

As noted above, I.69 invokes Law 3; thus another condition that would have to be met would be that these gravitational attractions are each other's Law 3 mutual reactions.

Possibly, when Newton asserts in his commentary to Law 3 that the law applies to "attractions" (his demonstrations in the scholium are instances of gravitational attractions), Newton considers it established that gravitation attraction between two bodies is a mutual Third Law action-reaction. Or perhaps he considers that it is demonstrated by his thought experiments in the scholium. And, as discussed in the note following the scholium after the Laws, laws of motion are not propositions and thus perhaps do not have to be proved.

But in any case, that Newton does assume this, or does consider it established, is apparent by his explicitly invoking it in Step 6 and, indeed, by his dropping the claim into his discussion in III.5 Corollary 1.

Step 3

"And the consequence of this (by Book I Proposition 69 and its corollaries) is that gravity towards all is proportional to the matter in the same bodies."

Therefore by I.69 (and its Corollary 1), gravity tending towards all the planets is proportional to the matter that the same bodies contain. The gravity tending towards all the planets is the absolute quantity of centripetal force. It is proportional to the matter in the *attracting* objects at the centers of forces.

Step 4

"Further, since all the parts of any planet you please *A* are heavy towards any planet you please *B*, ..."

This was established in III.6 Part 1. We use this in Step 7.

Step 5

"... and the gravity of any part is to the gravity of the whole as the matter of the part is to the matter of the whole, ..."

This was also established in III.6, in Part 3, and is used in Step 8.

Step 6

"... and to every action there is an equal reaction (by the Third Law of Motion), ..."

Assuming that the gravitation of *A* towards *B* and *B* towards *A* are mutual actions in the sense of Law 3, "the mutual actions of two bodies upon each other are always equal and directed to contrary parts" by that law.

Step 7

"... [therefore] the planet *B* will in turn gravitate towards all the parts of the planet *A*, ..."

Now, in Steps 7 and 8, Newton draws two conclusions from the conditions set up in Steps 4–6.

Step 4 told us that all parts of any planet *A* will gravitate towards any other planet *B*. Similarly, one could say that all parts of planet *B* (which could be "any planet *A*") will also gravitate towards planet *A* (which could be "any planet *B*"). So if Newton simply wanted to turn Step 4 around to say that all parts of *B* gravitated towards planet *A*, he would not have needed to invoke the Third Law.

But note that he is doing something slightly but significantly different. In Step 7 he wants to say that planet *B* (as a whole) will gravitate towards all the parts of planet *A* (that is, to each of *A*'s parts individually). This is not simply substituting *B* for *A*, but rather it is reversing the assertion of Step 4, taking the converse. For this he needs the justification he finds from the Third Law. If these attractions are each other's mutual Third Law reactions, they will be equal and opposite, and the converse of Step 4 will be established.

Step 8

"...and its gravity towards any particular part will be to its gravity towards the whole as the matter of the part to the matter of the whole."

That is, the weight of *B* as a whole towards any particular part of *A* will be to *B*'s weight towards the whole of *A* as the matter of the part of *A* is to the matter of the whole of *A*.

By Step 5, the gravity of a part of planet *A* towards any other planet *B* is to the gravity of the whole of *A* towards *B* as the matter of that particular part of *A* is to the matter of the whole of *A*.

Note again that we are not dealing with a simple substitution of *B* for *A* in the statement of Step 5. Rather, we have taken a converse. So again we need Law 3 to allow the reversal. Furthermore, only the invocation of Law 3 will allow us to establish that the gravities of *B* towards each part of *A* and each part of *A* towards *B* will be equal.

By Law 3, the gravity of the whole of *B* towards each of the parts of *A* is equal and opposite to the gravity of each part of *A* towards the whole of *B*.

Therefore the gravity of *B* towards any part of *A* will be to *B's* gravity towards the whole of *A* as the matter of *A's* part to the matter of *A's* whole. That is what was to have been proved.

Q.E.D.

[The corollaries to this proposition are not included, since they aren't used in Newton's central argument.]

Pause After III.7

Is This Universal Gravitation?

Newton's Own Doubts

Now it seems as if we really must have universal gravitation. We've said that planet *B* gravitates towards all parts of planet *A*, and that the gravity is as the quantity of matter in each part. Assuming as Newton does here that the attractions are mutual, the converse would also hold. Since every part must be attracting every part proportionally to its matter, it seems to follow that every particle attracts every particle.

But Newton wasn't willing to make that step here. He knew he needed something else before he could be satisfied. Yes, it does seem as if every particle attracts every other particle. But could it possibly be universally true? Our discoveries in the Phenomena and the first propositions of this book were that bodies attract inversely as the square of the distances between their centers. Is this consistent with particle-to-particle attraction? Would all the individual particle-to-particle attractions resolve to an attraction between the centers?

This isn't intuitively obvious. Consider a pebble dropped above the surface of the earth, attracted by every particle of the earth's matter inversely as the square of the pebble's distance from that particle. Doesn't it seem odd that the center of forces should be at the center of the earth? One might expect that the effective center of attraction would be a lot closer to the nearer attracting particles, considering that they are attracting much more vigorously. Instead we are being asked to believe that the effective center is halfway to the very farthest attracting particles, which must be exerting much less force, given that their force has fallen off as the square of their distances.

On 20 June 1686, the year before publication of *Principia*, Newton wrote to Halley:

> ... I never extended the duplicate proportion lower than to the superficies of the earth, and before a certain demonstration I found the last year have suspected it did not reach accurately enough down so low ...

(The "duplicate proportion" is the inverse square force law. "Superficies" means surface. The "certain demonstration" must refer to Proposition I.71 and its extensions I.74–I.76.)

Newton had worked hard on deriving the demonstrations that resolved this difficulty and he was proud of them. He goes on in the same letter to assert that Robert Hooke had only been guessing about how the inverse square law might apply to gravity. The letter continues,

> There is so strong an objection against the accurateness of this proportion that, without my demonstrations, to which Mr. Hooke is yet a stranger, it cannot be believed by a judicious philosopher to be any where accurate.

In the sketch to III.8, Newton describes his process of thought at this point, after proving III.7.

> After I had found that gravity in an entire planet arises and is compounded from the gravities to the parts, and that in the individual parts it is inversely proportional to the squares of the distances from the parts, I was in doubt whether that inverse duplicate proportion might apply accurately in the whole force compounded of many forces, or only approximately. For it might happen that the proportion that applies accurately enough at greater distances, would err noticeably near the surface of the planet because of the unequal distances and dissimilar positions of the particles. But at last, by Proposition 75 and 76 of Book I and their corollaries, I understood the truth of the proposition that is here in question.

His doubt was only removed by his success, "at last," in proving that it was possible that the results of III.7 could be consistent with the results of the first three propositions of this book, a result delivered by I.71 through I.76.

The Difficulty

The problem is this. Suppose that an outside particle is attracted to every particle of a sphere inversely as the square of the distance between particles. As the outside particle approaches the near particles on the sphere, the distance to those near particles gets smaller than any given amount and the attractive force becomes larger. One might expect that the force would grow beyond all bounds, approaching an infinite force as the distance approaches zero. The farther particles would add to this force, with lesser attraction.

One wouldn't expect that the force exerted by the near particles could ever be balanced out by particles beyond the center, or in any other way, such that we could be left with an attraction inversely as the square of the distance to the center.

But it is this very inverse square attraction to the center that we found to be the law of gravity in our first seven propositions of Book III.

Two conclusions would be possible for the judicious philosopher whom Newton thought would be unable to believe that, given particle-to-particle attraction, the law of inverse square to the center could hold at the surface.

One is that the inverse square law of gravity, while apparently true generally, doesn't apply at distances very near to the surface of a sphere. This is the suspicion that Newton says he had. But that would seem to violate the second Rule of Philosophizing.

The other possibility is that our theory that gravitational attraction holds between every two particles of matter was incorrect. But we have our evidence of the phenomena, and our argument for the continuity of the inverse square law between centers, and the rest of the argument for particle-to-particle attraction as we have developed it to this point.

Question for Discussion: We seem to be at an impasse. What plausibility could we find for particle-to-particle translating into center-to-center?

The Resolution

Proposition I.71 and its extensions give a rigorous proof that an inverse square force law will yield exactly the result we need, showing that no matter how close we get to the surface of the sphere, the force is always exactly inversely as the square of the distance to the center of the sphere—even though the outside particle is being attracted by every particle of the sphere inversely as the distance between *them.*

I.71 is also worth thoughtful attention in its own right, as a beautiful and amazing proposition. It seems so unlikely that a particle off to one side will not be more attracted to the closer side that we read the enunciation of this proposition with skepticism. And as the proposition unfolds one realizes that it is more astounding even than one might have guessed from the enunciation.

It turns out to depend not at all on the particles beyond the center attracting in a weaker way that balances out the more vigorous attraction of the nearer particles. Students trusting some intuition along that line to account for why particle-to-particle attraction could lead to a center-to-center experimental result should notice, in working through I.71, that even if we only considered strips of the surface on the near side, the effective attraction would *still* be to the center of the sphere of which these strips are a part.

Thus intuitions, including Newton's intuition all his life until he succeeded in proving these propositions, saying it would not work, turn out to be in error. And even our attempts at explaining it seem misplaced. The reality, as revealed mathematically, is astonishing.

Newton gives us the argument, and gives it to us not in algebra or calculus, where we lose the picture of what is happening and just have to trust the result that falls out of the manipulations of symbols, but in geometry, with the picture right before our delighted eyes.

We return to the question of the compatibility of our observations and the particle-to-particle theory of gravitation in the Pause After I.74.

Flashback to I.71–I.76 for Use in III.8

Once again, following Newton's advice, we dip back into Book I to establish conclusions we shall need in the proof of III.8. The three preceding pages (the Pause After III.7) discuss in some detail why we need this digression and why these propositions are well worth the attention they demand.

Book I Proposition 71

The same things being supposed [as in Proposition 70, i.e., that to the individual points of a spherical surface there tend equal centripetal forces decreasing in the duplicate ratio of the distances from the points], I say that a corpuscle placed outside the spherical surface is attracted to the center of the sphere with a force inversely proportional to the square of its distance from the same center.

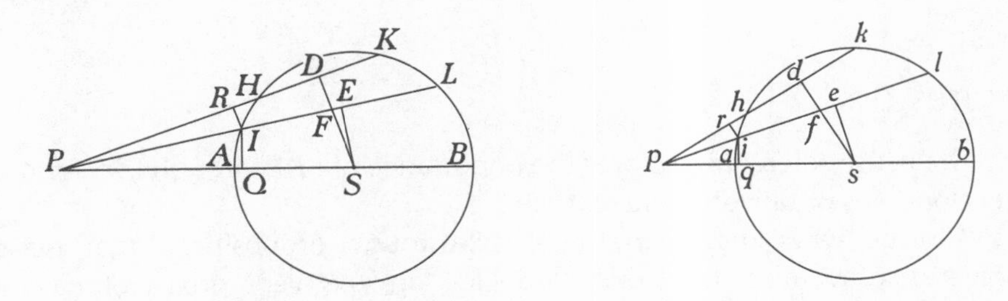

Let $AHKB$, $ahkb$ be two equal spherical surfaces described with centers S, s [and] diameters AB, ab; and P, p be corpuscles located externally on those diameters extended. From the corpuscles let the lines PHK, PIL, phk, pil be drawn, cutting off from the great circles AHB, ahb the equal arcs HK, hk and IL, il; and to them let the perpendiculars SD, sd, SE, se, IR, ir be dropped, of which SD, sd intersect PL, pl at F and f. Also, let the perpendiculars IQ, iq be dropped to the diameter. Let the angles DPE, dpe vanish, and because of the equals DS and ds, ES and es, the lines PE, PF and pe, pf, and the little lines DF, df may be taken as equals, that is, [magnitudes] whose ultimate ratio, when those angles DPE, dpe vanish at the same time, is that of equality. With these things set up thus, PI will be to PF as RI is to DF, and pf to pi as df or DF is to ri, and (*ex aequali*) $PI \times pf$ will be to $PF \times pi$ as RI is to ri that is (by Lemma 7 Corollary 3), as arc IH is to arc ih. Again, PI is to PS as IQ is to SE, and ps is to pi as se or SE is to iq, and (*ex aequali*) $PI \times ps$

is to $PS \times pi$ as IQ is to iq. And, these ratios being compounded, sq.$PI \times pf \times ps$ is to sq.$pi \times PF \times PS$ as $IH \times IQ$ is to $ih \times iq$, that is, as the circular surface which the arc IH will describe in the revolution of the semicircle AKB around the diameter AB, is to the circular surface which the arc ih will describe in the revolution of the semicircle akb around the diameter ab. And the forces with which these surfaces attract the corpuscles P and p along lines tending towards the surfaces themselves, are (by hypothesis) as those surfaces directly and the squares of the distances of the surfaces from the bodies inversely, that is, as $pf \times ps$ is to $PF \times PS$. And these forces are to the oblique parts of them that (constructed by the resolution of forces through Corollary 2 of the Laws) tend towards the centers along the lines PS, ps, as PI to PQ, and pi to pq, that is (because of the similar triangles PIQ and PSF, piq and psf), as PS to PF and ps to pf. Whence, *ex aequali*, it is brought about that the attraction of this corpuscle P towards S is to the attraction of the corpuscle p towards s as $\frac{PF \times pf \times ps}{PS}$ is to $\frac{pf \times PF \times PS}{ps}$, that is, as sq.$ps$ is to sq.PS. And by a similar argument, the forces by which the surface described by the revolution of the arcs KL, kl attract the corpuscles will be as sq.ps to sq.PS. And the forces of all the circular surfaces into which the two spherical surfaces can be divided, by taking sd always equal to SD and se equal to SE, will be in the same ratio. And, by composition, the forces of the whole spherical surfaces acting upon the corpuscles will be in the same ratio.

Q.E.D.

Notes on I.71

- The Pause After III.7 explores the importance of I.71 through I.75 for our development of universal gravitation.

In some ways these are the break-through propositions that make universal gravitation thinkable, and Newton was very proud of having discovered the way to make the proofs. Indeed, given the implausibility of the result, they would be among the most astonishing and thrilling of mathematical demonstrations, even if something as momentous as the theory of universal gravitation did not hang on their success.

- Proposition I.71 is the foundation of this sequence of propositions that culminate in I.75. The sequence establishes that if every particle attracts every particle by inverse square of the distance between the particles, the whole sphere will attract another sphere by inverse square of the distance between the centers—as if the spheres' masses were concentrated at the centers.

I.71 establishes that a spherical shell will attract an outside particle with a force that varies inversely as a square of the distance to the center of the ghost sphere of which this is the shell. The extension to the solid sphere is done by relatively easy steps. We take the conclusion found here about the spherical shell and nest successive shells inside one another to make up our solid sphere.

This conclusion about the solid sphere in turn allows us, in III.8, to establish universal gravitation.

Expansion of Newton's Sketch of I.71

"The same things being supposed [as in Proposition 70, i.e., that to the individual points of a spherical surface there tend equal centripetal forces decreasing in the duplicate ratio of the distances from the points], I say that a corpuscle placed outside the spherical surface is attracted to the center of the sphere with a force inversely proportional to the square of its distance from the same center."

Given:

To every individual point of a spherical surface there tend centripetal forces inversely as the squares of the distances from those points.

To Prove:

A corpuscle placed outside the spherical surface will be attracted inversely as the square of the distance from the center of the sphere whose surface this is.

Proof:

Before plunging into this proposition, let's get an overview of the plan of the proof.

The Plan:

We have two equal spherical shells, and two particles outside the spheres. We will show that the forces on the two particles are inversely as the squares of their distances from the centers of their respective spheres. Since the spheres are supposed equal, our result will compare particles attracted to the same sphere at different distances.

The key to the plan is to divide the spheres into bands, look at the ratio of forces on the bands, then add the bands up into the whole spheres. Let's see how we construct the bands, showing just one particle and sphere for simplicity.

From *P*, take a line through the center of the sphere *S*. Draw the tangent to the sphere from *P*, touching at *T*.

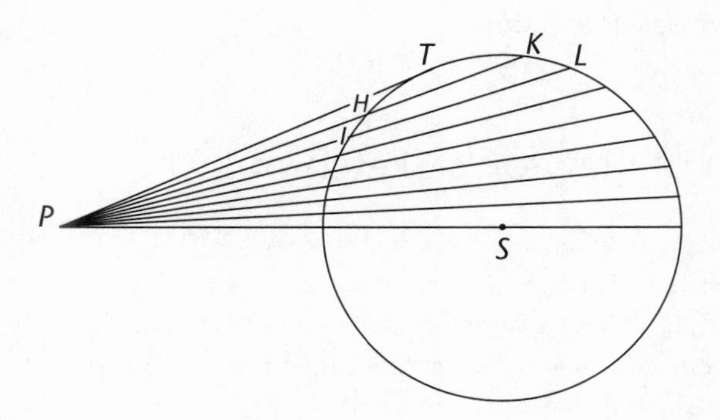

Now divide angle *TPS* into many angles, which need not be equal. Each angle defines arcs on a great circle, one on each side of the sphere: for example, *HT*, *KT*, and *IH*, *LK*.

Rotate each arc around the sphere pivoting on axis *AB* to make bands. This will cover the whole sphere with no gaps and no overlap.

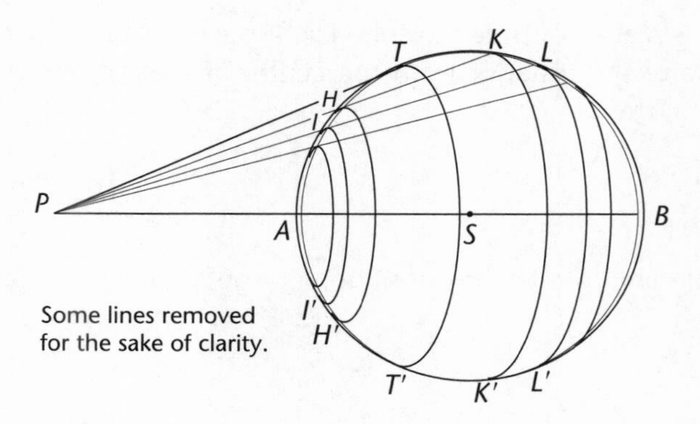

The force exerted on *P* by the surface areas of the whole sphere will be the sum of the forces exerted by each of the bands individually.

Now let's look at the details of the proof.

"Let *AHKB*, *ahkb* be two equal spherical surfaces described with centers *S*, *s* [and] diameters *AB*, *ab*; and *P*, *p* be corpuscles located externally on those diameters extended."

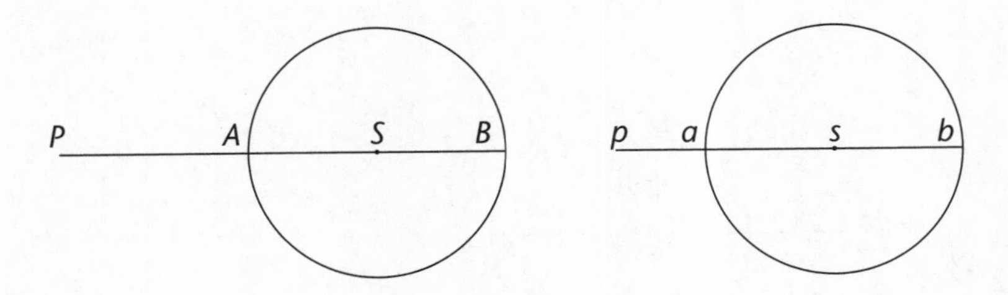

We are supposing that PS and ps are unequal distances. Because the spheres are equal, this could represent the same particle at different distances.

"From the corpuscles let the lines PHK, PIL, phk, pil be drawn, cutting off from the great circles AHB, ahb the equal arcs HK, hk and IL, il; and to them let the perpendiculars SD, sd, SE, se, IR, ir be dropped, of which SD, sd intersect PL, pl at F and f."

We start with a particular angle HPI making arcs HI and KL.

Construct $SD \perp PK$ and $SE \perp PL$.

Then we go to the second sphere and second distance. How do we construct our corresponding angles such as hpi? They are not simply made equal to the original angles such as HPI. Our new angle hpi will be constructed to cut off arc $\widehat{hk} = \widehat{HK}$ and $\widehat{il} = \widehat{IL}$. Thus, if we have a different distance ps, the angle will not be equal to angle HPI.

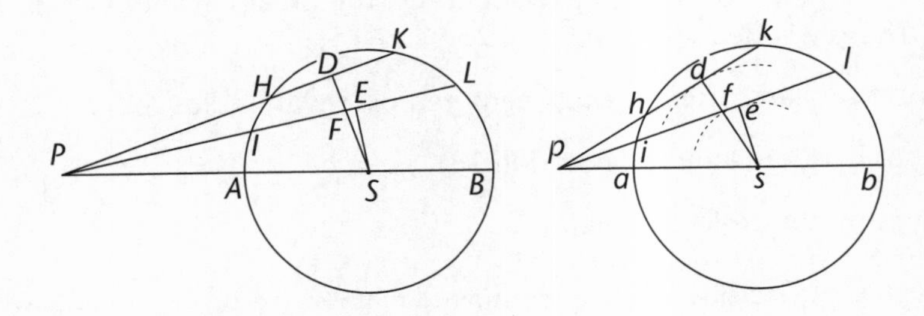

Construction of angle hpi:

We need to construct arcs $\widehat{il} = \widehat{IL}$ and $\widehat{hk} = \widehat{HK}$, where $\overline{pil}$ and $\overline{phk}$ are straight lines. Having constructed the arcs, we will have the angle hpi.

Suppose we had equal arcs HK and hk.

Because these are equal spheres, we have equal great circles AHB and ahb.

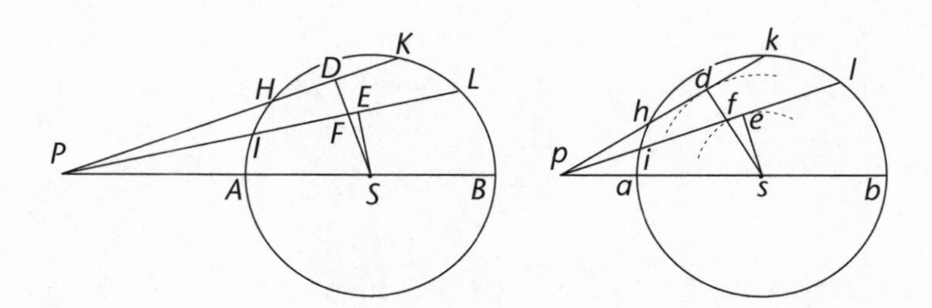

By Euclid III.29, in equal circles, equal circumferences are subtended by equal straight lines. Therefore chord $HK = hk$.

And by Euclid III.14, in a circle equal straight lines are equally distant from the center.

Therefore $SD = sd$.

Similarly, if arc $il = IL$, $se = SE$.

We'll go the other way for the construction.

Draw a circle with center *s* and radius *SD*.

From point *p*, by Euclid III.17, construct a line tangent to that circle, calling the point where the line touches the circle *d*.

By Euclid III.18, $sd \perp pd$.

Call the points where the tangent intersects the great circle of the sphere *h* and *k*. These will define arc *hk*.

Find arc *il* similarly, using a circle centered on *s* with radius *SE*.

Note that here similarly (by Euclid III.18), $se \perp pe$.

This gives us our desired angle *hpi*.

"Also, let the perpendiculars IQ, iq be dropped to the diameter."

We will do this when we are ready to use *IQ*.

"Let the angles DPE, dpe vanish, and because of the equals DS and ds, ES and es, the lines PE, PF and pe, pf, and the little lines DF, df may be taken as equals, that is, [magnitudes] whose ultimate ratio, when those angles DPE, dpe vanish at the same time, is that of equality."

Note first that we are supposing angles *DPE* and *dpe* to evanesce. This will carry through the proposition: each band we generate using these angles will evanesce.

F is the point at which *SD* intersects *PL* and similarly for the second sphere.

It has been shown that DS, ds and ES, es are always equal if the arcs are equal, which they are by construction.

SD, sd and SE, se are perpendiculars to the two lines making the angle. As the angles DPE and dpe vanish, the lines PK, PL and the lines pk, pl will coincide. Then the points E, F, D and points e, f, d will coincide. Then lines PE, PF and pe, pf will be equal and

$$SE \overset{\text{ult}}{=} SF; \quad se \overset{\text{ult}}{=} sf.$$

$$DF = SD - SF; \quad df = sd - sf.$$

Substituting,

$$DF \overset{\text{ult}}{=} SD - SE; \quad df \overset{\text{ult}}{=} sd - se.$$

Subtracting equals from equals, the differences will be ultimately equal:

$$DF \overset{\text{ult}}{=} df.$$

"With these things set up thus, PI will be to PF as RI is to DF, and pf to pi as df or DF is to ri, and (*ex aequali*) $PI \times pf$ will be to $PF \times pi$ as RI is to ri ..."

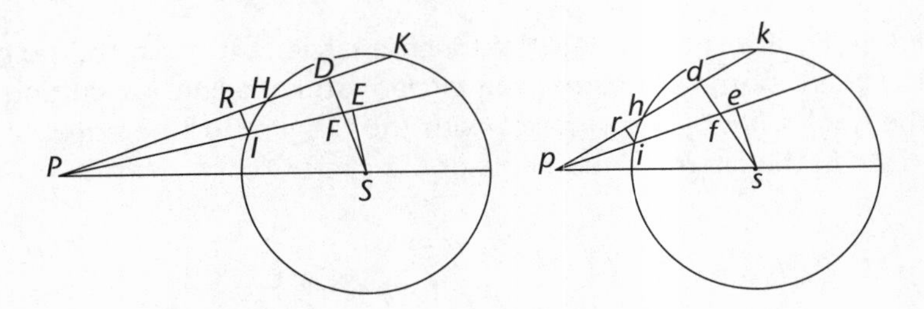

Let IR, ir be dropped perpendicular to PK, pk.

Since $SD \perp PK$ by construction,

In $\triangle PDF$, $IR \parallel FD$.

Therefore $\triangle PDF \sim \triangle PRI$. [Eu. VI.2 and 5]

Therefore by Euclid VI.4 and *alternando*,

$$PI : PF :: RI : DF. \qquad (1)$$

And by the same reasoning without *alternando*,

$$pf : pi :: df : ri.$$

But since, as argued in the previous bit of the sketch, $DF \overset{\text{ult}}{=} df$,

$$pf : pi \overset{\text{ult}}{=} DF : ri. \qquad (2)$$

Multiply simultaneous equations 1 and 2:

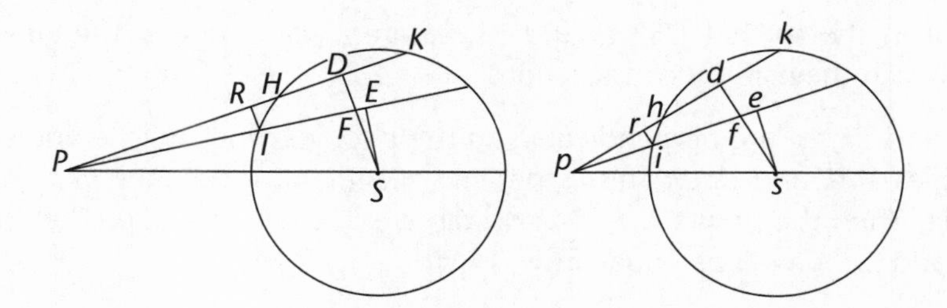

$$PI \times pf : PF \times pi \overset{ult}{=} RI \times DF : DF \times ri.$$

$$\frac{PI \times pf}{PF \times pi} \overset{ult}{=} \frac{RI}{ri}. \tag{3}$$

"... that is (by Lemma 7 Corollary 3), as arc IH is to arc ih."

Now Newton says that the ratios of Equation 3 are the same as that of the ratio of the arcs $\widehat{IH} : \widehat{ih}$.

What is the justification for this?

By Euclid III.32, if a straight line *HN* touch a circle, and from the point of contact *H* there be drawn across the circle a straight line *HK* cutting the circle, the angles *NHK* that it makes with the tangent will be equal to the angles *HMK* in the alternate segments of the circle.

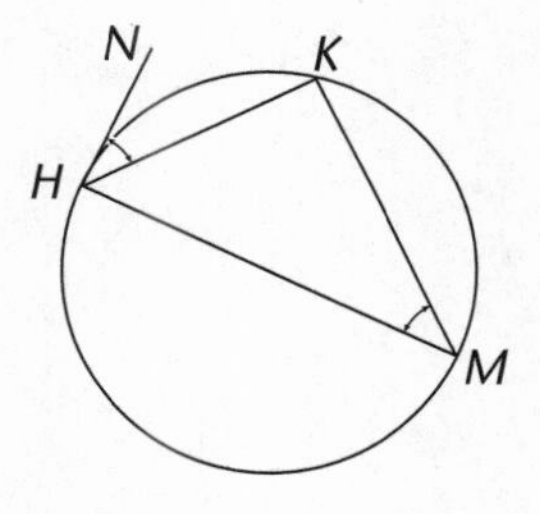

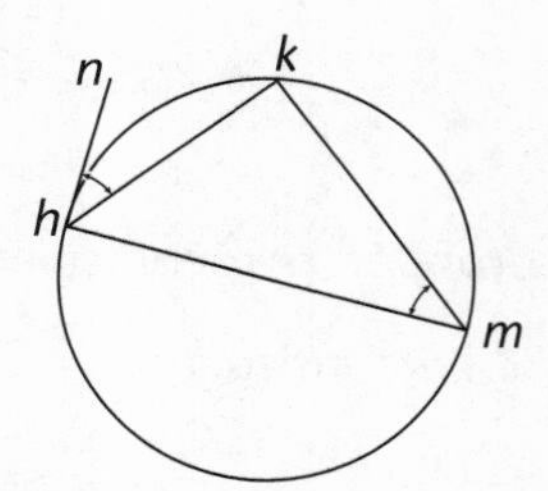

Angle NHK = angle HMK.

Angle nhk = angle hmk.

But $\widehat{HK} = \widehat{hk}$ by construction, so

Angle HMK = angle hmk. [Eu. III.21]

Therefore

$$\text{Angle } NHK = \text{angle } nhk. \tag{4}$$

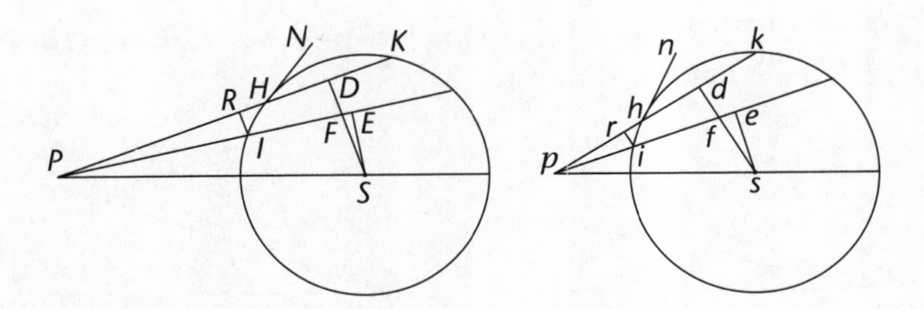

Consider chord IH.

Remember that we are supposing angle DPE to evanesce.

As angle $DPE \to 0$, $I \to H$.

As $I \to H$, IH approaches tangency at H.

Ultimately IH coincides with tangent HN.

By Euclid I.15, angle $RHI \overset{ult}{=}$ angle NHK and angle $rhi \overset{ult}{=}$ angle nhk.

By Equation 4, angle $NHK =$ angle nhk.

Therefore angle $RHI \overset{ult}{=}$ angle rhi.

Since $IR \perp PK$ and $ir \perp pk$ by construction,

Angle $HRI =$ angle hri.

Therefore $\triangle HRI \overset{ult}{\sim} \triangle hri$.

By Euclid VI.4,

$$RI : ri \overset{ult}{::} IH : ih. \tag{5}$$

Now we are ready to use Lemma 7 Corollary 3, which Newton cites in his sketch as the justification for this step:

$$IH : ih \overset{ult}{=} \widehat{IH} : \widehat{ih}. \tag{6}$$

So, picking up Equation 3:

$$\frac{PI \times pf}{PF \times pi} \overset{ult}{=} \frac{RI}{ri}.$$

Continuing the proportion with Equations 5 and 6:

$$\frac{PI \times pf}{PF \times pi} \overset{ult}{=} \frac{RI}{ri} \overset{ult}{=} \frac{IH}{ih} \overset{ult}{=} \frac{\widehat{IH}}{\widehat{ih}}.$$

$$\frac{PI \times pf}{PF \times pi} \overset{ult}{=} \frac{\widehat{IH}}{\widehat{ih}}. \tag{7}$$

"Again, PI is to PS as IQ is to SE, and ps is to pi as se or SE is to iq, and (*ex aequali*) $PI \times ps$ is to $PS \times pi$ as IQ is to iq."

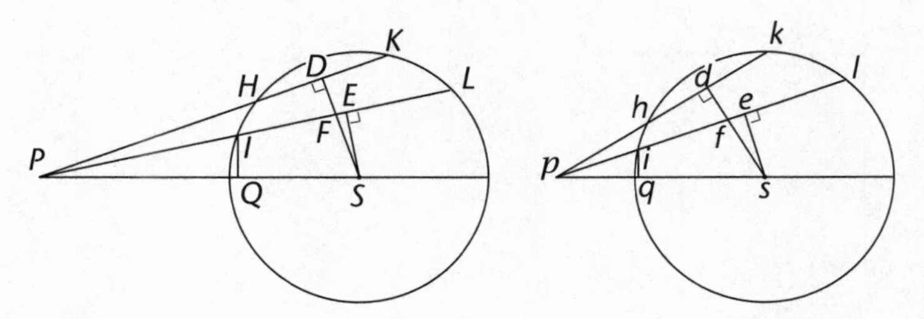

Drop $IQ \perp PS$.

$SE \perp PL$ by construction.

Therefore angle PQI = angle PES.

Angle IPQ = angle EPS. [identity]

Therefore $\triangle PQI \sim \triangle PES$.

And by Euclid VI.4,

$$PI : PS :: IQ : SE. \tag{8}$$

Similarly $\triangle pqi \sim \triangle pes$ and

$$ps : pi :: se : iq. \tag{9}$$

But se was constructed equal to SE.

Substituting into Equation 9,

$$ps : pi :: SE : iq. \tag{10}$$

Multiplying Equations 8 and 10:

$$PI \times ps : PS \times pi :: IQ \times SE : SE \times iq.$$

$$\frac{PI \times ps}{PS \times pi} = \frac{IQ}{iq}. \tag{11}$$

By Equation 7,

$$\frac{PI \times pf}{PF \times pi} \overset{ult}{=} \frac{\widehat{IH}}{\widehat{ih}}.$$

"And, these ratios being compounded, sq.$PI \times pf \times ps$ is to sq.$pi \times PF \times PS$ as $IH \times IQ$ is to $ih \times iq$, that is, as the circular surface which the arc IH will describe in the revolution of the semicircle AKB around the diameter AB, is to the circular surface which the arc ih will describe in the revolution of the semicircle akb around the diameter ab."

Multiplying Equation 7 by Equation 11:

$$\frac{PI^2 \times ps \times pf}{pi^2 \times PS \times PF} = \frac{IQ \times \widehat{IH}}{iq \times \widehat{ih}}.$$

Multiply numerator and denominator of right hand expression by 2π:

$$\frac{PI^2 \times ps \times pf}{pi^2 \times PS \times PF} \stackrel{ult}{=} \frac{IQ \times \widehat{IH} \times 2\pi}{iq \times \widehat{ih} \times 2\pi}. \qquad (12)$$

Now let I and H move around axis PQ of the sphere to make a surface strip.

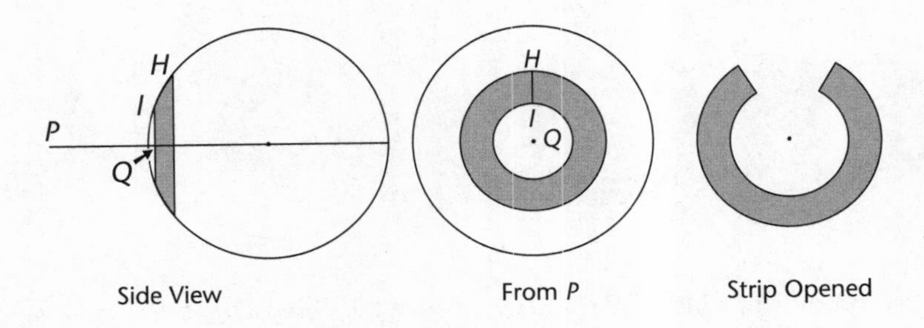

Side View From *P* Strip Opened

Newton has us take evanescent strips (he tells us that angle $DPE \to 0$), and so $IH \to 0$. Therefore the inner circumference or length of the strip $\stackrel{ult}{=}$ the outer length. At the limit, the surface area will be that of a rectangle; so surface area $\stackrel{ult}{=}$ length × width.

Q is the center of the circle made by I.

Therefore IQ is the radius r for the inner side of the strip.

Circumference $= 2\pi r$.

length $\approx 2\pi\, IQ$.

Surface area $\approx \widehat{IH} \times 2\pi\, IQ$.

So $IQ \times IH \times 2\pi$ in Equation 12 is ultimately equal to the surface area of the strip.

$$\frac{IQ \times IH \times 2\pi}{iq \times ih \times 2\pi}$$

is the ratio of the surface areas of the two strips.

But by Equation 12,

$$\frac{PI^2 \times ps \times pf}{pi^2 \times PS \times PF} \stackrel{ult}{=} \frac{IQ \times \widehat{IH} \times 2\pi}{iq \times \widehat{ih} \times 2\pi}.$$

Therefore

$$\frac{PI^2 \times ps \times pf}{pi^2 \times PS \times PF} \overset{ult}{=} \frac{\text{surface area of strip having width } IH}{\text{surface area of strip having width } ih}.$$

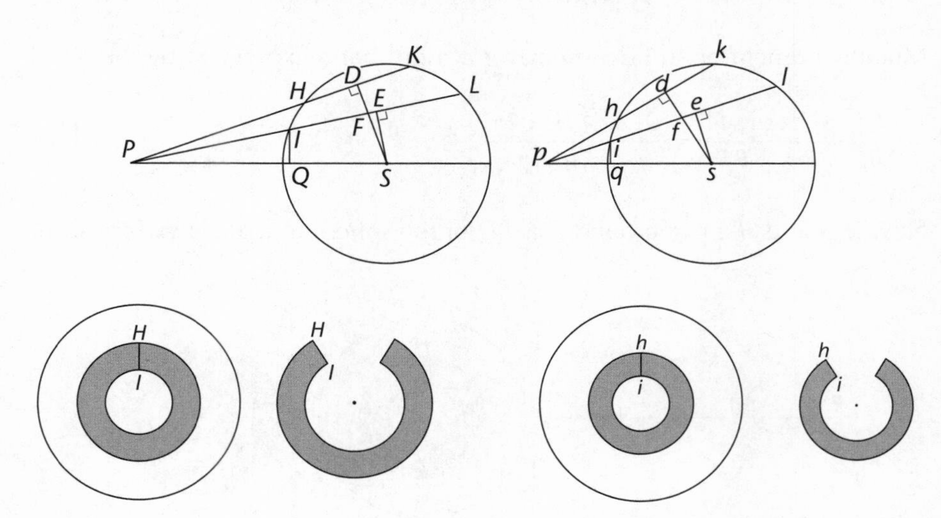

"And the forces with which these surfaces attract the corpuscles P and p along lines tending towards the surfaces themselves, are (by hypothesis) as those surfaces directly and the squares of the distances of the surfaces from the bodies inversely, that is, as $pf \times ps$ is to $PF \times PS$."

Let F_I be the force towards strip of width IH; let f_i be the force towards strip of width ih.

Since, by "the same things supposed," there tend equal centripetal forces to *every point of a spherical surface*, the forces will be directly as the area of the surfaces.

So, at equal distances, different areas:

$$\frac{F_I}{f_i} \overset{ult}{=} \frac{PI^2 \times ps \times pf}{pi^2 \times PS \times PF}. \tag{13}$$

And, by "the same things supposed," these forces are *decreasing as the square of the distances from these points*, the forces will be inversely as the squares of the distances.

So for each particle, that is, equal areas with different distances:

$$\frac{F_I}{f_i} = \frac{pi^2}{PI^2}. \tag{14}$$

Compounding Equations 13 and 14 to get ratio of forces for different areas and different distances:

$$\frac{F_I}{f_i} \overset{ult}{=} \frac{PI^2 \times ps \times pf}{pi^2 \times PS \times PF} \times \frac{pi^2}{PI^2}.$$

$$\frac{F_I}{f_i} \overset{ult}{=} \frac{ps \times pf}{PS \times PF}. \tag{15}$$

But this ratio of forces is the ratio of forces towards the strip $\widehat{IH}$, not towards the center S. We want to find what the force towards S will be under our assumption of particle-to-particle attraction.

"And these forces are to the oblique parts of them that (constructed by the resolution of forces through Corollary 2 of the Laws) tend towards the centers along the lines PS, ps, as PI to PQ, and pi to pq,..."

The force towards the evanescent strip $\widehat{IH}$ can be resolved into $\overrightarrow{PQ}$ and $\overrightarrow{QI}$.

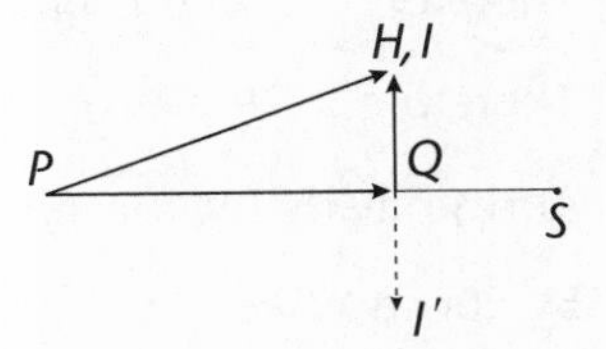

$\overrightarrow{PI}$ is the total force.

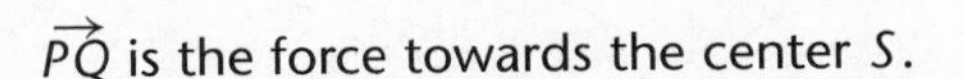
$\overrightarrow{PQ}$ is the force towards the center S.

The component $\overrightarrow{QI}$ is cancelled by the part $\overrightarrow{QI'}$ on the other side of line PQS, and similarly for every point H, I on the band above PQ there will be a point H', I' below it to cancel the QI component leaving a compounding of all the PQ components.

$$\frac{F_S}{F_I} = \frac{\text{components towards } S}{\text{total force towards } I} = \frac{PQ}{PI} \tag{16}$$

and

$$\frac{f_s}{f_i} = \frac{pq}{pi}. \tag{17}$$

"... that is (because of the similar triangles PIQ and PSF, piq and psf), as PS to PF and ps to pf."

As angle DPE vanishes, line PK approaches line PL. As this happens, line $SE \perp PL$ approaches line $SD \perp PK$. When angle DPE has vanished, we have the limiting condition, where E, D, and F are all in the same place. Thus as angle DPE evanesces, angle $PFS \rightarrow$ angle PES, that is,

Angle $PFS \overset{ult}{=}$ angle PES.

Since angle PES is a right angle by construction, angle PFS becomes a right angle.

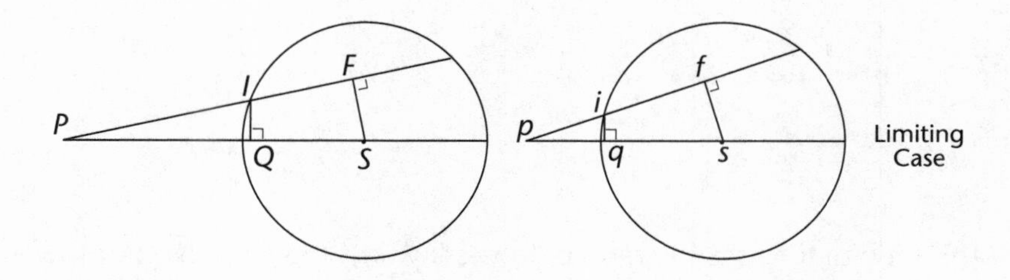

Angle IQP is also a right angle by construction.

Angle $IQP \overset{ult}{=}$ angle PFS.

Angle IPQ = angle FPS. [identity]

Therefore $\triangle PQI$ is ultimately equiangular with $\triangle PFS$.

Therefore $PF : PS \overset{ult}{::} PQ : PI$. [Eu. VI.4]

And similarly for the second sphere.

By Euclid VI.16,

$$\frac{PF}{PS} \overset{ult}{=} \frac{PQ}{PI} \quad \text{and} \quad \frac{pf}{ps} \overset{ult}{=} \frac{pq}{pi}. \tag{18}$$

"Whence, *ex aequali*, it is brought about that the attraction of this corpuscle P towards S is to the attraction of the corpuscle p towards s as $\frac{PF \times pf \times ps}{PS}$ is to $\frac{pf \times PF \times PS}{ps}$, that is, as sq.$ps$ is to sq.PS."

Dividing Equation 16 by Equation 17:

$$\frac{F_S/F_I}{f_s/f_i} = \frac{PQ/PI}{pq/pi}.$$

Substituting from Equations 18,

$$\frac{F_S/F_I}{f_s/f_i} \overset{ult}{=} \frac{PF/PS}{pf/ps}. \tag{19}$$

By Equation 15,

$$\frac{F_I}{f_i} \overset{ult}{=} \frac{ps \times pf}{PS \times PF}.$$

Multiplying Equations 19 and 15:

$$\frac{F_S/F_I}{f_s/f_i} \times \frac{F_I}{f_i} \overset{ult}{=} \frac{PF/PS}{pf/ps} \times \frac{ps \times pf}{PS \times PF}.$$

Rewriting:

$$\frac{F_S}{F_I} \times \frac{f_i}{f_s} \times \frac{F_I}{f_i} \overset{ult}{=} \frac{PF}{PS} \times \frac{ps}{pf} \times \frac{ps \times pf}{PS \times PF}.$$

Simplifying,

$$\frac{F_S}{f_s} = \frac{ps^2}{PS^2}.$$

The components of forces on P towards the center of the sphere, exerted by each strip, are inversely as the squares of the distances to the center.

Over the whole band, the other components, those not directed towards the center of the sphere, drop out.

Note that if we stopped with one strip, such as the strip made by rotating $\widehat{IH}$, having particle P attracted by all the matter in that strip, but dispensing with the rest of the spherical surface, the particle's accelerative force towards the strip would be inversely as the square of the distance not to the strip but to the center of the hypothetical sphere. Amazing! So it's not that the rest of the mass of the sphere somehow "balances out" the stronger effect of the nearer particle.

"And by a similar argument, the forces by which the surface described by the revolution of the arcs KL, kl attract the corpuscles will be as sq.ps to sq.PS. And the forces of all the circular surfaces into which the two spherical surfaces can be divided, by taking sd always equal to SD and se equal to SE, will be in the same ratio. And, by composition, the forces of the whole spherical surfaces acting upon the corpuscles will be in the same ratio."

Now take new arcs, for example KL, $K'L'$; rotate them for new bands; the same relation will hold. Repeat until the whole sphere is covered.

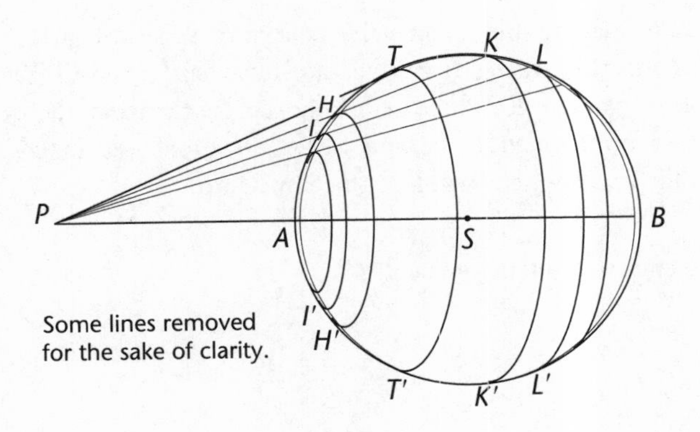

Since all the forces towards the centers for each band are inversely as the squares of the distances to the centers, by Euclid V.12, the forces over the whole sphere will be in the same ratio.

$$\frac{F_S}{f_s} = \frac{ps^2}{PS^2}.$$

The forces on a corpuscle placed outside the spherical surface are inversely proportional to the squares of the distances from the center.

Q.E.D.

Scholium

The surfaces of which the solids are composed are here not purely mathematical, but are orbs so tenuous that their thickness is as nothing: that is, they are evanescent orbs, of which the sphere ultimately consists, where the number of those orbs is increased and their thickness decreased *in infinitum*. Likewise, by the points of which the lines, surfaces, and solids are said to be composed, are to be understood equal particles of a size that may be disregarded.

Book I Proposition 74

The same things being supposed [as in Proposition 73, namely, that to the individual points of some given sphere there tend equal centripetal forces decreasing in the duplicate ratio of the distances from the points], I say that a corpuscle placed outside the sphere is attracted with a force inversely proportional to the square of its distance from the center of the sphere.

For let the sphere be divided up into innumerable concentric spherical surfaces, and the attractions of the corpuscle arising from the individual surfaces will be inversely proportional to the square of the distance of the corpuscle from the center (by Proposition 71). And, *componendo*, the sum of the attractions, that is, the attraction of the corpuscle to the whole sphere, will be in the same ratio.

Q.E.D.

[Newton gives no diagram with this proposition.]

Notes on I.74

- Now we nest enough spherical shells together to get a solid sphere, finding that the force on an outside corpuscle will still vary inversely as the square of the distance to the center.

What will still be left is to explore the situation where the outside corpuscle is a complete sphere itself, rather than a single corpuscle. This will be taken up in I.75.

Expansion of Newton's Sketch of I.74

Given:

Now not just the surface of a sphere, but the whole spherical body, each of whose point-particles (see previous scholium: "equal particles of a size that may be disregarded") attract the corpuscle with an inverse square force law.

To Prove:

The attraction to all these particles resolves into a force to the center of the sphere as inverse square of the distance of the corpuscle from the center of the sphere.

Proof:

"For let the sphere be divided up into innumerable concentric spherical surfaces, and the attractions of the corpuscle arising from the individual surfaces will be inversely proportional to the square of the distance of the corpuscle from the center (by Proposition 71)."

Suppose the sphere to be divided in thought into concentric spherical surfaces. These shells are imagined to have their thickness decreased and their number increased *in infinitum*. The solid sphere is imagined to be made up of these evanescent concentric shells.

For each of these surfaces, by Proposition 71, the forces attracting the outside corpuscle to each of the particles of the surface resolve to a force at the center attracting the corpuscle in the inverse square of its distance from the center.

"And, *componendo*, the sum of the attractions, that is, the attraction of the corpuscle to the whole sphere, will be in the same ratio."

Adding up the effects of all the concentric surfaces, the forces on the outside corpuscle of the whole sphere will be in the same ratio, by Euclid V.12.

Q.E.D.

Pause After I.74

• Since the attraction of an outside particle is inversely as the square of its distance from the center of the attracting sphere, it is *as if* the whole attracting force issued from one single corpuscle placed in the center of this sphere.

• We were concerned, before proving these propositions, that a particle-to-particle inverse square law might lead to near-infinite forces as a particle neared the surface of the body, since the distance between that particle and the surface particles was approaching zero.

For example, consider a pebble falling to the surface of the earth. Our observation is that the force on that pebble seems not very great; its acceleration is rather modest. Our concern was that this might call into question the whole particle-to-particle theory of gravity.

We may be seeing, now, why this theory of gravity is consistent with our observations, both of pebbles and of the motions of the planets and their moons. If the particle-to-particle attractions resolve into an attraction that acts inversely as the square of the distance to the *center*, then the pebble falling to the surface of a very large sphere may have very little force on it just because it is still a long way from the center.

• **Question:** Would there be more force on the pebble falling to the surface of the moon, since the distance to the center is smaller?

Book I Proposition 75

If to the individual points of a given sphere there tend equal centripetal forces decreasing in the duplicate ratio of the distances from the points, I say that any other similar sphere you please is attracted by the same [sphere] with a force inversely proportional to the square of the distance of the centers.

For the attraction of any particle you please is inversely as the square of its distance from the center of the pulling sphere (by Proposition 74), and therefore is the same as if the whole attracting force were to emanate from a single corpuscle placed in the center of this sphere. But this attraction is of the same magnitude as the attraction of the same corpuscle would in turn come out to be, if it were now pulled by the individual particles of the attracted sphere with the same force by which it attracts them. And this attraction of the corpuscle comes out to be inversely proportional to the square of its distance from the center of the sphere (by Proposition 74), and consequently it is in the same ratio to this equal attraction of the sphere.

Q.E.D.

[Newton gives no diagram with this proposition.]

Notes on I.75

• Again we have been given that all the particles of a whole sphere are attracting with an inverse square force law. I.74 showed that a *particle* outside this given attracting sphere will be attracted to the sphere's center by inverse square; now we prove that a *whole outside sphere* will be attracted to the given attracting sphere's center by inverse square.

• Note that this formulation is still considering only the effect of forces on one body towards another. There is no assumption of mutuality here. This proposition does not state or require or suppose that the attracting sphere is itself attracted to the other. The question of mutuality is taken up in the Corollary 2.

• In this series of propositions beginning with I.71, we have been adding forces on a single particle. We can do this using either accelerative force or motive force. But in the course of I.75, Newton starts adding forces on different particles. That brings up a problem. Suppose we had two particles side by side, each with a given accelerative force on it. If we add them together, we do not get twice the accelerative force; we have exactly the same accelerative force. But if we are considering the motive force, and each of the two original particles had a given motive force on it, when we added them together we would have twice the motive force. It is in these terms that we must start thinking in preparation for considering the masses of the two bodies and how each mass affects the resulting gravitational attraction. Corollary 1 will formulate the law of attraction considering the mass of the attracted body, and Corollary 2 will consider the mass of both.

Expansion of Newton's Sketch of I.75

Given:

All the particles of a whole sphere *T* are attracting the particles of a whole second sphere *S* with an inverse square law.

To Prove:

The combined effect of the attractions of all the individual particles of body S will be towards the center of body T and will vary inversely as the square of the distance between the two centers st.

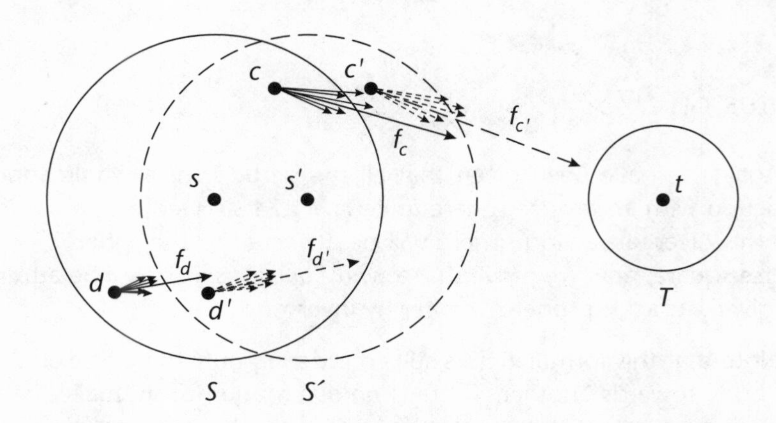

Proof:

"For the attraction of any particle you please is inversely as the square of its distance from the center of the pulling sphere (by Proposition 74),..."

A sum of all the little forces on point c within body S towards the various particles of body T adds up to a single vector f_c directed to center t of body T.

If body S now moves so that its center is at s', the new force $f_{c'}$ on point c' is augmented or diminished inversely as the square of its distance from center t of body T:

$$\frac{f_c}{f_{c'}} = \frac{tc'^2}{tc^2}.$$

The same will be true for all the other points d, etc., of body S. Therefore the attractions of all the individual particles of body S towards all the particles of body T resolve into many vectors, one per particle of S, all directed towards center t and all varying inversely as their distances from t.

"... and therefore is the same as if the whole attracting force were to emanate from a single corpuscle placed in the center of this sphere."

Therefore, it is as if all the attraction towards all the individual particles of T were directed to a single point body at center t.

But we need to know how the attraction varies with distance. The force on each particle may vary inversely as the square of the distance, but what about the combined forces on all the particles? We don't know how to resolve a cone of vectors into a relationship between centers. But Newton has a thought experiment for us to enable us to see how that cone would resolve, namely:

"But this attraction is of the same magnitude as the attraction of the same corpuscle would in turn come out to be, if it were now pulled by the individual particles of the attracted sphere with the same force by which it attracts them."

We do know how the forces would resolve and how they would vary if the vectors, instead of going from the many points of a sphere to a point outside, went from a center out to various points. That is the configuration I.74 addresses, and the result is an inverse square attraction to the center. If we leave these vectors the same magnitude they are, and change their direction (and therefore their sign), imagining them as an attraction of a particle at *t* towards all the particles of sphere *S*, we can determine that they will resolve to an attraction of point *t* towards the center of sphere *S* which, furthermore, is known, also by I.74, to vary inversely as the square of the distance between *t* and the center *s* of sphere *S*.

Note that this is only a mathematical fiction designed to give us insight into what we have, like finding an integral by doing a reverse differentiation. No claim or supposition is being made that body *T* is attracted to body *S*, let alone that this hypothetical point-body *t* is being attracted.

But just for the sake of reading the vectors backwards, imagine what the situation would be if a point body *t* were attracted towards each point *c*, *d*, etc. of body *S* with the forces $-f_c$, $-f_d$, etc.

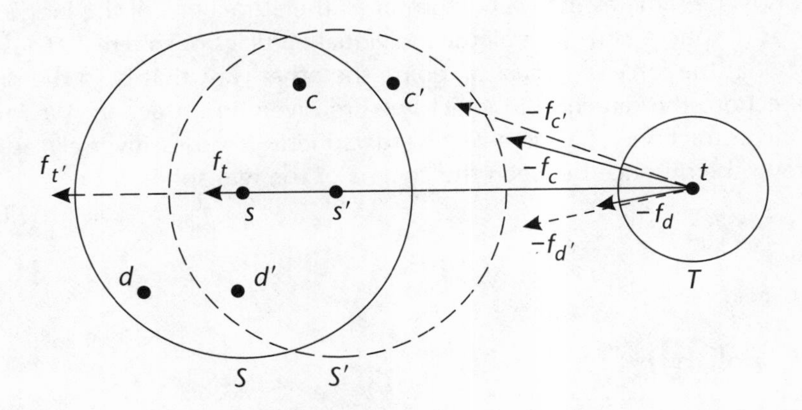

We now have an indefinitely large set of vectors, identical in number and magnitude, but opposite in direction, to those that urged points *c*, *d*, etc. towards point *t*. All of these new vectors act at point *t* urging it toward center *s* of body *S*.

"And this attraction of the corpuscle comes out to be inversely proportional to the square of its distance from the center of the sphere (by Proposition 74),..."

This set of vectors $-f_c$, $-f_d$, etc., adds up to a single vector f_t directed to center s. As body S moves, and s goes to s', f_t is varied inversely as the square of the distance of center t from center s.

$$\frac{f_t}{f_{t'}} = \frac{ts'^2}{ts^2}. \tag{1}$$

"... and consequently it is in the same ratio to this equal attraction of the sphere.

Q.E.D."

$$f_t = -f_c - f_d - \ldots$$

and

$$f_{t'} = -f_{c'} - f_{d'} - \ldots$$

Substituting into Equation 1,

$$\frac{-f_c - f_d - \cdots}{-f_{c'} - f_{d'} - \cdots} = \frac{ts'^2}{ts^2}.$$

Multiplying numerator and denominator of the left hand side by –1, we get

$$\frac{f_c + f_d + \cdots}{f_{c'} + f_{d'} + \cdots} = \frac{ts'^2}{ts^2}.$$

But this is the same as the vector sum of all the attractions of the individual particles of sphere S towards all the individual particles of sphere T. Thus we know what the force law must be going the other way, that is, in the original direction, the direction in which we are given an attraction. We know that this attraction, of sphere S towards sphere T, varies inversely as the square of the distance between the centers of the two spheres.

Q.E.D.

I.75 Corollary 1

The attractions of spheres towards other homogeneous spheres are as the pulling spheres applied to [i.e., divided by] the squares of the distances of their centers from the centers of the ones that they attract.

Expansion of Newton's Sketch of I.75 Corollary 1

Given:

Accelerative force on either sphere varies inversely as the square of the distance between the centers.

$$f_s \propto \frac{1}{st^2}$$

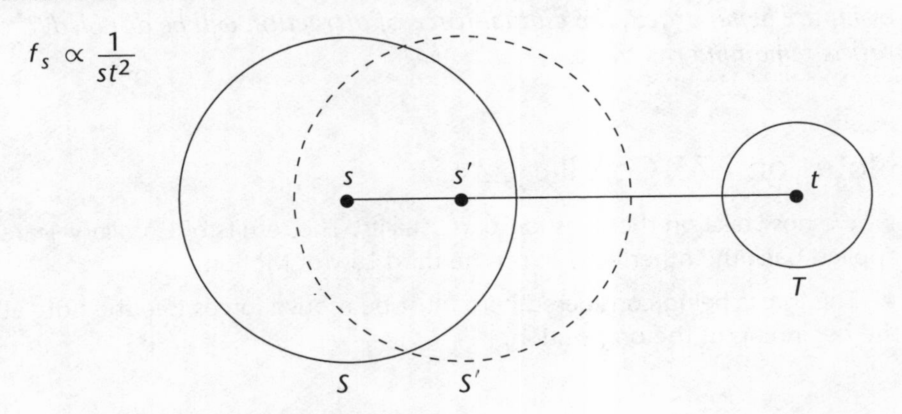

To Prove:

Accelerative force on *S* varies as mass of *T* divided by the square of the distance between the centers.

$$f_{s}(\text{on } S) \propto \frac{m_T}{st^2}$$

where *T* is the attracting sphere and *S* is the attracted sphere.

Proof:

By I.75, the accelerative force on sphere *S* towards sphere *T* varies inversely as the square of the distance between the centers of the spheres.

By the given of I.75, "to the individual points of a given sphere there tend equal centripetal forces." Since the total accelerative force is resolved of the forces to each of these several particles (because they are "points," they will be equal), the accelerative force on *S* will vary directly as the mass of *T* for a given distance.

Compounding,

$$f_s(\text{on } S) \propto \frac{m(\text{of } T)}{st^2}.$$

$$\frac{f_s(\text{on } S)}{f_s(\text{on } S')} \propto \frac{m(\text{of } T)/st^2}{m(\text{of } T')/s't'^2}.$$

I.75 Corollary 2

The same holds where the attracted sphere also attracts. For the individual points of the latter will pull those of the other with the same force by which they are in turn pulled by them, and consequently, since in all attraction (by Law 3), both the attracting point and the attracted point are being urged, the mutual force of attraction will be doubled, the ratios remaining the same.

Notes on I.75 Corollary 2

- We now take on the question of mutuality. The results of Corollary 1 are applied back the other way, using the third Law of Motion.
- The forces being considered here must be motive forces; see the note at the beginning of the proposition.

Expansion of Newton's Sketch of I.75 Corollary 2

"For the individual points of the latter will pull those of the other with the same force by which they are in turn pulled by them, . . ."

We are given that the attracted sphere also attracts: that is, we suppose it and explore what follows. That the particles of the latter will pull with the *same* force by which they are pulled must rely on the Third Law, even though Newton does not mention this law until the next step. (This "same force" is, presumably, a force of equal magnitude.)

But the fact that each particle attacts each other particle with a Third Law equal reaction between each pair of particles might not imply that the two unequal spheres have equal mutual attraction. For this reason, Newton brings us back to consideraton of the individual particles.

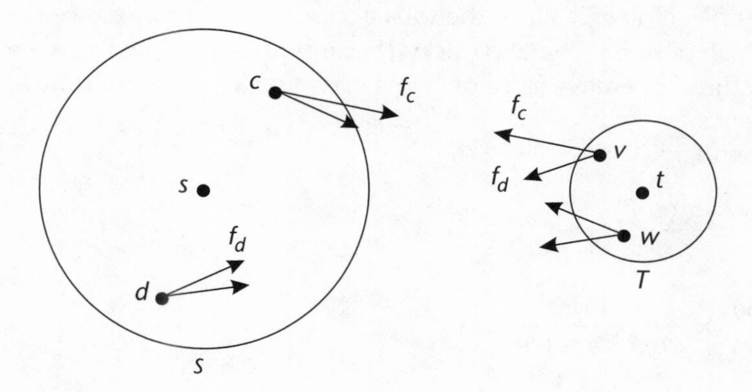

Suppose that sphere *S* contains *m* particles, and that sphere *T* contains *n* particles.

Body *S* has *n* forces on each of its *m* particles; body *T* has *m* forces on each of its *n* particles.

There will be a total of $m \times n$ forces on each body.

Therefore, the total forces on the two bodies are equal.

"... and consequently, since in all attraction (by Law 3), both the attracting point and the attracted point are being urged, ..."

The mysteriousness of the idea and claim of gravitational attraction obeying the Third Law has been noted elsewhere. Here we must take Newton's word for it. Gravitational attraction, he asserts, obeys the third Law of Motion. The particles are taken to be equal in quantity of matter, and the forces on each particle, going either direction, will be equal, and therefore the conclusion of the previous bit of the sketch will follow. Thus we may say that:

"... the mutual force of attraction will be doubled, the ratios remaining the same."

Assuming that the mutual forces are equal, the total mutual attraction will be double the attractive force of one on the other. But the variation ("the ratios") will be inversely as the square of the distance. The doubling, a constant factor, will not affect the proportionality.

Proposition 76

If spheres—being dissimilar in any manner (as regards density of matter and attractive force) in progression from the center to the circumference—are nonetheless similar everywhere in progression all around at every given distance from the center; and if the attractive force of any point decreases in the duplicate ratio of the distance of the attracting body: I say that the whole force by which one sphere of this kind attracts another is inversely proportional to the square of the distance of the centers.

[The sketch for I.76 is not quoted here; nor are its corollaries. They are summarized below.]

Notes on I.76 and Corollaries

• **I.76**: The proposition extends the conclusion of I.75 to spherical bodies that are not homogeneous, provided that the component spherical shells making up each body are themselves homogeneous.

Each of the spherical bodies may be made up of concentric spherical shells, each of which is homogeneous, but each of which may have a greater or lesser attractive force than the others. Newton might have said, referring to Definition 6, that the different spherical shells may have different absolute centripetal forces associated with them. Nonetheless, as long as the attractive force to each point of each shell decreases with the square of the distance, the total force toward the whole nest of spheres will also decrease with the square of the distance.

We should note that, at this point, no association has been made between the quantity of matter (mass) of the attracting body and the intensity of force toward that body. Remember: this proposition is from Book I, which is hypothetical and can only be applied to the real world when its conditions are shown to have been met. Although Newton does drop a hint here that density of matter might have something to do with the force, the proposition is much more generally applicable: the absolute force might be different for different colors, or it might be proportional to electrical charges, or to some other thing, and the proof would still hold.

It is only after we have proved III.6 that we can fully appreciate the role played by mass. In the application of this proposition, we would say that it doesn't matter how dense and massive, or how rarified and un-massive, each of the shells is: the attraction between spheres is still inverse-square center-to-center.

And with this last proposition of the sequence, our wonder at this amazing demonstration deepens yet further!

• **Corollaries**: In those that are relevant to this central argument, Newton looks ahead to what will emerge from his application of I.75 and I.76 in Proposition III.8.

Proposition 8

If the matter of two globes mutually gravitating to each other be everywhere homogeneous in regions that are equidistant from the centers, the weight of each globe towards the other will be inversely as the square of the distance between the centers.

After I had found that gravity in an entire planet arises and is compounded from the gravities to the parts, and that in the individual parts it is inversely proportional to the squares of the distances from the parts, I was in doubt whether that inverse duplicate proportion might apply accurately in the whole force compounded of many forces, or only approximately. For it might happen that the proportion which applies accurately enough at greater distances, would err noticeably near the surface of the planet because of the unequal distances and dissimilar positions of the particles. But at last, by Propositions 75 and 76 of Book I and their corollaries, I understood the truth of the proposition that is here in question.

Note on III.8

With this proposition, Newton concludes his argument for universal gravitation.

III.4 seems to establish that gravity operates on a center-to-center inverse square law down to the surface of the earth, at least as close as pendulums go.

III.7 seems to establish that gravity is particle-to-particle, in establishing that it is "part to part."

I.71, 74, and 75 resolve the perplexity the "judicious philosopher" might have had about these two results, both based on phenomena.

We now are in a position, calling on I.75, to state the conclusion to the brink of which III.7 had led us, but which without I.75 could not be fully accepted.

Expansion of Newton's Sketch of III.8

Given:

"If the matter of two globes mutually gravitating to each other be everywhere homogeneous in regions that are equidistant from the centers,..."

To Prove:

"...the weight of each globe towards the other will be inversely as the square of the distance between the centers."

This means that , as the distance between *S* and *T* changes, the weight of globe *S* towards globe *T* will vary inversely as the squares of the distances between their centers, and the weight of *T* towards *S* will also vary inversely as the squares of the distances between the centers.

Proof:

Part 1:

First we set up the conditions that allow us to invoke I.75.

Step 1

Proposition III.7 established that every planet gravitates towards all parts of every other planet.

Step 2

By III.1, 2, 5, and 6, and Rule 2, the accelerative force of gravity towards each planet is inversely as the square of the distances from the center of the planet.

Step 3

Given that matter is everywhere homogeneous in regions that are equidistant from the centers, it could be analyzed into concentric spheres of the same sort of matter. Here "same sort" means what it means in Rule 3: not matter in some regions having levity and others heaviness; nor different sorts falling at different rates.

Part 2: Application to Our World

Having met the conditions for I.75, we invoke it to conclude that, when two spheres are gravitating each towards the other, the accelerative force on either towards the other will be inversely as the square of the distance between the centers.

We are considering here the same two bodies but at different distances.

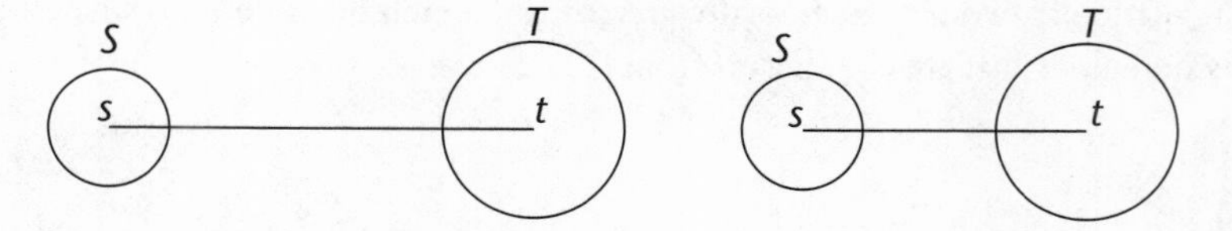

Since we are taking the same two bodies, the mass of the attracting body T cancels out.

$$f_a \text{ (on S)} \propto 1/st^2 \qquad (1)$$

"...the weight of each globe towards the other..."

Now we need to compare weights, that is, motive quantities of force instead of accelerative.

Take the mass of the attracted sphere and multiply it times both sides of Equation 1.

$$f_a \text{ (on } S) \times m \text{ (of } S) \propto \frac{m \text{ (of } S)}{st^2}.$$

By Definition 8 (elaborated in notes to II.24),

$$f_m \propto m \times f_a.$$

$$f_m \propto \frac{m \text{ (of } S)}{st^2}.$$

And weight varies as motive force, according to Newton's commentary to Definition 8 and elaborated in the expanded proof for II.24.

So the weight of S varies as the quantity of matter of S directly and the square of distance st inversely.

But the mass of S is constant, so

$$\text{Weight of } S \propto \frac{1}{st^2}.$$

Similarly,

$$\text{Weight of } T \propto \frac{1}{st^2}.$$

Q.E.D.

"After I had found that gravity in an entire planet arises and is compounded from the gravities to the parts, and that in the individual parts it is inversely proportional to the squares of the distances from the parts, I was in doubt whether that inverse duplicate proportion might apply accurately in the whole force compounded of many forces, or only approximately. For it might happen that the proportion that applies accurately enough at greater distances, would err noticeably near the surface of the planet because of the unequal distances and dissimilar positions of the particles. But at last, by Proposition 75 and 76 of Book I and their corollaries, I understood the truth of the proposition that is here in question."

The problem that might seem likely at near distances has been discussed in the notes to I.71, where Newton's comments on the difficulty of this problem, and the unlikeliness of the result, are quoted from a letter to Halley.

The fact that Newton discusses this obstacle here at length (apparently forgetting to offer any sketch for the proposition), and his speaking of "at last" understanding, give us a precious insight into his struggle. They suggest his sense of the difficulty of the problem, his sense of his own accomplishment in coming up with the solution given in the sequence of propositions we have just worked through, and some understated pride in having resolved the perplexity.

Pause Before III.8 Corollaries and III.13

We have finally established and explicitly asserted the principle that we now know as universal gravitation. All bodies attract all other bodies proportionally to their quantities of matter conjointly and inversely as the square of the distances between their centers.

What can we do with this? And can we check it in some dramatic way? Yes, we can, and we will, by using it to derive the elliptical orbits of the planets.

This is what Newton set out as the plan of *Principia*: to use the phenomena to discover the forces of nature, and to reason from these forces to demonstrate the rest of the phenomena.

In his Preface to the Reader (pages 3 and 4 in this book), Newton describes this revolutionary method thus:

> And on that account we present these [writings] of ours as the mathematical principles of philosophy. For the whole difficulty of philosophy appears to turn upon this: that from the phenomena of motion we investigate the forces of nature, and then from these forces we demonstrate the rest of the phenomena....

Then Newton states very specifically how he will demonstrate his method by giving an example of the procedure.

> In the third book ... we present an example of this procedure, in the unfolding of the system of the world. For there, from the celestial phenomena, using the propositions demonstrated mathematically in the preceding books, we derive the forces of gravity by which bodies tend to the sun and the individual planets.

We have established something very basic and of breath-taking originality about the forces of nature in "deriving the forces of gravity." But *Principia* undertakes to do more than that: it is Newton's demonstration of his deductive method.

> Then from the forces, using propositions that are also mathematical, we deduce the motions of the planets, of comets, of the moon, and of the sea. In just the same way it would be possible to derive the rest of the phenomena of nature from mechanical principles by the same manner of argument.

We have done the most difficult part. Can we hang in a little longer to reap the final rewards? Newton would like us to see him execute the finale, deriving the remaining phenomena.

We will cut the demonstration to a minimum, but that minimum will be a most dramatic and important exhibition, the elliptical orbits. Kepler had proposed elliptical orbits and equable description of areas based on a very different physics and view of gravity. Will Newton's new theory of gravity result in the same orbits and description of areas?

In order to execute this finale, we need to establish a couple more conclusions: we need to be able to work with the weights of the bodies at different distances from different central bodies. This will be done in the corollaries to III.8. These results will themselves be an application of the theory of gravitation and are important and revolutionary and exciting conclusions in their own right. The first, less a corollary to III.8 than the first step into the new unfolding of the phenomena of nature, shows us how to find the relative weight of a given body at the surface of various central bodies. This relative weight allows us to compare the absolute gravitational forces of those central bodies. Corollary 2 then determines the relative masses of Jupiter, Saturn, earth, and the sun. None of these conclusions could have been arrived at before we had the whole foundation that has been the subject of our study so far and that culminated in III.8.

In III.13 we use the gravitational theory, these first of the new conclusions, and the phenomena we already have to find that "the planets move as ellipses having their focus at the center of the sun, and by radii drawn to that center describe areas proportional to the times."

III.8 Corollary 1

Hence can be found and compared among each other the weights of bodies on different planets. For the weights of equal bodies revolving in circles about the planets are (by Book I Proposition 4 Corollary 2) as the diameters of the circles directly and the squares of the periodic times inversely; and the weights at the surfaces of the planets, or at any other distances from the planets you please, are greater or less (by this proposition) inversely in the duplicate ratio of the distances. Thus from the periodic times of Venus around the sun, 224 days and $16\frac{3}{4}$ hours, of the outermost satellite of Jupiter around Jupiter, 16 days and $16\frac{8}{15}$ hours, of the Huygenian satellite around Saturn, 15 days and $22\frac{2}{3}$ hours, and of the moon around the earth, 27 days 7 hours 43 minutes, compared with the mean distance of Venus from the sun and with the greatest heliocentric elongations of the outermost of Jupiter's satellites from the center of

Jupiter, 8′ 16″, of the Huygenian satellite from the center of Saturn, 3′ 4″, and of the moon from the center of the earth, 10′ 33″, by entering into a computation I have found that the weights of equal bodies at equal distances from the center of the sun, Jupiter, Saturn, and earth, to the sun, Jupiter, Saturn, and earth, respectively, are as 1, $\frac{1}{1{,}067}$, $\frac{1}{3{,}021}$, and $\frac{1}{196{,}282}$,* *respectively, and when the distances are increased or decreased, the weights are decreased or increased in the duplicate ratio. Thus the weights of equal bodies on the sun, Jupiter, Saturn, and earth, at distances of* 10000, 997, 791, *and* 109 *from their centers, and accordingly on their surfaces, will be as* 10000, 943, 529, *and* 435, *respectively. The magnitude of weights of bodies on the surface of the moon will be told in what follows.*

Notes on III.8 Corollary 1

- This corollary might better have been called a new proposition. It moves on from III.8 to the next stage of Newton's project. Corollary 2 is a true corollary of this demonstration, rather than of III.8.

- III.8 Corollary 1 finds the relative weight of equal bodies, first at given equal distances to Jupiter, Saturn, earth, and the sun, and then at unequal distances, namely at the surfaces of each of those central bodies.

  The relative weight of a given body (or equal bodies) at the surfaces of central bodies gives us a measure of the absolute gravitational force of those central bodies. It tells us the force being exerted from the center through the encircling regions relative to that exerted by the other central bodies.

  This will be used in III.13 to establish the amount by which the gravitational attraction of Jupiter would be expected to perturb Saturn's orbit around the sun.

- This corollary also allows us to determine the relative masses of Jupiter, Saturn, earth, and the sun. This calculation is done in Corollary 2.

---

* In the third edition, this number was printed as 1/169,282, which is not consistent with his stated heliocentric elongation of the moon. On the conjecture that two digits might have become transposed somewhere in the publishing process, the translator ventures the present correction. It is consistent with a heliocentric elongation of 10′ 30″.

Another possible source for this error has been identified by Robert Garisto, *Am. J. Phys.* **59** (1). On the basis of manuscript evidence, Garisto argues shows that a conjectured copying error, together with rounding in the intermediate calculations, gives exactly the number printed in the Third Edition. Inasmuch as a copying error is implicated on either account, we are justified in substituting a corrected figure.

Expansion of Newton's Sketch of III.8 Corollary 1

"Hence can be found and compared among each other the weights of bodies on different planets."

We are dealing here with equal attracted bodies. This could also be thought of as a standard unit mass "test body." We will compare their weights (motive gravitational attractions) towards different planets.

Part 1: Bodies Orbiting the Same Center

Step 1: Equal Bodies at Given Distances

"For the weights of equal bodies revolving in circles about the planets are (by Book I Proposition 4 Corollary 2) as the diameters of the circles directly and the squares of the periodic times inversely; ..."

Given for Step 1:

1. Equal bodies (same quantity of matter);

2. by observation, bodies are revolving in circles with observed diameters; centered on various planets;

3. by observation, certain periodic times.

To Prove for Step 1:

$$w \propto d/P^2,$$

where d represents the respective diameters of the circles, w the weights of the orbiting bodies, and P their periodic times.

Proof for Step 1:

By Definition 8 (elaborated in notes to Proposition II.24),

$$f_m \propto m \times f_a .$$

Since the bodies have equal quantities of matter, for them, $f_m \propto f_a$. That is, their weights will vary as the accelerative forces.

Since these bodies are moving in circles, we may invoke I.4 Corollary 2:

$$f_a \propto r/P^2 .$$

Or, by the previous conclusion,

Weights $\propto r/P^2$.

Since the circles' radii vary as the diameters,

Weights $\propto d/P^2$ for equal bodies revolving in circles.

Q.E.D.

Note that these are observed distances and periodic times, and that we are comparing hypothetical equal bodies.

Step 2: Equal Bodies at Any Distance

"... and the weights at the surfaces of the planets, or at any other distances from the planets you please, are greater or less (by this proposition) inversely in the duplicate ratio of the distances."

Given for Step 2:

Hypothetical equal bodies.

To Find for Step 2:

Their weights at different distances.

Proof for Step 2:

By the proposition, weight varies inversely as the square of the distance between the centers.

Step 1 gave us relative weights at some observed distances. Now using the proposition we can compute what the weight would be at any distance we please.

$$w \propto \frac{1}{r^2}.$$

Part 2: Gravitational Weights to Different Planets

To compare weights we must insist on the bodies being equal. (If we use accelerative forces, the bodies may be unequal, and that will not affect the accelerations, by III.6.) In this corollary we want to learn something about the central body by comparing the weights of these equal bodies (functioning as a unit-mass test body). In this part we compare the weights of these equal bodies first at observed distances from the different central bodies, then at equal distances from these different central bodies, then at the surfaces of the central bodies.

Step 1: Times and Distances

"Thus from the periodic times of Venus around the sun, 224 days and $16\frac{3}{4}$ hours, of the outermost satellite of Jupiter around Jupiter, 16 days and $16\frac{8}{15}$ hours, of the Huygenian satellite around Saturn, 15 days and $22\frac{2}{3}$ hours, and of the moon around the earth, 27 days 7 hours 43 minutes,..."

By observation, we are given periodic times for Venus around the sun and for the moons of Jupiter, Saturn, and earth around those central bodies. The following table summarizes the periodic times and converts them all to hours for ease of computation.

Venus	$224^d\,16\frac{3}{4}^h$	5392.75^h
Jupiter's moon	$16^d\,16\frac{8}{15}^h$	400.5^h
Saturn's moon	$15^d\,22\frac{2}{3}^h$	382.7^h
Earth's moon	$27^d\,7^h\,43^m$	655.71^h

"... compared with the mean distance of Venus from the sun and with the greatest heliocentric elongations of the outermost of Jupiter's satellites from the center of Jupiter, $8'16''$, of the Huygenian satellite from the center of Saturn, $3'4''$, and of the moon from the center of the earth, $10'33''$,..."

We now need to find the distances of each of these bodies from their central planets.

From the lemmita To Find the Mean Distance of an Inner Planet from the Sun, in Note 3 to Phenomenon 4, we know the distance of Venus from the sun relative to earth's distance from the sun.

In Phenomenon 1 and 2 we get the distances of the moons of Jupiter and Saturn in terms of the central planet's diameter, but what we want is those distances in the same units we are using for Venus' distance from the sun, namely in terms of earth's distance from the sun as 100,000.

We must calculate these distances from what we know, namely the distances of the planets from the sun and the heliocentric elongations of the moons from the central planets. The maximum heliocentric elongations are given in Phenomenon 1 and 2 and quoted again by Newton here; the heliocentric elongation of earth's moon is given here by Newton.

Later (in Corollary 2) Newton hints at how he arrived at the heliocentric elongation of earth's moon: he used a solar parallax of $10\frac{1}{2}''$ and calculated the maximum elongation that would be seen from the sun on that assumption. The correct parallax would be about $8'\,52'''$, making the elongation that much smaller.

By Phenomenon 4, distances from the sun are:

Venus	72,400
Jupiter	520,000
Saturn	951,000
Earth	100,000

Maximum heliocentric elongations of moons:

Jupiter's moon from Jupiter	8′ 16″
Saturn's moon from Saturn	3′ 4″
Earth's moon from Earth	10′ 33″

Using these numbers we will compute the distances of each satellite from the center of its orbit.

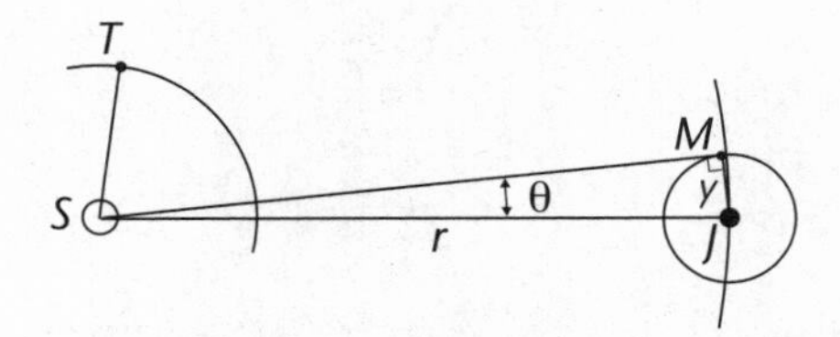

Let θ be the maximum heliocentric elongation.

$$\sin\theta = \frac{y}{r} = \frac{\text{distance of Jupiter's moon from Jupiter}}{\text{Jupiter's distance from the sun}} = \sin 8'16''$$

Venus from the sun		= 72,400
Jupiter's moon from Jupiter	$520{,}000 \times \sin 8'16''$	= 1,250
Saturn's moon from Saturn	$951{,}000 \times \sin 3'4''$	= 848.3
Earth's moon from earth	$100{,}000 \times \sin 10'33''$	= 306.9

These distances are relative to earth's orbit radius as 100,000, or, equivalently, to Venus's orbit radius as 72,400.

These distances are of unequal bodies at unequal distances around different centers. We have to sort this all out to be able to compare the weights. We do this by holding some things constant in successive calculations and move carefully to the most complex comparison.

Step 2 Accelerations at Observed Distances

"... by entering into a computation I have found that the weights of equal bodies at equal distances from the center of the sun, Jupiter, Saturn, and earth, to the sun, Jupiter, Saturn, and earth, respectively, are as 1, $\frac{1}{1{,}067}$, $\frac{1}{3{,}021}$, and $\frac{1}{196{,}282}$, respectively, and when the distances are increased or decreased, the weights are decreased or increased in the duplicate ratio."

First we will compute the accelerative forces on these unequal bodies towards their respective centers. By III.6, we know that the inequality of the attracted bodies isn't a problem for accelerative force. The accelerations will vary only with distance and with the quantity of matter in the attracting body. The quantity of matter in the attracted body will not affect its acceleration.

To compute the relative accelerative forces on these four bodies, use I.4 Corollary 2:

$$f_a \propto r/P^2$$

Using the periodic times Newton gave us earlier:

Venus	$224^d\,16\frac{3}{4}^h$	5392.75^h
Jupiter's moon	$16^d\,16\frac{8}{15}^h$	400.5^h
Saturn's moon	$15^d\,22\frac{2}{3}^h$	382.7^h
Earth's moon	$27^d\,7^h\,43^m$	655.71^h

$$F_{\text{Venus}} \propto 72{,}400 \div 5392.75^2 = 2.49 \times 10^{-3}.$$

$$F_{\text{Jupiter's moon}} \propto 1250 \div 400.5^2 = 7.79 \times 10^{-3}.$$

$$F_{\text{Saturn's moon}} \propto 848.3 \div 382.7^2 = 5.79 \times 10^{-3}.$$

$$F_{\text{Earth's moon}} \propto 306.9 \div 655.71^2 = 7.14 \times 10^{-4}.$$

For ease of comparison, multiply each relative force by 10^3:

Relative forces on:

Venus	2.49
Jupiter's moon	7.79
Saturn's moon	5.79
Earth's moon	0.714

These are the relative accelerative forces on Venus and the three moons at their respective distances from the centers of their own actual orbits.

Step 3: Accelerations at Venus's Distance as a Standard

The next step is to get relative accelerative forces at a standard distance.

Take the heliocentric distance of Venus as a standard distance. Suppose all these bodies (Venus and the three moons) were at the sun-Venus distance from their respective orbital centers.

Use the proposition to convert each body from a given distance to any distance:

$$f_a \propto 1/(\text{distances between centers})^2$$

These are the accelerations of a particular body as it moves closer to or farther from the center of its orbit, not a means of comparing the accelerations of Venus and the moons towards their respective central bodies with one another.

But we have the bodies' accelerations at each particular distance (namely, the distances of the various orbital radii) and want their accelerations at another distance, namely Venus's orbital radius.

$$f_a \propto 1/r^2$$

$$\frac{f_{\text{own orbit}}}{f_{\text{Venus's orbit}}} = \frac{(\text{radius Venus's orbit})^2}{(\text{radius own orbit})^2}.$$

We want to know the force on the body at Venus's orbit, so we solve for that.

$$f_{\text{Venus's orbit}} = \frac{f_{\text{own orbit}} \times (\text{radius own orbit})^2}{(\text{radius Venus's orbit})^2}.$$

Venus's distance is unchanged, so its relative force is still = 2.49.

$$f_{\text{Jupiter's moon}} = 7.79 \times 1{,}250.4^2 \div 72{,}400^2 = 2.32 \times 10^{-3}.$$

$$f_{\text{Saturn's moon}} = 5.79 \times 848.3^2 \div 72{,}400^2 = 7.95 \times 10^{-4}.$$

$$f_{\text{Earth's moon}} = 0.714 \times 306.9^2 \div 72{,}400^2 = 1.28 \times 10^{-5}.$$

For easier comparison, multiply each force by 10^4:

Force	Relative Value	Force Exerted Towards
f_{Venus}	24,900	sun
$f_{\text{Jupiter's moon}}$	23.2	Jupiter
$f_{\text{Saturn's moon}}$	7.95	Saturn
$f_{\text{Earth's moon}}$	0.128	earth

These are the accelerative forces on each attracted body if each were at the distance from its respective center that Venus is from the sun. When we state the conclusion in terms of equal bodies, these relative accelerations will also be relative weights towards these central bodies at this standard distance.

Put in terms of Venus's accelerative force as 1:

$$f_{\text{Venus}} = 24{,}900 \div 24{,}900 = 1 \text{ towards the sun.}$$
$$f_{\text{Jupiter's moon}} = 23.2 \div 24{,}900 = 1/1073 \text{ towards Jupiter.}$$
$$f_{\text{Saturn's moon}} = 7.95 \div 24{,}900 = 1/3132 \text{ towards Saturn.}$$
$$f_{\text{Earth's moon}} = 0.128 \div 24{,}900 = 1/194{,}531 \text{ towards the earth.}$$

Now we start dealing with equal bodies.

These different accelerative forces are also in the ratios we would get among the weights if equal bodies were placed at the same distance from the different attracting bodies.

So, by our calculations, the weights of equal bodies at equal distances from the centers of the sun, Jupiter, Saturn, and the earth are as 1, $^1/_{1,073}$, $^1/_{3,132}$, and $^1/_{194,531}$.

Newton's figures are slightly different: his values are 1, $^1/_{1,067}$, $^1/_{3,021}$, and $^1/_{196,282}$. It is not clear why we and Newton should not obtain exactly the same results if we both start with the same data. Perhaps there is a rounding discrepancy, or perhaps he used data he arrived at on another occasion (it is known that he did that for one value that went into these calculations).

These discrepancies are small, however, and our calculations have served our purpose of participating in Newton's reasoning. Let us employ his values in all further computations, so that we may continue to follow him in reaching his numerical conclusions.

"...and when the distances are increased or decreased, the weights are decreased or increased in the duplicate ratio."

If we move them in or out from here, the weights would vary inversely in the duplicate ratios of the distances.

Step 4: Weights on Surfaces of Attracting Bodies

"Thus the weights of equal bodies on the sun, Jupiter, Saturn, and earth, at distances of 10,000, 997, 791, and 109 from their centers, and accordingly on their surfaces, will be as 10,000, 943, 529, and 435, respectively."

The ratios 1, $^1/_{1,067}$, $^1/_{3,021}$, and $^1/_{196,282}$ will be the same at *any* distance as long as it is the same distance for all the pairs of attracting and attracted bodies. These ratios will work for any equal bodies at any equal distances.

Now, let's see what happens if the bodies are at different distances. Newton wants to see what the weights of the attracted bodies would be at the surfaces of the attracting bodies.

Since the preceding ratios hold for any equal bodies, this would tell us, for example, what we would weigh on Jupiter, Saturn, or the sun.

For a particular body at different distances, by the proposition:

Weight $\propto$ 1/(distance between centers)^2.

This will apply also to equal bodies.

Imagine equal bodies now at the *surface* of each of the attracting bodies.

Newton gives us the ratio of distances of these surfaces from the centers of the several bodies. These will be our new unequal distances. Note that since we are only concerned with the relative distances, the units are irrelevant. He chooses the units for convenience: the radius of the largest, the sun, is 10,000 units, making the smallest, earth, 109 units. These units have nothing to do with any other units we have used in this corollary.

Radius of Central Body

Sun	10,000
Jupiter	997
Saturn	791
Earth	109

weight $\propto 1/r^2$. [III.8]

$$\frac{\text{weight at distance a}}{\text{weight at distance b}} = \frac{(\text{distance b})^2}{(\text{distance a})^2}.$$

distance a = radius of attracting body;
distance b = standard distance, 10,000;
weight b = relative weight.

The relative weights of equal bodies at equal distances from these central planets will be*:

weight = 1 towards the sun.
weight = 1/1067 towards Jupiter.
weight = 1/3021 towards Saturn.
weight = 1/196,282 towards the earth.

* Newton's values

Then, for the sun:

$$\frac{\text{weight on sun surface}}{\text{relative weight}} = \frac{(\text{standard distance})^2}{(\text{distance to surface})^2} .$$

$$\frac{\text{weight on sun surface}}{1} = \frac{10{,}000^2}{10{,}000^2} .$$

Weight on sun's surface = 1.

For Jupiter:

$$\frac{\text{weight on Jupiter's surface}}{1/1{,}067} = \frac{10{,}000^2}{997^2} .$$

Weight on Jupiter's surface = 0.0943.

For Saturn:

$$\frac{\text{weight on Saturn's surface}}{1/3{,}021} = \frac{10{,}000^2}{791^2} .$$

Weight on Saturn's surface = 0.0529.

For earth:

$$\frac{\text{weight on earth's surface}}{1/196{,}282} = \frac{10{,}000^2}{109^2} .$$

Weight on earth's surface = 0.0429.

For more convenient comparison, multiply these values by 10,000:

Relative weights of unit mass at surface of the attracting bodies:

Sun	10,000
Jupiter	943
Saturn	529
Earth	429

Comments:

- You can now play around with this. For example, you can see that if you weigh 145 pounds on earth, you will weigh 179 pounds on Saturn, 319 pounds on Jupiter, and 3380 pounds on the sun.

This sort of calculation was something previously unheard of in the history of thought. It may be commonplace now, but in Newton's time, to be able to perform a thought experiment to imagine what one would weigh on the surface of a celestial body must have been mind-expanding, if not mind-boggling.

But for you, as well, it is special. This is not going to a science museum and standing on a scale that converts your weight to weight on the moon or on Jupiter. You have yourself, with much mental exercise and concentration, worked through the theory and the calculations to understand and produce this result. Your satisfaction cannot be compared with that of the passive museum-goer.

- Of more practical use, perhaps, the table of relative weights of equal bodies at equal distances from different attracting bodies allows us to find the relative quantities of matter in these various attracting bodies.

That is the work of Corollary 2.

"The magnitude of weights of bodies on the surface of the moon will be told in what follows."

Newton deals with this question later in Book III, in Proposition 37 Corollary 5, not included in this guidebook selection.

III.8 Corollary 2

The quantity of matter in the individual planets is also found. For the quantities of matter in the planets are as their forces at equal distances from their centers; that is, in the sun, Jupiter, Saturn, and the earth, they are as $1, \frac{1}{1{,}067}, \frac{1}{3{,}021}$, *and* $\frac{1}{196{,}282}$, *respectively. If the sun's parallax be set at greater or less than* $10''30'''$, *the quantity of matter in the earth will have to be increased or diminished in the triplicate ratio.*

Expansion of Newton's Sketch of III.8 Corollary 2

Proposition III.8 Corollary 1 found that there are different accelerative forces on equal bodies at the same distance from different central bodies.

Since the distance is the same and since accelerative force is independent of the mass of the *attracted* body, the difference in the force must be due to differences in the *attracting* body.

This is the difference in the *absolute* force of the attracting body, reflected in the accelerative forces as the measure of the efficacy of the cause. Proposition I.69 said that, where the accelerative forces vary with an inverse square law, the absolute forces are as the quantity of matter in the attracting body. Proposition III.7 said that in our world the power of gravity (here the accelerative force) is proportional to the quantity of matter in the attracting body.

Therefore the ratio of the accelerative forces is the ratio of the quantities of matter in the attracting bodies.

Part 2 Step 4 of Corollary 1 gives the table of relative weights of equal bodies at equal distances from various attracting bodies. These relative weights would be 1 towards the sun, 1/1067 towards Jupiter, 1/3021 towards Saturn, and 1/196,282 towards the earth.

The weights of equal bodies, or a unit-mass test body, will vary as the accelerative forces. We can now read off the relative quantities of matter of the attracting bodies: the sun will have 1,067 times the matter of Jupiter and 196,282 times the matter of the earth.

Proposition 13

The planets move in ellipses having their focus at the center of the sun, and by radii drawn to that center describe areas proportional to the times.

Hitherto we have reasoned about these motions from the phenomena. Now that the principles of the motions are known, we infer from these the celestial motions *a priori.* Because the weights of the planets towards the sun are inversely as the squares of the distances from the sun's center, if the sun were to be at rest and the remaining planets were not to act mutually upon each other, their orbits would come out elliptical, with the sun at the common focus. Also, the areas described would be proportional to the times (by Book I Proposition 1 and 11, and Proposition 13 Corollary 1). But the mutual actions of the planets upon each other are very slight (so that they can be disregarded), and (by Book I Proposition 66) they perturb the motions of the planets about a movable sun less than if those motions were carried out about a sun at rest.

Moreover, the action of Jupiter upon Saturn is not entirely to be disregarded. For the gravity to Jupiter is to the gravity to the sun (at equal distances) as 1 to 1,067, and therefore, at the conjunction of Jupiter and Saturn, since Saturn's distance from Jupiter is to Saturn's distance from the sun as about 4 to 9, the gravity of Saturn to Jupiter will be to the gravity of Saturn to the sun as 81 to 16 × 1,067, or about 1 to 211. And hence arises a perturbation of the orbit of Saturn at each conjunction of this planet with Jupiter that is perceptible enough that astronomers have difficulty with it. In accordance with the planet's changing position in these conjunctions, its eccentricity is now increased and now decreased, its aphelion is moved now forward and now back, and its mean motion is in turn accelerated and retarded. Nevertheless, nearly all the error in its motion around the sun arising from a force of this size (except for that in the mean motion) can be avoided by setting the lower focus of its orbit at the common center of gravity of Jupiter and the sun (by Book I Proposition 67): and therefore, where it is greatest, it hardly exceeds two minutes. And the greatest error in the mean motion hardly exceeds two minutes in a year. Further, in the conjunction of Jupiter and Saturn, the accelerative gravities of the sun towards Saturn, of Jupiter towards Saturn, and of Jupiter towards the sun, are approximately as 16, 81, and $\frac{16 \times 81 \times 3{,}021}{25}$, or 156,609, and accordingly, the difference of the gravities of the sun towards Saturn and of Jupiter towards Saturn is to the gravity of Jupiter towards the sun as 65 to 156,609, or 1 to 2,409. And to this difference is proportional the greatest efficacy of Saturn in perturbing Jupiter's motion, and consequently the perturbation of Jupiter's orbit is far less than that of Saturn's. The perturbations of the remaining orbits are much less again, except that the earth's orbit is perceptibly perturbed by the moon. The common center of gravity of the earth and the moon traverses an ellipse about the sun placed at the focus, and by a radius drawn to the sun describes areas proportional to the time on the same [ellipse]; the earth, however, revolves about this common center with a monthly motion.

Notes on III.13

• We are now at the culmination of this guidebook selection. In Newton's words, "having derived the forces of gravity by which bodies tend to the sun and the individual planets...from these froces we deduce the motions of the planets...." This is the demonstration of Newton's method. Earlier, in the Pause Before III.8 Corollaries and III.13, I described what was in store if we hung in a little longer. Now that we have actually arrived here, it may be useful to review some of the points discussed earlier to ensure that we understand the significance of this proposition.

Recall Newton's own description of his method, set forth in his preface:

> For the whole difficulty of philosophy appears to turn upon this: that from the phenomena of motion we investigate the forces of nature, and then from these forces we demonstrate the rest of the phenomena. And to this end are aimed the general propositions to which we have given careful study in the first and second books. In the third book, on the other hand, we present an example of this procedure, in the unfolding of the system of the world. For there, from the celestial phenomena, using the propositions demonstrated mathematically in the preceding books, we derive the forces of gravity by which bodies tend to the sun and the individual planets. Then from the forces, using propositions that are also mathematical, we deduce the motions of the planets, of comets, of the moon, and of the sea.

In the first two books we investigated the forces of nature, both by considering the Laws of Motion and by exploring consequences of various suppositions about centripetal forces and the motion of bodies. In the third book, as Newton says, he goes from the celestial phenomena, using the propositions demonstrated mathematically in the previous books, to derive the forces of gravity as gravity operates to effect the celestial mechanics.

Then, he says, from these forces, using propositions that are also mathematical, he deduces the motions of the planets, of comets, of the moon, and of the sea. We have completed the development of the forces of gravity in III.8 and are ready to move on to deduce the motions of the planets. This, as he says, is a demonstration of his procedure. The rest of Book III is occupied with examples of motions of planets, the moon, tides, comets, all demonstrated from this foundation we have laid. We'll take just one of his demonstration examples, the first one.

- Kepler had established that the planets moved in ellipses. His demonstration was based on his own very different foundations. The conclusion, however, was generally accepted as a true description. Would Newton's foundations lead to the elliptical orbits as well? This would be a test of his method and of the conclusions arrived at in the process of laying his foundations. Newton sets himself this test in III.13.

- Newton offered a proof in I.13 Corollary 1 that a body moving under an inverse square law would describe a conic section. And we do know from the development culminating in III.8 that bodies are moving under an inverse square force law. We know that the planets move in closed paths, so if an orbit is a conic section it must be an ellipse (the only closed conic section). Thus if we accept Newton's argument in I.13 Cor. 1 and if we look only at the sun and one planet, we should easily conclude that we could expect an ellipse.

But it isn't as simple as that, since we have demonstrated that all bodies are heavy towards all other bodies. So before we can be satisfied about whether Newton's system so far would predict planetary ellipses, we must investigate the perturbing effects of the many bodies in the vicinity.

- For summary and discussion of the relevant propositions from Book I that consider the interactions of multiple bodies, see Appendix B.

Expansion of Newton's Sketch of III.13

"The planets move in ellipses having their focus at the center of the sun, and by radii drawn to that center describe areas proportional to the times."

Using the principles of universal gravitation, Newton will now show that planets move in ellipses with their focus at the center of the sun; and that we will find equable description of areas around that focus as a geometric center. The former has come to be known as Kepler's First Law and the latter as Kepler's Second Law. Newton's proof here substantiates these discoveries of Kepler, or rather, substantiates his own method and foundations by arriving at the same result.

Proof:

"Hitherto we have reasoned about these motions from the phenomena."

We have not previously been able to determine the exact orbits of planets. From observations we have been able to see that orbits were approximately circular concentric, or circular eccentric, or elliptical close to circles. But we could not deduce the actual orbits from our mathematical foundations.

"Now that the principles of the motions are known, we infer from these the celestial motions *a priori*."

Newton refers here to his grand plan, as laid out in his Preface to the Reader, quoted above and quoted and discussed in the Pause After III.8. In doing so he draws our attention to the fact that we are now moving into the promised demonstration of his method.

"Because the weights of the planets towards the sun are inversely as the squares of the distances from the sun's center, if the sun were to be at rest and the remaining planets were not to act mutually upon each other, their orbits would come out elliptical, with the sun at the common focus."

III.8 tells us that the weight of each planet towards the sun is inversely as the square of its distance from the sun's center (assuming the other planets do not act on it).

Thus we have two conditions: (1) the weights of the planets are directed towards the sun as center of forces; (2) the weights are inversely as the squares of the distances from the sun.

This gives us the conditions we need in order to invoke I.13 Corollary 1 (the converse of Propositions I.11 through I.13).

By I.13 Corollary 1, if a body move with any velocity and at the same time be impelled by a centripetal force inversely as the square of the distance of places from the center, the body will move in a conic section having its focus at the center of forces.

Invoking I.13 Corollary 1 we know that the planets will move in conic sections around the sun at the focus. Since this is a closed orbit, as we know from Phenomenon 3, that conic must be an ellipse.

"Also, the areas described would be proportional to the times (by Book I Propositions 1 and 11, and Proposition 13 Corollary 1)."

By I.1, if we have a center of centripetal force operating on a body, the body will describe areas proportional to the times of description.

So far we have been assuming that the planet is not being directed to other planets than the sun, even though it has been asserted since III.5, and gradually more solidly proved from there on, that there is gravitation towards all planets. Thus we know that conclusions we have drawn for a system of two bodies cannot fully describe the forces on a planet.

"But the mutual actions of the planets upon each other are very slight (so that they can be disregarded),...."

For the most part, the actions of the planets on one another are small enough to be neglected, but Newton will consider exceptions. The disturbances of Jupiter and Saturn upon each other are significant, and so is the disturbance of earth's orbit by its moon.

"...and (by Book I Proposition 66) they perturb the motions of the planets about a movable sun less than if those motions were carried out about a sun at rest."

Furthermore, it turns out from I.66 (see Appendix B) that the disturbances from other planets will be less if we consider the sun's actual motion rather than maintaining the assumption that the sun is at rest. Newton gives an illustration of this in the case of Jupiter and Saturn.

"Moreover, the action of Jupiter upon Saturn is not entirely to be disregarded. For the gravity to Jupiter is to the gravity to the sun (at equal distances) as 1 to 1,067, and therefore, at the conjunction of Jupiter and Saturn, since Saturn's distance from Jupiter is to Saturn's distance from the sun as about 4 to 9, the gravity of Saturn to Jupiter will be to the gravity of Saturn to the sun as 81 to 16 × 1,067, or about 1 to 211."

Perturbation of Saturn's Orbit by Jupiter

By III.8 Corollary 2, at equal distances,

$$\frac{f_a \text{towards Jupiter}}{f_a \text{towards the sun}} = \frac{1}{1{,}067}$$

and we are given that, at the Jupiter-Saturn conjunction (the time when they are closest and so would perturb each others' orbits the most):

$$\frac{\text{distance from Jupiter to Saturn}}{\text{distance from sun to Saturn}} = \frac{4}{9}.$$

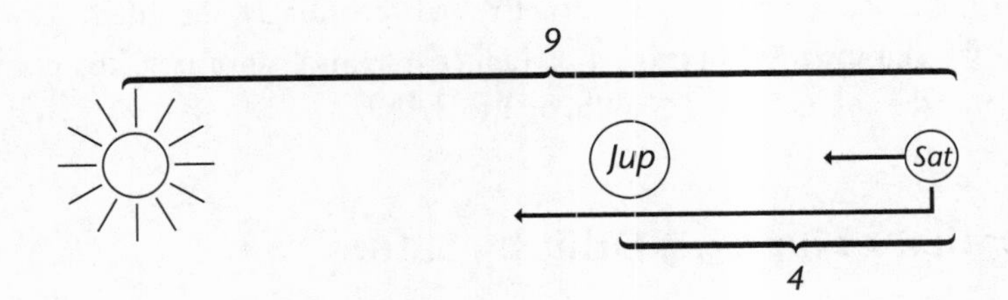

By III.2, $f \propto 1/r^2$.

Compounding to get the ratio of forces at different distances, that is, the Saturn-to-Jupiter distance and the Saturn-to-sun distance:

$$\frac{f_1}{f_2} = \frac{\text{distance}_2^{\,2}}{\text{distance}_1^{\,2}} \times \text{ratio of forces at the same distance.}$$

$$\frac{f_a \text{ Saturn towards Jupiter}}{f_a \text{ Saturn towards the sun}} = \frac{9^2}{4^2} \times \frac{1}{1{,}067} = \frac{1}{211}.$$

"And hence arises a perturbation of the orbit of Saturn at each conjunction of this planet with Jupiter that is perceptible enough that astronomers have difficulty with it. In accordance with the planet's changing position in these conjunctions, its eccentricity is now increased and now decreased, its aphelion is moved now forward and now back, and its mean motion is in turn accelerated and retarded."

The force accelerating Saturn towards Jupiter at conjunction is $\frac{1}{211}$ of the force accelerating it towards the sun. This is enough of a perturbation to be noticeable to astronomers, and astronomers had indeed been having trouble accounting for the motions of Saturn and Jupiter.

"Nevertheless, nearly all the error in its motion around the sun arising from a force of this size (except for that in the mean motion) can be avoided by setting the lower focus of its orbit at the common center of gravity of Jupiter and the sun (by Book I Proposition 67); and therefore, where it is greatest, it hardly exceeds two minutes. And the greatest error in the mean motion hardly exceeds two minutes in a year."

Newton says that if we could place the focus of Saturn's ellipse at the common center of gravity of Jupiter and the sun, this perturbation would be almost eliminated.

Proposition I.67 argues that, when several bodies are mutually attracted by an inverse square law, with two planets orbiting around a larger body, then the outer of the two orbiting planets describes areas more nearly proportional to the times when those areas are figured around the center of gravity of the other two bodies than it does around the larger body as center. (See Appendix B for the full statement of I.67.)

"Further, in the conjunction of Jupiter and Saturn, the accelerative gravities of the sun towards Saturn, of Jupiter towards Saturn, and of Jupiter towards the sun, are approximately as 16, 81, and $\frac{16 \times 81 \times 3{,}021}{25}$, or 156,609, and accordingly, the difference of the gravities of the sun towards Saturn and of Jupiter towards Saturn is to the gravity of Jupiter towards the sun as 65 to 156,609, or 1 to 2,409."

Perturbation of Jupiter's Orbit by Saturn

Now let's look at the perturbation of Jupiter towards Saturn.

By III.8 Corollary 2,

$$\frac{\text{mass of Saturn}}{\text{mass of the sun}} = \frac{1}{3{,}021}.$$

As given above,

$$\frac{\text{distance from Jupiter to Saturn}}{\text{distance from sun to Saturn}} = \frac{4}{9}.$$

In the Jupiter-Saturn conjunction,

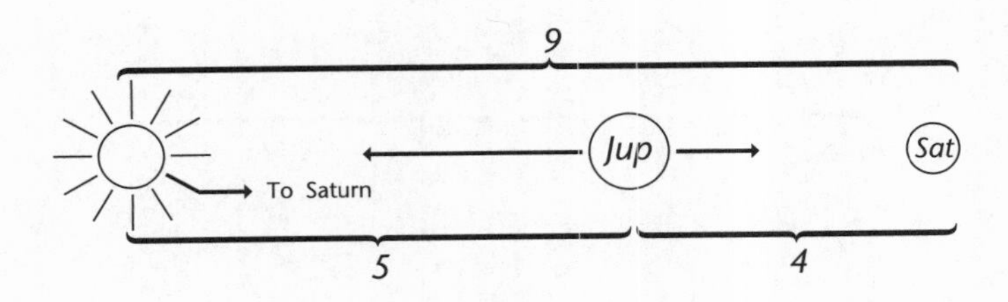

Since the Jupiter-Saturn distance is 4 to the sun-Saturn distance of 9, then the relative sun-Jupiter distance is 5.

By III.2, $F_a \propto 1/r^2$

By III.7, $F_a \propto m$

Therefore $F_a \propto m/r^2$.

$$f_{a\,\text{sun towards Saturn}} \propto \frac{\text{mass of Saturn}}{(\text{sun-Saturn distance})^2} = \frac{1}{81}.$$

$$f_{a\,\text{Jupiter towards Saturn}} \propto \frac{\text{mass of Saturn}}{(\text{Jupiter-Saturn distance})^2} = \frac{1}{16}.$$

$$f_{a\,\text{Jupiter towards the sun}} \propto \frac{\text{mass of the sun}}{(\text{Jupiter-sun distance})^2} = \frac{3{,}021}{25}.$$

To facilitate comparison, multiply each force by 16×81.

$$f_{a\,\text{sun towards Saturn}} = 16$$

$$f_{a\,\text{Jupiter towards Saturn}} = 81$$

$$f_{a\,\text{Jupiter towards the sun}} = \frac{3{,}021}{25} \times 16 \times 81 = 156{,}609$$

"And to this difference is proportional the greatest efficacy of Saturn in perturbing Jupiter's motion, and consequently the perturbation of Jupiter's orbit is far less than that of Saturn's."

At conjunction, the force on Jupiter towards the sun is perturbed by Saturn pulling Jupiter away from the sun, but this perturbation is reduced by Saturn pulling the sun towards it.

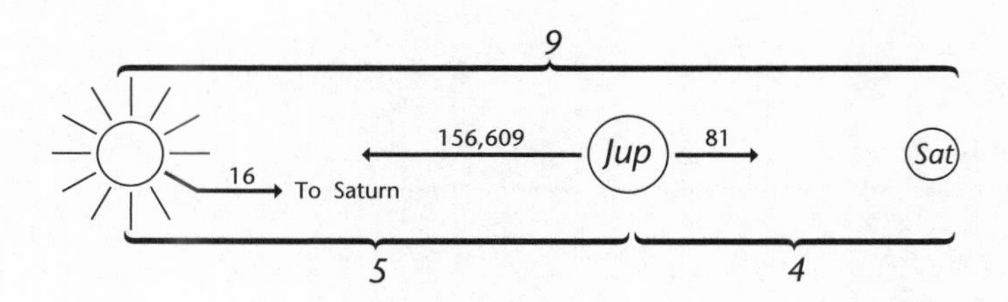

$$f_{a\,\text{sun towards Saturn}} = 16$$

$$f_{a\,\text{Jupiter towards Saturn}} = 81$$

$$f_{a\,\text{Jupiter towards the sun}} = 156{,}609$$

To get the perturbation we subtract the force pulling Jupiter towards Saturn (81) from the force pulling the sun towards Saturn (16). The force towards the sun (156,609) is reduced by 65.

Relatively, it's $^{65}/_{156{,}609}$, or $^{1}/_{2{,}409}$.

This is the greatest perturbation we will see of Jupiter's orbit by Saturn, and we see that it's much less than the perturbation of Saturn's orbit by Jupiter ($^{1}/_{211}$).

"The perturbations of the remaining orbits are much less again, except that the earth's orbit is perceptibly perturbed by the moon. The common center of gravity of the earth and the moon traverses an ellipse about the sun placed at the focus, and by a radius drawn to the sun describes areas proportional to the time on the same [ellipse]; the earth, however, revolves about this common center with a monthly motion."

Although earth's moon draws the earth off from an exact ellipse with the sun at the center, and perturbs equable description of areas around the sun, if we look at the common center of gravity of the earth and its moon, we find that that common center does follow an elliptical orbit with the sun at a focus while describing areas proportional to the times.

Pause After III.13

There is a common assumption that Newton depended on "Kepler's Laws" to prove his theories of celestial mechanics. You are now in a position to know that this is mistaken. Of course, Kepler's work pointed Newton in the right direction. But we have seen, as we worked through this key selection of propositions together, that Newton did not assume that Kepler was correct in asserting the ellipticity of the orbits or the area rule (equal areas swept out in equal times). Newton demonstrates both these conclusions in I.13 at the culmination of this sequence.

Newton claims (in a letter to Halley in June or July 1686) to be the first to establish the ellipticity of the planetary orbits satisfactorily. He established it on the foundation of his laws and mathematical propositions, and its accomplishment was a demonstration of the power of his system of rational mechanics, as he indicates in his prefaces to the first edition and to Book III.

General Scholium

The hypothesis of vortices is pressed by many difficulties. In order that any individual planet describe areas proportional to the time by a radius drawn to the sun, the periodic times of the parts of the vortex ought to have been in the duplicate ratio of the distances from the sun. In order that the periodic times of the planets be in the sesquiplicate ratio of the distances from the sun, the periodic times of the parts of the vortex ought to have been in the sesquiplicate ratio of the distances. In order that the smaller vortices about Saturn, Jupiter, and other planets be preserved in their circulations and float undisturbed in the sun's vortex, the periodic times of the parts of the solar vortex ought to have been equal. The rotations of the sun and the planets about their axes, which ought to have been consistent with the motions of the vortices, are in disagreement with all these ratios. The motions of the comets are in the highest degree regular, and observe the same laws as the motions of the planets, and cannot be explained by vortices. Comets are carried in highly eccentric motions into all parts of the heavens, which cannot happen unless vortices be removed.

Projectiles, in our air, feel only the resistance of our air. When the air is removed, as happens in Boyle's vacuum, the resistance stops, inasmuch as a slender feather and solid gold fall with equal velocity in that vacuum. And the account of the celestial spaces, which are above the sphere of the earth's exhalations, is the same. All bodies ought to move with complete freedom in these spaces, and for that reason the planets and comets revolve perpetually in orbits given in shape and position, following the laws set forth above. Though they will indeed carry on in their orbits by the laws of gravity, they nevertheless could by no means have attained the regular position of the orbits through these same laws.

The six principal planets revolve about the sun in circles concentric upon the sun, in the same direction of motion, approximately in the same plane. Ten moons revolve about the earth, Jupiter, and Saturn, in concentric circles, in the same direction of motion, very nearly in the planes of the orbits of the planets. And all these regular motions do not have their origin from mechanical causes, inasmuch as comets are carried freely in highly eccentric orbits, and to all parts of the heavens. In this kind of motion, the comets pass through the orbits of the planets with greatest ease and swiftness, and at their aphelia, where they are slower and delay for a longer time, they are at the greatest distance from each other, so that they pull each other least. This most elegant arrangement of the sun, planets, and comets could not have arisen but by the plan and rule of an intelligent and powerful being. And if the fixed stars be centers of similar systems, all these, constructed by a similar plan, will be under the rule of *One*, especially because the light of the fixed stars is of the same nature as the light of the sun, and all the systems send light into all mutually. And so that the systems of the fixed stars should not fall into each other mutually, he will have placed this same immense distance among them.

He governs everything, not as the soul of the world, but as lord of all things. And because of his dominion, he is usually called "Lord God *Παντοκράτωρ*."*
For "God" is a relative word, and is related back to servants, and "deity" is the absolute rule of God, not over his own body, as those believe for whom God is the world soul, but over servants. God most high is a being eternal, infinite, absolutely perfect; but a being without dominion, however perfect, is not the Lord God. For we say, "my God," "your God," "God of the Israelites," "God of gods," but we do not say, "my eternal," "your eternal," "eternal of the Israelites," "eternal of gods;" we do not say, "my infinite," or "my perfect." These names have no relation to servants. The word "God" everywhere signifies** the Lord; but not every lord is God. The absolute rule of a spiritual being constitutes God: true [absolute rule constitutes] the true [God]; the highest [absolute rule constitutes] the highest [God]; sham [absolute rule constitutes] a sham [God]. And from true absolute rule it follows that the true God is living, intelligent, and powerful; from the remaining perfections, that he is the highest, or in the highest

---

* [Newton's marginal note:] That is, "Universal Emperor."

** [Newton's marginal note:] Our Pocock derives the word "deus" from the Arabic word "du" (and in the oblique case, "di"), which signifies the Lord. And in this sense, princes are called "dii", Psalm 84:6 and John 10:45. And Moses is called "deus" of his brother Aaron, and "deus" of king Pharaoh (Exodus 4:16 and 7:1). And in the same sense the souls of dead princes used to be called "dii" by the people, but falsely, on account of the want of dominion.

degree perfect. He is eternal and infinite, omnipotent and omniscient; that is, he endures from eternity to eternity, and is present from infinity to infinity. He reigns over everything, and knows everything that happens or can happen. He is not eternity and infinity, but eternal and infinite; he is not duration and space, but he endures and is present. He endures always, and is present everywhere, and by existing always and everywhere, he has established duration and space. Since any single particle whatever of space is *always*, and any single indivisible moment whatever of duration is *everywhere*, surely the maker and lord of all things will not be *never*, *nowhere*. Every sentient soul is the same indivisible person at different times and in the different organs of perception and motion. Successive parts are given in duration, coexisting parts in space, neither [is given] in the human person or in his thinking principle: much less so in the thinking substance of God. Every person, *qua* sentient thing, is one and the same person throughout life in each and every organ of perception. God is one and the same God always and everywhere. He is omnipresent not in *power* alone, but also in *substance*. For power cannot subsist without substance. In him all things are contained* and moved, but without mutual effects [*passio*]. God is not affected by the motions of bodies, and these do not experience any resistance from God's omnipresence. It is universally acknowledged that the highest God exists necessarily, and by the same necessity he is *always* and *everywhere*. Hence also, the whole is entirely similar to himself, all eye, all ear, all brain, all force of perceiving, understanding, and acting, but in a manner by no means human, in a manner by no means corporeal, in a manner entirely unknown to us. Just as a blind man has no idea of colors, so we have no idea of the ways in which God most wise perceives and understands everything. He is entirely void of all body and corporeal form, and therefore cannot be seen, nor heard, nor touched; nor ought he to be worshipped under the image of any corporeal object. We have ideas of its** attributes, but we do not have the least knowledge of what the substance of any object is. We see only the shapes and colors of bodies, we hear only sounds, we touch only the external surfaces, we smell only odors, and taste flavors: we have no cognition of

---

* [Newton's marginal note:] This was the opinion of the ancients, such as Pythagoras (in Cicero, *On the nature of the gods* Book 1, Thales, Anaxagoras, Virgil (*Georgics* 4:220, and *Aeneid* 6:721), Philo (*Allegories*, beginning of Book 1), Aratus (*Phenomena*, at the beginning). So also the sacred writers, such as Paul (*Acts* 17:27–28), John 14:2, Moses (*Deuteronomy* 4:39 and 10:14), David (Psalm 139:7, 8, 9), Solomon (*1 Kings* 8:27), Job 22:12, 13, 14, Jeremiah 23:23–24. Moreover, idolaters used to make out that the sun, the moon, and the stars, people's souls, and other parts of the world, are parts of the highest god, and are therefore to be worshipped, but falsely.

** The word Newton uses here, 'eius', could be masculine or feminine or neuter; hence, it is not possible to tell whether Newton means God's attributes or those of bodies. In the context of the preceding sentence, this word would refer to God, but in the context of what follows, it would refer to bodies. Although the latter translation has been chosen, as being more consistent with the argument, the former is not impossible, and was adopted by Motte.

—Translator's note.

the inmost substances by any sense or act of reflection, and much less do we have an idea of the substance of God. We have cognition of him only through his properties and attributes, and through the wisest and best structures and final causes* of things, and marvel because of [his] perfections, and further, we revere and worship [him] because of his dominion. For we worship as servants, and God without dominion, providence, and final causes is nothing different from fate and nature. From blind metaphysical necessity, which is absolutely the same always and everywhere, no variation of things arises. The whole diversity of created things according to places and times could only have arisen from the ideas and will of a being existing necessarily. Moreover, God is said by way of allegory to see, to speak, to laugh, to love, to hate, to desire, to give, to receive, to rejoice, to become angry, to fight, to devise, to establish, to build. For every account of God is taken from human things through a certain likeness, not indeed perfect, but of a certain sort. And this much concerning God, to discourse of whom, at least from the phenomena, is the business of natural philosophy.

Hitherto I have set forth the phenomena of the heavens and of our sea through the force of gravity, but I have not yet assigned the cause of gravity. This force does indeed arise from some cause, which penetrates all the way to the centers of the sun and the planets, with no diminution of power, and which acts, not according to the quantity of the *surfaces* of the particles upon which it acts (as mechanical causes are wont to do), but according to the quantity of *solid* matter, and which acts at immense distances, extended everywhere, always decreasing in the duplicate ratio of the distances. Gravity towards the sun is compounded of the gravities towards the individual particles of the sun, and in receding from the sun decreases precisely in the duplicate ratio of the distances all the way to the orbit of Saturn, as is manifest from the planets' aphelia being at rest, and all the way to the aphelia of the comets, provided that those aphelia are at rest. The reason for these properties of gravity, however, I have not yet been able to deduce from the phenomena, and I do not contrive hypotheses. For whatever is not deduced from the phenomena is to be called a *hypothesis*, and hypotheses, whether metaphysical or physical, whether of occult qualities or mechanical ones, have no place in *experimental* philosophy. In this philosophy, propositions are deduced from the phenomena, and are rendered general by induction. Thus the impenetrability, mobility, and impetus of bodies, and the laws of motions and of gravity, came to be known. And it is enough that gravity really exists, and acts according to the laws set forth by us, and is sufficient [to explain] all the motions of the heavenly bodies and of our sea.

---

* "Final cause" is an Aristotelian term that denotes the end for the sake of which something happens. See Aristotle, *Physics*, II.3, 194b32. —Translator's note.

It would now be appropriate to add some remarks about a certain extremely subtle spirit pervading gross bodies and lying hidden in them, by whose force and actions the particles of bodies attract each other mutually at least distances, and stick together when brought into contact, and electrical bodies act at greater distances, both repelling and attracting neighboring corpuscles, and light is emitted, reflected, refracted, inflected, and heats bodies, and all perception is aroused, and the members of animals are moved by the will, that is, by vibrations of this spirit propagated through the solid filaments of the nerves from the external organs of perception to the cerebrum and from the cerebrum to the muscles. But these cannot be set forth in a few words, nor is there at hand a sufficient body of experiments by which the laws of action of this spirit are required to be accurately determined and shown.

APPENDIX A

Supplementary Proofs about Conics

Book I Lemma 12

All parallelograms circumscribed about any conjugate diameters of a given ellipse or hyperbola are equal among themselves.

This is proved in Conics.

Notes on Lemma 12:

Recall from Apollonius:

- **diameter:** A line through the center of the ellipse and meeting the curve in both directions. (See Apollonius Definition 4.)

- **conjugate diameters:** Take any diameter. Take the tangent where the diameter meets the curve (either one; they will be parallel). Then draw the *conjugate* diameter through the center parallel to this tangent. (See Apollonius Definitions 5 and 6, and I.47.)

  One set of conjugate diameters of the ellipse is the major and minor axes. These are the only ones that are perpendicular unless it's a circle, where all sets of conjugate diameters are mutually perpendicular. (See Apollonius Definitions 7 and 8.)

- **ordinate:** Drawn from a point on the curve to a particular diameter and parallel to the tangent where that diameter meets the curve. (See Apollonius Definition 5.)

Expansion of Newton's Sketch of Lemma 12

This will be demonstrated first for the ellipse and then for the hyperbola.

Given:

Let there be an ellipse with center *C* and conjugate diameters *ED*, *AB* and *GF*, *HI*.

Let the parallelograms *LKZT* and *RYOX* be circumscribed about the conjugate diameters. The sides of the parallelograms are tangents

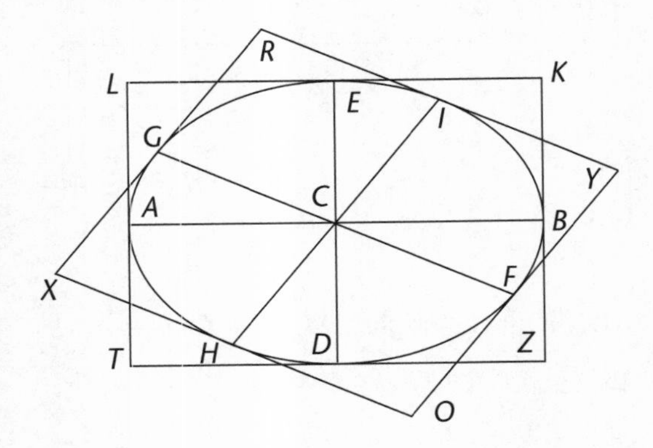

at the points where the diameters intersect the ellipse. (Neither parallelogram will be assumed to be built around major and minor axes.)

To Prove:

I say that parallelogram $LKZT$ = parallelogram $RYOX$.

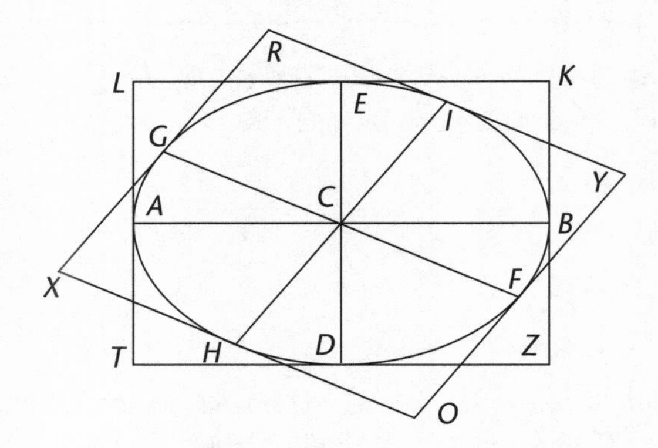

Proof:

Apollonius I.37 applied to Lemma 12, first use:

If a straight line XO touching an ellipse at H meets a diameter AB extended, intersecting at point V, and from the point of contact H a straight line HM is dropped ordinatewise to the diameter AB, intersecting the diameter in M, then the straight line CM cut by the ordinate HM from the center C of the ellipse multiplied by the straight line CV cut off by the tangent HV intersecting AB extended from the center of the ellipse will form an area equal to the square on the radius CA of the ellipse.

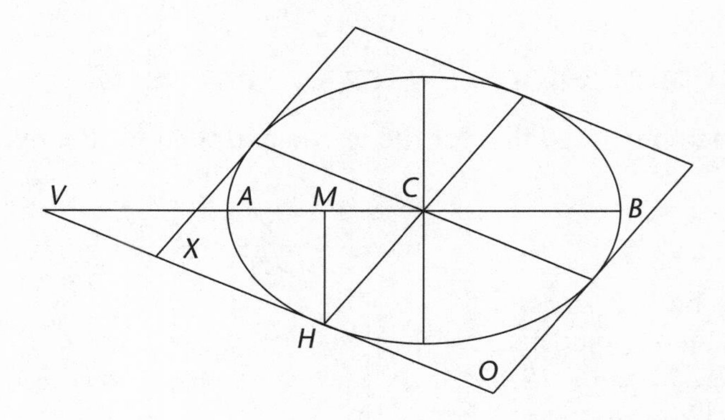

That is, rectangle $CM, CV = CA^2$

By Euclid VI.17,

$CV : AC :: AC : CM$ (1)

Where tangent XO intersects tangent TZ, call that point U.

Where diameter GF extended intersects tangent TZ, call that point P.

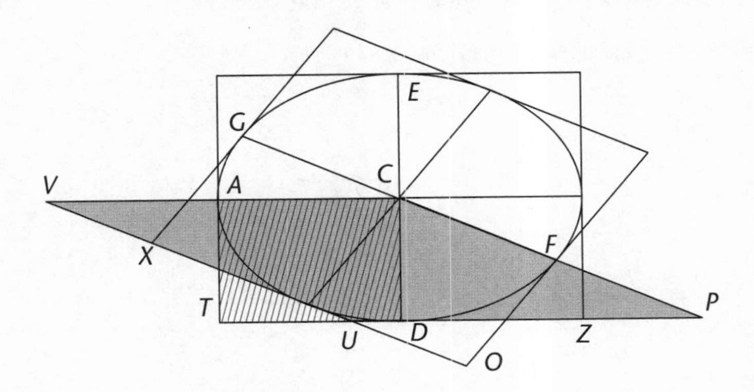

Figure $CVUP$ is a parallelogram because tangent $TZ \parallel$ diameter AB and tangent $XO \parallel$ diameter GF.

Figure $ACDT$ is a parallelogram because diameter $AC \parallel$ tangent TZ and tangent $AT \parallel$ diameter ED.

These parallelograms lie between the same parallels VC and TP; thus they are as their bases, by Euclid VI.1:

pllg $CVUP$: pllg $ACDT$ :: $CV : AC$ [Euclid VI.1]

Extend MH down to meet tangent TZ, intersecting it at S.

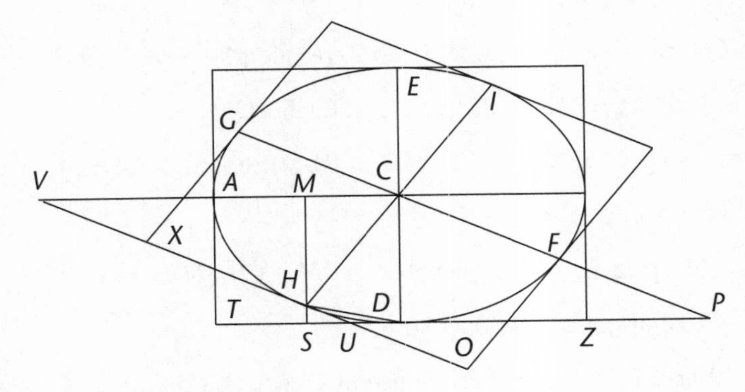

Parallelograms $ACDT$ and $MSDC$ lie between the same parallels AC and DT and therefore are as their bases.

pllg $ACDT$: pllg $MSDC$ :: $AC : MC$ [Euclid VI.1]

By Equation 1, $CV : AC :: AC : CM$

Substituting ratios of parallelograms:

pllg $CVUP$: pllg $ACDT$ :: pllg $ACDT$: pllg $MSDC$ (2)

By Euclid I.41, on the same base CD, between same parallels MS and CD, the area of parallelogram is twice that of the triangle.

pllg $MSDC = 2(\triangle CHD)$ [Euclid I.41]

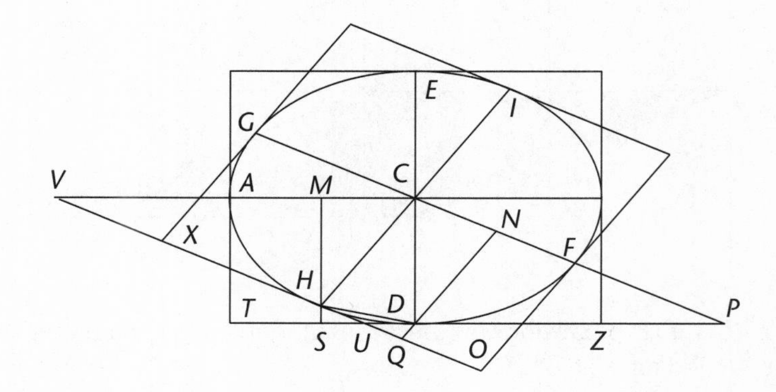

From D draw an ordinate to diameter GF intersecting in N.

Extend ND to meet tangent XO at Q.

Ordinate $DN \parallel$ diameter HI.

Therefore, $QN \parallel HI$.

Diameter $CN \parallel$ tangent XO.

Therefore, $CHQN$ is a parallelogram.

On base CH, between parallels CH and QN, lie pllg $CHQN$ and $\triangle CHD$.

Therefore, pllg $CHQN = 2(\triangle CHD)$ [Euclid I.41].

Therefore, pllg $CHQN$ = pllg $MSDC$ [both equal $2(\triangle CHD)$].

Substituting into Equation 2:

pllg $CVUP$: pllg $ACDT$ :: pllg $ACDT$: pllg $CHQN$ (3)

Apollonius I.37 applied, second use:

If a straight line TZ touching an ellipse at D meets a diameter GF extended, intersecting at P, and from the point of contact D to the diameter GF a straight line DN is dropped ordinatewise intersecting at N, then the straight line CN cut off by the ordinate from the center with the straight line CP cut off by the tangent on the diameter CF extended, from the center of the ellipse, will contain an area equal to that of the square on the radius of the ellipse CF.

That is, rectangle $CN, CP = CF^2$

$CP : CF :: CF : CN$ [Euclid VI.17]

On base CP between parallels CP and XO,

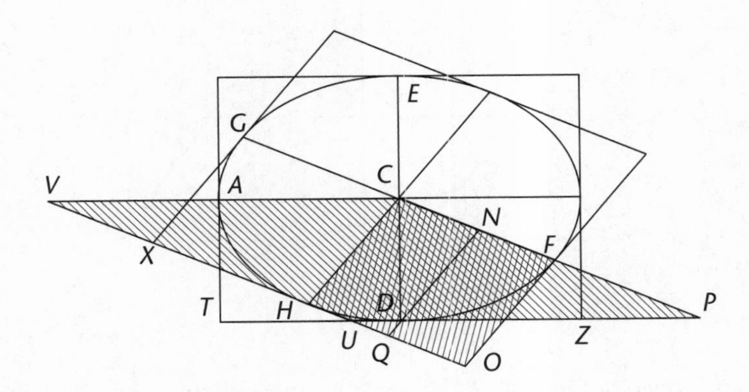

$CP : CF ::$ pllg $CVUP$: pllg $CHOF$ [Euclid I.35]

$CF : CN ::$ pllg $CHOF$: pllg $CHQN$ [Euclid I.35]

Therefore pllg $CVUP$: pllg $CHOF$ :: pllg $CHOF$: pllg $CHQN$ (4)

Comparing Equations 3 and 4, we see that

pllg $ACDT$ = pllg $CHOF$

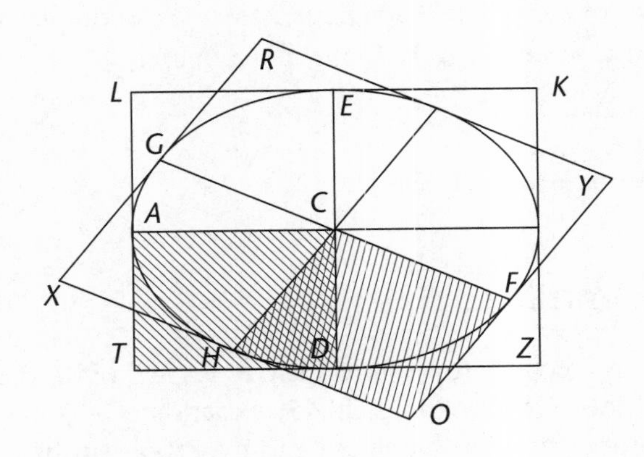

But pllg $ACDT = \frac{1}{4}$(pllg $KLTZ$)

and pllg $CHOF = \frac{1}{4}$(pllg $RYOX$)

Therefore pllg $LKTZ$ = pllg $RYOX$

Q.E.D.

For the hyperbola:

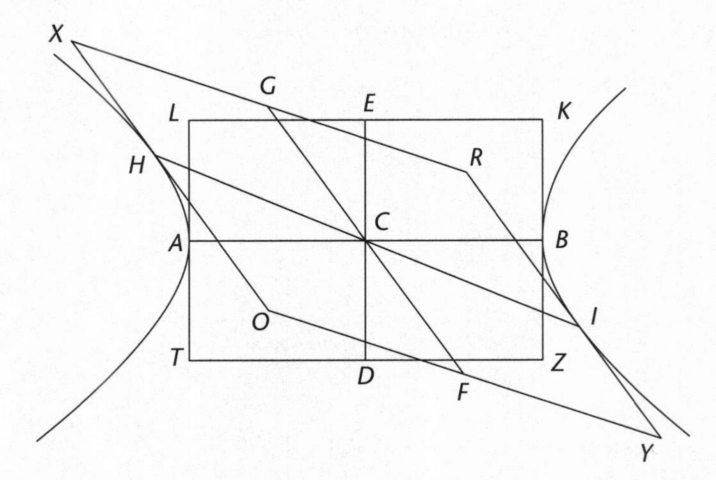

Construct parallelogram *LKZT* around conjugate diameters *AB* and *ED*.

Construct parallelogram *XRYO* around conjugate diameters *GF* and *HI*.

By the same reasoning as we used above for the ellipse, pllg *LKZT* = pllg *XRYO*. Apollonius I.37 applies to both ellipses and hyperbolas.

Book I Lemma 13

The latus rectum of a parabola belonging to any vertex you please is four times the distance of that vertex from the focus of the figure.

This is proved in Conics.

[Newton gives no diagram for this lemma.]

Notes on Lemma 13

• This lemma is proved for the principal vertex of a parabola in the translator's footnote* to Apollonius III.45, expanding and adapting what is proved in the proposition for the ellipse and hyperbola to the parabola. What we need, however, is the general case for the latus rectum of any point *P*.

• The construction used in this proof, and many of the intermediate conclusions, will be used for the proof of Lemma 14.

---

* Notes by R. Catesby Taliaferro are included in his translation published by Green Lion Press.

Expansion of Newton's Sketch of Lemma 13

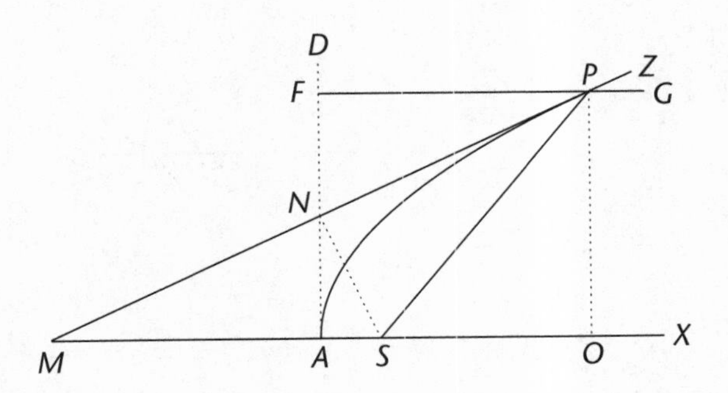

Let PM be tangent at P, a point on a parabola,intersecting the axis AX at M.

Let AD be tangent at the principal vertex A, intersecting PM at N.

Let diameter PG be drawn through P, parallel to AX and intersecting AD at F.

The tangent at A will be perpendicular to the principal axis AX.

Let PO be dropped ordinatewise from P to diameter AX.

Apollonius I.35 applied here: If a straight line PM touches a parabola, meeting the diameter AX outside the section at M, the straight line PO drawn from the point of contact ordinatewise to the diameter AX, will cut off on the diameter beginning from the vertex of the section A a straight line AO equal to the straight line AM between the vertex and the tangent.

That is, $MA = AO$. (1)

Since tangent AF at A is parallel to ordinate OP, and diameter PG is parallel to diameter AX,

Therefore $FP = AO$. [Eu. I.33]

Substituting from Equation 1, $MA = FP$.

Since diameter $FG \parallel AX$, cut by transversal AD, angle PFN = angle FAX by Euclid I.29.

But line AD is tangent to the principal vertex, therefore angle FAX is a right angle.

Therefore angle PFN is also a right angle. (2)

And by Euclid I.13, since angle FAX is a right angle, angle MAN is also a right angle.

Angle MAN = angle PFN.

Angle FNP = angle PFN. [Eu. I.15]

$\triangle MAN \cong \triangle PFN$. [Eu. I.26]

Therefore $MN = NP$. (3)

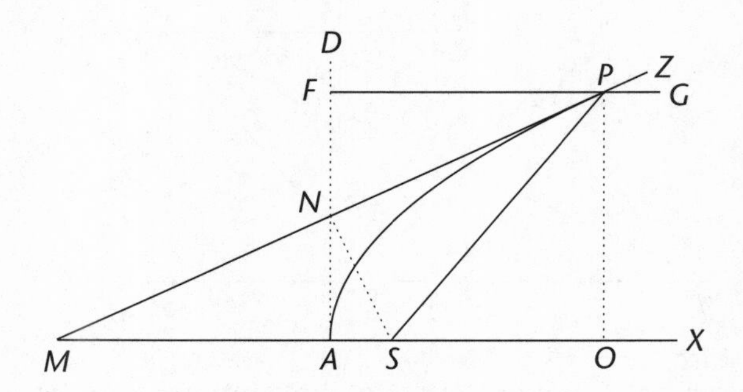

Apollonius III.48 says that for ellipses and hyperbolas the angle between the tangent at a point and the line from that point to one focus is equal to the angle between that tangent to that point and the line between that point and the other focus. This can be extended to the parabola, whose second focus is conceived to be at infinity, making the line to it parallel to the major axis. In the parabola this is another diameter through the point of contact. Applied here, it would say that the angle between the tangent ZPM and the diameter through P is equal to the angle between tangent MPZ and PS.

That is, angle ZPG = angle MPS. (4)

Because $GF \parallel OM$ with line PM as transversal,

Angle ZPG = angle PMS. [Eu. I.29]

Therefore angle MPS = angle PMS. (5)

Therefore by Euclid I.6, $MS = SP$. (6)

Connect SN.

Equations 3, 5, and 6 give us corresponding equality of two sides and the included angle in triangles MSN and PSN.

Therefore $\triangle MSN \cong \triangle PSN$. [Eu. I.4] (7)

Therefore angle MNS = angle PNS.

Since the two angles are made by a line NS set up on a straight line MP, the angles equal two right angles by Euclid I.13. But since they are equal, each is a right angle.

Therefore angle SNP is a right angle. (8)

[And generally, a line drawn from the point of intersection of tangents AD and PM, drawn to focus S, will be perpendicular to tangent PM. (8a)

Note this conclusion — Equation 8a — for use in Lemma 14.]

Now prove that $\triangle FPN \sim \triangle PNS$.

By Equation 4, angle NPS = angle ZPG.

And by Euclid I.15, angle FPN = angle ZPG.

Therefore angle NPS = angle FPN. (9)

By Equation 2, angle PFN is a right angle, and by Equation 8, angle SNP is a right angle,

Therefore angle PFN = angle SNP. [Euclid Post. 4]

But by Equation 9, angle NPS = angle FPN,

Therefore $\triangle FPN \sim \triangle PNS$.

$PF : PN :: PN : PS$. [Eu. VI.4]

By Euclid VI.19 porism,

$PF : PS :: PF^2 : PN^2$. (10)

Since by Equation 3, $MN = NP$,

$PM = 2PN$. (11)

Contrive a line L_P such that $PN : FP :: L_P : 2PM$. (12)

Apollonius I.49 applied here says: If a straight line PM touching a parabola at P meets the diameter AX at M, and through the point of contact P a parallel to the diameter is drawn $GF \parallel AX$, and from the vertex A a straight line AF is drawn parallel to an ordinate PO and it is contrived that as the segment of the tangent PN between the erected straight line AF and the point of contact P is to the segment FP of the parallel FG between the point of contact and the erected straight line AF, so is some straight line L_P to the double of the tangent $2PM$, then [L_P is the latus rectum of diameter FG at P].

Therefore L_P is the latus rectum of FG at P.

By Equation 12,

$PN : FP :: L_P : 2PM$.

Since by Equation 11, $PM = 2PN$,

$PN : FP :: L_P : 4PN$.

$L_P \times PF = 4PN^2$. [Eu. VI.16]

$L_P = 4PN^2/PF$. (13)

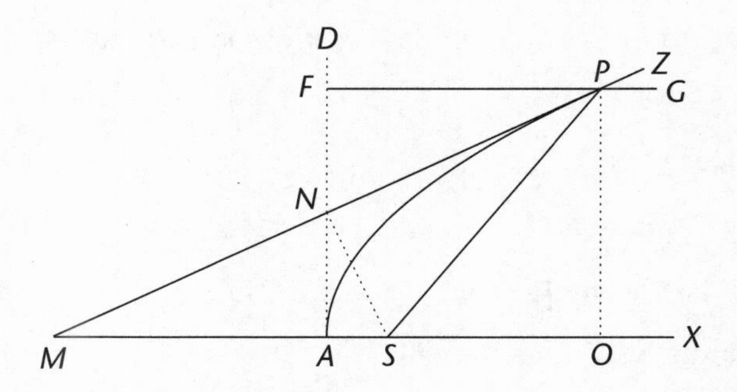

But by Equation 10, $PF : PS :: PF^2 : PN^2$.

$PS \times PF^2 = PF \times PN^2$. [Eu. VI.16]

$PS = PF \times PN^2 / PF^2$.

$PS = PN^2 / PF$.

Substituting into Equation 13,

$L_P = 4SP$.

Q.E.D.

Book I Lemma 14

The perpendicular drawn from the focus of a parabola to its tangent is the mean proportional between the distances of the focus from the point of tangency and from the principal vertex of the figure.

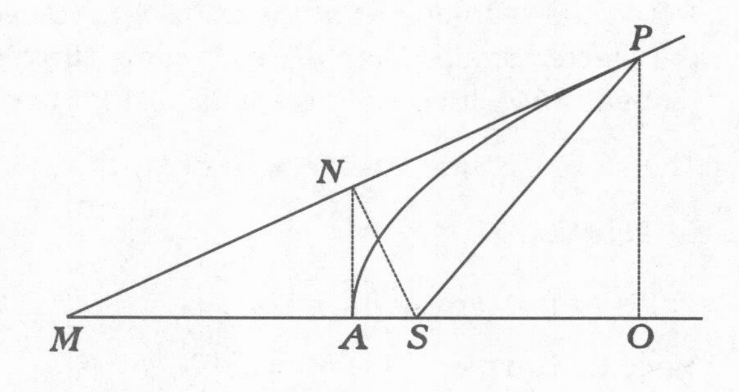

For let AP be a parabola, S its focus, A the principal vertex, P the point of tangency, PO applied ordinatewise to the principal diameter, PM the tangent meeting the principal diameter at M, and SN a line drawn from the focus perpendicular to the tangent. Let AN be connected, and because of the equals MS and SP, MN and NP, MA and AO, the straight lines AN and OP will be parallel, and hence the triangle SAN will have a right angle at A, and will be similar to the equal triangles SNM, SNP. Therefore, PS is to SN as SN is to SA.

Q.E.D.

Note on Lemma 14

Newton, having cited the conclusion of Lemma 13 on the authority of a writer on Conics, must now start from scratch to set up the conditions we used in our proof of Lemma 13 and prove the relationships deduced there. But since we proved Lemma 13 as we did, Lemma 14 will follow almost as a corollary.

Expansion of Newton's Sketch of Lemma 14

Given:

Parabola.

To Prove:

$SP : SN :: SN : SA$.

Proof:

Construction:

"For let *AP* be a parabola, *S* its focus, *A* the principal vertex, *P* the point of tangency, *PO* applied ordinatewise to the principal diameter, *PM* the tangent meeting the principal diameter at *M*, and *SN* a line drawn from the focus perpendicular to the tangent. Let *AN* be connected, ..."

We assume for this lemma the diagram and elements proved in Lemma 13. Construct as above, adding:

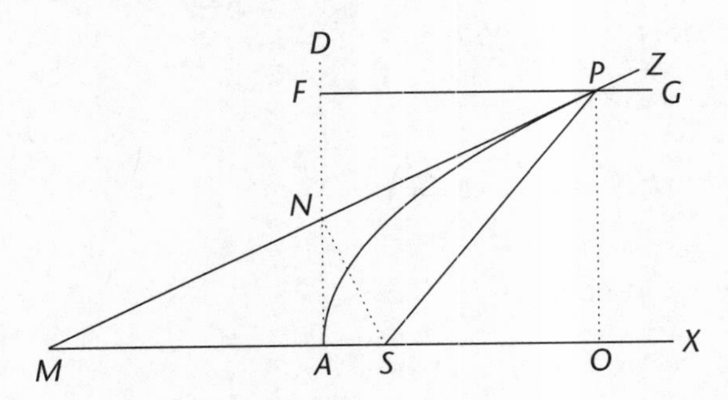

Let *AD* be tangent to the parabola at vertex *A*.

By Equation 8a of Lemma 13, a line drawn from the point of intersection of tangents *AD* and *PM*, drawn to focus *S*, will be perpendicular to tangent *PM*.

Therefore the *N* of Newton's Lemma 14 construction will be the intersection of the two tangents.

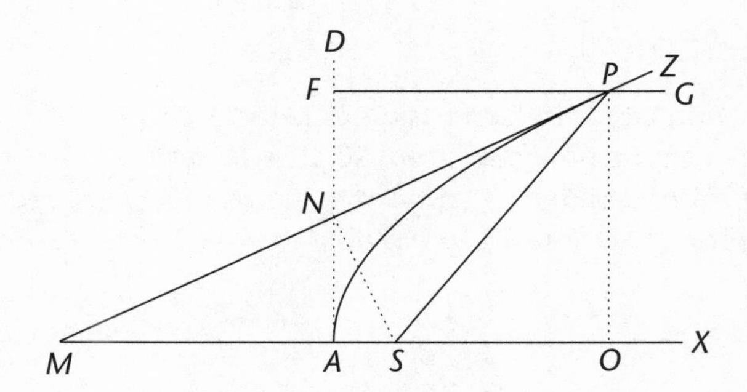

"… and because of the equals *MS* and *SP*, *MN* and *NP*, *MA* and *AO*, the straight lines *AN* and *OP* will be parallel, and hence the triangle *SAN* will have a right angle at *A*, and will be similar to the equal triangles *SNM*, *SNP*."

By Equation 6 of Lemma 13, $MS = SP$; by Equation 3 of Lemma 13, $MN = NP$; by Equation 1 of Lemma 13, $MA = AO$.

But since we showed above that *N* lies on tangent *AD*, we know that $AN \parallel PO$. Angle *SAN* is a right angle since it is the angle of the tangent at the principal vertex to the axis.

But Newton is using a different procedure here. He dropped *SN* perpendicular to *PM* and just connected *AN*. Now he shows that line *NA* cuts the sides *OM* and *PM* of $\triangle PMO$ proportionally, and so by Euclid VI.2, that line *AN* is parallel to the base *PO*. Since *PO* was dropped ordinatewise, it is parallel to the tangent at the principal vertex *A*, and so *AN* is that tangent. Then it follows that angle *SAN* is a right angle.

By Equation 7 of Lemma 13, $\triangle MSN \cong \triangle PSN$.

But $\triangle SAN \sim \triangle MSN$. [Eu. VI.8]

Therefore $\triangle SAN \sim \triangle PSN$.

"Therefore, *PS* is to *SN* as *SN* is to *SA*.

Q.E.D."

This follows by Euclid VI.4.

Q.E.D.

Lemma 14 Corollary 1

sq.*PS* is to sq.*SN* as *PS* to *SA*.

Expansion of Newton's Sketch of Corollary 1

This follows from the conclusion of the proposition by Euclid VI.19 porism.

Lemma 14 Corollary 2

And because SA is given, sq.SN varies as PS.

Expansion of Newton's Sketch of Corollary 2

Rewriting the conclusion of Corollary 1,

$$SN^2 = SA \times PS^2 / PS.$$

But SA is constant.

Therefore $SN^2 \propto PS$.

Lemma 14 Corollary 3

And the intersection of any tangent you please PM with the straight line SN, which is drawn perpendicular to it from the focus, lies upon the straight line AN, which touches the parabola at the principal vertex.

Expansion of Newton's Sketch of Corollary 3

This was proved in Lemma 13 as Equation 8a.

APPENDIX B
The "Three-Body Problem"

As discussed in our tracing of Newton's derivation of universal gravitation, he had reason to be suspicious of the full universality of the inverse square attraction, even *after* he had derived ellipses from the inverse square law. And one of his reasons for resisting this had to do with the question of whether the universe is stable.

Newton knew that our world exhibits stable elliptical orbits and equable description of areas. Kepler had shown this. If "gravity is given towards every planet" as Newton asserts in III.5 Corollary 1, and if this means, as it seems to, that every planet gravitates towards every other planet, this would lead to complicated relationships among many bodies. The stable ellipses that result from inverse square laws governing two-body systems would be thrown off by the effects of other bodies outside that system.

For example, the converse to I.9 would say that an inverse cube force law would lead to a spiral. Newton has not proved this converse, but it turns out that most inverse cube examples would lead to a spiral into the center or out away from it.

In Proposition 43–45 of Book I, Newton shows that when the line of apsides is moving we get an inverse cube component added to the inverse square component. In Propositions I.64–I.66, considering the effect of three or more bodies on one another, Newton shows that there will be movement of the lines of apsides. If one applied I.45 to this, one could conclude that there is an inverse cube component in addition to the inverse square component. Some inverse cube component to the combined force operating on a planet might be expected to make it spiral into the sun, or out away from it.

And yet we do find stable elliptical orbits in the planetary system. To justify his claim in III.5 Corollary 1 and its further development, Newton must account for that. Propositions I.65 and I.66 explore some reasons we might still find stable elliptical orbits. They argue that it is possible with many bodies, if they are suitably arranged, to get the bodies moving, at least approximately, in ellipses with equable description of areas. I.67 explores why we could still find equable description of areas when multiple bodies are influencing one another.

Book I Proposition 65

Many bodies, whose forces decrease in the duplicate ratio of distances from their own centers, can move among each other in ellipses, and, with radii drawn to the foci, can describe areas proportional to the times, very nearly.

In the above proposition, the case was demonstrated in which many motions are carried out in ellipses accurately [through forces directly proportional to the distances]. The more the law of forces departs from the law posited there, the more the bodies will perturb each others' motions, and it cannot happen that bodies, attracting each other according to the law posited here, move in ellipses accurately, except by preserving a certain proportion of distances from each other. In the following cases, however, the departure from ellipses will not be great.

Case 1. Suppose many smaller bodies to revolve about some very great one, at various distances from it, and to each let there tend absolute forces that are proportional to the same bodies. And because the common center of gravity of all of them (by the fourth corollary of the Laws) either is at rest or moves uniformly in a straight line, let us suppose that the smaller bodies are so small that the greatest body is imperceptibly distant from this center. That great one will [therefore] either be at rest or will move uniformly in a straight line without perceptible error. Further, the smaller bodies will revolve around this greatest one in ellipses, and by radii drawn to it will describe areas proportional to the times, unless errors be introduced either by a deviation of the greatest from that common center of gravity, or by the action of the smaller bodies on each other. And the smaller bodies can be diminished until this last error, and the mutual actions, be less than any given amount you please, and can therefore be diminished until the orbits match ellipses, and the areas correspond to the times, without any error that is not less than the given amount.

Case 2. Let us now imagine that the system—either that of the small bodies revolving about the greatest body, just described, or any other system you please consisting of two bodies revolving about each other—progresses uniformly in a straight line, and that it is meanwhile urged to the side by the force of another body, by far the greatest, and placed at a great distance. And because the equal accelerative forces, by which the bodies are urged along parallel lines, do not change the positions of the bodies among themselves, but cause the whole system to be carried along together, preserving the motions of the parts among themselves, it is manifest that no change of motion whatever of the attracted bodies among themselves arises, unless either from the inequality of the accelerative forces, or from the mutual inclination of the lines along which the attractions occur. Suppose, therefore, that all the accelerative attractions towards the greatest body are to each other inversely as the squares of the distances, and by increasing the distance of the greatest body until the differences of the straight lines in relation to their length, and the inclinations to each other, are less than any given [amount], the motions of the parts of the system among themselves will persevere without errors that are not less than any given [amount].

And because the system as a whole is attracted in the manner of a single body, on account of the slight distance of those parts from each other, the same system will move in the manner of a single body by this attraction. That is, by its center of gravity it will describe around the greatest body some conic section (namely, a hyperbola or parabola under a weak force, an ellipse under a stronger one), and by a radius drawn to the greatest body will describe areas proportional to the times, without any errors, except those which the distance of the parts, which are very small indeed and may be diminished at will, have the power to bring about.

Q.E.O.

By a similar argument it is possible to proceed to cases that are more compounded, *in infinitum*.

Q.E.O.

[Q.E.O. (Quod Erat Ostendendum): "That which was to have been shown."]

Note on I.65

• The I.65 brings in the complication of multiple bodies. If "gravity is given towards every planet" as Newton asserts in III.5 Corollary 1, and if this means, as it seems to, that every planet gravitates towards every other planet, then we would expect instability in the universe, as discussed above in the introduction to this appendix.

I.65 deals with three bodies and, with its corollaries 3 and 2 mentioned by Newton in this proposition, essentially restates I.3. I.65 Corollary 3 says that if we have ellipses or circles with equable description of areas, at least approximately, then either the central and orbiting bodies are not urged, or they are very little urged, by accelerative forces outside; or they are urged equally, by approximately parallel lines.

I.65 Corollary 1

In the second case, the nearer the greatest body of all approaches the system of two or more bodies, the more the motion of the parts of the system relative to each other will be disturbed, because the lines drawn from the greatest body to the latter ones are more inclined to each other, and the inequality of proportion is greater.

I.65 Corollary 2

They will, furthermore, be most greatly disturbed, supposing the accelerative attractions of the parts of the system towards the greatest body of all are not to each other inversely as the squares of the distances from that greatest body. This is especially so if the inequality of proportion is greater than the inequality of proportion of the distances from the greatest body. For if the accelerative force, acting equally and along parallel lines, does not perturb the motions relative to each other, it is necessary that the perturbation arise from the inequality of action, and that it be greater or less according as the inequality is greater or less. The excess of greater impulses acting on one body and not acting on the others, will necessarily change their relative positions. And this perturbation, added to the perturbation arising from the inclination and inequality of the lines, will make the whole perturbation greater.

Note on I.65 Corollary 2

• I.65 Corollary 2 says the motions will get "disturbed" or "confused"—this means we won't have ellipses and equable description of areas—if bodies are not accelerated by the same amount at the same distance.

• The stated condition might seem to require some proportion other than inverse square. However, what Newton is concerned about here is a hypothetical difference between the accelerative forces on two or more bodies of the system towards the large exterior body. A simple inverse square proportion would require the accelerative forces to be the same at the same distance. But what if they weren't? What if both were attracted with an inverse square law, but one were attracted a little more or a little less than another at the same distance?

The same argument would of course also hold if one body were attracted with an inverse square law and another were attracted in some other proportion to the distances.

I.65 Corollary 3

Whence, if the parts of this system move in ellipses or in circles without great perturbation, it is evident that the same bodies are either only slightly urged, or are urged equally and along lines that are approximately parallel, by accelerative forces tending towards other bodies.

Pause Before I.66

This is the proposition that is usually thought of as Newton's treatment of the famous "three body problem." It is not used in the argument leading to universal gravitation (though it is cited at its conclusion, in III.13), and is therefore not included in the main sequence of this book. Nor is it going to be given a full treatment here. However, since it has been the occasion of much misunderstanding, both in its own right and in its relation to the argument of the *Principia* as a whole, a brief account of it is included here.

A few remarks:

First, it must again be said of this proposition, as of all propositions in Book I, that it is a hypothetical proposition, which is to be applied to the phenomenal world only after its conditions have been carefully established. In this case, it could not have been used before III.5, is not cited until III.13, and does not fully come into play until the sequence of propositions beginning with III.21 (on the precession of the equinoxes), reaching a climax with III.35 and its lengthy scholium (on the moon's motions), and concluding with III.39. These propositions are, of course, of great interest, and well worth the considerable trouble involved in studying them; but they should be postponed until after the proper groundwork for them has been laid. The *Principia* is a carefully constructed argument, not just a grab-bag of miscellaneous proofs, and to skip the central argument in favor of propositions read out of sequence is to deprive oneself of seeing the full force of Newton's achievement, as he intended it to be seen.

Second, it does not, as is sometimes claimed, "solve the three body problem." Newton was unable to solve that problem, and in fact a perfectly general solution is not possible (as was discovered much later). It does, however, set out a way of analyzing, aproximately, the effects of perturbing forces on the earth's rotation, the tides, the moon's orbit, and other phenomena. Qualitatively, it is brilliant, but Newton never managed to get the numbers to come out right, and finally (in the scholium to III.35, pages 871–872 of the Cohen translation), had to resort to the ancient Greek device of the epicycle to complete his lunar theory.

The proposition will not be presented in its entirety, since that would distract too much from the aim of this book. The following is a summary, with a few select quotations, which will give readers an idea of the potential of Newton's completed gravitational theory. The corollaries following the proposition are also summarized, sketching out the proofs and their significance. The aim of the summaries is to allow those who wish to go farther into Newton's treatment to figure out where to begin.

Book I Proposition 66

[The enunciation of this proposition is somewhat convoluted, and does not give an idea of what the proposition and its corollaries are really about. Briefly put, it states that planetary systems are better behaved when the central body is allowed to wobble around under the forces exerted towards the planets. The main accomplishment of the proposition, on the other hand, is the geometrical analysis of forces and the application of this analysis to various special cases. A translation of Newton's construction follows.]

This is fairly evident from the demonstration of the second corollary of the preceding proposition, but the point is made by a more explicit and more broadly cogent argument as follows.

Case 1. Let the smaller bodies P and S revolve in the same plane about the greatest body T, and of these, let P describe the inner orbit PAB and S describe the outer orbit ESE.

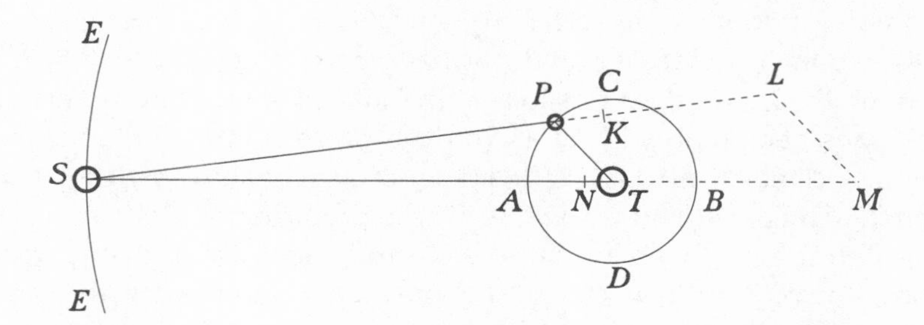

Let SK be the mean distance of bodies P and S, and let the accelerative attraction of body P towards S, at the mean distance, be set out by the same [SK]. Let SL be taken to SK in the duplicate ratio of SK to SP, and SL will be the accelerative attraction of body P towards S at any distance you please SP. Join PT, and draw LM parallel to it, meeting ST at M, and the attraction SL is resolved (by corollary 2 of the Laws) into the attractions SM, LM. And thus the body P will be urged by a threefold accelerative force. One force tends to T, and arises from the mutual attraction of the bodies T and P.... The second force is that of the attraction LM, which, because it tends from P to T, being added to the previous force, coincides with it.... But now the third force SM, in pulling body P along a line parallel to ST, will compose with the prior forces a force which is no longer directed from P to T, and which all the more deviates from this direction the greater this third force is to the prior forces, other things being the same....

Let the accelerative attraction of body T towards S be set out by the line SN.... Now if the attraction SN were less than the attraction SM, it would itself remove the part SN of the attraction SM, and only the part MN would remain, by which the proportionality of times and areas and the form of the orbit would be perturbed.

This completes Newton's construction.

To prove the proposition itself, he then argues that if SN were equal to SM—that is, if the force on the central body T were the same as the force on the inner orbiting body P, the perturbing effect would be zero (by

Corollary 6 of the Laws). On the other hand, if *SN* were zero, while *SM* remained the same, the perturbing force *MN* would be the same as *SM*, and its effect would be considerable. Therefore, the perturbations are less if the central body is acted on by the orbiting bodies.

Note that by removing the component *SN* by which *T* is pulled toward *S*, Newton in effect views the two bodies *T* and *P* as if *T* were at rest and *MN* were a perturbing force on *P* that goes toward *S* when $SP > ST$ and goes away from *S* when $SP < ST$. It isn't really ever a repulsive force, but represents the difference between two unequal attracting forces. An observer on *T* thinks that *P* is being pushed away from *S*, although the truth is that *T* is being pulled toward *S* more strongly than *P* is.

I.66 Corollary 1

This corollary extends the conclusion of the proposition to more than three bodies.

I.66 Corollaries 2–5. Speed and curvature.

Here Newton looks at the effective perturbing force *NM* (ignoring the component *LM*) and shows that it pulls the body *P* forward (toward *A*) between *C* and *A*, backward between *A* and *D*, and again forward between *D* and *B* and backward between *B* and *C*. The result is that the areas described are greater in proportion to the times near *A* and *B* (Cor. 2); that the velocity of *P* is also greater near *A* and *B* (Cor. 3); that the orbit (other things being the same) is less curved near *A* and *B* (Cor. 4); and that *P* is therefore closer to *T* near *A* and *B* (Cor. 5). At the same time, Newton points out that, under the influence of its own gravitational attraction toward *T*, the body *P* is moving on an eccentric ellipse that has its own greatest and least distances, fast and slow places, and changing curvature. The effects described in Corollaries 2–5 are modifications to the properties that the orbit would have had if it were not disturbed.

I.66 Corollaries 6–9. Motions of the apsides.

Here Newton begins to consider the effects of the combined forces *LM* and *MN*. When *P* is at *C* or *D*, the force *MN* is zero, and *LM* is attractive, pulling *P* toward *T*. And when *P* is at *A* or *B*, the force *MN* always pulls *P* away from *T*. Corollaries 6–9 all depend on these simple relations.

Corollary 6 provides a way of relating the forces to the distance of *P* from *T* and the periodic time of its orbit, assuming that the orbit is approximately circular and using I.4 and its corollaries.

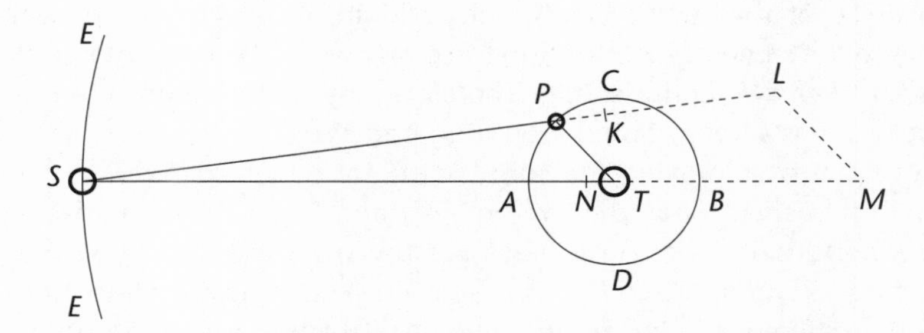

In Corollaries 7–9, Newton begins to consider the orbit, not as an approximately concentric circle, but as an eccentric ellipse that is nearly circular. These corollaries are about the motion of the apsides and related matters.

Corollary 7 uses I.45 to determine the motion of the apsides of the orbit of *P*. In I.45, Newton proved that if the forces decrease in a ratio *greater* than the squares of the radii, then the apsides will move *forward;* if less, then backward. Here, *LM* is roughly proportional to *PT,* and when this is added to the inverse square force on *P,* an increase in *PT* will decrease the force by a ratio *less* than the squares of the distances, so the apsides will move backward, clockwise in the diagram.

At *A,* on the other hand, the forces all line up, and the situation looks like this:

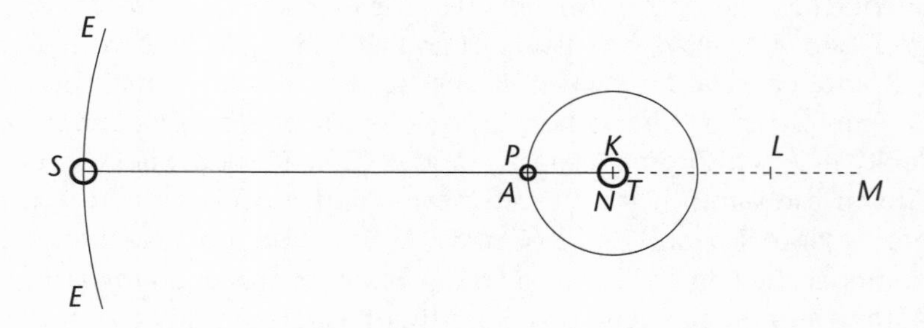

L must be beyond *K* (which now coincides with *N* and *T*), because $SL:ST :: ST^2:SA^2$, and likewise, *M* is beyond *L.* Force *MN* has the effect of pulling *P* toward *S,* while force *LM* pulls *P* toward *T,* as before. Therefore, the net force on *P* is the difference, *TL* or *KL*. But if *S* is sufficiently far away, *KL* becomes approximately twice *PK* or *PT* (we won't go into this here, but it can be proved from the way these points have been defined). And therefore, the force on *P* is the resultant of the inverse square force pulling in toward *T* and the outward (subtractive) force *KL* which is directly as the distance. So as the distance increases, the force will decrease by a ratio *greater* than the squares of the distances, and the apsides will move *forward* (counterclockwise, in the diagram).

On the average, however, since *KL* (the subtractive force near *A* and *B*) is about twice *PT* (see above), while *PT* is approximately equal to *LM* (the additive force near *C* and *D*, assuming *S* to be sufficiently distant, the subtractive force wins out, and so the net motion of the apsides is forward (counterclockwise). To help us visualize this, Newton writes,

> Moreover, the truth of this corollary and the preceding one will be more easily understood by conceiving of a system of two bodies *T, P,* surrounded on all sides by many bodies *S, S, S,* and so on, standing on the orbit *ESE*. For by the actions of all of these the action of *T* will be diminished on all sides, and would decrease in a ratio greater than the duplicate ratio of the distances.

Corollary 8 notes that because of the relative strengths of the two forces described in Corollary 7, the apsides move forward more quickly at *A* and *B* and move backward more slowly at *C* and *D*.

Corollary 9 states that for the same reasons, the eccentricity of the orbit increases as the apsides move from *C* to *A* and from *D* to *B,* and decreases as the apsides move from *A* to *D* and from *B* to *C*.

I.66 Corollaries 10–22.

Corollaries 10–12 consider the orbit *PA* as not lying in the same plane as orbit *EST,* and they examine the motion of the plane of the orbit and of the nodes, that is, of the points where orbit *PAB* intersects the plane of orbit *EST*.

Corollary 10 shows how the inclination of the plane of orbit *PA* is affected by the disturbing force. Of the two components of the disturbing force (*LM* and *MN*), *LM* is always parallel to *PT,* and therefore can't divert *P* from its orbital plane. Of force *NM,* Newton writes,

> When the nodes are in the syzygies [i.e., at *A* and *B*], because it acts in the same plane of the orbit, it does not disturb these motions, but where the nodes are in the quadratures [i.e., at *C* and *D*], it disturbs them most greatly, and, always drawing the body *P* from the plane of its orbit, it decreases the inclination of the plane when the body is passing from the quadratures to the syzygies, and in turn increases the inclination when the body is passing from the syzygies to the quadratures.

Corollary 11 takes up the motion of the nodes. As in the previous corollary, the disturbing effect happens only when the nodes are in the neighborhood of *C* and *D*. Imagine the body *P* moving from *A* towards *D* (where the node is at the beginning of the motion), and suppose that at this part of its orbit it is above the plane of the paper. Since the force *NM* is directed toward *S,* it will deflect *P* downward toward the plane of the paper. Therefore, *P* will reach the plane of the paper before it gets to *D,* and hence the node has moved backward (counterclockwise). The same will happen from *B* to *C*. So the motion of the nodes is always backward, but varies in speed.

Corollary 12 states that these effects are greater at *A* than at *B,* because the disturbing force toward *S* is greater at *A.*

Up to this point, the diagram has shown *T* as the central body, with *P* and *S* orbiting around it. In Corollary 13, Newton notes that none of the arguments involved any assumptions about the sizes of the three bodies, and that the same things would therefore hold true if *S* were a very large body, with *P* and *T* orbiting around it as a system.

In Corollary 14, Newton supposes *S* to be very distant, and states that, other things remaining the same, the disturbing forces are approximately proportional to the absolute force of *S* directly and the inverse triplicate ratio of the distance *ST.* This is a consequence of a proportion which Newton states (without derivation) at the beginning of the corollary. A derivation of this proportion is given by Cohen in the introductory matter to his translation of *Principia* (pp. 355–359), so we won't go into it here.

This corollary has two sub-corollaries of its own. First, Newton applies I.4 corollaries 2 and 6 to prove that the disturbing forces will be inversely as the square of the periodic time. Second, assuming that the absolute force of *S* is proportional to its magnitude, the disturbing forces will be directly proportional to the cube of the apparent diameter of *S.*

Corollary 15 says that the scale of the system may be changed, as well as the forces of *S* and *T,* and all the effects will be similar and proportional.

Corollary 16 shows how to use Lemma 10 Cor. 2 to calculate the effect of changing the sizes, forces, and distances of the bodies.

Corollary 17 shows how to calculate the relative magnitude of the disturbing forces, using I.4 cor. 2.

Corollaries 18–22 lay the groundwork for explaining the tides and the slow change in the orientation of the earth's axis (the motion of precession).

In Corollary 18, Newton asks us to imagine a ring of bodies *P,* coalescing into a fluid ring encircling *T,* and argues that the conclusions of Corollaries 6–12 (on the motions of apsides and orbital inclination and nodes) will apply to this ring.

In Corollary 19, the ring is brought down to earth, as it were: globe *T* is expanded so as to reach out to *ABCD,* and the fluid ring of Corollary 18 is included in a channel cut around the circumference of *T.* The body *S* plays the role of the moon. Since, by Cor. 1, the water moves fastest at *A* and *B,* and slowest at *C* and *D,* it will pile up into a high tide between *A* and *D* and between *B* and *C.* The upward and downward forces *LM* and *KL* reinforce this tidal motion.

In Corollary 20, the ring surrounding (and attached to) *T* is made rigid, and Newton considers the effect of the disturbing force on the ring as a whole. As in Cor. 11, the nodes will tend to move backward. Extending this to the globe itself, Newton writes:

> The same holds for a globe without a ring, which in the region of the equator is a little higher than near the poles, or consists of a somewhat denser matter. For the excess of matter in the region of the equator takes the place of that ring.

As a result of this force, the axis of the globe's rotation will be slowly moved backwards in a circular motion.

The rest of the corollary considers the effect of increasing the (gravitational) force of the central body on the conclusions of this and the preceding corollary.

Corollaries 21 and 22 extend the conclusion of Corollary 20 to a globe that is narrower at the equator, or is less dense there, and they show how the composition or form of the globe could be studied by measuring the precession of the axis.

Book I Proposition 67

The same laws of attractions being supposed, I say that the outer body S, by radii drawn to the common center of gravity O of the inner bodies P, T, will describe areas more proportional to the times, and an orbit more nearly approaching the shape of an ellipse having its focus at that same center, than it can describe about the inmost and greatest body T, by radii drawn to it.

For the attractions of body S towards T and P make up its absolute attraction, which is directed more to the common center of gravity O of bodies T and P than to the greatest body T, and which is more nearly inversely proportional to the square of the distance SO then to the square of the distance ST, as will easily become evident to anyone thinking the matter through.

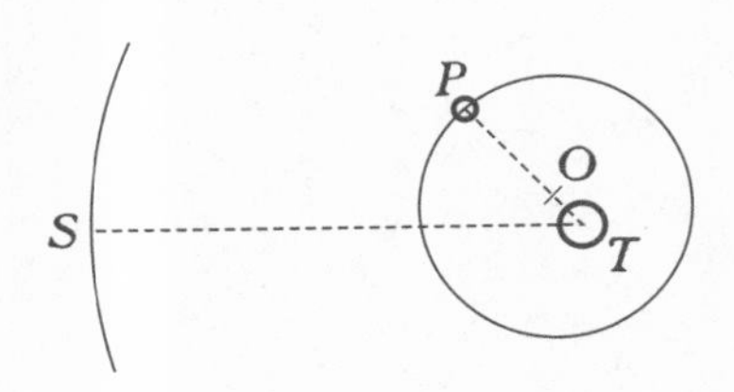

Proposition I.67 argues that, when several bodies are mutually attracted by an inverse square law, with two planets orbiting around a larger body, then the outer of the two orbiting planets describes areas more nearly proportional to the times when those areas are figured around the center of gravity of the other two bodies than it does around the larger body as center.

Bibliography: Primary Sources

Apollonius of Perga, *Conics Books I–III,* translation by R. Catesby Taliaferro. Green Lion Press, 1998.

Archimedes, *On Spirals,* translation by R. Catesby Taliaferro. University of Chicago and Encyclopedia Brittanica Inc., 1952.

Aristotle, *On the Heavens,* translation by W.K.C. Guthrie. Harvard University Press, 1939.

Aristotle, *Physics,* translation by Joe Sachs. Rutgers University Press, 1995.

Bacon, Francis, *The Advancement of Learning,* published in 1605. J.M. Dent & Sons, 1973, London, Everyman's Library edition.

Copernicus, Nicolaus, *Revolutions of Heavenly Spheres,* translation by C.G. Wallis. Encyclopedia Britannica, 1952.

Descartes, Rene, *Meditations on First Philosophy,* translation by Haldane and Ross. *Philosophical Works of Descartes,* Dover Publications, 1955.

Descartes, *Le Monde,* The Janus Library [no date or other publisher identification], edited by Michael Sean Mahoney.

Descartes, *Principles of Philosophy,* translation by Valentine R. Miller and Reese P. Miller. Kluwer Academic Publishers, 1991.

Euclid, *Elements,* translation by Thomas L. Heath. Green Lion Press, 2002.

Galileo (Galileo Galilei), *Dialogue Concerning the Two Chief World Systems,* translation by Stillman Drake. University of California Press, 1967.

Galileo, *De Motu,* translation by I.E. Drabkin. University of Wisconsin Press, 1960.

Galileo, *Two New Sciences,* translation by Stillman Drake. Wall & Thompson, 1989.

Gilbert, William, *De Magnete,* translation by P. Fleury Mottelay. Dover Publications, Inc., 1958.

Huygens, Christian, *The Pendulum Clock,* translation by Richard J. Blackwell. The Iowa State University Press, 1986.

Kepler, Johannes, *Epitome of Copernican Astronomy,* translation by C.G. Wallis. Encyclopedia Britannica, 1952.

Kepler, *New Astronomy,* translation by William H. Donahue. Cambridge University Press, 1992.

Leibniz, Gottfried Wilhelm, *Philosophical Essays,* translations by Roger Ariew and Daniel Garber. Hackett Publishing Company, 1989.

Newton, Isaac, *Correspondence of Isaac Newton,* edited by H.W. Turnbull. Cambridge University Press, 1960.

Newton, *The Mathematical Papers of Isaac Newton,* edited by D.T. Whiteside. Latin text with translation by D.T. Whiteside. Cambridge University Press. Vols I–VIII, 1967–1981.

Newton, *Philosophiae Naturalis Principia Mathematica,* Latin Text, Third Edition with Variant Readings and Notes. Koyre, Alexandre and I. Bernard Cohen, editors, Harvard University Press, 1972.

Newton, *The Principia,* translation by I. Bernard Cohen and Anne Whitman. University of California Press, 1999.

Newton, *Theory of the Moon's Motion* (1702). See Kollerstrom under Secondary Sources (below).

Ptolemy (Claudius Ptolemaeus), *The Almagest,* translation by R. Catesby Taliaferro. University of Chicago and Encyclopedia Brittanica, 1952.

Bibliography: Selected Secondary Sources

Aiton, E.J., *The Vortex Theory of Planetary Motions*. American Elsevier, Inc., 1972.

Bart, Robert S., *Notes to Accompany the Reading of Newton's Principia Mathematica*. Second Edition, St. John's College, 1968.

Brackenridge, J. Bruce, *The Key to Newton's Dynamics*. University of California Press, 1995.

Chandrasekhar, S., *Newton's Principia for the Common Reader*. Clarendon Press, Oxford, 1995.

Cohen, I. Bernard, *Introduction to Newton's 'Principia'*. Harvard University Press, 1971.

Cohen, I. Bernard, and George E. Smith. *The Cambridge Companion to Newton*, Cambridge University Press, 2002.

Crowe, Michael J., *Theories of the World from Antiquity to the Copernican Revolution*. Dover Publications, 1990.

De Gandt, François, *Force and Geometry in Newton's Principia*. Princeton University Press, 1995.

Guicciardini, Niccolò, *Reading the Principia*. Cambridge University Press, 1999.

Kollerstrom, Nicholas, *Newton's Forgotten Lunar Theory*. Green Lion Press, 2000.

Westfall, Richard S., *Never at Rest: A Biography of Isaac Newton*. Cambridge University Press, 1984.

Wilson, Curtis, *Astronomy from Kepler to Newton*. Variorum Reprints, London, 1989.

Yoder, Joella G., *Unrolling Time*. Cambridge University Press, 1988.

Index

BIOGRAPHICAL NOTE

After earning her BA at St. John's College in Annapolis in 1965, Dana Densmore worked as a systems programmer at MIT for the Apollo Project and the Space Shuttle. One of her responsibilities there was coordination of all the on-board guidance, navigation, and control programs for the Lunar Lander. Subsequently she helped design a state-of-the-art on-board guidance computer system for advanced space missions, specializing in design of the real-time multiprocessing operating system. She also helped develop and market a proprietary methodology for software engineering design for large real-time guidance and control applications. She left MIT in 1977 to work in industry. As a senior product planner for a major international computer manufacturer she researched and evaluated advanced and emerging cybernetic technologies.

In 1987 she was appointed Tutor at St. John's College, Santa Fe, where she led tutorials and discussions in the St. John's College great books curriculum. Among subjects she taught were ancient, medieval, and modern literature and philosophy; ancient Greek mathematics and astronomy; Galileian and Newtonian physics; non-Euclidean geometry; electrodynamics; and special relativity. In 1993 she received her MA from St. John's for a thesis on the cause of gravity in Newton's *Principia.* While leading students through the *Principia,* she developed a series of notes and expansions which became the nucleus of this book. In 1994, she left St. John's to finish the book and to found the Green Lion Press, of which she is now Chief Editor.

William H. Donahue, co-director and technical manager of Green Lion Press, has been an independent scholar in history of science since 1981. He received his BA from St. John's College, Annapolis, in 1967, and his Ph.D. from the University of Cambridge, England, in 1973. His dissertation, *The Dissolution of the Celestial Spheres, 1595–1650,* was published by Arno Press. From 1973 to 1976, he taught in the Great Books curriculum at St. John's College in Santa Fe. He is translator of Kepler's *Astronomia nova* (published by Cambridge University Press) and Kepler's *Optics* (published by Green Lion Press). He has done research on the Kepler manuscripts in St. Petersburg, Russia and has published numerous articles in the *Journal for the History of Astronomy* and the *British Journal for the History of Science.*